MW01101162

A CHECKLIST OF NORTH AMERICAN AMPHIBIANS AND REPTILES

A CHECKLIST OF NORTH AMERICAN AMPHIBIANS AND REPTILES

The United States and Canada

SEVENTH EDITION

Volume 1—Amphibians

M. J. Fouquette, Jr.

School of Life Sciences, Arizona State University, Tempe, Arizona, USA

and

Alain Dubois

Département Systématique & Evolution, Muséum National d'Histoire Naturelle, Paris, France

Library of Congress Control Number:		2014902312
ISBN:	Hardcover	978-1-4931-7034-0
	Softcover	978-1-4931-7035-7

This book was printed in the United States of America.

Front cover photos by Suzanne L Collins, used with permission.

Rev. date: 02/12/2014

To order additional copies of this book, contact:
Xlibris LLC
1-888-795-4274
www.Xlibris.com
Orders@Xlibris.com
551050

CONTENTS

PREFACE FOR VOLUME 1

by the senior author

The most recent edition of a herpetological checklist for North America (Schmidt, 1953) appeared about 60 years ago. As Schmidt pointed out in that publication, preparing such a listing is time-consuming, and does not produce any new science, so it is understandable why committees charged with preparing such a list were not able to make much progress. As a consequence, Schmidt took over the work himself, as a labor of love and a project that he viewed to have great value. I am in much the same position. Since my retirement I have endeavored to keep up with the rapid taxonomic and nomenclatural changes in herpetology, and lamented the fact that there is no checklist for North America in print with current names and information on the species of amphibians and reptiles inhabiting this continent. Some may suggest that published paper checklists are no longer useful, but I would argue the opposite. A checklist is a nomenclatural document, listing all the species of some larger taxon, usually with synonymies of other names used for each species, current for that point in time. Today there are numerous field guides to the herpetofauna of all or parts of North America, with descriptions, photographs, keys, and range maps. The species names are generally included, but the common names are emphasized, as these are meant to be documents to be used more by the non-technical public to recognize species of their locality. There are also electronic checklists of various types for amphibians or reptiles (see further comments below). But to keep track of nomenclatural changes and their dates, a hard-copy text seems a necessity. When changes are made in an online checklist in 2001, then updated or modified in 2003, the original change of 2001 may disappear and there is no record of that event. If we have a series of paper-printed checklists published at intervals, each is a snapshot of current nomenclature, and the dates of changes are easily determined.

To assist in this endeavor I have teamed up with Alain Dubois for amphibians, and Kenneth Williams for reptiles. Dubois' broad expertise in herpetological nomenclature and taxonomy exceeds mine, and while I have the greater familiarity with the North American herpetofauna and associated literature, he is probably a better taxonomist and certainly understands animal nomenclature and the associated rules as well as or better than perhaps any other active herpetologist. In addition, he has better access to the early European literature. Original descriptions of much of the North American fauna were by European naturalists, and a large portion of the historical specimens are housed in European museums, and especially the MNHN in Paris.

For amphibians, the division of labor is along these lines. I prepared a draft of the accounts initially and turned them over to Alain. He found errors, suggested alternate possibilities of arrangement, solved any nomenclatural questions, and perhaps challenged some of my initial ideas. He turned this back to me and I either accepted or rejected some of his suggestions, or we continued to confer about others until we worked out a satisfactory result. Alain did much of the early work on synonymy listings, especially for familial and generic series names. We agreed that the final taxonomic decisions would be mine. Many/most of the taxonomic options we have agreed on. But there are some where a taxonomic option used herein is mine alone, and in disagreement with his. So we have agreed to note that **not all taxonomic opinions herein are necessarily in agreement with those of the junior author.**

As we progressed with the amphibian volume, Alain encountered some health-related problems, and also was involved with numerous other commitments of his museum position, teaching, and such, and found he was unable to devote the time necessary to complete the checklist within our previously agreed time-frame. Thus, during the last several months, the completion of the first volume became mostly my responsibility, and the failure to include some important synonyms, or possible errors that may be found should be attributed to me, as Alain is a perfectionist in his contributions, and would have ensured that all synonymies were complete, and the manuscript was error-free. I should also note that Alain did continue to contribute, especially in responding to nomenclatural questions, after he had to withdraw from active participation in the checklist. And he reviewed a semi-final version of the manuscript and made some corrections and additions.

A few words of explanation may be in order here. Regarding our adherence to the International Code of Zoological Nomenclature (hereafter, "the *Code*"), we firmly hold to that document as the only authority for nomenclature of animals (with one exception, to be explained). Every so often some of us become dissatisfied with the *Code,* and introduce an alternative, which may have some advantages, but overall is unsatisfactory as a substitute. The current alternative system of nomenclature that enjoys a significant degree of popularity is called the *Phylocode.* It purports to be based on the evolutionary history of animals, so that its rules are designed to name taxa that conform to evidence of their phylogenies, providing names for the major clades, in a looser system not requiring hierarchical ranking. We consider the Linnean system of a hierarchy of taxa to be one of the finest concepts that has been contributed to taxonomy. The system has been codified into a complex of rules for the proper naming of of animal taxa, at least for species, genera, and family-series. While the best authority, the *Code* is not perfect. Probably the greatest problem with the *Code* is that it stops there, and does not govern naming taxa above the family-series ranks. My co-author (Dubois, 2005a, 2005c, 2006a, 2009a, 2011; Dubois & Raffaëlli, 2012) has advanced several suggestions for including logical rules for naming of these higher ranks, and it would be wise for the next version of the *Code* to include a set of rules such as Dubois has proposed, to cover nomenclature of higher ranks. I feel the proposals of Dubois are a good starting point, but we badly need modifications to the *Code* to include the higher categories, whether similar to those of Dubois or not. Otherwise, alternate proposals, such as the *Phylocode,* may attract more dissatisfied zoologists. In addition to the need to include rules for nomenclature of higher ranks, we feel that the *Code* should allow for more ranks at the genus-and species-series level; at other levels there can be virtually an infinite number of additional subcategories, to fit the needs of the taxonomist. But taxonomic freedom is limited by the *Code* restricting the possible ranks at the genus-series to genus and subgenus, and at the species-series to species and subspecies (plus rather cumbersome possibilities of "aggregations of species" and "aggregations of subspecies"). Dubois (2011) has argued strongly for these and other necessary modifications to make the *Code* a better working document. And quite recently Dubois *et al.* (2013) have called for better handling of electronic publication in the *Code*. These three matters seem to demand most immediate attention of the International Committee.

But aside from its few glaring faults, the *Code* has served zoology well through the last 250 years, and I feel we should not desert it for another system. Accordingly, we follow the *Code* in all matters of nomenclature in this work. The exception noted above, is for those higher taxonomic levels not currently regulated by the *Code*. Individual zoologists who may unilaterally decide to follow their own rules rather than adhere to those of the *Code,* can potentially lead us into chaos; the *Code* is the only universal set of rules regulating our nomenclature, so we must adhere to it, and urge changes where they are needed. For higher rank names not regulated by the *Code,* we follow the rules proposed by Dubois (see citations above), and implore the International Committee on Zoological Nomenclature to adopt such changes in the *Code* that will remedy that problem.

This may seem like material more appropriate to the Introduction, than to this Preface, but I am including it here to indicate that these are my opinions in this matter. My co-author's opinions about such matters are well documented, whereas mine are not, so I need to separate myself from him in this section and indicate that we share a similar view on nomenclatural matters, and it is not a matter of my just "going along" with him.

We have strived in our effort, to make this checklist as nearly error-free as possible. But we realize that we may miss some important synonym for a given taxon, or perhaps citation of a given page number may be wrong, or other such error, including typos, may have crept in and not been caught by us. If so, our readers may notice these, and we hope they will expeditiously report them to us so that they may be corrected for the next edition.

Finally, I dedicate my portion of this work to the memory of Leonard Stejneger, Thomas Barbour, and Karl P. Schmidt, for being the giants upon whose shoulders we may stand. I hope that future editions of this checklist will continue, but hopefully at shorter intervals. At the present rate of taxonomic and nomenclatural changes in herpetology, I personally feel a new edition is needed approximately every four to five years, but certainly no more than every ten years. At this writing, it is our intent to produce the next edition by 2017 or 2018.

M.J.F., November, 2013.
jack.fouquette@asu.edu

PREFACE for Volume 1

by the Junior Author

The senior author has already adequately introduced this work, so I will take this opportunity to address a matter of importance to all works of this nature:

A plea for nomenclatural accuracy in faunistic checklists

Taxonomy and nomenclature

Faunistic and taxonomic checklists are reference works that can potentially be used and referred to by many biologists in various disciplines: not only taxonomy and biogeography, but also physiology, ethology, ecology, conservation biology, etc. Because taxonomy is an ever-evolving science, all faunistic lists, whatever the taxonomic group and the geographical area concerned, are bound to evolve and to be regularly updated. A North American physiologist working in the years 1920-1950 and using leopard frogs as material would have considered it equivalent to work on any specimens from most regions of the United States, because all populations of leopard frogs east of the Sierra Nevada were referred to the same species then called *Rana pipiens*, whereas nowadays several distinct species are recognized in different parts of the country and different results may be expected if specimens from different regions are used as scientific material.

Changes regularly occur in the classification of all groups of organisms, for several reasons: 1) new field work providing new data on the biology, characters and distribution of organisms (e.g., new data on mating call, on variability, or on hybrid zones and their dynamics), and even resulting in the recognition of new taxa which had escaped the attention of previous naturalists and researchers (e.g., the recent discovery of *Urspelerpes brucei* Camp *et al.*, 2009); 2) new methods and techniques of analysis of the characters of the organisms (*e.g.*, cytogenetics, protein electrophoresis or nucleic acid sequencing); 3) new taxonomic concepts and theories, and resulting changes in the dominant taxonomic paradigms (*e.g.*, the transition from phenetic to cladistic or synthetic conceptions of biological classification).

Taxonomy being in permanent evolution, nomenclature also has to change. But there is an important difference here. *Taxonomy* is the science of classification of organisms into formal units, the *taxa* (singular *taxon*), that nowadays are largely, although not only, based on our hypotheses of cladistic relationships among taxa, whereas nomenclature is simply a system, regulated by the *International Code of Zoological Nomenclature* (the "*Code*"; ICZN, 1999), of allocation of scientific names or *nomina* (singular *nomen*) to these taxa, but does not intervene into the recognition or definition of these taxa. Taxonomy, being a complex scientific discipline, must remain free of any regulatory constraints such as *a priori* rules or regulations. Like for all sciences, its harmonious development requires respect for what the "Preamble" of the *Code* calls "*the freedom of taxonomic thought or actions.*" In contrast, nomenclature is not a science but a technique, a set of more or less arbitrary conventions which scientists belonging in a large community agree to follow. In order to function as a unique international system of communication between biologists, nomenclature cannot depend on vague "consensus" or "usage", but must be strictly regulated by a set of Rules; this is the function of the *Code*. Strict respect of the *Code* by all zootaxonomists is crucial if communication about organisms among them as well as with the rest of society is to be unambiguous and reliable.

Nomenclature is a specialized field which should not be tackled superficially by people lacking the appropriate formation and background to do so. The Rules of the *Code* should be followed scrupulously but, the terminology used in this text being imprecise and sometimes misleading (for example, the term "name" is used indiscriminately in this text for very distinct concepts). For more clarity, it is better to use a more precise and accurate terminology for the concepts and tools of the *Code* (Dubois, 2011).

Nomenclatural stability vs. nomenclatural accuracy
Faunistic and taxonomic checklists tend to replace each other, often though new editions, and there may exist a temptation for their authors to try and change them minimally, i.e., only through additions, mostly of new taxa, but without modifying the information that was already present in the previous edition, even if this information is known to be partly wrong or defective. The reason given for this attitude can be qualified as a "cult for stability". The idea behind it is that stability is more important than accuracy. This mostly applies to the nomenclatural information of the checklist: better to keep a wrong nomen for a taxon because the latter is "well-known" than correct it.

This philosophy has been followed in a number of successive editions of various checklists, but it is not recommended, for a simple reason: nomina are not here to please a particular group of biologists (e.g., those working on the North American fauna), but for the universal communication among all biologists and users of the results of taxonomic research. Nomina play a crucial role in the storage and retrieval of taxonomic information about the organisms of the planet, and as a key to the literature where are stored all the data that biologists have gathered on these organisms over centuries. In order for checklists to play properly this role, the taxonomic allocation and the validity of their nomina must be properly established through a strict respect of the Rules of the *Code*, and not through an approximate use of vague and undefined concepts like "consensus" or "usage".

The Rules of the *Code* were devised in such a way that, in most cases, the valid nomen of a taxon, within a given taxonomic frame, is unambiguously and automatically established through their rigorous application, which leaves no place for interpretations, opinions, debates or decisions. Usually, in front of a given nomenclatural problem, two specialists on opposite sides of the planet, and working seriously, must come to the same final conclusion regarding the proper nomenclature to apply to any given classification. The cases where choices have to been made and non-automatic decisions taken are rare. They occur mainly in two situations. First, whenever, after the original publication, an ambiguity remains about the status of a nomen, this ambiguity must be removed through a first-reviser action (*e.g.*, a subsequent designation of lectotype or of type-species, or the fixation of precedence between two simultaneous competing nomenclatural acts). Second, whenever the long-established nomenclatural stability of a *really* well-known nomen (i.e., used by hundreds of authors of all kinds, not only taxonomists, and for a long time) is threatened by the sudden rediscovery of a long forgotten nomen, the case must be brought to the knowledge of the International Commission on Zoological Nomenclature ("the Commission"), which may use its Plenary-Powers to suspend the Rules in that particular case. Apart from these two particular situations, no taxonomist is entitled to make personal decisions that would not respect the Rules of the *Code*, as this would result in nomenclatural instability and unreliability, and ultimately in miscommunication among biologists.

Nomina, once appropriately published, never "die", they remain for ever available and potentially valid (if they are not junior synonyms or homonyms of older nomina), except by action of the Commission. Therefore, even if a mistake has been repeated hundreds of times by careless workers, there will always come a day when a more serious taxonomist will discover the error and care for correcting it. For this reason, nomenclatural corrections should be published as soon as possible after the mistake has been published, in order to avoid it becoming "entrenched in the literature." Faunistic checklists, as well as taxonomic revisions or nomenclatural updates, are privileged places where nomenclatural corrections should be carried out. Caring for nomenclatural accuracy in checklists is always beneficial because it avoids the establishment of a wrong usage that could appear to some as deserving to be preserved against the need of a subsequent change that will unavoidably result from the subsequent discovery of the nomenclatural mistake.

Resistance to corrections to please some people or "tradition" is counterproductive, not only because it just delays the implementation of a correct nomenclature for the taxa at stake, but also, more generally, because it entertains among some taxonomists an attitude of carelessness and neglect, if not deliberate ignorance, about the Rules of the *Code*, which in the long run could be a strong recipe for nomenclatural confusion, if not chaos.

Besides, and especially nowadays, with an increase in the speed of transfer of information among scientists through the Internet and the World Wide Web, a correction needs only a short time to be spread and to replace a wrong usage, for example the common situation of a nomen that has been used for several decades for a taxon but that in fact applies to another one. Contrary to a widespread belief, this was already the case one century ago: for example the replacement of the well-known amphibian generic nomen *Ixalus* Duméril & Bibron, 1841 by *Philautus* Gistel, 1848, required by the fact that the former was found to be an invalid junior homonym, took only 15 years (1905-1920) to become universal (Dubois & Ohler, 2001), and more recently the replacement of *Ololygon* Fitzinger, 1843 by *Scinax* Wagler, 1830 was almost instantaneous worldwide (Dubois, 1995).

For further discussion or examples of some of these points, see: Bonaparte, 1838c; Dubois, 1984c, 1987b, 2000, 2005a, 2006a, 2007b, 2010, 2012; Dubois *et al.,* 2005; Dubois & Raffaëlli, 2009, 2012; Frost, 1985; Nemésio & Dubois, 2012; and Ng, 1994.

A. Dubois, November, 2013

INTRODUCTION

What is a checklist?

Minimally, a faunistic **checklist** provides a listing of all the species (and usually subspecies) of some given taxon, which occur in a given geographic region, or perhaps are represented in the collection of a museum, with the current names and the current assignment to higher ranks. At least the source of each name is almost always given, and often a synonymy, which may be exhaustive or abbreviated. It is meant to provide a current inventory of the taxonomy and nomenclature for the fauna of that region, and if necessary, provide some fine-tuning. This checklist provides such an inventory of the extant herpetofauna of North America (as defined below), including more extensive synonymies than in prior editions, and also provides reference to phylogenetic relationships of the taxa, based on recent analyses. This checklist is the next edition in the series begun by Stejneger & Barbour (1917-1943; five editions, published through Harvard University), and the most recent sixth edition by Schmidt (1953; published by the American Society of Ichthyologists and Herpetologists, or ASIH). Ours is considerably different (improved) from the earlier editions and it is in no way associated with or endorsed by Harvard or ASIH. We follow the rules and recommendations of the International Code of Zoological Nomenclature (International Commission on Zoological Nomenclature, 1999), which we henceforth refer to as the *Code*. Where the *Code* does not provide guidance, such as for nomenclature of taxa above the level of the family-group or family-series, we follow the suggestions outlined in detail by Dubois (2005a), unless otherwise indicated. For the most part the earlier checklists adhered to the *Code*, but in some instances they may have disregarded some of the rules.

For amphibians, there have also been checklists of species worldwide. Gorham (1974) listed all amphibian species as known up to 1970. Synonyms were given in an abbreviated format, but no literature citations were given. Frost (1985) supplied a checklist of all amphibian species as known up to that time, with complete synonymies and literature citations, a tremendous advance in providing taxonomic and nomenclatural information for the Class. He has continued and expanded that endeavor in the online listing noted below (Frost, 2009-2013).

Traditionally a **catalog** (or catalogue) contains more information about species of a region than does a checklist, usually including species descriptions, figures, and often keys and perhaps maps or other kinds of data. For the American continents, but especially North America, there is the Catalogue of American Amphibians and Reptiles, currently published by the Society for the Study of Amphibians and Reptiles (SSAR). Several new accounts are published each year, and most species of North America are now represented, and many of Middle and South America. Eventually it is hoped all species in the Americas will have accounts, but presently there are still many species from North America for which accounts are lacking, and relatively few species from further south are represented. Unfortunately, this catalogue has been publishing accounts for over forty years and is still less than half completed, and many of the earlier accounts are seriously out of date. The website for Catalogue information is http: //www.herplit.com/SSAR/catalogue /caar.html. Hereafter, that publication is referred to as "CAAR."

In today's technological world there are also online catalogs and listings available on the internet which may give a herpetologist the most current listing of species, with synonymies, for all the amphibians and reptiles of the world, which may be more than a person needs, if his work is restricted to North America, or a portion thereof. For amphibians there is the extremely useful website, Amphibian Species of the World http://research.amnh. org/herpetology/ amphibia/), managed by Frost (2009-2013). That site has been very useful for our research on amphibians; in many cases we used it as a guide to some of the synonyms and literature we may otherwise have missed. In referencing that site, we are including the years in which we accessed it for information as cited here; we do not give a specific citation for each time we may have checked data that are presented there. For the most part, we have seen all the literature we cite, and exceptions usually include a note after the citation, indicating we did not see it and on which work we depended. For reptiles, there is The TIGR Reptile Database (http://reptile-database.org/), supported by the systematics working group of the German Herpetological Society (DGHT), and edited by Peter Uetz *et al.* (1995-). Both sites are updated frequently. The Center for North American Herpetology has a website (http://www.cnah.org) where it maintains an updated list of North American species, and a searchable database of recent publications on North American herpetology, along with other very useful and helpful information.

It should also be noted that there were some publications prior to the six editions of the checklist which served a similar function. One of these was Harlan (1827). The classic Holbrook (1836-1842) five-volume set was more of a catalog, as was Garman (1883), though Garman (1884a) was more of a checklist. Cope (1875b) also published a checklist for North America based on specimens in the US National Museum, and later (Cope 1889b, 1900) published his more extensive catalogs. Yarrow (1883a) also published a checklist. Regarding Holbrook's work, it has a confusing history. As detailed by Worthington & Worthington (1976), volumes 1 and 2 have two versions, distinguishable only by number of pages and presence or absence of certain species accounts. Volume 1 was originally published in 1836, volume 2 in 1838, then the second *version* of both were published later, probably in 1839 (we assume that year in citing those versions). In the first edition, there were only four volumes. The second *edition* was published in 1842, in five volumes (see Worthington & Worthington, 1976, for details). The second edition was reprinted in 1976 by the SSAR.

Taxonomic coverage

Unlike previous editions, this checklist is presented in two volumes. The decision to do so came rather late in the preparation, but it seemed advisable primarily for two reasons. First, the two initial co-authors had finished most of the amphibian portion at that time, and decided it should not be necessary to wait until we completed the reptile portion as well, in order to make that part available to the herpetological community. Secondly, by publishing in this manner, we could add or change co-authors for the reptiles, without affecting the authorship of the amphibian portion. At this writing, Kenneth L. Williams has been added to the authorship of the reptile volume, and the senior author will remain the same, and Alain Dubois has commitments that do not allow for him to continue on the reptile volume, but his advice will be sought during its completion. Additionally, we find the content is now too great to be conveniently contained in a single volume. So we will publish the first volume, covering the Amphibia, followed by the second volume, covering Reptilia a few months later.

The term amphibian needs little explanation; as used here. We refer specifically to salamanders and frogs and toads occurring north of the US-Mexican border. Amphibians also include caecilians (Gymnophiona), but none of these occurs within the US and Canada. However, we must define our use of the term "reptile."

Traditionally, the Class Reptilia includes lizards, snakes, amphisbaenians, turtles, crocodilians, and tuataras, along with other fossil groups, such as dinosaurs. Phylogenetically, this would also include mammals and birds. Recently there has been suggestion that the Reptilia be revised to include only lizards, snakes, amphisbaenians, and tuataras, and relegating turtles and crocodilians to separate taxa at the Class level. We hold to the traditional definition of the Reptilia, and the justification and other details are explained in the Introduction to the second volume of this work.

In our counting of total numbers of families, genera, or species included in a higher taxon of amphibians, we rely on the latest figures given by Frost (2009-13) on his website, or as modified by any difference in classification we may employ. Of course these numbers will change with time, and will quickly become outdated, but should be useful approximations for the next few years.

Geographic coverage

For purposes of this publication, North America is defined as the United States and Canada. The first five editions by Stejneger & Barbour (1917-1943) also included Baja California in conformation with the existing checklists for North American birds, but the sixth edition (Schmidt, 1953) restricted coverage to the US and Canada. We considered using the title, "Checklist of Amphibians and Reptiles of the United States and Canada," to be more definitive, but decided that the title should remain the same as for the first six editions, but we have added the subtitle for clarification. There are checklists available for Mexico and portions thereof, and inclusion of the Mexican herpetofauna would have taken more than twice as long to research. Moreover, it has become a useful biogeographic practice to regard North America as the US and Canada, Middle America (or Mesoamerica) as Mexico and Central America, and then South America, as three separate geographic entities (*e.g.*, Duellman, 2001a, b). None of the previous checklists has included US territories, such as Puerto Rico, US Virgin Islands, or Guam, nor does this edition. For that same reason, Hawaii and Alaska were not included, as they were not states at the time those editions were published. As they have now been states for over 50 years, their herpetofauna are included in this edition, even though most/all Hawaiian species are introduced, and Alaska's herpetofauna is meager. As in all previous editions, this checklist covers only extant species; though we do include species which may have recently become extinct. We include introduced species, at least where they have established breeding populations, in their normal taxonomic position, rather than placing them in a separate section as was done in previous editions. **An asterisk (*) precedes the name of any taxon which is non-native to North America, and any statement about their introduction into this area.**

Nomenclatural philosophy

In all nomenclatural matters we adhere to the *Code,* as noted above. We again remind the reader that Rules of the international *Code* do not currently apply to names above the Family-series, so a taxonomist is essentially free to use a nomen that he/she prefers for orders or classes. Frost *et al.* (2006a; appendix 6) used some rather vague guidelines for choosing nomina of higher taxa, but, as shown by Dubois (2008b, 2009a), they are imprecise and inconsistent and cannot be used for an automatic allocation of the valid nomen to a higher taxon, similar to the automatic allocation of nomina to taxa allowed by the *Code* for lower taxa. The only detailed rules that have been proposed for such an automatic nomenclatural allocation of nomina in zoology are those of Dubois (2004, 2005a, c, 2006a-b, 2009a). Those rules are consistent with those of the *Code* in that they rely on the *original* content of the taxon named, not on the rank; nor on subsequent usages, restrictions, emendations, etc. Those rules are followed for the few higher taxa covered in this work. Thus, the nomen Amphibia is credited to de Blainville (1816a) who was the first to use it (in the form "Amphybiens") for a taxon including all anamniote tetrapods then known, and who used Reptilia (as "Reptiles") for all amniotes then known, excluding the Aves and Mammalia. Similarly, Batrachia was created (as "Batraciens") by Brongniart (1800) for a taxon including all frogs and salamanders then known, but excluding the caecilians and all "reptiles." Anura was created (as "Anoures") by Duméril (1805) for a subtaxon of Batrachia including all frogs but excluding all salamanders, and Urodela (as "Urodèles") for its sister-taxon, salamanders. The latter nomen is to be preferred to Caudata, coined by Scopoli (1777) for a taxon including salamanders but also lizards and crocodiles.

In one departure from the *Code* (merely one of terminology), we prefer the use of the term "series" for the aggregate rankings around a major category, rather than the term "group." We have adopted the suggestion of Dubois (2000) in referring to the three levels of taxa regulated by the Code as "species-series," "genus-series," and "family-series," instead of "species-group," etc., to avoid any confusion that the term "group" may have because of usage in another sense. For example, the term "species-group" is widely used as an informal monophyletic set of species, essentially below the level of subgenus.

Arrangement of taxa

Initially, we began arranging the taxa in phylogenetic order, utilizing the latest studies indicating phylogenetic relationships among the species. We also planned to have an extensive index that would serve as an alphabetical listing for those having trouble finding a given species in our listings. But after getting feedback from some colleagues, we decided to rearrange our listing, following checklist tradition in making it more alphabetical. However, we ignore the alphabetical arrangement in our listings of orders. In this volume, we list salamanders first, then anurans. In the reptile volume, we list crocodylians first, then turtles, then lizards, snakes, and amphisbaenians. But within each order, we alphabetize the taxa of major subdivisions, whether they be suborders, infraorders, superfamilies, epifamilies, or other. Then within each of those taxa, we alphabetize the families and subfamilial divisions, then the genera, and within genera the subgenera, and within those the species, then the subspecies. Where there are subdivisions, such as subfamilies, subgenera, or subspecies, we treat the nominotypical one first, then the others in alphabetical order. The checklist editions by Stejneger & Barbour were all alphabetical. That of Schmidt (1953) was largely alphabetical, though it was sometimes apparently partially phylogenetically arranged, but without explanation.

For taxonomic arrangement, we generally follow the most recent taxonomy for salamanders (Dubois & Raffaëlli, 2012; with modifications for higher taxa by Shen *et al.,* 2013), and for frogs the latest molecularly-based higher classification of Pyron & Wiens (2011), which is mostly congruent with that of Frost *et al.* (2006a) and other recent studies, particularly for North American taxa. We have made some modifications of genus-series taxonomy where we feel such changes better serve herpetological classification.

Layout of accounts

For each taxon, the account includes a consistent set of items. The scientific name of the taxon is given, in **boldface**, preceded by the rank assigned (above species level) and followed by the author we accept for that taxon. We follow traditional nomenclatural practice and Recommendation 22A.3 of the *Code* in placing the species author in parentheses if the current genus used for the species is different from that of the original description. For the unregulated, higher ranked taxa, as well as the family-series, Small Caps are used.

Synonymy: We here refer to our listings of nomina as "synonymies," but in effect they are really "paronymies" (Dubois, 2000; see that article for definitions). For species-series and genus-series taxa, a rather complete synonymy is included, listing significant protonyms (original names which apply to the same taxon) and aponyms (subsequent listings or modifications of protonyms) that we found, but not necessarily all the *lapses* and misspellings, though many are included. For family-series and higher, the synonymy is more abbreviated, but unlike previous checklist editions, the higher taxa are accorded proper authorship and at least a limited synonymy listing. We also depart from the usual mode of citation for authors than is traditionally found in synonym listings in checklists. Rather than merely giving an abbreviated citation, we cite using the Author-Year system, then give the full citation in the Literature Cited section at the end. This same arrangement is seen in CAAR accounts. We feel this is one of the special values of this new checklist, so that a reader may immediately have the full name of an old or unfamiliar journal, the full title of an article, and full pagination (invaluable if you need to request an interlibrary loan of a journal article).

The format of our synonymies may be unfamiliar to many readers. We adopt the format suggested by Dubois (2000), in which true synonyms are not indented, and aponyms of each name are indented below that name. The true synonyms have no punctuation between the nomen and the author; the aponyms are separated from the author's name by a colon.

Misapplied names are sometimes appended to the end of the synonymy or noted in a commentary. Synonyms and aponyms are listed in chronological sequence, as is logical, appropriate, and traditional.

In most cases we adopt the most recent taxonomy proposed. There are some exceptions. In reptiles, for example, Hoser (2009) proposed splitting the genus *Crotalus* into several genera. His justification for this seems unacceptable so we have ignored that proposal (see that account in the Reptile volume for explanations). There have been other proposals with a strong phylogenetic basis for subdividing a genus, and proposing alternative names for the subdivisions. In some such cases we may suggest retention of the original generic name for all the species and consider the subdivided units as subgenera. This is explained in detail and justified where it is pertinent.

Distribution: The geographic range given for species or subspecies is usually rather abbreviated, sometimes more so than in previous editions. More definitive information on ranges is available from the many field guides that include range maps, or maps in the CAAR accounts. We try to cite an available map in each distribution paragraph for a taxon. We also include the Canadian provinces and U.S. states in which each species or subspecies has been reported, using the official abbreviation for those political subdivisions (refer to the Appendix for such abbreviations). Where a taxon has been introduced there by human intervention from another source, the abbreviation is preceded by an *asterisk. If a form is found in all or nearly all states, we may instead list only those states where the form has **not** been reported (*e.g.*, "**U.S.**: all except AK, HI, ID, MT, ND.").

Common name: Common names used are the English vernacular names as listed by Crother (2012) and Collins & Taggart (2009) for family-series and below; for higher taxa, we follow those sources for many names, but may follow others for some higher taxa, with explanation. Or if no common name is given by those sources, we may suggest one. Where the sources specified here differ, both names are given, and identified, respectively as SECSN (Standard English & Current Scientific Names; Crother) or C&T (Collins & Taggart). For forms that occur in Canada, the French name used there is also given, as found in Mazerolle, *et al.* (2012).

Content: For taxa above the species-series, the content is given in terms of taxa that are currently included in that taxon, by our interpretation, or for many of the higher taxa that may include many extralimital forms, simply an approximation of the numbers of included taxa as listed by Frost (2009-13). If a species has subspecies that are generally recognized, we list the number, then include an account for each one that *occurs in our region.* If there are no subspecies, we may not use this paragraph.

Commentary: This paragraph is used to indicate phylogenetic relationships of the taxon to others according to recent studies, as well as any other comments about literature, conservation, or other aspects.

Terminology and abbreviations

Dubois (2005a) has argued that terms including the word "type" (such as type-specimen, holotype, type locality) should be replaced in the *Code* by more meaningful terms that do not suggest typological philosophy. Readers of this checklist may not be familiar with that terminology, hence we use both the traditional terms and his alternative terminology in this work.

The following abbreviations are employed in synonymies:

H—Holotype (*Code*, 1999) or **holophoront** (Dubois, 2005a).

S—Syntypes (*Code*, 1999) or **symphoronts** (Dubois, 2005a).

L—Lectotype (*Code*, 1999) or **lectophoront** (Dubois, 2005a).

N—Neotype (*Code*, 1999) or **neophoront** (Dubois, 2005a).

O—Name-bearing type(s) (*Code*, 1999) or **onomatophore(s)** (Simpson, 1940).

OT—Type locality (*Code*, 1999) or **onymotope** (Dubois, 2005a).

NS—Type species (*Code*, 1999) or **nucleospecies** (Dubois, 2005a).

NG—Type genus (*Code*, 1999) or **nucleogenus** (Dubois, 2005a).

NN—Replacement name or **nomen novum** (*Code*, 1999) or **alloneonym** (Dubois, 2000, 2005a).

NU—Nomen nudum (*Code*, 1999) or **gymnonym** (Dubois, 2000).

ND—Nomen dubium according to Art. 23.x.x of the *Code* (1999).

NO—Nomen oblitum according to Art. 23.9.2 of the *Code* (1999).

NP—Nomen protectum according to Art. 23.9.2 of the *Code* (1999).

OS—Objective synonym (*Code*, 1999) or **isonym** (Dubois, 2000, 2005a), names having the same O.

SS—Subjective synonym (*Code*, 1999) or **doxisonym** (Dubois, 2000, 2005a) names refering to the same taxon, with different O.

IS—Incorrect subsequent spelling (*Code*, 1999) or **ameletonym** (Dubois, 2000).

UE—Unjustified emendation (*Code*, 1999) or **autoneonym** (Dubois, 2000, 2005a).

mo.—Fixation (of onomatophore) by **monotypy** (*Code*, 1999) or **monophory** (Dubois, 2005a).

od.—Fixation (of onomatophore) by **original designation.**

pd.—Fixation (of onomatophore) by **present designation.**

sd.—Fixation (of onomatophore) by **subsequent designation.**

The onomatophore(s) (type specimens) of a species or subspecies are listed to the extent we have been able to discover them; generally we simply give the information as we have found it cited, without reference to publications where that information was seen. More recently described taxa include the museum number of the holophoront (holotype), but older descriptions often did not designate a particular specimen or even where it was deposited, so that information may have been made available in a subsequent publication, by a more recent author, or may remain unknown. If the onomatophore consists of several symphoronts (syntypes, sometimes called co-types in older literature), the number of specimens catalogued under a museum number is indicated in parentheses following the number (*e.g.*, USNM 101010 (3) indicates a **S** of three specimens catalogued under that museum number; if the museum number is not followed by a number in parentheses, either there is but a single specimen or the number of specimens is unknown to us). Museum abbreviations are contained in an Appendix, which is repeated in both volumes, whether or not any specimens from that museum were cited in that volume. Except for rare instances, we have not examined any type material. Information on museum numbers is taken from the literature, often from several sources.

Regarding the type locality (onymotope), the only way the *Code* allows *restricting* the locality from which the type specimen (onomatophore; Simpson, 1940) of a species (H or S) was collected is through designation of a lectotype from a known locality among several localieties of origin of the syntypes. If a species description did not state a precise collection locality for the specimen(s) intended as the name-bearer, it is not proper to arbitrarily "restrict" the collecting locality of that/those specimen(s) unless there is evidence to support a more specific site of collection. Thus, if the original description gave the collecting locality as "Carolina," it is not valid to arbitrarily restrict that to "Charleston, South Carolina," just because we know the species occurs there, or the author lived there. This has been pointed out earlier (*e.g.*, Dubois & Ohler, 1995; Dubois & Raffaëlli, 2009). Thus, in *synonymies,* we note the locality given by the author as the onymotope (type locality), but any subsequent, *invalid* restriction may only be noted in comments. Sometimes there has been a valid restriction of the type locality for a species, through investigation by a subsequent author, where appropriate data are cited, and we recognize such valid restrictions. Also, it may sometimes be discovered that the author of a nomen erred in giving the locality of collection, and the onymotope (type locality) may be validly "corrected."

In citing the type locality (onymotope), we do not usually give it as an exact quotation from the original author, but include the gist of that locality, modified to a standard form for this checklist: country, state or province, county or equivalent, specific locality, latitude and longitude if given, elevation if given, habitat if given. When "altitude" was given, that is corrected to "elevation;" this follows the defined distinction that *elevation* refers to how high the ground surface is above sea level, whereas *altitude* refers to how high an object (such as a bird or an airplane) is above the ground.

In citing authors who recognize a nomen as a junior synonym, we give the author that first declared that synonymy (if we have been able to discover that), or if we feel that such a synonymy might be controversial, and there are subsequent authors that confirm that synonymy we may also list them. Also, we try to give the aponym as recognized in at least the first and the most recent North American checklist editions (*i.e.,* Stejneger & Barbour, 1917, and Schmidt, 1953), and where different we may also cite an intervening edition. We may cite spelling errors or emendations, but especially where the spelling might not be recognized as referring to the same nomen. We also cite some of the popular field guides that have used any of the aponyms.

Often, an incorrect name is given for a taxon in the literature. This may be an error, where the author may have misidentified the taxon, or it may have been the name in use at the time and later changed (such as when that population is described as new). Strictly speaking, these uses of a name are not synonyms, rather they are the misapplication of a name (sometimes it becomes a misapplication after the fact). Thus, at the end of a synonymy listing for a taxon, we may also list "misapplied names", indicating those names used in the wrong context, though not actually synonyms.

Another convention we may occasionally use in synonymies is that for a name that may be considered a synonym but is nomenclaturally **unavailable** for any reason (*e.g.,* is a junior homonym, or a *nomen nudum*), the name is "placed between quotation marks."

Literature citation

As noted above, we abandon checklist tradition, where in synonymies an abbreviated citation is usually given with no further citation in the bibliographic listing, and instead we give an author-year citation there, and a full citation of the work in the Literature Cited. Also, we try to list the author as it was printed in the work itself (although given names are abbreviated to initials throughout). So if Palisot de Beauvois' name appeared in a work merely as M. de Beauvois, we cite it as "de Beauvois" and list it with no initials, as none were given in the work (the M. of course is the abbreviation for "Monsieur"). Or sometimes we may add initials, if well known, but placed in brackets to indicate material that was not originally present. Another example is with a less-known Polish work, usually cited as authored by "Jarocki." However, in his 1822 work, the name as given on the title page is "Jarockiego," so that is how we list it. We have endeavored to read the pertinent portion of every reference we cite (at least one of us), but there were some which we did not see, in which case in the Literature Cited, the reference has a note in brackets that we did not see it, and usually where we saw it cited. Sometimes a work actually appears later (or occasionally earlier) than the year printed on it (probably less than 3% of our cited works). When this information is known, we use the convention of citing the actual publication year as determined externally, followed by the internal printed year in quotation marks and parentheses. For example, Peters 1873 ("1872") means that the date indicated on the journal in which Peters' article appeared was printed as 1872, but we have external evidence that the issue did not actually appear until 1873. This, of course, could be important in determining priority of any new names given in the work. In some cases, notably in articles published in the 19th century in Proceedings of the Academy of Natural Sciences of Philadelphia, the journal volume is typically indicated as the year(s) of the academy meetings for that volume, then at the bottom of the title page, the year of publication. We have found that the citation often given lists the year as that of the meeting where the paper was presented, rather than the year of publication. We use the year of publication of the article, as that is the year that determines priority of any new names in that article.

A few words about identification of references published by the same author in one year: When an author publishes more than one work in the same year, we use the usual method to distinguish among those, by appending a letter to each year. Even though a letter may appear earlier in the alphabet, that does not necessarily mean the work was published earlier. We often simply use "a" for the first work that we cited for that year, so "b" may identify a work that we cited later but may have been published earlier. Also note that we have marked in our files hundreds of such citations, applying the letters as we list them. Then we actually cite only those that apply to amphibians in Volume 1, so the Literature Cited for this volume might include say Smith (1966a, 1966c, and 1966f) because those were cited in this volume; Smith (1966b, 1966d, and 1966e) would be cited in Volume 2.

In citing opinions and other publications of the International Commission on Zoological Nomenclature, some authors cite authorship of these as "Anonymous" (*e.g.*, Dubois, 2005a), but for this work we decided to cite these in the text as "ICZN (year)," and in the Literature Cited with the full "International Commission..." as the author.

Comparing numbers of taxa in previous editions.

New taxa are being described or proposed every year. At the level of genus and species, this is largely because we find new forms, and as technology improves we are able to discover even more cryptic species. As more species are found in a genus, there is a tendancy to split the genus so it is not so speciose, thus the number of genera also grows. Table 1 compares the numbers of genera and species listed by the first North American checklist, the most recent one, and this one. It is also interesting to compare salamanders and anurans in the same manner. Table 2 has the same kind of comparison for the numbers listed for the two orders.

Between the first and sixth editions, 36 years, the number of genera increased by a factor of 29% and the number of species remained the same. During that same timespan, the number of introduced genera increased by 100%, and the number of introduced species also grew by 100%. Between the sixth edition and this one, 60 years, the number of genera increased by a factor of 7.5% and the number of species increased by a factor of 115%. During that period, the number of introduced genera increased by 100%, and the number of introduced species grew by 150%. Treating these in terms of numbers per year, between the first and the 6th, the number of genera grew by 0.25 per year, while the number of species grew by 0.0 per year. During the later period, genera increased by 0.05 per year while species grew by 2.68 per year. Genera of introduced forms grew by 0.03 per year during the earlier timespan, and by 0.03 per year in the more recent timespan. Introduced species for the earlier period grew by 0.03 per year, and for the more recent period by 0.05 per year.

To compare these numbers for the two orders of amphibians, we turn to Table 2. Here we can see that for salamanders, during the earlier 36 year period, the number of genera grew by 15% and the number of species grew by 3.9%. For anurans, for the earlier 36 years the number of genera grew by 27%, and the number of species decreased by 4.6%. For the later 60 year period, the number of salamander genera decreased by 11.5% and the number of species increased by 143%. Comparing the numbers per year, for salamanders, during the earlier period, the number of genera per year was 0.17, while species increased only 0.08 per year. For anurans, during the earlier period, number of genera increased by 0.08. per year, and during the later period, by 0.10 per year. And the number of species decreased by 0.08 per year in the earlier timespan, and increased by 0.80 per year in the later one.

	1st ed., 1917	*6th ed., 1953*	*7th ed., 2014*
Number N.A. genera	31	40	43
Number N.A. species	140	140	301
Introduced genera	1	2	4
Introduced species	1	2	5

Table 1. Historical comparison of taxon numbers listed in three editions of the North American checklist, for Amphibians: Stejneger & Barbour (1917), Schmidt (1953) and the present work. Number of North American genera and Number of North American species *include* introduced forms listed. Then numbers of introduced genera and species listed in the three editions are given separately.

	1st ed., 1917	*6th ed., 1953*	*7th ed., 2014*
SALAMANDERS			
Number N.A. genera	20	26	23
Number N.A. species	76	79	192
ANURANS			
Number N.A. genera	11	14	20
Number N.A. species	64	61	109

Table 2. Historical comparison of taxon numbers of salamanders and frogs & toads in three editions of the North American checklist, for Amphibians (as cited in Table 1). All numbers include introduced forms.

So the number of genera seems to have increased slightly in salamanders then decreased slightly. The reduction in number of genera in the later period reflects the synonymy of some genera with others (such as *Manculus* with *Eurycea* and *Leurognathus* with *Desmognathus*), while in anurans the number of genera has increased regularly to a slight extent. The number of species of salamanders increased only slightly in the earlier 36 years, but increased explosively in the later 60 year period. This is largely a reflection of the many new cryptic species added since molecular technology has been introduced, as well as increasing the status of many former subspecies to species rank. In anurans, there is a gradual and regular increase in number of genera, with little synonymy of genera. And for species, there was a decrease in numbers during the earlier period, due mainly to some species being reduced in rank to subspecies, but in the later period there has been again a rather explosive increase in numbers. Note that all introduced amphibians are anurans—no species of salamanders has been introduced to North America.

ACKNOWLEDGEMENTS

M. J. Fouquette acknowledges the following for their assistance:

ASU and the School of Life Sciences has allowed the senior author to occupy space and use library and computer facilities after retirement, as well as making office and staff support available.

ASU interlibrary loan librarians, and especially Victoria McLaughlin and Danielle Schumacher, helped in obtaining many hundreds of pages of literature not otherwise readily available, and René Tanner, worked to find some obscure references for me.

Darrel Frost responded to questions concerning the AWS. His website provided a starting point for our search for amphibian synonyms.

Kraig Adler responded to some questions on history of some North American herpetological literature, and to questions about early literature. He also commented on parts of the manuscript.

Roger Bour, and other colleagues at MNHN, helped with certain old French literature.

Christopher A. Phillips responded to questions about specimens in the UIMNH collection, from the Illinois State Biological Survey collection.

Bert Hölldobler at ASU helped with transliteration and translation of some older German literature.

Roy McDiarmid discussed parts of the manuscript with me and made valuable suggestions.

The online Biodiversity Heritage Library was extremely useful in providing access to a great number of publications, mostly prior to 1923. In particular, Bianca Crowley and Grace Costantino provided assistance with articles that were not downloading properly.

An anonymous reviewer read an early copy of the manuscript and made several suggestions which led to improvements.

David Wake reviewed in great detail a final version of the manuscript, with a focus on the salamander section. He pointed out errors, alerted me to overlooked literature, and made many valuable suggestions. He disagrees with some of our nomenclatural decisions at the higher levels, not currently covered by the *Code*.

Ken Dodd reviewed a final version of the manuscript, with a focus on the anuran section, and made valuable suggestions, most of which we adopted; however, he does not completely agree with our terminology or nomenclatural decisions.

And finally I thank Elena and Mari for understanding when I was working 12-hour days on this manuscript for extended periods.

Class **Amphibia** de Blainville, 1816

Amphybiens de Blainville, 1816*a*: "107" [115] [*nec* Amphibia Linnaeus, 1758: 12, 194; *nec* Amphibiesa Garsault, 1764: 18; *nec* Amphibiens De Blainville, 1816*a*: "111" [119]; *nec* Amphibia Kirby, 1835: 411; *nec* Amphibia de Queiroz & Gauthier, 1992: 474].

Amphibiens: De Blainville, 1816*b*: 246.

Amphibien: De Blainville, 1818: 1368.

Amphibia: Macleay, 1821: 263; Gray, 1925: 213; Bonaparte, 1850: [2].

Amphibii: Desmarest, 1856: 150.

Distribution: Cosmopolitan, except Antarctica and higher northern latitudes; absent from most oceanic islands.

Common name: Amphibians.

Content: 3 orders, containing 62 families, with 484 genera and 6771 species (derived from Frost, 2009-13). Two orders, 20 families, and 301 species occur in North America.

Commentary: Retention of the nomen Amphibia de Blainville, 1816, for this taxon rests on the set of Rules proposed by Dubois (2006a) for the nomenclature of higher taxa. Frost *et al.* (2006) proposed other nomenclatural interpretations, but these do not rely on clear formalized Rules. Dubois (2004, 2009a) argued that de Blainville's (1816*a*) use of the nomen Amphybiens to include the taxa we regard as today's extant amphibians, and excluding all other vertebrates, requires that we credit him as the author of this nomen. Macleay (1821)—not Gray (1825)—subsequently provided the first latinized aponym of this nomen, as Amphibia, using virtually the same name for a taxon of the same content. Other authors used the nomina Amphibia or Batrachia for taxa with the same or different content, and many other names were used in the early years of zoology for this or other related taxa. Early use of the name "Amphibia" by Linnaeus is considered an unavailable homonym for a group including more reptiles and fishes than what we now consider amphibians. See Frost *et al.* (2006) for a more extensive list of references (called "synonymy," but which not only includes synonyms, but other nomenclatural interpretations, many of which are questionable.

Many recent authors (*e.g.*, Dubois, 2004, 2005b; Frost *et al.*, 2006) place salamanders and frogs in a superordinal or subclass taxon Batrachia, as the sister taxon to Gymnophiona (extralimital). We generally follow a modification of the classification indicated by the results of the massive study by Frost *et al.* (2006) for amphibians, but their use of many new unranked names for taxa above the family-series, without proper consideration for older nomina for the same taxa, results in a high number of ranks which seem unnecessary for a Linnean classification. For salamanders we generally follow the recent revised classification of Dubois & Raffaëlli (2012), which is similar to that of Pyron & Wiens (2011), which we follow for anurans. We use a further modification of the Frost *et al.* (2006) classification to fit the simpler system as proposed by Dubois (2005b), which does not need the additional rankings. We also eliminate redundant taxa where not needed for balance.

North America contains only 3.5% of the species diversity of this Class.

Superorder **Batrachia** Brongniart, 1800

Batraciens Brongniart, 1800: 82.

Batrachii: Latreille, 1800: xxxvii.

Batrachia: Ross & Macartney, *in* Cuvier, 1802: tab. 3; Bonaparte, 1850: [2].

Batracii*:* Duméril, 1805: 90.

Batracian: Barnes, 1826: 268.

Batrachi: Wagler, 1828: 859.

Batracia*:* Swainson, 1839: 86.

Distribution: As for the Class.

Content: Two orders (Urodela, the salamanders, and Anura, the frogs and toads), containing 58 families, 472 genera, and 6585 species. Of these, both orders, 20 families, 43 genera, and 301 species are in North America.

Commentary: Retention of the nomen Batrachia Brongniart, 1800, for this taxon rests on the set of rules proposed by Dubois (2006a) for nomenclature of higher taxa. Dubois (2009a) discussed the alternative interpretation of Frost *et al.* (2006), which does not rely on clear formalized rules. Most modern molecular studies support the sister relationship of frogs and salamanders, making this superorder appropriate for amphibian classification, including the highly resolved study of Shen *et al.* (2013).

North America contains only 3.6% of the species diversity of this taxon.

Order **Urodela** Duméril, 1805

Urodèles Duméril, 1805: 91.

Urodeli: Fischer, 1813: 58.

Urodelia: Rafinesque, 1815: 78.

Urodela: Gray, 1825: 215; Zittel, 1888.

Urodeles: Gray, 1842a: 111.

Distribution: Palearctic in Europe and Asia, northern Africa, and North America, south to northern Neotropics.

Common name: Salamanders. *French Canadian name:* Salamandres.

Content: Ten families, containing 63 genera with about 600 species (derived from Dubois & Raffaëlli, 2012). Nine families, 23 of the genera, and 192 of the species occur in North America. Thus, North America contains about 32% of the species diversity of urodeles.

Commentary: Both nomina Urodela and Caudata have been commonly used for this order in the past. As nomina above the family-series are not regulated by the *Code*, the choice for them should logically be based on a rationale for allocation of nomina to higher taxa. As stated in the *Introduction*, we here follow the rationale of Dubois (2006a). As shown by Dubois (2004) and recognized by Frost *et al.* (2006), the taxon named Caudata by Scopoli (1777) differed in content, including three genera of reptiles and only one of salamanders, so should not be used, as the nomen does not apply as a synonym of the salamander order. Scopoli (1777) also recognized a nomen Ecaudata for the single genus *Rana*, so that nomen applies to the order of frogs. As pointed out by Dubois (2004), Duméril (1805) used the nomen Caudati as an equivalent of his French nomen Urodèles. Although Duméril (1805) used the term "family" for this taxon the 151 nomina he included are not based on generic names and are not available in the family-series. They should be referred to the class-series (*i.e.*, all ranks above the family-series) of nomina (for details see Dubois & Raffaëlli, 2012). Duméril's taxon that he called Caudati or Urodèles was used as a higher taxon, including only the salamanders, so both these nomina apply to this order. Similarly, both Duméril's (1805) nomina Ecaudati and Anoures apply to the order of frogs. The nomen Caudati Duméril, 1805, being a junior homonym of Caudata Scopoli, 1777, and the nomen Ecaudata Scopoli, 1777, having been ignored by most authors, Dubois (2004, 2009a) and Dubois & Raffaëlli (2012) supported rejection of both nomina, Caudata and Ecaudata, and retention of the nomina Urodela and Anura, credited to Duméril (1805), for the two orders of batrachians. Dubois & Raffaëlli (2012: 109) also showed that, besides priority, the use of Urodela for the order of salamanders was supported by usage, as this nomen has been used more in the literature than Caudata for salamanders. They also pointed out that Zittel (1888) selected Urodela in preference to Caudata, as first reviser. This is but one example in which changes in the *Code* to include regulation of higher ranked taxa is deemed necessary, as has been strongly recommended by Dubois (2004, 2005a, c, 2006a, b, 2009a); until such incorporation occurs, it is likely that different nomenclatures will be used by different authors, which may result in poor communication, even among specialists.

Dubois & Raffaëlli (2012) introduced use of suborder and infraorder rankings in their new ergotaxonomy of the salamander order, and we mostly follow that arrangement. In an earlier working taxonomic classification of amphibians, Dubois (2005b) presented an arrangement using epifamily and superfamily ranks, and we revive those to some extent here, as new evidence necessitates changes. Also, we minimize use of redundant ranks, so that where a suborder or superfamily contains only one family, there is no infraorder or epifamily used, unless needed for balance. Other arrangements of families based on recent molecular studies (*e.g.*, Pyron & Wiens, 2011; Dubois & Räfaelli, 2012; Shen *et al.*, 2013) vary in interpretation of relationships and taxonomic arrangement. The most recent study, Shen *et al.* (2013) seems to indicate the best arrangement of families, based on their fully resolved analysis. Thus we use a simplified classification following that study.

North America contains 21.2% of the species diversity of the Order.

Suborder Imperfectibranchia Hogg, 1838

Imperfectibranchia Hogg, 1838: 152.

Imperfectibranchia: Dubois & Raffaëlli, 2012: 136.

Distribution: Eastern North America and Asia.

Common name: None in general use.

Content: This taxon includes two families. One is North Anerican and is treated below, and the other, Hynobiidae, is extralimital, in Asia.

Commentary: Shen *et al.* (2013) demonstrated that this taxon is the sister to all other salamander families. They used the name Cryptobranchoidea to refer to this taxon, without designating a rank. We designate the suborder rank for the taxon. Cryptobranchoidea was introduced by Fitzinger (1826: 41), but at the family-series level, hence is not available at the ordinal-series level. The name Imperfectibranchia seems to be the oldest appropriate name containing these families, and was used by Dubois & Raffaëlli (2012) at the infraorder rank.

Family **Cryptobranchidae** Fitzinger, 1826

Cryptobranchoidea Fitzinger, 1826: 41.—**NG**: *Cryptobranchus* Leuckart, 1821.

Cryptobranchidae: Cope, 1889b: 18.

Cryptobranchiidae: Cope, 1889b: 30.

Cryptobranchinae: Regal, 1966: 405.

Menopomatidae Hogg, 1838: 152.—**NG**: *Menopoma* Harlan, 1825.

Menopomina Bonaparte, 1839c: 16.

Menopomidae: Cope, 1875b: 12.

Menopomida: Smith, 1877: 19.

Andriadini Bonaparte, 1839b: [259].—**NG**: *Andrias* Tschudi, 1837.

Andriadina: Bonaparte, 1839c: 125.

Andriadidae: Bonaparte, 1845b: 6.

Andriantidae: Bonaparte, 1850: [2].

Andriantina: Bonaparte, 1850: [2].

Protonopsidina Bonaparte, 1839c: 125.—**NG**: *Protonopsis* Le Conte, 1824.

Protonopsina: Bonaparte, 1845b: 6.

Protonopsidae: Gray, 1850: 6.

Protonopseidae: Bonaparte, 1850: [2].

Protonopseina: Bonaparte, 1850: [2].

Megalobatrachi Fitzinger, 1843: 34.—**NG**: *Megalobatrachus* Tschudi, 1837.

Megalobatrachinae: Kuhn, 1965: 99.

Salamandropes Fitzinger, 1843: 34.—**NG**: *Salamandrops* Wagler, 1830.

Sieboldiidae Bonaparte, 1850: 1.—**NG**: *Sieboldia* Gray, 1838.

Sieboldiina: Bonaparte, 1850: [2].

Distribution: Eastern North America, Japan and China.

Common Name: Giant salamanders.

Content: Two genera, one in North America. The genus *Andrias* occurs in Asia.

Commentary: Frost *et al.* (2006) included this family in the Superfamily Cryptobranchoidea Noble, 1931, with the Family Hynobiidae (extralimital). They also recognized an infraordinal taxon, "Perennibranchia Latreille, 1825" that includes the Cryptobranchoidea, Proteidae, and Sirenidae, which they found to be closely related, paedogenic taxa. Results of studies by Zhang & Wake (2009) and Pyron & Wiens (2011), Shen *et al.* (2013), and others, disagree with that relationship, so it is not followed here.

Genus **Cryptobranchus** Leuckart, 1821

Cryptobranchus Leuckart, 1821: 259.—**NS** (mo.): *Cryptobranchus salamandroides* Leuckart, 1821. Placed on the official list of generic names (ICZN, 1926).

Cryptobranchus: Stejneger & Barbour, 1917: 7; Schmidt, 1953: 11.

Urotropis Rafinesque, 1822a: 3.—**NS**: *Urotropis mucronata* Rafinesque, 1822, od.—**SS:** Brame (1972: 27), *fide* Frost (2009-13), with *Cryptobranchus.*

Protonopsis Le Conte, 1824: 57.—**NS**: *none, see comment.*—**Comment**: Frost (2009-13) stated that the NS of this genus is *Salamandra horrida* Barton, 1808, mo. Actually, no Latin specific nomen was associated with this genus in the original publication, and two earlier species descriptions (*Salamandra alleganiensis* Sonnini & Latreille, 1802; *S. horrida* Barton, 1808) were associated with this genus. So Le Conte's name has no original NS, and its nomenclatural availability rests only on Le Conte's (1824) descriptive notes.—**SS:** Harlan (1827: 320) with *Menopoma*: Wagler (1830: 209) with *Salamandrops.*

Protonophis: Tschudi, 1838: 68.—**IS.**

Abranchus Harlan, 1825c: 233 (*nec* Boie, 1824).—**NS**: *Salamandra alleganiensis* Sonnini & Latreille, 1801: 406, mo.—**SS:** Wagler (1830: 209) with *Protonopsis.*—**Homonymy:** Junior primary homonym of *Abranchus* Boie, 1824 (Mollusca), which was a NN for Van Hasselt's *Abrancha.*

Menopoma Harlan, 1825d: 270.—**NN** for *Abranchus* Harlan, 1825c: 233.—**OS:** Eichwald (1831: 164), with *Cryptobranchus.*
Salamandrops Wagler, 1830: 209.—**NS** (mo.): *Salamandra gigantea* Barton, 1808.—**SS:** Gray (1850: 53) with *Protonopsis.*

Distribution: Eastern United States.

Common name: Hellbenders.

Content: One species, with two subspecies. Endemic to North America.

Commentary: Dundee (1972) provided a CAAR generic account. As the source of the generic nomen he cited Leuckart, as 1821, in Isis öder Encyklopädische Zeitung von Oken, noting that other synonymies give 1821 in Isis von Oken, but that the original source was the former, and the latter was only an abstract. However, the former journal was published from 1817-1819 (vol. 1-5), then the latter replaced it beginning with vol. 6 in 1820. He cites the volume as 1 (6); we found that volume and number, dated 1817, seemed to be an unrelated subject, and was neither titled nor with an author, and pagination was 41-48. Dundee gave pages only slightly different from the Isis von Oken reference cited in other synonymies. Thus, we cite the reference of Leuckart (1821) that seems to be the original source, and has been cited in synonymies by all other authors we have read.

Regarding the origin of the name *Abranchus,* we have seen references indicating the situation cited above; however, none of these has cited the Boie or Van Hasselt work, and we have been unable to find the citations, so these are not listed in the Literature Cited section.

Cryptobranchus alleganiensis (Sonnini & Latreille, 1802)

Synonymy: See the subspecies *C. a. alleganiensis,* below.

Distribution: Eastern North America; see range map in Petranka (1998) and Dundee (1972). **U.S.**: AL, AR, GA, IL, IN, KY, MD, MO, MS, NC, NY, OH, PA, SC, TN, VA, WV.

Common Name: Hellbender.

Content: Two subspecies.

Commentary: Dundee (1972) provided a CAAR account, including the genus, species, and subspecies. This species is federally listed as endangered. The St. Louis Zoo has a successful captive breeding program.

Cryptobranchus alleganiensis alleganiensis (Sonnini & Latreille, 1802)

Salamandra alleganiensis Sonnini & Latreille, 1802: 406.—**O**: **H**: MNHN unnumbered, shown in plate in front of page 253 in Sonnini & Latreille (1801); now lost.—**OT**: U.S., North Carolina, Michell County, North Toe River, probably vicinity of Davenport's plantation, 1 mile south of the mouth of the Brushy Creek and 4 miles east-northeast of the Spruce Pine Creek.—**Comments**: 1) This nomen is often (*e.g.*, Harper, 1940; Crother, 2012) credited to Daudin (1803), but it was first published by Sonnini & Latreille (1802: 406) in the index of the Latin nomina of their species, at the end of the fourth volume of their work; this nomen clearly refers there to the description of "la salamandre des monts Alléganis" in page 253 of the second volume of their book (Sonnini & Latreille, 1801), and the plate preceding it. 2) The OT given by Sonnini & Latreille (1801: 253) for this species (USA, Virginia, Allegheny Mountains) was shown by Harper (1940: 720-721) to be incorrect. The OT given above is the original one but "corrected", not "restricted." 3) the nomen was placed on the official list of specific names (ICZN, 1956b).

Salamandra alleghaniensis: Harlan, 1825c: 222, 225;—**IS**

Salamandra alleganensis: Gray, 1825: 217.—**UE.**

Salamandra alleghanensis: Tschudi, 1838: 96.—**IS.**

Triton alleghaniensis: Daudin, 1803: 231.—**IS.**

Triton alleganiensis: Oppel, 1811: 81.

Triton alleghaniensis: Harlan, 1825c: 225.—**IS.**

Abranchus alleghaniensis: Harlan, 1825c: 233' Gray, 1850: 53.—**IS.**

Abranchus alleganensis: Gray, 1825: 217.—**UE.**

Menopoma alleghaniensis: Harlan, 1825d: 271; Baird *in* Heck & Baird, 1851: 253.—**IS.**

Menopoma alleghaniense: Knauer, 1878: 96.—-**IS/UE.**

Menopoma allegheniense: Davis & Rice, 1883: 26.—**IS/UE.**

Salamandrops alleghaniensis: Wagler, 1830: 209.—**IS.**

Cryptobranchus alleghaniensis: Van der Hoeven, 1838: 384.—**IS.**

Cryptobranchus alleganiensis: Cope, 1889b: 38; Stejneger & Barbour, 1917: 7; 1943: 4; Bishop, 1943: 59; Frost, 1985: 560.

Cryptobranchus alleganiensis alleganiensis: Schmidt, 1953: 12; Brame, 1967: 5; Cochran & Goin, 1970: 9; Nickerson & Mays, 1973: 2; Gorham, 1974: 19;

Conant, 1975: 240; Behler & King, 1979: 270; Frost, 1985: xx; Conant & Collins, 1998: 418; Petranka, 1998: 140; Raffaëlli, 2007: 65.

Salamandra horrida Barton, 1808: 8.—**O**: fate unknown, probably lost.—**OT**: U.S. ("in the great lakes of our country, in the waters of the Ohio, and Susquehanna, and other parts of the United States").—**Comments**:1) by today's *Code* this would be a *nomen nudum* as the only description is the length and habitat/distribution; however at that time the greater length of the specimen was sufficient to provide availabllity. 2) the restriction of OT by Schmidt (1953: 11) to the Muskingum River (Ohio) is invalid, not being associated with a neotype designation, or otherwise providing evidence the O came from there; Schmidt cited Barton (1808), but the report of the Muskingum River was in a later work (Barton, 1814). Also see Comment in Dundee (1972: 3).

Salamandra horrida: Barton, 1814: 5. Here Barton described in some detail, with a plate, the animal he named earlier. There can be no doubt of the synonymy.

Protonopsis horrida: Barnes, 1826: 278. Dundee (1972) notes that Barnes credits Barton (1808) but none of Barton's works mentions that binomial. Barnes probably was referring to use of the specific epithet, but preferred this generic treatment.

Abranchus horridus: Gray, 1831a: 109.

Salamandra gigantea Barton, 1808: 8.—**OS** for *S. horrida,* suggested in a footnote.

Salamandra gigantea: Cuvier, 1817: 101; Gray, 1825: 217; Bory de Saint-Vincent, 1842: 233.

Molge gigantea: Merrem, 1820: 187.

Salamandrops gigantea: Wagler, 1830: 209; Gray, 1850: 53.

Salamandra gigantia: Griffith & Pidgeon, 1831: 410.

Menopoma giganteum: Van der Hoeven, 1833: 304 (*fide* Frost, 2009-2013; not seen).

Menopoma gigantea: Tschudi, 1838: 96; Gray, 1831a: 410.

Salamandra (Menopoma) gigantea: Schlegel, 1858: 61.

Salamandra maxima Barton, 1808: 8.—**OS** for *S. horrida,* suggested in a footnote.

Cryptobranchus salamandroides Leuckart, 1821: 260.—**NN** for *Salamandra gigantea* Barton, 1808.

Cryptobranchus salamandroides: Fitzinger, 1826: 66; Eichwald, 1831: 164; Gray, 1850: 54.

Urotropis mucronata Rafinesque, 1822a: 4.—**O**: **H:** mo., fate unknown, probably lost.—**OT**: U.S., Kentucky, Kentucky River.

Eurycea mucronata: Rafinesque, 1832c: 121.

Menopoma fusca Holbrook, 1842: 5: 99.—**O**: **H**: specimen of plate 33; fate unknown, probably lost.—**OT**: U.S., North Carolina, Buncomb County, Ashville ("waters of French Broad ... of Ashville"). Given as Tennessee, Knoxville, in error (Dundee, 1972).

Protonopsis fusca: Gray, 1850: 54.

Menopoma fusca: Baird *in* Heck & Baird, 1851: 253.

Menopoma fuscum: Boulenger, 1882b: 82; Yarrow, 1883a: 20.

Cryptobranchus fuscus: Cope, 1889b: 43.

Cryptobranchus terassodactylos Wellborn, 1936: 63.—**O**: **H**: ZMB 9639 (Bauer *et al.*, 1993: 290).—**OT**: North America.—**SS:** Bauer *et al.*,1993: 290.—**Comment**: The restriction by Schmidt (1953: 11) to "Allegheny Mountain in Virginia" is invalid, not being associated with evidence that the H came from there.

Distribution: Eastern United States, with a disjunct population in central Missouri. See Dundee (1972) for map and more detailed range information. **U.S.**: AL, GA, IL, IN, KY, MD, MO, MS, NC, NY, OH, PA, SC, TN, VA, WV.

Common name: Eastern Hellbender.

Commentary: Crowhurst *et al.* (2011) recognized populations of this subspecies as a conservation unit, genetically distinct from two others in the *bishopi* subspecies, and suggested that further study may support recognition of the three as separate species.

Cryptobranchus alleganiensis bishopi Grobman, 1943

Cryptobranchus bishopi Grobman, 1943: 6.—**O**: **H**: UMMZ 68930.—**OT**: U.S., Missouri, Carter County, Current River at Big Spring Park.

Cryptobranchus bishopi: Bishop, 1943: 63; Stejneger & Barbour, 1943: 4; Collins, 1991: 42; Dubois & Raffaëlli, 2012: 112.

Cryptobranchus alleganiensis bishopi: Schmidt, 1953: 12; Brame, 1967: 5; Cochran & Goin, 1970: 9; Nickerson & Mays, 1973: 2; Gorham, 1974: 19; Conant, 1975: 241; Behler & King, 1979: 270; Conant & Collins,

1998: 419; Petranka, 1998: 140; Raffaëlli, 2007: 66; Crowhurst *et al.*, 2011: 637.

Distribution: Black River drainage in Arkansas and Missouri; see Dundee (1972) for detailed range information. **U.S.**: AR, MO.

Common name: Ozark Hellbender.

Commentary: Collins (1991) argued that this population should be ranked at the species level, based on geographic isolation from *C. a. alleganiensis*. We disagree; there is no evidence of reproductive isolation, and little significant morphological differentiation — molecular phylogenetic studies might help decide this question. Crowhurst *et al.* (2011) found two genetically distinct units and suggested that further study may elevate them to separate species, but we here retain this population as a subspecies, pending further study.

Suborder **Meantes** Linnaeus, 1767

Meantes Linnaeus, 1767: [xxxvi, Addendum].

Meantes: Stejneger & Barbour, 1917: 24; Dubois & Raffaëlli, 2012: 153.

The remainder of the synonymy is seen in the Family Sirenidae, below.

Distribution: As for the family.

Content: A single family, which is often thought to represent the most primitive living salamanders (*e.g.*, Zhang & Wake, 2009); however, the analysis of Shen *et al.* (2013) demonstrates that Cryptobranchoids are more primitive, and this taxon is the sister to all remaining families.

Family **Sirenidae** Gray, 1825

Sirenina Gray, 1825: 215.—**NG**: *Siren* Linnaeus, 1767.

Sirenea: Hemprich, 1820: xix.

Sirenidae: Hogg, 1838: 152; Bonaparte, 1840a: 395; 1850: [2]; Cope, 1889b: 223.

Sirenida: Knauer, 1878: 95.

Sirenoidia: Dubois, 2005b: 20.

Sirenoidea: Dubois, 2005b: 20.

"Chirodysmolgae" Ritgen, 1828: 277. Unavailable family-series name, not formed from a generic nomen.

Sirenes Fitzinger, 1843: 35.—**NG**: *Siren* Linnaeus, 1767.

"Trachystomata" Stannius, 1856: 4 (*fide* Frost, 2009-13, not seen).—Name unavailable above family-series, as it was explicitly coined for the family-series; however, it is not available as a family-series name because it is not formed from a genus name. As it is unavailable, it does not compete in homonymy or synonymy for the same name by Cope, 1866.

Distribution: Atlantic coast, gulf states, and Mississippi valley.

Common name: Sirens.

Content: Two genera with four species, all of which are North American.

Commentary: Gray (1825) was first to employ the name, but included Cryptobranchids and Proteids as well as Sirens. Hogg (1838) used the modern suffix, and Cope (1889b) first used the name in the restricted sense as now used. Goin & Goin (1962) proposed that sirens be treated as part of an order Trachystomata Cope, 1866, along with some fossil taxa, but this was rejected by Estes (1965); more recently Gao & Shubin (2001) found this family to be a sister to Proteidae, and even more recently Zhang & Wake (2009) presented evidence that this family is basal to all other salamander families, while proteids are a sister taxon of amphumioids; hence, placement in a separate suborder seemed appropriate. However, the analysis of Pyron & Wiens (2011) indicates this family is basal to other salamandroids, but not closely related to amphiumoids or cryptobranchoids. The more recent and completely resolved molecular analysis of Shen *et al.* (2013) demonstrates that this taxon is sister to the salamandroids (in the sense treated here). Martof (1974a) provided a CAAR family account.

Genus **Pseudobranchus** Gray, 1825

Pseudobranchus Gray, 1825: 216.—**NS**: *Siren striata* Le Conte, 1824.

Parvibranchus Hogg, 1839: 270.—**NN** for *Pseudobranchus* Gray.

Distribution: Florida, Georgia, and South Carolina.

Common name: Dwarf Sirens.

Content: Two species.

Commentary: Martof (1972) provided a generic account in the CAAR.

Pseudobranchus axanthus Netting & Goin, 1942

Synonymy: See under the subspecies, *P. a. axanthus,* below.

Distribution: Most of peninsular Florida. Petranka (1998) has a recent range map, including subspecies distributions. **U.S.**: FL.

Common name: Southern Dwarf Siren.

Content: Two subspecies

Commentary: Originally described by Netting & Goin (1942) as a subspecies of *P. striatus,* but Moler & Kezer (1993) presented evidence for full species status.

Pseudobranchus axanthus axanthus Netting & Goin, 1942

Pseudobranchus striatus axanthus Netting & Goin, 1942: 183.—**O**: **H**: CM 200339.—**OT**: U.S., Florida, Alachua County, about 5 miles southeast of Gainesville, eastern edge of Payne's Prairie, where Prairie Creek enters the River Styx.

Pseudobranchus striatus axanthus: Stejneger & Barbour, 1943: 36; Schmidt, 1953: 15; Brame, 1967: 5; Cochran & Goin, 1970: 6; Gorham, 1974: 19; Conant, 1975: 249; Behler & King, 1979: 272; Conant & Collins, 1998: 430.

Pseudobranchus axanthus: Moler & Kezer, 1993: 44.

Pseudobranchus axanthus axanthus: Collins, 1997: 9; Petranka, 1998: 480; Raffaëlli, 2007: 38.

Distribution: Northeast and central Florida. **U.S.**: FL.

Common name: Narrow-striped Dwarf Siren.

Commentary: This is the same race listed in the CAAR account by Martof (1972) as *P. striatus axanthus,* but elevation of *axanthus* to full species status included the former subspecies, *P. striatus belli,* now recognized as a subspecies of *P. axanthus.*

Pseudobranchus axanthus belli Schwartz, 1952

Pseudobranchus striatus belli Schwartz, 1952: 1.—**O**: **H**: UMMZ 10600.—**OT**: U.S., Florida, Dade County, 23.1 miles west of Miami, on the Tamiami Trail.

Pseudobranchus striatus belli: Brame, 1967: 5; Cochran & Goin, 1970: 8; Gorham, 1974: 19; Conant, 1975: 250; Behler & King, 1979: 272; Conant & Collins, 1998: 431.

Pseudobranchus axanthus belli: Collins, 1997: 9; Petranka, 1998: 480; Raffaëlli, 2007: 39.

Distribution: Southern peninsular Florida (Everglades). **U.S.**: FL.

Common name: Everglades Dwarf Siren.

Commentary: In elevating *P. s. axanthus* to full species status, Moler & Kezer (1993) included *P. s. belli,* so it becomes a race of *P. axanthus.*

Pseudobranchus striatus (Le Conte, 1824)

Synonymy: See synonymy of the subspecies, *P. s. striatus,* below.

Distribution: Southern South Carolina, southern Georgia, northern Florida. Petranka (1998) has a recent range map showing two subspecies distributions. **U.S.**: FL, GA, SC.

Common name: Northern Dwarf Siren.

Content: Three subspecies are currently recognized.

Commentary: Martof (1972) provided an account of the species in the CAAR, but the status of some of the subspecies he listed has changed.

Pseudobranchus striatus striatus (Le Conte, 1824)

Siren striata Le Conte, 1824: 53.—**O**: none stated, but indicated in the cabinet of the Lyceum, and based on the animal in Pl. 4.—**OT**: none given, but by inference (Stejneger & Barbour, 1917) one of Le Conte's plantations in Liberty or Floyd County, Georgia. Restricted (Harper, 1935: 280) to vicinity of Riceborogh, Liberty County, Georgia. Harper pointed out that the distant Floyd County reference was undoubtedly an error or lapsus, and the restriction seems a valid one.

Pseudobranchus striatus: Gray, 1825: 216; Cope, 1889b: 230; Stejneger & Barbour, 1917: 24; Schmidt, 1953: 15; Frost, 1985: 617.

Siren striata: Knauer, 1878: 95.

Pseudobranchus striatus striatus: Netting & Goin, 1942: 193; Stejneger & Barbour, 1943: 36; Cochran & Goin, 1970: Gorham, 1974: 19; 8; Conant, 1975: 250; Behler & King, 1979: 271; Conant & Collins, 1998: 430; Petranka, 1998: 482; Raffaëlli, 2007: 39.

Distribution: Coastal southern South Carolina and southeastern Georgia. **U.S.**: GA, SC.

Common name: Broad-striped Dwarf Siren.

Pseudobranchus striatus lustricolus Neill, 1951

Pseudobranchus striatus lustricolus Neill, 1951c: 39.—**O**: **H**: ERA-WTN 14215 (may now be FSM).—**OT**: U.S., Florida, Levy County, 7.8 miles east of Otter Creek.

Pseudobranchus striatus lustricolus: Schmidt, 1953: 15; Brame, 1967: 5; Cochran & Goin, 1970: 8; Gorham, 1974: 19; Conant, 1975: 250; Behler & King, 1979: 272; Conant & Collins, 1998: 430; Petranka, 1998: 482; Raffaëlli, 2007: 39.

Distribution: Gulf hammock region of Florida. Martof's (1972) range map includes a distribution for this form. **U.S.**: FL.

Common name: Gulf Hammock Dwarf Siren.

Commentary: Petranka (1998) indicated that, although he recognized this subspecies, identification and definition of the range are problematic, and no range map was presented.

Pseudobranchus striatus spheniscus Goin & Crenshaw, 1949

Pseudobranchus striatus spheniscus Goin & Crenshaw, 1949: 277.—**O**: **H**: CM 29015.—**OT**: U.S., Georgia, Lee County, 7 miles south of Smithville.

Pseudobranchus striatus spheniscus: Schmidt, 1953: 16; Brame, 1967: 5; Cochran & Goin, 1970: 8; Gorham, 1974: 19; Conant, 1975: 250; Behler & King, 1979: 271; Conant & Collins, 1998: 430; Petranka, 1998: 482; Raffaëlli, 2007: 39.

Distribution: Southwestern Georgia and the Florida panhandle. **U.S.**: FL, GA.

Common Name: Slender Dwarf Siren.

Genus **Siren** Österdam, 1766

Siren Österdam, 1766: 1.—**NS**: *Siren lacertina* Österdam, 1766 (sd.). Placed on the official list of generic names (ICZN, 1926; see also in facsimile—ICZN, 1958: 339-340).

Siren: Linnaeus, 1767: [xxxvi], Gray, 1825: 216; Barnes, 1826: 281; Cope, 1889b: 225.

Distribution: Gulf and Atlantic coasts of southeastern U.S.

Common name: Sirens.

Content: Two species.

Commentary: Martof (1974b) provided a generic CAAR account. Authorship of the generic nomen is often accorded to Linnaeus, but Österdam is the proper author (see discussion in Dubois & Raffaëlli, 2012: 100-101).

Siren intermedia Barnes, 1826

Synonymy: See under *S. intermedia intermedia.*

Distribution: Coastal drainages, North Carolina to eastern Texas and the Mississippi valley. Petranka's (1998) range map shows subspecies distributions. **U.S.**: AL, AR, FL, GA, IL, IN, KY, LA, MI, MO, MS, NC, OK, SC, TN, TX.

Common name: Lesser Siren.

Content: Three subspecies are recognized.

Commentary: Original description seems to have been by Barnes (1826), though this was apparently overlooked by subsequent authors who believed the first description was in Harlan's (1827) checklist, referring to manuscript notes attributed to Le Conte, who later (Le Conte, 1828) provided further detail. Cope (1889b: 226) treated this species as a synonym of *S. lacertina.* Martof (1973a) provided a CAAR account, with a range map indicating subspecies distributions, though Petranka (1998) has a more recent range map.

Siren intermedia intermedia Barnes, 1826

Siren intermedia Barnes, 1826: 269.—**O**: not designated but indicated to be "in the Cabinet of the Lyceum," apparently lost.—**OT**: Southern states. Corrected (Harper, 1935: 279) to vicinity of Riceborough, Liberty County, Georgia. Schmidt (1953: 14) restricted the OT to Liberty County, but without explanation, apparently without realizing Harper's correction, so Schmidt's action is invalid.

Siren intermedia: Le Conte, 1827: 322; 1828: 133; Wagler, 1830: 210; Harper, 1935: 277; Frost, 1985: 618.

Siren intermedia intermedia: Goin, 1942: 211; Stejneger & Barbour, 1943: 35; Viosca, 1949: 9; Schmidt, 1953: 14; Brame, 1967: 5; Cochran & Goin, 1970: 6; Gorham, 1974: 20; Conant, 1975: 248; Behler & King, 1979: 273; Conant & Collins, 1998: 429; Petranka, 1998: 485; Raffaëlli, 2007: 36.

Distribution: Coastal plain, North Carolina to southeastern Alabama. **U.S.**: AL, FL, GA, MS, NC, SC.

Common name: Eastern Lesser Siren.

Commentary: Le Conte (1827) referred to what seems now to be Barnes' type material in the Lyceum, as well as to other specimens (syntypes?) in the ANSP.

Siren intermedia nettingi Goin, 1942

Siren intermedia nettingi Goin, 1942: 211.—**O**: **H**: CM 7850.—**OT**: U.S., Arkansas, Lawrence County, Imboden.

Siren intermedia nettingi: Stejneger & Barbour, 1943: 35; Viosca, 1949: 9; Schmidt, 1953: 15; Brame, 1967: 5; Cochran & Goin, 1970: Gorham, 1974: 20; 6; Conant, 1975: 248; Behler & King, 1979: 273; Garrett & Barker, 1987: 72; Liner, 1994: 15; Conant & Collins, 1998: 429; Petranka, 1998: 485; Dixon, 2000: 51; Raffaëlli, 2007: 37.

Distribution: Eastern Texas, Louisiana and Mississippi, plus the Mississippi valley. **U.S.**: AL, AR, IL, IN, KY, LA, MI, MO, MS, OK, TN, TX.

Common name: Western Lesser Siren.

Commentary: In the early literature, the two species were often confused, leading to reports of this form as *S. lacertina* (*e.g.*, Davis & Rice, 1883) or *vice versa*.

Siren intermedia texana Goin, 1957

Siren intermedia texana Goin, 1957: 37.—**O**: **H**: TCWC 10567.—**OT**: U.S., Texas, Cameron County, 7 miles north of Brownsville.

Siren intermedia texana: Brame, 1967: 6; Cochran & Goin, 1970: 6; Gorham, 1974: 20; Conant, 1975: 248; Behler & King, 1979: 273; Petranka, 1998: 485; Conant & Collins, 1998: 429; Raffaëlli, 2007: 37.

Siren texana: Dixon, 2000: 51.

Misapplication of names:

Siren intermedia nettingi: Flores-Villela & Brandon, 1992: 289-291.

Distribution: Rio Grande drainage in Texas and adjacent Mexico. **U.S.**: TX.

Common name: Rio Grande Siren (subspecies not recognized by Tilley *et al.,* 2012).

Commentary: Flores-Villela & Brandon (1992), argued that the holotype was not distinguishable from *S. i. nettingi.* They tentatively suggested the larger specimens may be *S. lacertina,* but this is improbable unless there were an intentional introduction from the more eastern range. Dixon

(2000) argued that some specimens are larger than other known *S. intermedia,* and claimed species distinctness for the Rio Grande population. Pending resolution of this question, we tentatively treat the population as distinct at the level of subspecies. See further commentary of Frost (2009-13), under *S. intermedia.*

Siren lacertina Österdam, 1766

Siren lacertina Österdam, 1766: 2.—**O**: none designated.—**OT**: U.S., "Carolinae paludosis." Placed on the official list of specific names (ICZN, 1956b).

Siren lacertina: Sonnini & Latreille, 1801: 259; 1830a: 259; Barton, 1808; Fitzinger, 1826: 66; Baird, 1859e: 29; Knauer, 1878: 95; Boulenger, 1882b: 87; Cope, 1889b: 226; Werner, 1909: 13; Stejneger & Barbour, 1917: 24; 1943: 35; Strecker & Williams, 1928: 8; Schmidt, 1953: 14; Brame, 1967: 5; Cochran & Goin, 1970: 5; Gorham, 1974: 20; Conant, 1975: 247; Behler & King, 1979: 273; Frost, 1985: 618; Liner, 1994: 15; Conant & Collins, 1998: 426; Petranka, 1998: 489; Raffaëlli, 2007: 37.

Syren lacertina: Custis *in* Freeman & Custis, 1807: 23.—**IS** (genus).

Sirene lacertina: Oken, 1816: 187 (*fide* Frost, 2009-13; not seen).—**IS** (genus).

Muraena siren Gmelin, 1789: 1136.—**NN**: for *Siren lacertina* Österdam.—**SS**: Schinz, 1822: 188 (*fide* Frost, 2009-13, not seen).

Phanerobranchus dipus Leuckart, 1821: 260.—**NN** for *S. lacertina* Österdam.

Distribution: Coastal drainages of Florida and southeastern Alabama, to Maryland. Petranka (1998) has a range map. **U.S.**: AL, FL, GA, MD, NC, SC, VA.

Common name: Greater Siren.

Commentary: Schmidt (1953: 14) invalidly restricted the type locality to vicinity of Charleston, South Carolina. Martof (1973b) provided an account in the CAAR. As with the generic nomen, authorship of this species nomen is often cited as Linnaeus, but Österdam is the proper author (Dubois & Raffaëlli, 2012).

Suborder Pseudosauria de Blainville, 1816

Pseudosauriens de Blainville, 1816a: 119.

Pseudosauria: Dubois & Raffaëlli, 2012: 138.

Distribution: All temperate and tropical continental areas except Australia.

Content: Seven families, containing 54 genera and 685 species; 296 species are in North America (42.3% of the species diversity of the suborder). This suborder should probably be divided into two infraorders, based on the two distinctive clades found by Shen *et al.* (2013). One of these would include the families Amphiumidae, Plethodontidae, Proteidae, and Rhyacotritonidae, while the other would include the families Ambystomatidae, Dicamptodontidae and Salamandridae.

Commentary: Shen *et al.* (2013) referred to this taxon as Salamandroidea. This nomen has been used in the family series and is thus not available as an ordinal series name. Dubois & Raffaëlli (2012) used the name Pseudosauria de Blainville as an infraorder to include the same taxa, and we follow that usage, with modification of rank.

Family **Ambystomatidae** Gray, 1850

Ambystomina Gray, 1850: 32.—**NG**: *Ambystoma* Tschudi, 1838.

Ambystomidae: Hallowell, 1855a: 11; Stejneger & Barbour, 1943: 8; Schmidt, 1953: 16.

Ambystominae: Cope, 1859a: 122.

Amblystomidae: Cope, 1863: 54.—**IS**

Amblystomida: Knauer, 1878: 98.

Amblystomatinae: Boulenger, 1882b: 1, 31.—**IS**

Amblystomatidae: Garman, 1884a: 37.—**IS**

Ambystomoidea: Noble, 1931: 471.

Ambystomatoidea: Tihen, 1958: 1. As a superfamily.

Ambystomatoidia: Dubois, 2005b: 19. As an epifamily.

Siredontina Bonaparte, 1850: [2].—**NG**: *Siredon* Wagler, 1830.—**SS**: Boulenger, 1882b: 38, by implication.

"Acholotida" Stannius, 1856: 4. *fide* Frost (2009-13), not seen. Unavailable family-series name for *Siredon*.

Distribution: Southern Canada and Alaska, south through the Mexican Plateau.

Common name: Mole salamanders.

Content: One genus with 33 species, of which 16 species occur in the US or Canada.

Commentary: Edwards (1976) felt that the genera *Ambystoma* and *Dicamptodon* were divergent enough that each should be in a different family, thus split Ambystomatidae into two families. Good & Wake (1992) removed the genus *Rhyacotriton* to its own family as well, leaving only the nominotypical genus in this family. Frost (2009-13) retained *Dicamptodon* in the Ambystomatidae. Here we follow Edwards (1976) in including *Ambystoma* and *Dicamptodon* in separate families.

North America contains 48.5% of the species diversity of this family.

Genus **Ambystoma** Tschudi, 1838

Synonymy: See under subgenus *Ambystoma,* below.

Distribution: Southernmost Alaska, western coastal and southern Canada, the US except the desert southwest, south through the Mexican Plateau.

Common name: Mole salamanders.

Content: 33 species, of which 16 occur in the US and Canada.

Commentary: At least seven names may have been valid contenders for the genus, all published before the name *Ambystoma.* But those names have been placed on the Official Index of Rejected and Invalid Generic Names, while the name *Ambystoma* Tschudi was placed on the Official List of Generic names in Zoology. Cope (1887b) described a genus *Linguaelapsus,* and transferred *A. annulatum* to that genus. Tihen (1958) recognized other species related to *A. annulatum* and proposed they be included in *Linguaelapsus.* Freytag (1959) recommended the taxon be recognized at the genus level. Tihen (1969) provided a complete nomenclatural history for the genus in his CAAR account.

Jones *et al.* (1993) confirmed the monophyly of the subgenus *Linguaelapsus,* and two other groups of species seem to form monophyletic clades, so we recognize four subgenera for the species of the genus *Ambystoma,* following Dubois & Raffaëlli (2012).

Subgenus **Ambystoma** Tschudi, 1838

"Axolotus" Jarockiego, 1822: 179.—**NS:** *Gyrinus mexicanus* Shaw & Nodder, 1789 **(**Smith & Tihen, 1961: 215, sd.**)**. Placed on the official list of rejected and invalid names (ICZN, 1963c).—**SS**: Smith & Tihen, 1961: 215.

"Axolotus": Cuvier & Latreille, 1831: 89.—**NS:** *Siren pisciformis* Shaw, 1802 **(**mo**)**. Homonymy with *Axolotus* Jarockiego.

"Axolot": Bonaparte, 1832: 77.—**NS**: *Siren pisciformis* Shaw, 1802 **(**Smith & Tihen, 1961: 174, sd.**)**. On official list of rejected and invalid names (ICZN, 1963c).

Axolotyl: Van der Hoeven, 1833: 18.—**IS** of *Axolotus* Cuvier & Latreille, 1831.

Axolotes: Owen, 1844: 23.—**UE.**

"Axolotus": Gray, 1850: 49.—**NS**: *Axolotus mexicanus* Shaw & Nodder, 1789 **(**mo.**)**.—Homonymy with *Axolotus* Jarockiego.—**SS**: Boulenger, 1882b: 44, by implication.

Axoloteles Wood, 1863: 183.—**NS**: *Axoloteles guttatus* Wood, 1826 **(**mo.**)**.—**SS**: Smith *et al.*, 1970: 363.—**UE.**

Acholotes: Cope, 1868a: 184.—**UE**

"Philhydrus" Brookes, 1828: 16.—**NS:** *Siren pisciformis* Shaw, 1802 **(**mo.**)**. On official list of rejected and invalid names (ICZN, 1963c).—**SS**: Smith & Tihen, 1961: 174.

Phylhydrus: Swainson, 1839: 94.—**IS**

"Phillhydrus": Gray, 1831b: 108.—**NS**: *Siren pisciformis* Shaw, 1802 **(**mo.**)**. On official list of rejected and invalid names (ICZN, 1963c).—**SS**: Gray, 1831b: 108, with *Siredon*.

Phyllidrus: Agassiz, 1842-1846: 6.—**IS**

Siredon Wagler, 1830: 209.—**NS**: *Axolotus axolotl* Cuvier & Latreille, 1831 (mo.) (*Gyrinus mexicanus* Shaw & Nodder, 1789).

Sirenodon Wiegmann, 1832: 204.—**NS**: *Siredon axolotl* Wagler, 1830 **(**mo.**)**.—**SS**: Leunis, 1860: 148, by implication. On official list of rejected and invalid names (ICZN, 1963c).

Stegoporus Wiegmann, 1832: 204.—**NN** for *Sirenodon* Wiegmann & Ruthe, 1832. On official list of rejected and invalid names (ICZN, 1963c).

Hemitriton Van der Hoeven, 1833: 305.—**NN** for *Hypochthon* Merrem, *Menobranchus* Harlan, and *Siredon* Wagler.

Ambystoma Tschudi, 1838: 92.—**NS**: *Lacerta subviolacea* Barton, 1809 (od.). Priority over *Xiphonura* Tschudi, 1838, by Baird, 1850a: 283 (first reviser). On official list of generic names (ICZN, 1963c).

Amblystoma: Agassiz, 1842-1846: 2; Knauer, 1878: 98.—**IS**

Ambistoma: Jan, 1857: 55.—**IS**

Ambystoma: Stejneger & Barbour, 1917: 8; Schmidt, 1953: 17; Tihen, 1958: 3 (as subgenus).

Salamandroidis Fitzinger, 1843: 33.—**NS**: *Lacerta subviolacea* Barton, 1809, od.—**SS**: Gray, 1850: 35.

Salamandroides: Gray, 1850: 35.—**UE.**

Limnarches Gistel, 1848: xi.—**NN** for *Ambystoma* Tschudi, 1838.

Xiphoctonus Gistel, 1848: xi.—**NN** for *Xiphonura* Tschudi, 1838.

Plagiodon Duméril, Bibron, & Duméril, 1854c: 101.—**NN** for *Ambystoma* Tschudi, 1838.

Desmiostoma Sager, 1858: 428.—**NS**: *Desmiostoma maculatus* Sager, 1858 (mo.).—**SS**: Cope, 1868a: 180.

Camarataxis Cope, 1859a: 122.—**NS**: *Ambystoma maculatum* Hallowell, 1858 (od.), (but see comments by Gehlbach, 1966).—**SS**: Strauch, 1870: 61.

Pectoglossa Mivart, 1868: 698.—**NS**: *Plethodon persimilis* Gray, 1859 (mo.).—**SS**: Strauch, 1870: 69.

Rhyacosiredon Dunn, 1928: 85.—**NS**: *Amblystoma altamirani* Duges, 1895 (od.).—**SS**: Brandon, 1989: 18.

Plioambystoma Adams & Martin, 1929: 505.—**NS:** *Plioambystoma kansense* Adams & Martin, 1929 (mo.).—**SS**: Brame, 1972: 122 (unpublished) *fide* Frost (2009-13).

Bathysiredon Dunn, 1939: 1.—**NS**: *Siredon Dumerilii* Duges, 1870 (od.).—**SS**: Tihen, 1958: 44 (as subgenus).

Lanebatrachus Taylor, 1941b: 180.—**NS**: *Lanebatrachus martini* Taylor, 1941 (od.).—**SS**: Brame, 1972: 124 (unpublished, *fide* Frost, 2009-13).

Ogallalabatrachus Taylor, 1941b: 181.—**NS**: *Ogallalabatrachus horarium* Taylor, 1941 (od.).—**SS**: Brame, 1972: 124 (unpublished, *fide* Frost, 2009-13).

Distribution: Most of the U.S. and southern Canada, south to the Mexican Plateau. Representatives have been found in every state except Alaska and Hawaii, and possibly Nevada, and widely distributed in the southern tier of Canadian provinces.

Common name: No name in general use. Perhaps Typical Mole Salamanders is appropriate.

Content: Three of the 33 species of the genus are in this subgenus, all inhabiting North America.

Commentary: Tihen (1969) provided a CAAR generic account.

Ambystoma (Ambystoma) gracile (Baird, 1859)

Siredon gracilis Baird, 1859d: 13, pl. 44, fig. 2.—**O**: **S**: USNM 4080 (2 larvae).—**OT**: U.S., Oregon, Cascade Mountains, near latitude 40° N.

Ambystoma gracile: Dunn, 1926b: 136; Stejneger & Barbour, 1933: 5; 1943: 9; Frost, 1985: 554; Stebbins, 2003: 155; Raffaëlli, 2007: 93.

Ambystoma gracile gracile: Dunn, 1944: 129; Schmidt, 1953: 19; Snyder, 1963: 1; Cochran & Goin, 1970: 13; Gorham, 1974: 28; Behler & King, 1979: 291; Stebbins, 1985: 36; Petranka, 1998: 54.

Ambystoma (Ambystoma) gracile: Tihen, 1958.

Amblystoma paroticum Baird, *in* Cope, 1868a: 200.—**O**: **H**: USNM 4708 (Yarrow, 1883a: 151, sd.). Cochran, 1961a: 5, reported 2 syntypes under that number, apparently in error.—**OT**: Can., British Columbia, Chilliwack Lake (originally reported as "Chiloweyuck, Washington Territory."—**SS**: Dunn, 1926b: 135.

Chondrotus paroticum: Cope, 1887b: 88.

Ambystoma paroticum: Stejneger & Barbour, 1917: 11.

Amblystoma decorticatum Cope, 1886b: 522.—**O**: **H**: USNM 14493.—**OT**: Can., British Columbia, Port Simpson.

Chondrotus decorticatum: Cope, 1887b: 88; 1889b: 107.

Ambystoma decorticatum: Stejneger & Barbour, 1917: 9; 1943: 9.

Ambystoma gracile decorticatum: Dunn, 1944: 129; Schmidt, 1953: 19; Snyder, 1963: 2; Brame, 1967: pp; Cochran & Goin, 1970: 13; Gorham, 1974: 28; Behler & King, 1979: 291; Stebbins, 1985: 36; Petranka, 1998: 54.

Distribution: Coastal northern California through British Columbia to southeastern Alaska. Petranka (1998) has a range map, showing the subspecies. Shaffer (2005) has an updated map of distribution in the U.S. Fisher et al. (2007) have a map for Canada. **Can.**: BC. **U.S.**: AK, CA, OR, WA.

Common name; Northwestern Salamander. *French Canadian name:* Salamandre foncée.

Content: The 2 subspecies indicated in the synonymy have been recognized for many years; however, Titus (1990) presented evidence that these are probably not appropriate, so we do not recognize them here. CNAH also does not recognize any subspecies.

Commentary: Snyder (1963) provided a CAAR account.

Ambystoma (Ambystoma) maculatum (Shaw, 1802)

"Salamandra punctata" de la Cepède, 1788b: 237.—**O**: not stated, probably originally in MNHN.—**OT**: U.S., Carolina (invalid restriction by Schmidt, 1953). Suppressed on the official list of rejected/invalid names, de la Cepède's work considered nonbinomial (ICZN, 2005). Frost (2009-13) gives further information on this name.

"Salamandra punctata": Bonnaterre, 1789: 63. Junior homonym of *S. punctata* de la Cepède, 1788, or perhaps a subsequent use.—**NO.**

Ambystoma punctata: Baird, 1850a: 283; *in* Heck & Baird, 1851: 254.

Ambistoma punctatum: Jan, 1857: 55.—**IS**

Amblystoma punctatum: Cope, 1868a: 175; Knauer, 1878: 98; Davis & Rice, 1883: 26.

Ambystoma punctatum: Hay, 1892: 435.

Lacerta maculata Shaw, 1802a: 304.—**O**: none known to exist; based on Catesby, 1743: pl. 10, fig. 10.—**OT**: U.S., Carolina (invalid restriction by Schmidt, 1953: 20).

Salamandra maculata: Green, 1818: 350; Wied, 1865: 129.

Siredon maculatus: Baird, 1850a: 291.

Camarataxis maculata: Cope, 1859a: 123.

Amblystoma maculatum: Boulenger, 1882b: 38; Garman, 1884a: 36.

Ambystoma maculatum: Stejneger, 1902b: 239; Stejneger & Barbour, 1917: 10; 1943: 10; Strecker & Williams, 1928: 5; Viosca, 1949: 9; Schmidt, 1953: 20; Brame, 1967: 11; Cochran & Goin, 1970: 10; Gorham, 1974: 29; Conant, 1975: 255; Behler & King, 1979: 294; Frost, 1985: 556; Conant & Collins, 1998: 439; Petranka, 1998: 76; Raffaëlli, 2007: 91.

Ambystoma (Ambystoma) maculatum: Tihen, 1958: 3.

Salamandra venenosa Daudin, 1803: 229.—**O**: none known to exist, probably originally in MNHN.—**OT**: U.S., Pennsylvania, Philadelphia County, Philadelphia.

Lacerta subviolacea Barton, 1809: 109.—**O**: none known to exist, probably originally in ANSP.—**OT**: U.S., Pennsylvania, Philadelphia County, a few miles from Philadelphia.

Salamandra subviolacea: Harlan, 1827: 317.

Ambystoma subviolaceum: Tschudi, 1838: 92.

Salamandroidis subviolacea: Fitzinger, 1843: 33.

Ambystoma carolinae Gray, 1850: 35. Based on a fictitious *Lacerta carolinae* Linnaeus, 1766. Stejneger (1902b) suggested this was meant as a NN for *Lacerta punctata* Linnaeus, 1766.

Ambystoma carolina Wood, 1863: 182.—**UE.**

Salamandra argus Gray, 1850: 35. Proposed as a junior synonym of *A. carolinae* Gray, 1850.—**NU.**

Ambystome argus Duméril, Bibron, & Duméril, 1854c: 103.

Salamandra margaritifera Duméril, Bibron, & Duméril, 1854c: 105.—**NN**.

Distribution: Eastern U.S. and adjacent southeastern Canada. Petranka (1998) has a range map. **Can.**: NB, NS, ON, PE, QC. **U.S.**: AL, AR, CT, DE, GA, IL, IN, LA, MA, MD, ME, MI, MO, MS, NH, NJ, NY, NC, OH, OK, PA, RI, TX, VA, WI, WV.

Common name; Spotted Salamander. *French Canadian name:* Salamandre maculée.

Commentary: Anderson (1967c) provided a CAAR account. Note: *Salamandra palustris* Bechstein, 1800, is often listed as a synonym for this species but it is not; see Frost's (2009-13) notation in the synonymy of *A. maculatum.*

Ambystoma (Ambystoma) talpoideum (Holbrook, 1838)

Salamandra talpoidea Holbrook, 1838b: 117, pl. 29.—**O**: none known to exist.—**OT**: U.S., sea islands on the borders of South Carolina.

Ambystoma? talpoideum: Gray, 1950: 36 (questioned whether genus correct).

Amblystoma talpoideum: Cope, 1868a: 172; 1889b: 52; Davis & Rice, 1883: 26.

Ambystoma talpoideum: Hay, 1892: 582; Stejneger & Barbour, 1917: 11; 1943: 11; Viosca, 1949: 9; Schmidt, 1953: 21; Cochran & Goin, 1970: 11; Gorham, 1974: 29; Conant, 1975: 251; Behler & King, 1979: 296; Frost, 1985: 557; Conant & Collins, 1998: 432; Petranka, 1998: 96; Raffaëlli, 2007: 92.

Ambystoma (Ambystoma) talpoideum: Tihen, 1958: 38.

Distribution: Southeastern U.S.; Shoop (1964), Petranka (1998) and Trauth (2005b) have range maps. **U.S.**: AL, AR, FL, GA, IL, KY, LA, MO, MS, NC, OK, SC, TN, TX.

Common name; Mole Salamander.

Commentary: Shoop (1964) provided a CAAR account. Petranka (1998) suggested further studies may show the Atlantic coast and Gulf coast populations to be distinct species.

Subgenus **Heterotriton** Gray, 1850

Heterotriton Gray, 1850: 49.—**NS (**mo.**)**: *Salamandra ingens* Green, 1831.—**SS**: Cope, 1859a: 123.

Distribution: Much of central and northwestern U.S. and adjacent Canada.

Common name: No current name in use. Tiger Salamanders seems appropriate.

Content: Twenty of the 33 species of the genus are in this subgenus, of which three are in North America.

Ambystoma (Heterotriton) californiense Gray, 1853

Ambystoma californiense Gray, 1853a: 11.—**O**: **H**: BM, not known to exist.—**OT**: U.S., California, Monterey, in a well. as corrected by Schmidt (1953: 23), from vicinity of San Francisco.

Amblystoma californiense: Yarrow, 1875: 516.

Ambystoma tigrinum californiense: Cope, 1889b: 86; Dunn, 1940: 157; Stejneger & Barbour, 1943: 12; Schmidt, 1953: 23; Gehlbach, 1967: 2; Cochran & Goin, 1970: 16; Gorham, 1974: 29; Stebbins, 1985: 35.

Ambystoma californiensis: Behler & King, 1979: 289.—**UE.**

Ambystoma californiense: Shaffer & McKnight, 1996: 426; Irschick & Shaffer, 1997: 45; Petranka, 1998: 47; Raffaëlli, 2007: 80.

Ambystoma (Heterotriton) californiense: Dubois & Raffaëlli, 2012, by implicaton.

Distribution: West-central California. See range map in Petranka (1998). Gehlbach (1967) also has a map as a subspecies of *A. tigrinum*. **U.S.**: CA.

Common name; California Tiger Salamander.

Commentary: Schmidt (1953) modified the type locality to vicinity of San Francisco, apparently without evidence, so probably invalid. Considered a subspecies of *A. tigrinum* for many years, but Shaffer & McKnight (1996) presented genetic evidence that this is a distinct species, basal to the rest of the *tigrinum* complex. The species is federally listed as endangered. Gehlbach (1967) provided a CAAR account as a subspecies of *A. tirgrinum.* Dubois & Raffaëlli (2012) placed this species alone in a supraspecies *californiense.*

Ambystoma (Heterotriton) mavortium Baird, 1850

Synonymy: See under subspecies, *A. m. mavortium,* below.

Distribution: Central and northwest U.S. and adjacent Canada. Petranka's (1998) range map of *A. tigrinum* shows distribution of the species and its subspecies; note that *A. t. tigrinum* should not be included. **Can.**: AB, BC, MB, SK. **U.S.**: AR, AZ, CA, CO, KS, MN, MT, ND, NE, NM, OK, SD, TX, UT, WA, WY.

Common name; Western Tiger Salamander (SECSN) or Barred Tiger Salamander (C&T).

Content: Five subspecies are usually recognized. These were previously considered subspecies of *A. tigrinum* (Schmidt, 1953; Gehlbach, 1967; Petranka, 1998). Another population, *Amblystoma velasci* Dugés, 1891, is known from central Mexico, and may represent an extralimital subspecies of this species, or may be a distinct species (see Gehlbach, 1967). Relationships among these populations are still not well understood. Lannoo & Phillips (2005) apparently treat *A mavortium* as part of *A. tigrinum,* as there is no account for this species, and the map for *A. tigrinum* seems to include the ranges of both species.

Commentary: This species is a member of the *tigrinum* clade (Jones *et al.,* 1993). Irschick & Shaffer (1997) found this to be a species separate from *A. tigrinum,* but could not resolve how to treat the populations that have been designated as subspecies. Dubois & Raffaëlli (2012) included this species in a supraspecies *mexicanum.* Listed in British Columbia as Threatened.

Ambystoma (Heterotriton) mavortium mavortium Baird, 1850

"Ambystoma mavortia" Baird, 1850a: 284.—**NU**.

Ambystoma mavortia Baird, 1850b: 292.—**O**: **H**: possibly USNM 3990 (Yarrow, 1883a: 149, sd.); questioned by Cochran, 1961a: 21.—**OT**: U.S., New Mexico (included what is today Arizona); perhaps the Rio Grande valley between Santa Fe, NM, and El Paso, TX (Gehlbach, 1967).—**SS**: (Garman, 1884a: 36; Cope, 1889b: 68; Dunn, 1940: 158), with *A. tigrinum*; removed from synonymy by Shaffer & McKnight, 1996.

> *Ambystoma mavortium*: Gray, 1850: 37; Shaffer & McKnight, 1996: 430.
>
> *Amblystoma mavortium*: Baird, 1859b: 20; Cope, 1868a: 184; Yarrow, 1875: 516; Coues, 1875: 631.
>
> *Ambystoma tigrinum mavortium*: Dunn, 1940: 158; Stejneger & Barbour, 1943: 12; Schmidt, 1953: 22; Cochran & Goin, 1970; Gorham, 1974: 30; Conant, 1975: 256; 15; Behler & King, 1979: 298; Stebbins, 1985: 35; Garrett & Barker, 1987: 54; Shaffer & McKnight, 1993: 419; Liner, 1994: 9; Conant & Collins, 1998: 440; Petranka, 1998: 109; Dixon, 2000: 54; Lemos-Espinal *et al.*, 2004: 6, 79.
>
> *Ambystoma mavortium mavortium*: Raffaëlli, 2007: 81.
>
> *Ambystoma (Heterotriton) mavortium*: Dubois & Raffaëlli, 2012, by implication.

Amblystoma proserpine Baird & Girard, 1852d: 173.—**O**: **S:** USNM, including USNM 4082 (3), od., and the authors mentioned several other specimens in the onymatophore-series (but did not cite catalog numbers).—**OT**: the extant specimens and 3 others, U.S., Texas, Bexar County, 4 miles east of San Antonio at Salado Creek. Other specimens from "on the route from Montgomery, Mexico." Restricted to U.S., Texas, Bexar County, San Antonio, by Smith & Taylor, 1950a: 360.—**SS**: Strauch, 1870: 64; Cope, 1889b: 68; Dunn, 1940: 161.

> *Amblystoma prosperpina*: Baird, 1859e**:** 29.—**IS**
>
> *Ambystoma tigrinum proserpine*: Smith & Taylor, 1948: 14.

Siredon lichenoides Baird & Girard, 1852b: 68.—**O**: **S:** USNM 4061 (2) (Cochran, 1961a: 25, sd.).—**OT:** U.S., New Mexico, Santa Fe County, Spring Lake, at the head of Santa Fe Creek.—**SS**: Smith, 1877: 43, with *A. tirgrinum*.—**SS**: Hartmann, 1879: 76, and Fowler & Dunn, 1917: 8, with *A. mavortium*.

Siredon lichenoides: Baird, 1859b: 20; 1859d: 13; Desor, 1870: 268 (*fide* Frost, 2009-13, not seen).

Amblystoma trisruptum Cope, 1868a: 194.—**O**: **H**: USNM 4068 (Yarrow, 1883a: 150, sd.).—**OT**: U.S., New Mexico, Mora County, Ocate Rivere.—**SS**: Garman, 1884a: 36, with *A. tigrinum*; Dunn, 1940: 158, with *A. tigrinum mavortium*.

Distribution: The easternmost subspecies, from Texas to Nebraska, with fragmented populations in Texas. Petranka's (1998) map marked *A. t. mavortium* shows the range of this race, as does Gehlbach's (1967) map marked "2". **U.S.**: CO, KS, ND, NE, NM, OK, TX, UT.

Common name: Barred Tiger Salamander.

Commentary: Individuals of this subspecies have been introduced into non-native habitats in southern Arizona (Collins, 1981; Espinoza *et al.*, 1970). Gehlbach (1967: 2) provided a CAAR account of the subspecies as a race of *A. tigrinum.* Stebbins (2003: 153) treated this species as a synonym of *A. tigrinum,* but did not recognize subspecies.

Ambystoma (Heterotriton) mavortium diaboli Dunn, 1940

Ambystoma tigrinum diaboli Dunn, 1940: 160.—**O**: **H**: UMMZ 50156,—**OT**: U.S., North Dakota, Ramsey County, Devil's Lake.

Ambystoma tigrinum diaboli: Stejneger & Barbour, 1943: 12; Schmidt, 1953: 22; Brame, 1967: 12; Cochran & Goin, 1970: 15; Gorham, 1974: 29; Behler & King, 1979: 299; Stebbins, 1985: 35; Shaffer & McKnight, 1993: 419; Petranka, 1998: 109.

Ambystoma mavortium diaboli: Raffaëlli, 2007: 83.

Ambystoma (Heterotriton) mavortium diaboli: Dubois & Raffaëlli, 2012, by implication.

Distribution: The Dakotas, between the ranges of *A. tigrinum* and *A. m. melanostictum,* north into Canada where the range is more extensive. Petranka's (1998) map marked *A. t. diaboli* shows the range of this form, as does Gehlbach's (1967) map marked "7". **Can.**: MB, SK. **U.S.**: MN, ND, SD.

Common name: Gray Tiger Salamander. *French Canadian name:* Salamandre tigrée grise.

Commentary: Gehlbach (1967: 3) provided a CAAR account as a subspecies of *A. tigrinum.*

Ambystoma (Heterotriton) mavortium melanostictum (Baird, 1860)

Siredon lichenoides melanosticta Baird *in* Cooper, 1860: 306.—**O**: **H**: stated as USNM 4073, actually USNM 7043 (Gehlbach, 1966: 881).—**OT**: U.S., Nebraska, between Ft. Union and Ft. Benton; but corrected (Gehlbach, 1966: 881) to U.S., Montana, Valley County, near Frazer in the Missouri River valley, 100 miles west of Ft. Union, North Dakota (based on museum data for the corrected H).—**SS**: Bishop, 1942: 256, as *tigrinum*.

Ambystoma tigrinum melanostictum: Bishop, 1942: 256; Schmidt, 1953: 22; Brame, 1967; Cochran & Goin, 1970: 16; Gorham, 1974: 30; Conant, 1975: 256; Behler & King, 1979: 298; Conant & Collins, 1998: 441; Gehlbach, 1967; Stebbins, 1985: 34-35; Shaffer & McKnight, 1993: 419; Petranka, 1998: 109.

Ambystoma mavortium melanostictum: Raffaëlli, 2007: 82

Ambystoma (Heterotriton) mavortium melanostictum: Dubois & Raffaëlli, 2012, by implication.

Ambystoma tigrinum slateri Dunn, 1940: 159.—**O**: **H**: CPS 2489, now USNM 108982.—**OT**: U.S., Washington, Grant County, 5 miles southeast of Coulee Dam.—**SS**: Bishop, 1942: 25, with *A. t. melanostictum*.

Ambystoma tigrinum slateri: Stejneger & Barbour, 1943: 13.

Distribution: Westernmost and northernmost ranging subspecies, with fragmented populations mainly from Kansas to Canada, plus Washington and Idaho. Gehlbach's (1967: 1) map marked "5" and Petranka's (1998) map marked *A. t. melanostictum* shows the range of this form. **Can.**: AB, BC, SK. **U.S.:** MT, ND, SD.

Common name: Blotched Tiger Salamander. *French Canadian name:* Salmandre tigrée à éclaboussures.

Commentary: Gehlbach (1967: 3) provided a CAAR account of the subspecies, as a race of *A. tigrinum*.

Ambystoma (Heterotriton) mavortium nebulosum Hallowell, 1854

Ambystoma nebulosum Hallowell, 1854f: 209.—**O: S** (sd.) including USNM 4702a (Cochran, 1961a: 7), and ANSP 1294 (Lowe, 1955a: 244).—**L**: USNM 4702a, designated by Lowe, 1955a: 244.—**OT:** U.S., New Mexico; corrected by Hallowell, 1854d: 144, and 1858b: 352, to what is now U.S., Arizona, Coconino County, San Francisco Mountain, near Flagstaff.—**SS**: Smith, 1877: 42.

Ambystoma tigrinum nebulosum: Dunn, 1940: 158; Stejneger & Barbour, 1943: 13; Schmidt, 1953: 23; Brame, 1967; Gehlbach, 1967: pp; Cochran & Goin, 1970: 16; Gorham, 1974: 30; Behler & King, 1979: 298; Stebbins, 1985: 35; Shaffer & McKnight, 1993: 419; Petranka, 1998:109; Dixon, 2000: 54.

Ambystoma (Ambystoma) tigrinum nebulosum: Tihen, 1958:36.

Ambystoma mavortium nebulosum: Raffaëlli, 2007: 83.

Ambystoma (Heterotriton) mavortium nebulosum: Dubois & Raffaëlli, 2012, by implication.

Ambystoma maculatum Hallowell, 1858a: 215 (repeated, Hallowell, 1858b: 355)(nec *Ambystoma maculatum* Shaw, 1802**)**.—**O**: **H** (sd. Hallowell, 1858b: 355), originally ANSP, but could now be USNM 14481 **(**Gehlbach, 1966: 881).—**OT:** U.S., New Mexico territory (now either New Mexico or Arizona). Secondary junior homonym of *Lacerta maculata* Shaw, 1802 (= *Ambystoma maculatum*).—**SS**: with *Ambystoma tigrinum utahense* by Lowe, 1955a: 246; with *A. t. nebulosum* by Gehlbach, 1966: 881; with *A. mavortium* by Cope, 1868a: 184.

Ambystoma tigrinum utahense Lowe, 1955a: 246.—**O**: **H**: MVZ 29481.—**OT**: U.S., Utah, Uintah County, Lapoint.

Distribution: Most of Arizona, eastern New Mexico, central Colorado, Utah. Gehlbach's (1967) map marked "3" and Petranka's (1998) map marked *A. t. nebulosum* show the range of this form. **U.S.**: AZ, CO, NM, UT.

Common name: Arizona Tiger Salamander.

Commentary: Gelhbach (1967: 2) provided a CAAR account of the subspecies, as a race of *A. tigrinum.*

Ambystoma (Heterotriton) mavortium stebbinsi Lowe, 1954

Ambystoma tigrinum stebbinsi Lowe, 1954: 243.—**O**: **H**: UAZ 665.—**OT**: U.S., Arizona, Santa Cruz County, southwest side of Huachuca Mountains, J.A. Jones ranch in Parker Canyon, elev. ca. 5000 ft.

Ambystoma (Ambystoma) tigrinum stebbinsi: Tihen, 1958: 36.

Ambystoma tirgrinum stebbinsi: Cochran & Goin, 1970: 16; Behler & King, 1979: 299; Jones *et al.*, 1988: 633; Petranka, 1998: 109.

Ambystoma mavortium stebbinsi: Raffaëlli, 2007: 83.

Ambystoma (Heterotriton) mavortium stebbinsi: Dubois & Raffaëlli, 2012, by implication.

Ambystoma tigrinum utahense Lowe, 1955a: 246.—**O**: **H** (od.), MVZ 29481.—**OT**: U.S., Utah, Uintah County, Lapoint.

Ambystoma tigrinum utahense: Brame, 1967: 12; Cochran & Goin, 1970: 16.

Distribution: Known only from the vicinity of the type locality. **U.S.**: AZ.

Common name: Sonoran Tiger Salamander.

Commentary: Gehlbach (1967) regarded this form as indistinguishable from *A. t. nebulosum,* but Jones *et al.* (1988) supported recognition of the subspecies. Gorham (1974) listed it as a synonym of *A. tigrium.* The race is federally listed as endangered (under the name *Ambystoma tigrinum stebbinsi*).

Ambystoma (Heterotriton) tigrinum (Green, 1825)

"Siren operculata" de Beauvois, 1799: 281.—**O**: specimen in figs 1 & 2, not now known to exist.—**OT**: U.S., New Jersey, in a swamp near the Delaware River, not far from the middle ferry opposite Philadelphia. On official index of rejected/invalid names (ICZN, 1963d).—**SS**: Smith & Tihen, 1961: 214.—**Comment:** Only of academic interest, as the name is now invalidated, but it appears from de Beauvois' article that he was translating a portion of Peale's forthcoming catalogue, and perhaps authorship should be given as "Peale *in* de Beauvois."

"Proteus neocaesariensis" Green, 1818: 358.—**O**: not stated, probably ANSP but not now known to exist.—**OT**: not stated, probably in vicinity of U.S., New Jersey, Princeton. On official index of rejected/invalid names (ICZN, 1963d).—**SS**: Say, 1818: 406, with *Siren operculata*; Smith & Tihen, 1961: 214, with *A. tigrinum.*

Proteus neocaesareanus: Gray, 1850: 45.—**IS**

"Axolotus philadelphicus" Jarockiego, 1822: 179.—**NN** for *Siren operculata* de Beauvois, 1799. On official list of rejected/invalid names (ICZN, 1963d).—**SS**: Smith & Tihen, 1961: 214.

Salamandra tigrina Green, 1825: 116.—**O**: none known to exist; Yarrow (1883a) listed USNM 3970, but it is a paratype, not from the type locality (*fide* Frost, 2009-13).—**OT**: U.S., New Jersey, Burlington

County, near Moore's town [= Moorestown]. On official list of specific names (ICZN, 1963d).

Triton tigrinus: Holbrook, 1842: 79.

Ambystoma tigrina: Baird, 1850a: 284.

Amblystoma tigrinum: Boulenger, 1882b: 43; Cope, 1889b: 68; Stejneger & Barbour, 1917: 12; Irschick & Shaffer, 1997: 44; Raffaëlli, 2007: 80.

Ambystoma tigrinum tigrinum: Dunn, 1940: 158; Stejneger & Barbour, 1943: 12; Schmidt, 1953: 21; Cochran & Goin, 1970: 15; Gorham, 1974: 29; Conant, 1975: 255; Behler & King, 1979: 298; Frost, 1985: 557; Garrett & Barker, 1987: 55; Conant & Collins, 1998: 440; Petranka, 1998: 108; Dixon, 2000: 54.

Ambystoma (Ambystoma) tigrinum: Tihen, 1958: 3.

Ambystoma (Heterotriton) tigrinum: Dubois & Raffaëlli, 2012, by implication.

Salamandra ingens Green, 1831: 254.—**O**: **H:** ANSP 1309 **(**Fowler & Dunn, 1917: 8, sd.**)**.—**OT**: U.S., Louisiana, freshwater stream near New Orleans.—**SS**: Cope, 1868a: 179.

Triton ingens: Holbrook, 1842: 85.

Heterotriton ingens: Gray, 1850: 33.

Ambystoma ingens: Hallowell, 1858b: 353.

Salamandra lurida Sager, 1839: 323.—**O**: **H:** USNM 39442 **(**Cochran, 1961a: 24, sd.**)**.—**OT**: U.S., Michigan, Wayne County, Detroit, *fide* Cochran, 1961a: 24. Preoccupied by *Salamandra lurida* Rafinesque, 1832, which is a *nomen nudum, fide* Brame, as stated by Frost, 2009-13.—**SS**: Cope, 1868a: 179.

Ambystoma lurida: Baird, 1950a: 284.

Ambystoma episcopus Baird, 1850a: 284.—**O**: not stated or known to exist.—**OT**: U.S., Mississippi, Kemper County.—**SS**: Cope, 1868a: 179.

Siredon Harlanii Duméril, Bibron, & Duméril, 1854c: 181.—**O: S**: MNHN 4777, and one other lost.—**L**: MNHN 4777 (Gehlbach, 1966: 882).—**OT**: U.S., Caroline (stated as U.S., New Mexico, Spring Lake, by Schmidt, 1953: 23, in error).—**SS**: Bishop, 1945: 24, with *A. tigrinum mavortium*; Gehlbach, 1966: 882, with *A. t. tigrinum*.

Ambystoma bicolor Hallowell, 1858a: 215.—**O**: **H**: ANSP 10584 (Fowler & Dunn, 1917: 8, sd.).—**OT**: U.S., New Jersey, Cape May County, near Beesley's Point.—**SS**: Garman, 1884a: 36.

Amblystoma bicolor: Boulenger, 1882b: 42,

Amblystoma conspersum Cope, 1859a: 123.—**O**: **H**: ANSP 10589 (Fowler & Dunn, 1917: 10, sd.).—**OT**: U.S., Pennsylvania, Chester County, Londongrove.—**SS**: Dunn, 1940: 156.

Amblystoma conspersum: Boulenger, 1882b: 42.

Ambystoma conspersum: Stejneger & Barbour, 1917: 9; Fowler & Dunn, 1917: 10.

Amblystoma obscurum Baird, *in* Cope, 1868a: 192.—**O**: **H**: USNM 3994 (Cochran, 1961a: 5, sd).—**OT**: U.S., Iowa, Polk County, Fort Des Moines.—**SS**: Garman, 1884a: 36.

Amblystoma xiphias Cope, 1868a: 192.—**O**: **H**: USNM 14470 (ex USNM 4135; Cochran, 1961a: 6, od.).—**OT**: U.S., Ohio, Franklin County, Columbus.—**SS**: Dunn, 1940: 156.

Ambystoma xiphias: Stejneger & Barbour, 1917: 12.

Amblystoma copeianum Hay, 1885: 209.—**O**: **H**: USNM 14112 (Cochran, 1961a: 5, sd.).—**OT**: U.S., Indiana, Marion County, Irvington, near Indianapolis.—**SS**: Dunn, 1940: 156.

Ambystoma sopeanum: Brimley, 1907: 153.—**IS**.

Distribution: Much of eastern North America (see range map in Gehlbach, 1967, or Petranka, 1998, for *A. t. tigrinum* only; or Odum & Corn, 2005a, for the species in the U.S.), but absent from New England, and in Canada only at one locale. **Can.**: ON. **U.S.**: AL, AR, FL, GA, IA, IL, IN, KS, KY, LA, MI, MN, MO, NC, NE, OH, SC, SD, TN, TX, WI.

Common name; Eastern Tiger Salamander. *French Canadian name:* Salmandre tigrée.

Commentary: Three earlier synonyms that apparently apply to this species have been placed on the Official Index of Rejected and Invalid Species Names in Zoology. The name was applied for many years to a broader taxon including *A. mavortium* and *A. velasci* (extralimital), until partitioning by Shaffer & McKnight (1996). The species now is equivalent to the nominotypical subspecies listed by Schmidt (1953) and in Gehlbach's (1967) CAAR account, as well as Petranka (1998). The other subspecies listed by those authors are now treated as races of *A. mavortium.* For Canadian herpetologists, Fisher et al. (2007) did not distinguish between this species and *A. mavortium,* listing both under the name *A. tigrinum.* This species was placed alone in a supraspecies *tigrinum* by Dubois & Raffaëlli (2012). Listed federally in Canada as Endangered (Great Lakes population), and in Ontario as Extirpated.

Subgenus **Linguaelapsus** Cope, 1887

Linguaelapsus Cope, 1887b: 88 (genus).—**NS**: *Ambystoma annulatum* Cope, 1886 (Dunn & Dunn, 1940: 71, sd.).—**SS**: Günther, 1901c: 295. Use of subgenus rejected by Shaffer *et al.*, 1991: 286, as rendering *Ambystoma* paraphyletic.

Linguaelapsus: Tihen, 1958: 3 (subgenus); Jones *et al.*, 1993 (subgenus, by implication); Freytag, 1959: 80 (genus).

Distribution: Much of the eastern half of the U.S.

Common name: No name in general use. Perhaps Flatwoods Salamanders might be appropriate.

Content: 6 of the 16 North American species of the genus are placed in this subgenus.

Commentary: Tihen (1958) placed several species of *Ambystoma* in a taxonomic group, reviving Cope's name, at a subgeneric level. Freytag (1959) recognized this group of species as a full genus, but few have followed that taxonomic assessment.

Ambystoma (Linguaelapsus) annulatum Cope, 1886

Amblystoma annulatum Cope, 1886b: 88.—**O**: **H**: USNM 11564 (Cochran,1961a: 18, sd.).—**OT**: not stated, sd. (Schmidt, 1953: 18) U.S., Arkansas, Garland County, vicinity of Hot Springs; validity of this designation depends on locality data with the H.

Linguaelapsus annulatum: Cope, 1887b: 88.

Linguaelapsus annulatus: Cope, 1889b: 115.

Ambystoma annulatum: Stejneger & Barbour, 1917: 8; 1943: 8; Schmidt, 1953: 18; Anderson, 1965: 1; Cochran & Goin, 1970: 12; Gorham, 1974: 28; Conant, 1975: 252; Behler & King, 1979: 288; Frost, 1985: 554; Conant & Collins, 1998: 434; Petranka, 1998: 37; Raffaëlli, 2007: 100.

Ambystoma (Linguaelapsus) annulatum Tihen, 1958: 3; Tihen, 1969: 1.

Linguaelapsus annulatum: Freytag, 1959: 88.

Distribution: Southern Missouri, northwestern Arkansas, eastern Oklahoma. Anderson (1965) and Petranka (1998) have range maps for the species; Trauth (2005a) has a more recent map. **U.S.**: AR, MO, OK.

Common name; Ringed Salamander.

Commentary: Anderson (1965) provided a CAAR account.

Ambystoma (Linguaelapsus) barbouri Kraus & Petranka, 1989

Ambystoma barbouri Kraus & Petranka, 1989: 95.—**O**: **H**: UMMZ 182844.—**OT**: U.S., Kentucky, Pendleton County, in a first order tributary of Harris Creek which flows parallel to U.S. Hwy. 27, 4.6 km S of the Licking River.

Ambystoma (Linguaelapsus) barbouri: Jones *et al.*, 1993: 100, by implication.

Ambystoma barbouri: Kraus, 1996: 1; Petranka, 1998: 40; Conant & Collins, 1998: 436; Raffaëlli, 2007: 99.

Distribution: Western West Virginia, through northern Kentucky and central Tennessee to southern Indiana and southwestern Ohio. See maps in Kraus (1996) and Watson & Pauley (2005). **U.S.**: IN, KY, OH.

Common name; Streamside Salamander.

Commentary: Jones, *et al.* (1993) found *barbouri* to be a sister species of *A. texanum,* and explicitly supported monophyly of the subgenus, implying that *A. barbouri* is a member of the subgenus. Kraus (1996) provided a CAAR account. Robertson *et al.* (2006) indentified this species as a primary ancestor of populations of unisexual *Ambystoma.*

Ambystoma (Linguaelapsus) bishopi Goin, 1950

Ambystoma cingulatum bishopi Goin, 1950: 300.—**O**: **H**: CM 29137.—**OT**: U.S., Florida, Escambia County, about five miles north of Pensacola.

Ambystoma cingulatum bishopi: Schmidt, 1953: 18; Conant, 1958: 209; Martof, 1968: 2; Cochran & Goin, 1970: 13; Gorham, 1974: 28.

Ambystoma bishopi: Pauly, Piskurek, & Shaffer, 2007: 424.

Ambystoma (Linguaelapsus) bishopi: Tihen, 1958; Dubois & Raffaëlli, 2012: 48.

Distribution: Florida panhandle, southwestern Georgia to southern Alabama. **U.S.**: AL, FL, GA.

Common name; Reticulated Flatwoods Salamander.

Commentary: Described as a subspecies of *A. cingulatum,* and so treated by Schmidt (1953), and perhaps regarded as conspecific with *A. cingulatum* by Martof (1968) in his CAAR account. Likewise, Petranka (1998) did not mention it, but probably regarded it as conspecific with *A. cingulatum.* Pauly *et al.* (2007) found evidence it should be elevated to species status, and that it was related to the other species placed in the subgenus or genus *Linguaelapsus* by Tihen (1958) and Freytag (1959). This species is federally listed as endangered.

Ambystoma (Linguaelapsus) cingulatum Cope, 1867

Amblystoma cingulatum Cope, 1868a: 205.—**O**: **H**: USNM 3786, not listed by Cochran, 1961, assumed lost.—**OT**: U.S., South Carolina, Jasper County, Grahamville.—**N**: USNM 129396 (Goin, 1950); but Frost, 2009-13, cites statements that the H still exists, thus designation of N invalid.—**OT**, for N: U.S., South Carolina, Jasper County, Robertsville.

Chondrotus cingulatus: Cope, 1889b: 99.

Ambystoma cingulatus: Brimley, 1907: 153.

Ambystoma cingulatum: Stejneger & Barbour, 1917: 8; 1943: 9; Schmidt, 1953: 18; Brame, 1967; Martof, 1968; Conant, 1975: 252; Behler & King, 1979: 289; Frost, 1985: 554; Conant & Collins, 1998: 435; Petranka, 1998: 50; Raffaëlli, 2007: 101.

Ambystoma angulatum Wright & Wright, 1932: 10.—**IS**, *lapsus*.

Ambystoma cingulatum cingulatum: Goin, 1950: 307; Conant, 1958: 209; Cochran & Goin, 1970: 12; Gorham, 1974: 28.

Ambystoma (Linguaelapsus) cingulatum: Tihen, 1958: 3.

Linguaelapsus cingulatus Freytag: 1959: 88.

Amblystoma lepturum Cope, 1886b: 524.—**O**: **H**: USNM 14583 (Cochran, 1961a: 5, sd.).—**OT**: U.S. Designated by Schmidt, 1953: 18 as South Carolina, Jasper County, but without justification (Cope indicated the collector and locality were unknown), thus the restriction is invalid.—**SS**: Schmidt, 1953: 18.

Linguaelapsus lepturum: Cope, 1889b: 116.

Distribution: Southern South Carolina and southern Georgia, northern Florida, southern Alabama and eastern Mississippi. An updated map is available in Pulis & Means (2005). **U.S.**: AL, FL, GA, MS, SC.

Common name; Frosted Flatwoods Salamander.

Content: A. bishopi was originally described as a subspecies of *A. cingulatum* by Goin (1950) but no subspecies are currently recognized.

Commentary: Martof (1968) provided a CAAR account. This species is federally listed as threatened.

Ambystoma (Linguaelapsus) mabeei Bishop, 1928

Ambystoma mabeei Bishop, 1928a: 157.—**O**: **H**: USNM 75058.—**OT**: U.S., North Carolina, Harnett County, near Dunn, low grounds of the Black River.

Ambystoma mabeei: Stejneger & Barbour, 1933: 5; 1943: 10; Schmidt, 1953: 20; Brame, 1967: 11; Hardy and Anderson, 1970: 1; Cochran & Goin, 1970: 14; Gorham, 1974: 28; Conant, 1975: 253; Behler & King, 1979: 292; Frost, 1985: 555; Conant & Collins, 1998: 436; Petranka, 1998: 68; Raffaëlli, 2007: 99.

Ambystoma (Linguaelapsus) mabeei: Tihen, 1958: 3; 1969: 1.

Linguaelapsus mabeei: Freytag, 1959: 88.

Distribution: Atlantic coastal North and South Carolina into southeastern Virginia. Mitchell (2005) has a range map. **U.S.**: NC, SC, VA.

Common name: Mabee's Salamander.

Commentary: Hardy & Anderson (1970) provided a CAAR account.

Ambystoma (Linguaelapsus) texanum (Matthes, 1855)

Salamandra texana Matthes, 1855: 266.—**O**: none known to exist.—**OT**: U.S., Texas, Fayette County, Rio Colorado and Cumming's Creek bottom.

Amblystoma texanum: Baird, 1859e: 29; Boulenger, 1882b: 50.

Chondrotus texanus: Cope, 1889b: 104.

Ambystoma texanum: Stejneger & Barbour, 1917: 12; 1943: 11; Strecker & Williams, 1928: 6; Schmidt, 1953: 21; Anderson, 1967a: 1; Cochran & Goin, 1970: 12; Gorham, 1974: 29; Conant, 1975: 253; Behler & King, 1979: 297; Frost, 1985: 557; Garrett & Barker, 1987: 53; Conant & Collins, 1998: 436; Petranka, 1998: 103; Dixon, 2000: 53; Raffaëlli, 2007: 98.

Ambystoma (Linguaelapsus) texanum: Tihen, 1958, 3; 1969: 1.

Linguaelapsus texanus: Freytag, 1959: 88.

Amblystoma microstomum Cope, 1861b: 123.—**O**: **H**: ANSP 1285 (Fowler & Dunn, 1917: 10, sd.), or ANSP 2286 (Malnate, 1971: 348).—**OT**: unclear. U.S., Ohio, in original, but possibly U.S., Indiana, Wabash River (see Fowler & Dunn, 1917: 11, and Anderson, 1967a).

Chondrotus microstomus: Cope, 1889b: 101.
Ambystoma microstomum: Strecker, 1915: 56; Fowler & Dunn 1917: 11.

Ambystoma schmidti Taylor, 1939a: 263.—**O**: **H**: EHT-HMS 3999 (now UIMNH 25042; see Frost, 2009-2013).—**OT**: "10 miles east of San Martín (Asuncion) at Rancho Guadalupe, México, Mex." but in error.—**SS**: Kraus & Nussbaum, 1989.
Ambystoma (Linguaelapsus) schmidti: Tihen, 1958: 3.
Linguaelapsus schmidti: Freytag, 1959: pp.

Ambystoma nothagenes Kraus, 1985: 14.—**O**: **H**: UMMZ 176237.—**OT**: U.S., Ohio, Erie County, Kelley' Island.—**SS**: Bogart *et al.*, 1987: 2200.
Ambystoma nothogenes: Frank & Ramus, 1995: 27.—**IS.**

Distribution: Eastern Texas, north to southern Iowa, east to Alabama, and up the Mississippi valley to Ohio. In Canada only on Pelee Island in Lake Erie (Ontario). Petranka (1998) has a range map. **Can.**: ON. **U.S.**: AL, AR, IA, IL, IN, KS, LA, MI, MO, MS, OH, OK, TN, TX, WV.

Common name; Small-mouthed Salamander (SECSN) or Smallmouth Salamander (C&T). *French Canadian name:* Salamandre à petite bouche.

Commentary: Schmidt (1953) restricted the type locality to Rio Colorado bottom land, but without citing any evidence, so that restriction is invalid. One of the species included in the subgenus *Linguaelapsus* by Tihen (1958). Another species he included was *Ambystoma schmidti* Taylor, 1939a, which has been synonymized with *A. texanum* (Kraus & Nussbaum, 1989). Interestingly, Schmidt (1953) did not list *A. schmidti* either as a valid species or as a synonym; perhaps it was because the type locality was erroneously cited as in Mexico. Anderson (1967a) provided a CAAR account. Listed as Endangered in Ontario.

Subgenus **Xiphonura** Tschudi, 1838

Xiphonura Tschudi, 1838: 95.—**NS** (mo.): *Salamandra jeffersoniana* Green, 1827.—**SS**: Baird, 1850a: 283.
Xiphonura: Dubois & Raffaëlli, 2012: 147.

Distribution: Most of eastern U.S. and adjacent Canada, plus northwestern U.S. and adjacent Canada, and peninsular Alaska.

Common name: No name in use.

Content: Four species, all of which are in North America.

Ambystoma (Xiphonura) jeffersonianum (Green, 1827)

Salamandra jeffersoniana Green, 1827: 4.—**O**: **S:** probably included USNM 3968 (Uzzell, 1967a, sd.).—**OT**: U.S., Pennsylvania, Washington County, Cannonsburg, near Chartier's Creek in the vicinity of Jefferson College.

Xiphonura jeffersoniana: Tschudi, 1838: 93.

Salamandra jeffersoniana: Holbrook, 1842: 5: 51.

Ambystoma jeffersoniana: Baird, 1850a: 283; *in* Heck & Baird, 1851: 254.

Amblystoma jeffersonianum: Cope, 1868a: 195.

Amblystoma jeffersonianum jeffersonianum: Cope, 1875b: 26; 1889b: 89.

Ambystoma jeffersonianum: Stejneger & Barbour, 1917: 9; 1943: 9; Dunn, 1918b: 458; Schmidt, 1953: 19; Cochran & Goin, 1970: 13; Gorham, 1974: 28; Conant, 1975: 254; Behler & King, 1979: 291; Frost 1985: 555; Conant & Collins, 1998: 437; Petranka, 1998: 58; Raffaëlli, 2007: 95.

Ambystoma (Ambystoma) jeffersonianum: Tihen, 1958: 39.

Ambystoma (Xiphonura) jeffersonianum: Dubois & Raffaëlli, 2012: 147.

Salamandra granulata DeKay *in* Holbrook, 1842: 5: 63.—**O**: **S**: including specimen of pl. 20, not now known to exist, but Cochran, 1961a: 24, listed USNM 39281 (2) as syntypes of *S. granulata* DeKay (sd.).—**OT**: U.S., northern districts of New York. See Adler, 1976: xxxviii, discussion of authorship; also noted synonymy is questionable.—**Comment**: Duméril, *et al.*, 1854c: 81, suggested this may be a synonym of *Plethodon glutinosus*.

Ambystoma fuscum Hallowell, 1858b: 355. (description repeated, Hallowell, 1858a: 216).—**O**: **H**: ANSP 1379 (Fowler & Dunn, 1917: 10, sd.).—**OT**: U.S., Indiana, Jefferson County, Hanover, near Hanover College.—**SS**: Cope, 1868a: 197.

Amblystoma jeffersonianum var. *fuscum*: Cope, 1868a: 197.

Plethodon persimilis Gray, 1859: 230.—**O**: not stated, probably BM.—**OT**: uncertain; given in error as Siam, Laos Mountains (Mivart 1868: 699). "Designated" (invalidly) as vicinity of New York city (Schmidt, 1953: 19).

Pectoglossa persimilis: Mivart, 1868: 698.

Amblystoma persimile: Strauch, 1870: 65.

Amblystoma persimilis: Bourret, 1927: 254.

Amblystoma platineum Cope, 1868a: 198.—**O**: **S**: USNM 7145, 4688. But Cochran (1961a: 5) reported USNM 3998 and USNM 3988 (now 39444); also ANSP 1299, regarded as holotype (holophoront) by Fowler & Dunn, 1917: 10 (see discussion by Uzzell, 1964: 292).—**OT**: U.S., Ohio, Cuyahoga County, Cleveland.—**SS**: not determined, but resurrected by Uzzell (1964) for a hybrid lineage (see discussion by Lowcock *et al.*, 1987, who advised that the population represented by this name cannot be named under the *Code*, because of its hybrid nature).

Amblystoma jeffersonianum platineum: Davis & Rice, 1883: 26.

Ambystoma tremblayi Comeau, 1943: 124.—**O**: unclear, probably originally in Section of Biology, Université Laval, Quebec, Canada; Brame, 1959: 20, considered lost.—**OT**: not stated; Brame, 1959: 20 (sd.), designated Can., Quebec, Quebec County, Cap Rouge.—**Comment:** Brame (1959) argued the name is a *nomen nudum*, but Frost (2009-13) notes the description provides a diagnosis. Uzzell (1964) suggested this is a gynogenetic lineage, but Lowcock *et al.* (1987) argued it is not a lineage but a hybrid swarm.

Ambystoma tremblayi: Brame, 1967: 12.

Distribution: Western New England to Indiana and Kentucky; restricted range in southern Ontario. Petranka (1998) has a range map for the species. **Can.:** ON. **U.S.**: KY, IN, MD, NY, OH, PA, VA, WV.

Common name; Jefferson Salamander. *French Canadian name:* Salamandre de Jefferson.

Commentary: Lowcock *et al.* (1987) considered *A. tremblayi* a hybrid swarm of *A. jeffersonianum X A. laterale,* so partly synonymous with both species. *A. jeffersonianum* and *A. laterale* hybridize extensively, confusing the identity of these species. Uzzell (1967a) provided a CAAR account, and accounts (1967c, d) of *A. platineum* and *A. tremblayi*, regarded here as hybrid populations. A member of the *jeffersonianum* clade (Jones *et al.*, 1993). Listed as Endangered in Ontario. Also see Bogart & Klemens (2008) and references cited therein, regarding status of pure diploid, bisexual populations, versus diploid and polyploid unisexual hybrid populations of this species and the following one.

Ambystoma (Xiphonura) laterale Hallowell, 1857

Ambystoma laterale Hallowell, 1857a: 6.—**O**: **H**: ANSP 1377 (Fowler & Dunn, 1917: 10, sd.).—**OT**: U.S., Michigan, Marquette County, Marquette, southern border of Lake Superior.

Amblystoma jeffersonianum laterale: Davis & Rice, 1883: 26.

Ambystoma jeffersonianum laterale: Cope, 1889b: 92.

Ambystoma laterale: Minton, 1954; Brame, 1967; Cochran & Goin, 1970: 14; Gorham, 1974: 28; Conant, 1975: 252; Behler & King, 1979: 292; Frost, 1985: 555; Petranka, 1998: 63; Raffaëlli, 2007: 96.

Ambystoma (Ambystoma) laterale: Tihen, 1958: 39.

Ambystoma (Xiphonura) laterale: Dubois & Raffaëlli, 2012: 48.

Amblystoma platineum: Cope, 1868a: 198. See this nomen in the synonymy of *A. jeffersonianum*.

Ambystoma tremblayi: Comeau, 1943: 124. See this nomen in the synonymy of *A. jeffersonianum*.

Ambystoma nothagenes Kraus, 1985: 14.—**O**: **H** (od.), UMMZ 176237.—**OT**: U.S., Ohio, Erie County, Kelley's Island.—**SS**: Bogart *et al.*, 1987: 2200; they considered this name applied to non-gynogenetic hybrids, *A. texanum X A. laterale,* thus not a valid species name.

Misapplied names:

Ambystoma jeffersonianum: Stejneger & Barbour, 1943: 9-10 (part); Schmidt, 1953: 19 (part).

Distribution: Northeastern U.S. and southeastern Canada; hybridizes with *A. jeffersonianum* where the two overlap. Petranka (1998) has a range map. **Can.:** MB, NB, NE, NS, ON, QC. **U.S.:** CT, DE, IL, MA, MD, MI, MN, NH, NJ, NY, OH, PA, RI, VA, VT, WI; may now be extirpated from RI.

Common name; Blue-spotted Salamander. *French Canadian name:* Salmandre à points bleus.

Commentary: See commentary for *A. jeffersonianum*; some names are synonyms for that species as well as this. Schmidt (1953: 19) considered this a synonym of *A. jeffersonianum.* Uzzell (1967b) provided a CAAR account. Listed as endangered by New Jersey, special concern in Massachusetts, and threatened in Connecticut.

Ambystoma (Xiphonura) macrodactylum Baird, 1850

Synonymy: See the subspecies, *A. m. macrodactylum,* below.

Distribution: Northwestern U.S and southwestern Canada and peninsular Alaska. Petranka (1998) has a range map showing subspecies distributions. **Can.**: AB, BC. **U.S.**: AK, CA, ID, MT, OR, WA.

Common name; Long-toed Salamander. *French Canadian name:* Salamandre à longs doigts.

Content: Five subspecies are recognized.

Commentary: Ferguson (1963) provided a CAAR account.

Ambystoma (Xiphonura) macrodactylum macrodactylum Baird, 1850

Ambystoma macrodactyla Baird, 1850b: 292.—**O**: **S**: USNM 4042 exists, others (ANSP) apparently lost.—**OT**: U.S., Oregon, Clatsop County, Astoria.

Ambystoma macrodactylum: Gray, 1850: 37; Stejneger & Barbour, 1917: 10; 1943: 10; Bishop, 1943: 139; Schmidt, 1953: 20; Frost, 1985: 555.

Amblystoma macrodactylum: Cope, 1868a: 198; 1889b: 95.

Ambystoma macrodactylum macrodactylum: Mittleman, 1948a: 92; Brame, 1967: 11; Cochran & Goin, 1970: 17; Gorham, 1974: 28; Behler & King, 1979: 293; Stebbins, 1985: 37; 2003: 157; Petranka, 1998: 71; Raffaëlli, 2007: 97.

Ambystoma (Ambystoma) macrodactylum: Tihen, 1958: 39.

Ambystoma (Xiphonura) macrodactylum macrodactylum: Dubois & Raffaëlli, 2012: 48.

Amblystoma epixanthum Cope, 1884: 16.—**O**: **S**: ANSP 3880-3883 (Fowler & Dunn, 1917: 10, sd.); reported as ANSP 13880-13883 by Malnate, 1971: 347, who noted 13882 sent to MCZ; Barbour & Loveridge, 1929: 210, reported **S** MCZ 4900, from ANSP.—**OT**: U.S., Idaho, Ellmore County, Atlanta, South fork of Boise River. Slevin (1928: 29) reported the OT as swamp near head of South Boise River, south side of the Sawtooth Mountain range, Idaho.—**SS**: Slevin, 1928: 29 and Stejneger & Barbour, 1933: 6.

Ambystoma epixanthum: Stejneger & Barbour, 1917: 9; Fowler & Dunn, 1917: 11.

Ambystoma stejnegeri Ruthven, 1912b: 517.—**O**: **H**: USNM 48598.—**OT**: U.S., Iowa, Davis County, Bloomfield; reported as error by Mittleman, 1948a: 94, in treating as synonymy with *A. macrodactylum*.

Ambystoma stejnegeri: Stejneger & Barbour, 1917: 11.

Distribution: Coastal Oregon to British Columbia. **Can.**: BC. **U.S.**: OR, WA.

Common name: Western Long-toed Salamander. *French Canadian name:* Salmandre à longs doigts de l'Ouest.

Ambystoma (Xiphonura) macrodactylum columbianum Ferguson, 1961

Ambystoma macrodactylum columbianum Ferguson, 1961: 313.—**O**: **H**: USNM 142228.—**OT**: U.S., Oregon, Union County, 0.5 mile N. Anthony Leakes (SW 1/4, Sec. 7, R37E, T78), elev. 7100 feet.

Ambystoma macrodactylum columbianum: Brame, 1967: 11; Cochran & Goin, 1970: 14; Gorham, 1974: 28; Behler & King, 1979: 293; Stebbins, 1985: 293; 2003: 157; Petranka, 1998: 71; Raffaëlli, 2007: 97.

Ambystoma (Xiphonura) macrodactylum columbianum: Dubois & Raffaëlli, 2012: 48.

Distribution: Northeastern Oregon and Washington, northern Idaho, western British Columbia. Petranka (1998) has a range map. **Can.**: BC. **U.S.**: ID, OR, WA.

Common name: Eastern Long-toed Salamander. *French Canadian name:* Salamandre à longs doigts de l'Est.

Commentary: A. m. krausei is a more eastern subspecies, and *A. m. columbianum* ranges farther north, so the common names are not very appropriate.

Ambystoma (Xiphonura) macrodactylum croceum Russell and Anderson, 1956

Ambystoma macrodactylum croceum Russell & Anderson, 1956: 137.—**O**: **H**: MVZ 63734.—**OT**: U.S., California, Santa Cruz County, Rio Del Mar.

Ambystoma macrodactylum croceum: Brame, 1967: 11; Cochran & Goin, 1970: 14; Gorham, 1974: 28; Behler & King, 1979: 293; Stebbins, 1985: 37; 2003: 157; Petranka, 1998: 71; Raffaëlli, 2007: 98.

Ambystoma croceum: Collins, 1991: 43.

Ambystoma (Xiphonura) macrodactylum croceum: Dubois & Raffaëlli, 2012: 48.

Distribution: Known only from the vicinity of the type locality. Petranka (1998) shows the disjunct range on his map. **U.S.**: CA.

Common name: Santa Cruz Long-toed Salamander.

Commentary: Some may argue this disjunct population should be given species status, as there is no gene flow between it and the nearest *A. m. sigillatum* across the mountains on the other side of the state. The subspecies is federally listed as endangered.

Ambystoma (Xiphonura) macrodactylum krausei Peters, 1882

Ambystoma krausei Peters, 1882b: 145.—**O**: **H**: ZMB 10364 (Bauer, *et al.*, 1993: 290, sd.).—**OT**: U.S., Montana, Flathead River.—**SS**: Mittleman, 1948a: 94.

Ambystoma macrodactylum krausei: Mittleman, 1948a: 94; Brame, 1967: 11; Cochran & Goin, 1970: 15; Gorham, 1974: 28; Behler & King, 1979: 293-294; Stebbins, 1985: 38; 2003: 157; Petranka, 1998: 71; Raffaëlli, 2007: 97.

Ambystoma (Xiphonura) macrodactylum krausei: Dubois & Raffaëlli, 2012, by inference.

Distribution. Northeastern Idaho, western Montana, eastern British Colombia, western Alberta. Petranka (1998) has a range map. **Can.**: AB, BC. **U.S.**: ID, MT.

Common name. Northern Long-toed Salamander. *French Canadian name:* Salamandre à longs doigts du Nord.

Ambystoma (Xiphonura) macrodactylum sigillatum Ferguson, 1961

Ambystoma macrodactylum sigillatum Ferguson, 1961: 316.—**O**: **H**: USNM 142212.—**OT**: U.S., Oregon, Klamath County, 100 yards W. of the boat landing in Eagle Cove of Crater Lake.

Ambystoma macrodactylum sigillatum: Brame, 1967: 11; Cochran & Goin, 1970: 15; Gorham, 1974: 28; Behler & King, 1979: 294; Stebbins, 1985: 37; 2003: 157; Petranka, 1998: 71; Raffaëlli, 2007: 98.

Ambystoma (Xiphonura) macrodactylum sigillatum: Dubois & Raffaëlli, 2012: 48.

Distribution: Northern California to southwestern Oregon. **U.S.**: CA, OR.

Common name: Southern Long-toed Salamander.

Ambystoma (Xiphonura) opacum (Gravenhorst, 1807)

Salamandra opaca Gravenhorst, 1807: 431.—**O**: None known to exist.—**OT**: U.S., New York.

Amblystoma opacum: Gray, 1850: 36; Knauer, 1878: 98; Davis & Rice, 1883: 26; Cope, 1889b: 54.

Ambystoma opacum: Baird, 1850a: 283; Dunn, 1918b: 456; Stejneger & Barbour, 1917: 10; 1943: 11; Bishop, 1924: 88; 1928a: 157; Strecker & Williams, 1928: 7; Viosca, 1949: 9; Schmidt, 1953: 21; Brame, 1967: 11; Cochran & Goin, 1970: 11; Gorham, 1974: 29; Conant, 1975: 251; Behler & King, 1979: 295; Frost, 1985: 556; Conant & Collins, 1998: 433; Petranka, 1998: 88; Raffaëlli, 2007: 94.

Ambystoma opaca: Baird *in* Heck & Baird, 1851: 254.

Salamandra opaca: Duméril, Bibron, & Duméril, 1854c: 66.

Ambystoma (Ambystoma) opacum: Tihen, 1958: 39.

Ambystoma (Xiphonura) opacum: Dubois & Raffaëlli, 2012: 48.

Salamandra fasciata Green, 1818: 350.—**O**: **S**: ANSP 1420-1423 (Fowler & Dunn, 1917: 9, sd.).—**OT**: not stated, probably vicinity of Princeton, New Jersey. Schmidt (1953: 21) restricted it to U.S., New Jersey, vicinity of Princeton, probably on the basis that Green lived there.—**SS:** Duméril, Bibron, & Duméril (1854c: 67, with "?").

Ambistoma fasciatum: Jan, 1857: 55.—**IS** of generic nomen.

Salamandra gravenhorstii Fitzinger, 1826: 66.—**NN** for *S. opaca* Gravenhorst, 1807.

Salamandroides gravenhorstii: Fitzinger, 1864: pl. 97, fig. 173.

?*"Salamandra armigera"* Duméril, Bibron, & Duméril, 1854c: 106.—**NU**, attributed to Valenciennes, under the name *Salamandra semi-fasciata*, and stated it could be a variety of *Ambystome argus*.

Distribution: Much of the eastern U.S. Petranka (1998) has a range map. **U.S.**: AL, AR, CT, DC, DE, FL, GA, IL, IN, KY, LA, MA, MD, MI, MO, MS, NC, NJ, NY, OH, RI, SC, TN, TX, VA, WV.

Common name; Marbled Salamander.

Commentary: Anderson (1967b) provided a CAAR account.

Family **Amphiumidae** Gray, 1825

Amphiumidae Gray, 1825: 216.—**NG**: *Amphiuma* Garden, 1821

Amphiumoidea: Fitzinger, 1828: 24 (explicit family)

Amphiumina: Bonaparte, 1840a: 101, as subfamily of Amphiumidae.

Amphiumidae: Bonaparte, 1850: [2]; Cope, 1875b: 25; 1889b: 213; Blackburn & Wake, 2011: 46.

Amphiumina: Bonaparte, 1850: [2].

Amphiumida: Jan, 1857: 19; Knauer, 1878: 96.

Amphiumoideae: Stejneger, 1907: 3. Explicit superfamily.

Amphiumoidea: Dunn, 1922: 426; Dubois, 2005b: 19; superfamily.

Muraenopses Fitzinger, 1843: 34.—**NG**: *Muraenopsis* Fitzinger, 1843.

Distribution: Mississippi valley and Gulf coast from eastern Texas through Florida, Atlantic coast from Florida to Virginia.

Common name: Amphiumas; Congo (or Conger) Eels, and Lamper Eels are vernacular names in parts of the range.

Content: One genus with three species, all in North America.

Commentary: Frost's (2009-13) synonymy implicitly corrected the nomenclatural history given in the comment by Salthe (1973a) in his CAAR family account. Frost *et al.* (2006) included this family and the Plethodontidae in a new suprafamilial taxon, Xenosalamandroidei; that nomen is considered invalid for the family-series because it is not based on a generic name. Several studies have shown that this family is a sister family of Plethodontidae (*e.g.,* Pyron & Wiens, 2011; Shen *et al.*, 2013). Both of those studies also showed the clade formed by the two familes is a sister to the Rhyacotritonidae, and that overall clade is a sister to the Proteidae.

Genus **Amphiuma** Garden, 1821

Amphiuma Garden *in* Smith, 1821: 333, 599.—**NS** (mo.): *Amphiuma means* Garden, 1821.

Amphiuma: Barnes, 1826: 279; Wagler, 1830: 209; Knauer, 1878: 96; Cope, 1889b: 215.

Chrysodonta Mitchill, 1822: 503.—**NS** (mo.): *Chrysodonta larvaeformis* Mitchill, 1822 (= *Amphiuma means* Garden, 1821).—**SS**: Cuvier, 1827: 2.

Sirenoidis Fitzinger, 1843: 34.—**NS (mo.)**: *Amphiuma didactylum* Cuvier, 1827 (= *Amphiuma means* Garden, 1821).—**SS**: Gray, 1850: 54.

Muraenopsis Fitzinger, 1843: 34.—**NS** (mo.): *Amphiuma tridactylum* Cuvier, 1827.—**SS**: Ryder, 1879: 15.

Distribution: As for Family.

Common name: Amphiumas. The vernacular name Conger Eel or Congo Eel, is often used by the layman in many localities.

Content: Three species, all North American.

Commentary: Garden described the genus in correspondence in 1773, which was published by Smith (1821) along with other letters by early naturalists. Salthe (1973a) provided a CAAR generic account.

Amphiuma means Garden, 1821

Amphiuma means Garden *in* Smith, 1821: 333, 599.—**O**: none stated; Lönnberg (1896) may have rediscovered the **H**: ZMUU 15.—**OT**: not stated but from context perhaps either Charleston, South Carolina, or eastern Florida; restricted to Charleston, South Carolina (Schmidt, 1953).

Amphiuma means: Wagler, 1830: 209; Knauer, 1878: 96; Boulenger 1882b: 83; Cope, 1889b: 216; Werner, 1909: 76; Stejneger & Barbour, 1917: 6; Baker, 1947: 11; Viosca, 1949: 9; Cochran & Goin, 1970: 19; Gorham, 1974: 27; Conant, 1975: 245; Behler & King, 1979: 285; Frost, 1985: 559; Conant & Collins, 1998: 425; Petranka, 1998: 132; Raffaëlli, 2007: 171.

Amphiuma means means: Goin, 1938: 128; Stejneger & Barbour, 1943: 3; Baker, 1945: 58; Schmidt, 1953: 27: Conant, 1958: 205.

Sireni simile Garden *in* Smith, 1821: 591. Noted this nomen was used by Linnaeus.

Chrysodonta larvaeformis Mitchill, 1822: 503.—**O**: **S:** (2) deposition unknown.—**OT**: U.S., Georgia, Chatham County (vicinity of Savannah).—**SS**: Wagler, 1830: 7, with *A. didactylum*.

Amphiuma didactylum Cuvier, 1827: 4.—**O**: not stated, probably originally in MNHN.—**OT**: U.S., Louisiana, New Orleans, Georgia, and South Carolina. Given as a synonym of *A. means*.

Amphiuma didactylum: Wagler, 1830: 209; 1933a: (not paginated) Pl. 19, Fig. 1.

Sirenoidis didactylum: Fitzinger, 1843: 34

Distribution: Louisiana east of the Mississippi through Florida, Atlantic coast to Virginia, below the fall line. See maps in Salthe (1973b) and Petranka (1998). **U.S.**: AL, FL, GA, LA, MS, NC, SC, VA.

Common name: Two-toed Amphiuma.

Content: No subspecies are currently recognized, but *A. tridactylum* was previously considered a race of this species.

Commentary: Salthe (1973b) provided a CAAR account. Alexander Garden (a botanist living in Charleston, SC), in a letter to Linnaeus, published by James Edward Smith (1821), in the first volume of a two-volume work in which were printed a number of items of correspondence among several early naturalists, described (p. 333; June 20, 1770) this species to Linnaeus, and wondered if it could be related to a legless lizard (*Anguis*), or a link between lizards and snakes. He had at first thought it was a *Siren,* but after examination suspected it to be a different taxon. He stated he was sending the specimen to Linnaeus. In a letter (p. 591; December 10, 1772) Garden later noted that Linnaeus had used the nomen *Sireni simile* for the species (apparently in a missing letter). Then in another letter (May 15, 1773; p. 599), he used his name *Amphiuma means* in referring to the specimen he had sent to Linnaeus. Publication of the letters by Smith constitutes the final step for making the nomen available, carrying the authorship of Garden, and the date of Smith's volume (some 30 years after Garden's death). Bonett *et al.* (2009) showed this to be a sister species to *A. pholeter*.

Amphiuma pholeter Neill, 1964

Amphiuma pholeter Neill, 1964: 62.—**O**: **H**: FSM 17655 (ex WTN US2675).—**OT**: U.S., Florida, Levy County, 4.5 miles NE by E of Rosewood.

Amphiuma pholeter: Cochran & Goin, 1970: 20; Gorham, 1974: 27; Conant, 1975: 246; Behler & King, 1979: 286; Frost, 1985: 559; Petranka, 1998: 134; Conant & Collins, 1998: 426; Raffaëlli, 2007: 170.

Distribution: Gulf coast from southwestern Alabama to northwestern Florida and adjacent Georgia. Petranka (1998) has a range map. **U.S.**: AL, FL, GA.

Common name: One-toed Amphiuma.

Content: No subspecies have been described.

Commentary: Means (1996) provided a CAAR account.

Amphiuma tridactylum Cuvier, 1827

"Syren quadrapeda" Custis *in* Freeman & Custis, 1807: 23.—**O**: not stated nor known to exist.—**OT**: U.S., Louisiana, 21 miles below Natchitoches, in a pond near the Red River. Suppressed for purposes of priority but not homonymy.

Amphiuma tridactylum Cuvier, 1827: 7.—**O**: **H**: not stated, but MNHN 7821, *fide* Frost, 2009-13.—**OT**: U.S., Louisiana, New Orleans.

Amphiuma tridactylum: Wagler, 1830: 209; 1933a: (unpaginated) Pl. 19, Fig. 2; Knauer, 1878: 96; Stejneger & Barbour, 1917: 6; Strecker & Williams, 1928: 4; Baker, 1947: Viosca, 1949: 9; 12; Hill, 1954: 214; Cochran & Goin, 1970: 19; Gorham, 1974: 27; Conant, 1975: 246; Behler & King, 1979: 286; Frost,1985: 559; Conant & Collins, 1998: 425; Petranka, 1998: 136; Raffaëlli, 2007: 171.

Muraenopsis tridactyla: Fitzinger, 1843: 34.

Amphiuma tridactyla: Boulenger, 1882b: 82.

Amphiuma means tridactylum: Goin, 1938: 128; Bishop, 1943: 54; Stejneger & Barbour, 1943: 3; Baker, 1945: 59; Schmidt, 1953: 28; Conant, 1958: 208.

Distribution: Eastern Texas to western Alabama, Mississippi River valley north to southeastern Missouri and southwestern Kentucky, with questionable records from Indiana and Illinois. Petranka (1998) has a range map. **U.S.**: AL, AR, IL?, IN?, KY, LA, MO, MS, OK, TN, TX.

Common name: Three-toed Amphiuma.

Commentary: Most authors have treated this as a distinct species, but Cope (1889b: 216) treated it as a synonym of *A. means.* For several years this species was treated as a subspecies of *A. means,* following Goin's (1938) report of intergrades in Louisiana. But Baker (1947) found contradictory evidence and returned the taxon to full species status; Hill (1954) supported this action with further evidence. Bonett *et al.* (2009) provided genetic data showing this species is distinct and is a sister to the other two species. Salthe (1973c) provided a CAAR account.

Family **Dicamptodontidae** Tihen, 1958

Ambystomina Gray, 1850: 32.—**NG**: *Ambystoma* Tschudi, 1838.

Ambystomidae: Hallowell, 1857a: 11; Noble, 1931: 472; Stejneger & Barbour, 1943: 8; Schmidt, 1953: 16.

Amblystomidae: Cope, 1863: 54.

Dicamptodontinae Tihen, 1958: 3 (as a subfamily).—**NG**: *Dicamptodon* Strauch, 1870.

Dicamptodontidae: Edwards, 1976; Good & Wake, 1992; Petranka, 1998; Frost *et al.*, 2006.

Distribution: Coastal northern California to Washington and adjacent British Colombia; northern Idaho.

Common name: Pacific Giant Salamanders.

Content: One genus with four species, all in North America. There are also several fossil genera referred to this family.

Commentary: Recognition of the family has been controversial, and the genus is often treated within the family Ambystomatidae. However, Good and Wake (1992) presented evidence beyond that of Tihen (1958) and Edwards (1976) and recommended that the family be recognized as distinct from ambystomatids, as well as the Rhyacotritonidae. Frost *et al.* (2006: 37) and Dubois & Raffaëlli (2012) also recognized it as a separate family.

Genus **Dicamptodon** Strauch, 1870

Dicamptodon Strauch, 1870: 68.—**NS**: *Triton ensatus* Eschscholz, 1833 (mo.).

Chondrotus Cope, 1887b: 88.—**NS**: *Amblystoma tenebrosum* Baird & Girard, 1852 (od.).—**SS**: Hay, 1892: 427.

Distribution: As for the family.

Common name: Pacific Giant Salamanders. *French Canadian name:* Grandes salamandres.

Content: 4 species, all of which occur in the US or Canada.

Commentary: Anderson (1969) provided a generic CAAR account. Frost (2009-13) has a more complete synonymy. Nussbaum (1976) provided a good nomenclatural history for the genus and its species to that time. Steele *et al.* (2005) studied phylogenetic relationships among the species.

Dicamptodon aterrimus (Cope, 1868)

Amblystoma aterrimum Cope, 1868a: 201.—**O**: **H**: USNM 5242 (Cochran, 1961a, sd.).—**OT**: North Rocky Mountains; data with H—"crossing of Bitter Root River, north Rocky Mountains (Montana)" (Cochran, 1961a).

Amblystoma aterrimum; Boulenger, 1882b: 49; Stejneger & Barbour, 1917: 8.

Chondrotus aterrimus: Cope, 1887b: 88; 1889b: 109.

Dicamptodon aterrimus: Dunn, 1923: 39; Stejneger & Barbour, 1923: 7; Daugherty *et al.*, 1983: 679; Frost, 1985: 561; Petranka, 1998: 146; Stebbins, 2003: 159; Steele *et al.*, 2005: 95; Raffaëlli,2007: 73.

Distribution: North-central Idaho. Lohman & Bury (2005) have a map. **U.S.**: ID.

Common name: Idaho Giant Salamander.

Commentary: Stejneger & Barbour (1943: 8), Schmidt (1953: 16), and Gorham (1974: 30) treated this as a synonym of *D. ensatus,* as did most subsequent authors, until the 1980's. Steele *et al.* (2005) found this to be a basal sister taxon to the other species.

Dicamptodon copei Nussbaum, 1970

Dicamptodon copei Nussbaum, 1970: 506.—**O**: **H**: USNM 166784.—**OT**: U.S., Washington, Cowlitz County, Marratta Creek 85 m upstream from bridge on state highway 504; SW 1/4 sec. 3, T.9R.4 E.; 46° 17' N, 122° 18' W; 840 m elevation.

Dicamptodon copei: Behler & King, 1979: 300; Frost, 1985: 561; Stebbins, 1985: 38; 2003: 160; Petranka, 1998: 147; Raffaëlli, 2007: 74.

Distribution: Western Washington; see map in Petranka (1998). **U.S.**: WA.

Common name: Cope's Giant Salamander.

Commentary: Nussbaum (1983) provided a CAAR account. Steele *et al.* (2005) found this to be a sister species to the pair *D. ensatus* + *D. tenebrosus.*

Dicamptodon ensatus (Eschscholtz, 1833)

Triton ensatus Eschscholtz, 1833: 6.—**O**: none known to exist.—**OT**: U.S., California, vicinity of the Bay of San Francisco. Probably near Ft. Ross, Sonoma County (Nussbaum, 1976: 4).

Dicamptodon ensatus: Strauch, 1870: 69; Stejneger & Barbour, 1923: 7; 1943: 8; Schmidt, 1953: 16; Gorham, 1974: 30; Frost, 1985: 561; Stebbins, 1985: 38; 2003: 468; Petranka, 1998: 150; Raffaëlli, 2007: 74.

Ambystoma ensatum: Grinnell & Camp, 1917: 139.

Distribution: Confined to coastal area north and south of San Francisco, California. See map in Petranka (1998). **U.S.**: CA.

Common name: California Giant Salamander.

Commentary: Anderson (1969) provided a CAAR account. Steele *et al.* (2005) found this to be a sister species to *D. tenebrosus.*

Dicamptodon tenebrosus (Baird & Girard, 1852)

Amblystoma tenebrosum Baird & Girard, 1852a: 174.—**O**: **H**: USNM 4710 (Yarrow, 1883a: 152, sd.; confirmed by Cochran, 1961a).—**OT**: U.S., Oregon.

Xiphonura tenebrosa: Girard, 1858: 14.

Amblystoma tenebrosum: Cope 1868a: 202; Boulenger, 1882b: 49.

Chondrotus tenebrosus: Cope, 1887b: 88.

Ambystoma tenebrosum: Stejneger & Barbour, 1917: 11.

Dicamptodon tenebrosus: Good, 1989: 728; Petranka, 1998: 152; Stebbins, 2003: 159; Raffaëlli, 2007: 73.

Distribution: Coastal northern California to extreme southwestern British Columbia. Petranka (1998) has a range map. **Can.**: BC. **U.S.**: CA, OR, WA.

Common name: Coastal Giant Salamander (SECSN) or Pacific Giant Salamander (C&T). *French Canadian name:* Grande Salamandre du Nord.

Commentary: Stejneger & Barbour (1943: 8) and Schmidt (1953: 16) treated this as a synonym of *D. ensatus,* as did most subsequent authors (*e.g.,* Brame, 1967: 10; Gorham, 1974: 30), until the late 1980's. Steele *et al.* (2005) found this a sister species to *D. ensatus.* Listed in British Columbia as Threatened.

Family **Plethodontidae** Gray, 1850

Synonymy: See under Tribe Plethodontini, below.

Distribution: Southern Alaska, much of Canada and most of the US, south into the South American tropics; southern Europe; Korea.

Common name: Lungless salamanders.

Content: 27 genera with 547 species, of which 13 genera and 255 species are in North America (most of the others are neotropical). Taxonomic grouping of genera was rather stable for some thirty years or so, with a subfamily Desmognathinae recognized for those forms with a unique jaw mechanism, and Plethodontinae for all others. Three tribes were recognized within the latter subfamily (Wake, 1966, followed by Petranka, 1998). New evidence, mostly molecular (Vieites *et al.*, 2011; Wake, 2012), suggested a new arrangement, and that is followed here, with minor modification.

Commentary: Pyron & Wiens (2011) supported the findings of several other studies indicating a sister relationship of this family to Amphiumidae, further supported by Shen *et al.* (2013). Dubois & Raffaëlli (2012) followed this and other studies with similar results. Vieites *et al.* (2011) suggested a new taxonomy for the family, with two subfamilies, the first with four tribes (though in the paragraph above the classification they claimed only three), the other with five tribes. Wake (2012) clarified and somewhat modified that arrangement.

North America contains 22% of the species diversity of this family.

Subfamily **Hemidactyliinae** Hallowell, 1857

Synonymy: See under Tribe Hemidactyliini, below.

Distribution: Most of the U.S. and parts of southern Canada, south into Baja California and Mexico to tropical South America.

Content: Wake (2012) recognized four tribes and we follow that classification here. Another more comprehensive analysis in preparation supports Wake's (2012) classification (Wake, *in litt.*). Dubois & Raffaëlli (2012) recognized only two tribes (but several subtribes). The subfamily contains 19 genera and 446 species, 166 of which are found in North

America. North America thus contains about 37.2% of the species diversity of the subfamily.

Commentary: Hallowell (1857a) regarded the species *Hemidactylium scutatum* as so distinctive that it merited its own family. It was later treated for many years in a melting-pot version of Plethodontidae until Wake (1966) placed it in a Tribe Hemidactyliini, along with genera now regarded as spelerpines, in the subfamily Plethodontinae. Then Chippindale *et al.* (2004) again placed *Hemidactylium* in its own subfamily Hemidactylinae, and later studies (especially Vieites *et al.,* 2011 and Wake, 2012) included the taxa much as treated here. The largest tribe, Bolitoglossini, is entirely extralimital in the neotropics.

Tribe **Batrachosepini** Wake, 2012

"Batrachosepini" Dubois, 2008b: 71. *Nomen nudum* (see Dubois, 2012).

"Batrachosepita" Dubois, 2008b: 73. *Nomen nudum* (see Dubois, 2012)

"Batrachosepini" Vieites *et al.,* 2011: 11, 633.—**NG:** *Batrachoseps* Bonaparte, 1839; *nomen nudum* (see Dubois, 2012).

Batrachosepini Wake, 2012, 75.—**NG:** *Batrachoseps* Bonaparte,1839.

Batrachosepina: Dubois & Raffaëlli, 2012: 115.—**OS**.

Distribution: As for the genus.

Common name: Slender Salamanders.

Content: A single genus, with 22 species.

Commentary: Wake (1966) placed these species in a tribe Bolitoglossini, with the neotropical plethodontids, within the Plethodontinae. The revised treatment of Vieites *et al.* (2011) and Wake (2012) places the Slender Salamanders in a separate tribe, in the subfamily containing the neotropical plethodontids and spelerpines. Pyron & Wiens (2011) found *Batrachoseps* as a distinctive clade within their subfamily Bolitoglossinae, and a sister taxon to the neotropical bolitoglossines, indicating it could be a subtribe of a tribe Bolitoglossini, or a tribe of the Hemidactylinae. We treat it here, following Wake (2012).

Genus **Batrachoseps** Bonaparte, 1839

Synonymy: See under Subgenus *Batrachoseps,* below.

Distribution: California to Oregon and northern Baja California.

Common name: Slender Salamanders.

Content: 22 species; all occur in North America.

Commentary: Jackman *et al.* (1997) implied a splitting of the genus, recognizing a subgenus *Plethopsis* for a small portion of the genus, and the nominotypical subgenus for the remainder. Wake (2012) followed that treatment, and we follow it here.

Subgenus **Batrachoseps** Bonaparte, 1839

Batrachoseps Bonaparte, 1839e: [267].—**NS** (od.): *Salamandrina attenuata* Eschscholtz, 1833.

*Batrachoseps***:** Jackman *et al.*, 1997: 884 (implied as subgenus).

Distribution: Much of California to southwestern Oregon.

Content: Nineteen of the 22 described species of the genus are in this subgenus; others may soon be described in this subgenus as well. The species of the subgenus are generally recognized as forming five species-groups (Jockush & Wake, 2002: 363). These are the **B. attenuatus group:** *B. attenuatus,* and possibly other undescribed cryptic species; the **B. gabrieli group**: *B. gabrieli* (perhaps best treated as part of the *pacificus* group); the **B. nigriventris group**: *B. bramei, B. gregarius, B. nigriventris, B. relictus* (the relationships of the latter now questionable [Jockusch *et al.*, 2012]), *B. simatus, B. stebbinsi*; the **B. pacificus group**: *B. gavilanensis, B. incognitus, B. luciae, B. major, B. minor, B. pacificus*; and the former **B. relictus group** (which Jockusch *et al.*, 2012, renamed the **diabolis Group**): *B. altasierrae, B. diabolicus, B. kawia, B. regius. B. relictus* has been moved to the *nigriventris* group, hence the name of the group needed to change.

Batrachoseps (Batrachoseps) altasierrae Jockush, Martínez-Solano, Hansen & Wake, 2012

Batrachoseps altasierrae Jockush, Martínez-Solano, Hansen & Wake, 2012: 13.—**O: H:** MVZ 59909.—**OT:** California, Kern County, 2.4 km southeast of Alta Sierra, in the Greenhorn Mountains (35.732 N, 1118.538 W).

Misapplied names. Some specimens now treated as this species, have been treated as part of other species in the past:

Batrachoseps attenuatus: Grinnell & Camp, 1917: 137.

Batrachoseps attenuatus attenuatus: Dunn, 1926b: 232.

Batrachoseps relictus: Brame & Murray, 1968: 5.
Batrachoseps pacificus relictus: Yanev, 1980: 535.
Batrachoseps relictus: Jockusch, Wake & Yanev, 1998: 13.

Distribution: Higher elevations (900-2440 m) in the southern Sierra Nevada, in Kern and Tulare Counties, California, and allopatric to other members of the genus.

Common name: Greenhorn Mountains Slender Salamander (SECSN).

Commentary: This species was formerly treated within *B. attenuatus* or *B. relictus*. Some of the paratypes of *B. relictus* were referred to this population as a distinct species by Jockusch *et al.* (2012), who also showed a sister relationship of this species with *B. kawia*.

Batrachoseps (Batrachoseps) aridus Brame, 1970

Batrachoseps aridus Brame, 1970: 2.—**O**: **H**: LACM 56271.—**OT**: U.S., California, Riverside County, Hidden Palm Canyon, a tributary of Deep Canyon, (10.5 miles by road S of the intersection of state Highways 111 and 74, town of Palm Desert), NW end of Santa Rosa Mountains, from slopes on western side of the Coachella Valley, elevation approximately 2800 feet.—**SS**: Wake & Jockusch, 2000: 116.

Batrachoseps aridus: Behler & King, 1979: 306; Frost, 1985: 571; Petranka, 1998: 219; Hansen & Wake, 2005a: 666.
Batrachoseps major aridus: Wake & Jockusch, 2000: 117; Stebbins, 2003: 189.
Batrachoseps (Batrachoseps) major aridus: Jockusch & Wake, 2002, by implication; Raffaëlli, 2007: 199.

Distribution: Known only from the vicinity of the type locality and another canyon about 4 1/2 miles from there. Hansen & Wake (2005a) provided a map. **U.S.**: CA.

Common name: Desert Slender Salamander.

Commentary: Hansen & Wake (2005a) treated this form as a full species, though reiterating the evidence that it is not phylogenetically distinguishable from *B. major*. It is more often treated as a subspecies of *B. major,* and is apparently sister to that taxon. It is federally listed (as *B. aridus*) as endangered. We treat this highly restricted population as a full species, based on its distribution within that of its sister taxon.

Batrachoseps (Batrachoseps) attenuatus (Eschscholtz, 1833)

Salamandrina attenuata Eschscholtz, 1833: 1.—**O**: none stated nor known to exist.—**OT**: U.S., California, vicinity of the Bay of San Francisco.

Batrachoseps attenuatus: Bonaparte, 1839e: [267]; Fitzinger, 1843: 33; Knauer, 1878: 98; Cope, 1889b: 127; Stejneger & Barbour, 1917: 13; Gorham, 1974: 30; Behler & King, 1979: 307; Frost, 1985: 572; Stebbins, 1985: 59; 2003: 194; Petranka, 1998: 221.

Batrachoseps attenuata: Baird, 1850a: 288.

Batrachoseps attenuatus attenuatus: Dunn, 1926b: 224; Campbell, 1931: 133; Stejneger & Barbour, 1943: 22; Schmidt, 1953: 40; Cochran & Goin, 1970: 31.

Batrachoseps (Batrachoseps) attenuatus: Jockusch & Wake, 2002: 363; Raffaëlli, 2007: 194.

Batrachoseps caudatus Cope, 1889b: 126.—**O**:**H** (od.) USNM 13561.—**OT**: U.S., Alaska, Hassler Harbor (probably Annette Island *fide* Stejneger & Barbour 1917: 13. Frost (2009-13) suggested this is best regarded as museum records error.—**SS**: Wake *et al.* 1998: 13.

Batrachoseps caudatus: Stejneger & Barbour, 1917: 13.

Batrachoseps attenuatus caudatus: Dunn, 1926b: 232; Campbell, 1931: 133; Stejneger & Barbour, 1943: 22.

Distribution: Northwestern California to southwestern Oregon. Petranka (1998) and Boundy (2000) provided range maps. **U.S.**: CA, OR.

Common name: California Slender Salamander.

Content: No subspecies are currently recognized, but for a while all known species of the genus were considered subspecies of *B. attenuatus* (Dunn, 1926b). *B. leucopus* Dunn, 1922, was also considered a subspecies for a while, but is now treated as a synonym of *B. major*.

Commentary: Boundy (2000) provided a CAAR account. Hansen & Wake (2005l) found little decline in this species as currently recognized. The taxon as recognized here may be composed of up to five distinct species (Martinez-Solano *et al.*, 2007). The members of this complex are a sister taxon to the rest of the subgenus.

Batrachoseps (Batrachoseps) bramei Jockush, Martínez-Solano, Hansen & Wake, 2012

Batrachoseps bramei Jockush, Martínez-Solano, Hansen & Wake, 2012: 8.—**O: H:** MVZ 217944.—**OT:** California, Tulare County, Packsaddle Canyon adjacent to Kern River, elev. 1137 m (35.945 N, 118.476 W).

Misapplied name:

Batrachoseps simiatus: Brame & Murray, 1968:15-18.

Distribution: Known only from a restricted area around the upper Kern River Canyon, allopatric to all other members of the genus.

Common name: Fairview Slender Salamander (SECSN).

Commentary: Members of this species were formerly treated within *B. simatus*. Jockusch *et al.* (2012) recognized this as a distinct species, and their analysis indicates that the species consists of four divergent clades within the *nigroventris group.* Its sister relationships are unclear but allozymes suggest it is a sister to the *simatus-relictus-gregarius* clade, within that species group.

Batrachoseps (Batrachoseps) diabolicus Jockusch, Wake, & Yanev, 1998

Batrachoseps diabolicus Jockusch, Wake, & Yanev, 1998: 7.—**O**: **H**: MVZ 95446.—**OT**: U.S., California, Mariposa County, Hell Hollow, at the junction with the Merced River at Lake McClure on California Highway 49.

Batrachoseps (Batrachoseps) diabolicus: Jockusch & Wake, 2002: 363; Raffaëlli, 2007: 195.

Batrachoseps diabolicus: Stebbins, 2003: 194.

Distribution: Western foothills of Sierra Nevada of California, from Merced River drainage to the American River, below 300 m. See range maps in Jockusch *et al.* (1998) and Hansen & Wake (2005b). **U.S.**: CA.

Common name: Hell Hollow Slender Salamander.

Batrachoseps (Batrachoseps) gabrieli Wake, 1996

Batrachoseps gabrieli Wake, 1996: 1.—**O**: **H**: MVZ 196449.—**OT**: U.S., California, Los Angeles County, San Gabriel Mountains, above Soldier Creek in the upper San Gabriel River drainage, approximately 1 km ESE Crystal Lake; SW Section 28, R9W, T3N; 34° 18' 47" N, 117° 49' 57" W; approximately 1500 m elevation, under cover on a steep talus slope.

Batrachoseps (Batrachoseps) gabrieli: Jockusch & Wake, 2002: 363; Raffaëlli, 2007: 199.

Batrachoseps gabrieli: Petranka, 1998: 225; Stebbins, 2003: 189.

Distribution: Southwestern California mountains, Los Angeles and San Bernardino Counties. Wake (1996) and Hansen *et al.* (2005) included range maps. **U.S.**: CA.

Common name: San Gabriel Mountains Slender Salamander (SECSN) or San Gabriel Slender Salamander (C&T).

Batrachoseps (Batrachoseps) gavilanensis Jockusch, Yanev, & Wake, 2001

Batrachoseps gavilanensis Jockusch, Yanev, & Wake, 2001: 69.—**O**: **H**: MVZ 155642.—**OT**: U.S., California, San Benito Co., 0.5 miles (0.8 km) south of cement plant on San Juan Creek Road; 36° 49' 13" N, 121° 31' 30" W, elevation ca. 120 m.

Batrachoseps (Batrachoseps) gavilanensis: Jockusch & Wake, 2002: 263; Raffaëlli, 2007: 196.

Batrachoseps gavilanensis: Stebbins, 2003: 190.

Distribution: Parts of Santa Cruz & San Luis Obispo Counties, California. Jockusch *et al.* (2001) and Hansen & Wake (2005m) have distribution maps. **U.S.**: CA.

Common name: Gabilan Mountains Slender Salamander.

Commentary: References have been made to members of this species prior to its description, as *B. attenuatus* (*e.g.*, Brame & Murray, 1968), *B. attenuatus attenuatus* (*e.g.*, Stebbins, 1951; Hendrickson, 1954), or *B. pacificus* (*e.g.*, Yanev, 1980).

Batrachoseps (Batrachoseps) gregarius Jockusch, Wake, & Yanev, 1998

Batrachoseps gregarius Jockusch, Wake & Yanev, 1998: 2.—**O**: **H**: MVZ 224581.—**OT**: U.S., California, Madera-Mariposa County line, Sierra National Forest, Westfall Picnic Ground east of Highway 41.

Batrachoseps (Batrachoseps) gregarius: Jockusch & Wake, 2002: 363; Raffaëlli, 2007: 200.

Batrachoseps gregarius: Stebbins, 2003: 187.

Distribution: West slope of Sierra Nevada and Greenhorn Mountains, California. Jockusch *et al.* (1998) included a range map. **U.S.**: CA.

Common name: Gregarious Slender Salamander.

Commentary: Hansen & Wake (2005c) presented a species account. Some of the analyses by Jockusch *et al.* (2012) indicate this species may be a sister to *B. bramei.*

Batrachoseps (Batrachoseps) incognitus Jockusch, Yanev, & Wake, 2001

Batrachoseps incognitus Jockusch, Yanev, & Wake, 2001: 67.—**O**: **H**: MVZ 100059.—**OT**: U.S., California, San Luis Obispo County, Near Rocky Butte, 14.7 km NE Highway 1 on San Simeon Creek Road; 35.683° N, 121.076° W, elevation ca. 900 m.

Batrachoseps (Batrachoseps) incognitus: Jockusch & Wake, 2002: 363; Raffaëlli,2007: 197.

Batrachoseps incognitus: Stebbins, 2003: 191.

Distribution: Santa Lucia Mountains, southwestern Monterey County to San Luis Obispo County line, California. See map in Jockusch *et al.* (2001) or Hansen & Wake (2005d). **U.S.**: CA.

Common name: San Simeon Slender Salamander.

Batrachoseps (Batrachoseps) kawia Jockusch, Wake, & Yanev, 1998

Batrachoseps kawia Jockusch, Wake, & Yanev, 1998: 11.—**O**: **H**: MVZ 94134.—**OT**: U.S., California, Tulare County, Kaweah River, west side of the South Fork.

Batrachoseps (Batrachoseps) kawia: Jockusch & Wake, 2002: 363; Raffaëlli, 2007: 195.

Batrachoseps kawia: Stebbins, 2003: 192.

Distribution: Apparently limited to a restricted range in the Kaweah River Drainage, Tulare County, California. Jockusch *et al.* (1998) included a range map. **U.S.**: CA.

Common name: Sequoia Slender Salamander.

Commentary: Hansen & Wake (2005e) noted that recent check of known sites found no salamanders.

Batrachoseps (Batrachoseps) luciae Jockusch, Yanev, & Wake, 2001

Batrachoseps luciae Jockusch, Yanev, & Wake, 2001: 61.—**O**: **H**: MVZ 104941.—**OT**: U.S., California, Monterey County, Monterey, Don Dahvee Park; 36° 35' 22" N, 121° 53' 38" W, elevation ca. 24 m.

Batrachoseps (Batrachoseps) luciae: Jockusch & Wake, 2002: 363; Raffaëlli, 2007: 197.

Batrachoseps luciae: Stebbins, 2003: 190.

Distribution: Parts of Santa Cruz and San Luis Obispo Counties, California. Jockusch *et al.* (2001) have a distribution map. **U.S.**: CA.

Common name: Santa Lucia Mountains Slender Salamander.

Commentary: Hansen & Wake (2005f) noted no apparent decline in populations of this species.

Batrachoseps (Batrachoseps) major Camp, 1915

Batrachoseps major Camp, 1915: 327.—**O**: **H**: MVZ 611.—**OT**: U.S., California, Los Angeles County, Sierra Madre, 1000 feet elevation.

Batrachoseps attenuatus major: Dunn, 1926b; 234

Batrachoseps major: Stejneger & Barbour, 1917: 14; Brame & Murray, 1968: 22; Gorham, 1974: 30; Behler & King, 1979: 307.

Batrachoseps pacificus major: Campbell, 1931: 133; Stejneger & Barbour, 1943: 23; Schmidt, 1953: 41; Cochran & Goin, 1970: 32; Yanev, 1980: 531; Stebbins, 1985: 57; Liner, 1994: 9; Petranka, 1998: 229.

Batrachoseps (Batrachoseps) pacificus major: Jackman, Applebaum, & Wake, 1997: 883.

Batrachoseps (Batrachoseps) major: Jockusch & Wake, 2002: 363.

Batrachoseps major major: Wake & Jockusch, 2000: 116; Stebbins, 2003: 188.

Batrachoseps (Batrachoseps) major major: Raffaëlli, 2007: 198.

Batrachoseps catalinae Dunn, 1922: 62.—**O**: **H**: USNM 57335.—**OT**: U.S., California, Los Angeles County, Santa Catalina Island. Synonymy with *B. attenuatus* by Dunn, 1926b: 241, and with *B. major* by Dunn, 1926b: 239.

Batrachoseps attenuatus catalinae: Dunn, 1926b: 239.

Batrachoseps pacificus catalinae: Campbell, 1931: 133; Stejneger & Barbour, 1943: 23; Schmidt, 1953: 41; Cochran & Goin, 1970: 32.

Batrachoseps leucopus Dunn, 1922: 60.—**O**: **H**: USNM 64319.—**OT**: Mexico, Baja California del Norte, Los Coronados, North Island.—**SS**: Dunn, 1926b: 241.

Batrachoseps leucopus: Stejneger & Barbour, 1923: 9.
Batrachoseps attenuatus leucopus: Dunn, 1926b: 241; Campbell, 1931: 133; Stejneger & Barbour, 1943: 23; Schmidt, 1953: 40; Cochran & Goin, 1970: 31.
Batrachoseps pacificus leucopus: Zweifel, 1958: 3.

Distribution: Coastal Southern California mountains into Baja California and some of Channel Islands. Jockusch & Wake (2002) have a range map of several species, including this one. **U.S.**: CA.

Common name: Southern California Slender Salamander (SECSN), or Garden Slender Salamander (C&T).

*Content***:** Wake & Jockusch (2000) recognized two subspecies. Martínez-Solano *et al.* (2012) found non-monophyly in two lineages more closely related to other species than to each other, in material identified as *B. major,* using mtDNA data, but nuclear data indicated a sister relationship between the two. Data for *B. aridus,* often treated as a subspecies, were not very helpful due to unavailability of fresh material, and poor condition of the older material available.

Commentary: Hansen & Wake (2005n) noted populations of this species have been extirpated from much of their historical distribution.

Batrachoseps (Batrachoseps) minor Jockusch, Yanev, & Wake, 2001

Batrachoseps minor Jockusch, Yanev, & Wake, 2001: 65.—**O**: **H**: MVZ 155968.—**OT**: U.S., California, San Luis Obispo County, from along the Santa Rita—Old Creek Road, 8.5 km SW intersection with Vineyard Road; 35° 31' 20" N, 120° 46' 40" W; elev. ca. 400 m.
Batrachoseps (Batrachoseps) minor: Jockusch & Wake, 2002: 363; Raffaëlli, 2007: 197.
Batrachoseps minor: Stebbins, 2003: 191.

Distribution: Southern Santa Lucia Mountains, San Luis Obispo County, California. Jockusch *et al.* (2001) provided a range map. **U.S.**: CA.

Common name: Lesser Slender Salamander.

Commentary: Hansen & Wake (2005g) noted significant declines in abundance of this species throughout its range.

Batrachoseps (Batrachoseps) nigriventris Cope, 1869

Batrachoseps nigriventris Cope, 1869b: 98.—**O**: **S** (od.), USNM 6734 and ANSP 1865.—**OT**: U.S., California, Kern County, Fort Tejon.—**Comment**: Dunn (1926b: 232) also cited ANSP 481-482 as **S**.

Batrachoseps nigriventris: Cope, 1889b: 129; Frost, 1985: 572; Petranka, 1998: 226; Stebbins, 2003: 186.

Batrachoseps (Batrachoseps) nigriventris: Jockusch & Wake, 2002: 363; Raffaëlli, 2007: 200.

Distribution: Coastal southern California, and Sierra Nevada slopes. Petranka (1998) and Hansen & Wake (2005o) have range maps. **U.S.**: CA.

Common name: Black-bellied Slender Salamander.

Commentary: Prior to the mid-1970s (e.g., Gorham, 1974: 30) this species was usually regarded as a synonym of B. attenuatus. Petranka (1998: 226) incorrectly credited authorship to Yanev. Jockusch & Wake (2002) found the species of the *nigriventris*-group to be a sister taxon to all members of the *pacificus* and *relictus* groups. Within the group, Jockusch *et al.* (2012) indicated this species is perhaps most closely related to *B. bramei. Hansen & Wake (2005o) noted no notable declines in this species.*

Batrachoseps (Batrachoseps) pacificus (Cope, 1865)

Hemidactylium pacificum Cope, 1865b: 195.—**O**: **H**: USNM 6733 (sd., Dunn, 1926b: 237).—**OT**: U.S., on the coast of southern California, Santa Barbara; corrected to one of the northern Channel Islands (Van Denburgh, 1905).

Batrachoseps pacificus: Cope, 1869b: 98; 1889b: 129; Boulenger, 1882b: 59; Stejneger & Barbour, 1917: 13; Gorham, 1974: 31; Behler & King, 1979: 308; Frost, 1985: 572; Stebbins, 2003: 187.

Batrachoseps attenuatus pacificus: Dunn, 1926b: 236.

Batrachoseps pacificus pacificus: Campbell, 1931: 133; Stejneger & Barbour, 1943: 23; Schmidt, 1953: 41; Cochran & Goin, 1970: 32; Stebbins, 1985: 57; Petranka, 1998: 228-229.

Batrachoseps (Batrachoseps) pacificus: Jockusch & Wake, 2002: 363; Raffaëlli, 2007: 199.

Distribution: Channel Islands and coastal central and southern California. Petranka (1998) and Hansen & Wake (2005p) have range maps. **U.S.**: CA.

Common name: Channel Islands Slender Salamander.

Batrachoseps (Batrachoseps) regius Jockusch, Wake, & Yanev, 1998

Batrachoseps regius Jockusch, Wake, & Yanev, 1998: 9.—**O**: **H**: MVZ 94029.—**OT**: U.S., California, Fresno County, South bank of the North Fork, Kings River, 1.6 km N of Kings River (by road).

Batrachoseps (Batrachoseps) regius: Jockusch & Wake, 2002: 363; Raffaëlli, 2007: 195.

Batrachoseps regius: Stebbins, 2003: 192.

Distribution: Kings River system and Kaweah River drainage, Fresno and Tulare Counties, California. Jockusch *et al.* (1998) provided a range map. **U.S.**: CA.

Common name: Kings River Slender Salamander.

Commentary: Hansen & Wake (2005h) reported populations seem stable.

Batrachoseps (Batrachoseps) relictus Brame & Murray, 1968

Batrachoseps relictus Brame & Murray, 1968: 5.—**O**: **H**: LACM 34360.—**OT**: U.S., California, Kern County, Bakersfield, 150 yards above the junction of State Hwy. 173 and the road turnoff to Democrat Hot Springs Resort, above the upper dirt road, in the Kern River Canyon (25.7 mi. by road NE of intersection of Niles and Baker Streets), 2400 feet elevation.

Batrachoseps pacificus relictus: Yanev, 1980: 535; Stebbins, 1985: 58; Petranka, 1998: 229.

Batrachoseps relictus: *Gorham, 1974: 31;* Behler & King, 1979: 309; Collins, 1997: 5; Jockusch, Wake & Yanev, 1998:13; Stebbins, 2003: 193.

Batrachoseps (Batrachoseps) relictus: Jockusch & Wake, 2002: 363; Raffaëlli, 2007: 196.

Distribution: Lower Kern River drainage, Kern County, California, and in the Breckenridge Mountains, from 480-2000+ m elev. Jockusch *et al.* (2012) provided a range map. **U.S.**: CA.

Common name: Relictual Slender Salamander.

Commentary: We treat this as a full species, based mainly on Jockusch *et al.* (1998) and Jockusch & Wake (2002), though it is sometimes treated as a subspecies of *B. pacificus*. Hansen & Wake (2005i) note that the species has apparently been extirpated from the type locality, as none have been seen there since 1971.

Batrachoseps (Batrachoseps) simatus Brame & Murray, 1968

Batrachoseps simatus Brame & Murray, 1968: 15.—**O**: **H**: LACM 34527.—**OT**: U.S., California, Kern County, 1 to 1.5 miles southwest of Democrat Hot Springs Resort turnoff, on the steep slopes to the south side of the Kern River Canyon above State Hwy. 178 (24.7 mi NE of the intersection of Niles and Baker Streets, in Bakersfield, or about 2 mi NE of Cow Flat Creek).

Batrachoseps simatus: Gorham, 1974: 31; Behler & King, 1979: 309; Frost, 1985: 573; Stebbins, 1985: 56; 2003: 185; Petranka, 1998: 231.

Batrachoseps (Batrachoseps) simatus: Jackman, Applebaum & Wake, 1997: 883; Raffaëlli, 2007: 201.

Distribution: Kern River drainage in Kern and Tulare Counties, California. See range map in Jockusch *et al.* (2012). **U.S.**: CA.

Common name: Kern Canyon Slender Salamander.

Commentary: Hansen & Wake (2005k) noted that there may be another restricted population currently included as this species which may be another undescribed cryptic species; this was later described as *B. bramei* (Jockusch *et al.*, 2012). Jockusch *et al.* (2012) found this to be a sister species to *B. relictus*.

Batrachoseps (Batrachoseps) stebbinsi Brame & Murray, 1968

Batrachoseps stebbinsi Brame & Murray, 1968: 18.—**O**: **H**: MVZ 81835.—**OT**: U.S., California, Kern County, Piute Mountains, southern Sierra Nevada, 3 miles west of Paris Loraine (sometimes called Loraine), at 2500 ft. elevation.

Batrachoseps stebbinsi: Gorham, 1974: 31; Behler & King, 1979: 310; Frost, 1985: 573; Stebbins, 1985: 54; 2003: 185; Petranka, 1998: 232.

Batrachoseps (Batrachoseps) stebbinsi: Jockusch & Wake, 2002: 363; Raffaëlli, 2007: 202.

Distribution: Scattered localities in the Piute Mountains and Tehachapi Mountains, California. Petranka (1998) has a range map. **U.S.**: CA.

Common name: Tehachapi Slender Salamander.

Commentary: According to Jockusch *et al.* (2012), this is a sister species to *B. gregarius* + *B. bramei. Listed federally as of concern, and by California as threatened.*

Subgenus **Plethopsis** Bishop, 1937

Plethopsis Bishop, 1937: 93.—**NS** (mo.): *Plethopsis wrighti* Bishop, 1937.

Plethopsis: Jackman *et al.*, 1997: 884. Treated implicitly as a subgenus.

Distribution: Northern California to northern Oregon.

Common name: None in common use; Northwestern Slender Salamanders seems appropriate.

Content: Three of the 22 described species of the genus are in this subgenus.

Batrachoseps (Plethopsis) campi Marlow, Brode, & Wake, 1979

Batrachoseps campi Marlow, Brode, & Wake, 1979: 3.—**O**: **H**: MVZ 122993.—**OT**: U.S., California, Inyo County, W slope of the Inyo Mountains, Long John Canyon, 3.2 km (2 mi) N, 5.3 km (3.3 mi) E Lone Pine, elevation 1695 m (5560 ft).

Batrachoseps campi: Frost, 1985: 572; Stebbins, 1985: 54; 2003: 184; Petranka, 1998: 224.

Batrachoseps (Plethopsis) campi: Jackman, Applebaum, & Wake, 1997: 884, implied; Jockush & Wake, 2002: 363; Raffaëlli, 2007: 193.

Distribution: Inyo Mountains, Inyo County, California. Petranka (1998), Jockusch *et al.* (2001) and Stebbins (2003) have range maps. **U.S.**: CA.

Common name: Inyo Mountains Salamander.

Commentary: Jockusch (2001) provided a CAAR account.

Batrachoseps (Plethopsis) robustus Wake, Yanev, & Hansen, 2002

Batrachoseps (Plethopsis) robustus Wake, Yanev, & Hansen, 2002: 1017.—**O**: **H**: MVZ 219115.—**OT**: U.S., California, Tulare County, 3.4 km south-southeast Sherman Pass Road on USFS rd 22S19; approximately 4.4 km (air) north-northwest Sirretta Peak; approximately 0.9 km (air) southeast Round Meadow on Kern Plateau, (T22S, R34E Sect. 31), 35.9604° N, 118.3514° W; 2775-2800 m elevation.

Batrachoseps robustus: Stebbins, 2003: 184.

Batrachoseps (Plethopsis) robustus: Raffaëlli, 2007: 193.

Distribution: Kern Plateau and Scodie Mountains, Kern and Tulare Counties, California. Wake *et al.* (2002) provided a range map. **U.S.**: CA.

Common name: Kern Plateau Salamander.

Commentary: Hansen & Wake (2005j) noted that most populations appear stable but no salamanders have been seen in the Scodie Mountains of Kern County since an extensive wildfire.

Batrachoseps (Plethopsis) wrighti (Bishop, 1937)

Plethopsis wrighti Bishop, 1937: 93—**O**: **H**: USNM 102445.—**OT**: U.S., Oregon, Clackamas County, 8.7 miles southeast of Sandy, in woods bordering Mt. Hood highway.

Plethopsis wrighti: Stejneger & Barbour, 1943: 22; Dubois, 2007b: 67, correcting an unjustified emendation.

Batrachoseps wrighti: Stebbins & Lowe, 1949: 128; Schmidt, 1953: 41; Cochran & Goin, 1970: 31; Gorham, 1974: 31; Behler & King, 1979: 310; Frost, 1985: 573; Stebbins, 1985: 53; 2003: 183; Petranka, 1998: 234.

Batrachoseps wrightorum: Applegarth, 1994: 5, 31; Collins, 1997: 6; Collins & Taggart, 2002: 7; Michels & Bauer, 2004: 84.—**UE** (see Commentary).

Batrachoseps (Plethopsis) wrighti: Jackman, Applebaum, & Wake, 1997: 884, implied; Jockush & Wake, 2002: 363; Raffaëlli, 2007: 192.

Distribution: Cascade Mountains of northwestern Oregon. Kirk (1991) and Petranka (1998) have range maps. **U.S.**: OR.

Common name: Oregon Slender Salamander.

Commentary: Kirk (1991) provided a CAAR account. Applegarth (1994) and Michels & Bauer (2004: 84) emended the termination of the name, and discussed prior arguments about it. The form *wrighti* was originally proposed for the name, honoring A. H. and M. R. Wright, and Applegarth and Michels & Bauer noted the *Code* provides for such a correction. But the *Code* does not provide for the specific "correction" they made, as pointed out by Brandon-Jones *et al.* (2007) and Dubois (2007b). Rather, the nomen as originally proposed, is to be considered the correct one. And the nomen, *Batrachoseps wrightorum* Applegarth, 1994, remains an available name, barring its being considered a *nomen nudum,* or a junior homonym. Miller *et al.* (2005) found two separate molecular clades of the species, and suggested they may be worthy of designation as different conservation units; more study is needed before suggesting these may be separate cryptic species.

Tribe **Hemidactyliini** Hallowell, 1857

"Mycetoglossina" Bonaparte, 1850: [2].—**NG**: *Mycetoglossus* Bonaparte, 1839. Bonaparte (1839e: 267) attributed the genus name to Bibron (no year). Placed on the official list of rejected/invalid names, (ICZN, 1997b)

Hemidactylidae Hallowell, 1857a: 11.—**NG**: *Hemidactylium* Tschudi, 1838, mo. Placed on official list of family-series names (ICZN, 1997b).

Hemidactyliini: Wake, 1966: 50; Blackburn & Wake, 2011: 46; Wake, 2012: 78.

Hemidactylinae: Chippindale *et al.*, 2004: 2819; Macey, 2005: 201; Dubois, 2005b: 19; Vieites *et al.,* 2011: 633; Wake, 2012: 76.

Hemidactyliina: Dubois & Raffaëlli, 2012: 139.

Distribution: Endemic to North America in the eastern US and southeastern Canada, with a discontinuous distribution.

Common name: Four-toed Salamanders.

Content: Monotypic.

Genus **Hemidactylium** Tschudi, 1838

Hemidactylium Tschudi, 1838: 59, 94.—**NS**: *Salamandra scutata* Schlegel, 1838, mo.—Placed on official list of generic names (ICZN, 1997b).

Cotobotes Gistel, 1848: xi. Substitute name for *Hemidactylium* Tschudi, 1838.

Desmodactylus Duméril *et al.*, 1854c: 117. Substitute name for *Hemidactylium* Tschudi, 1838.

Dermodactylus: David, 1871: 95.—**IS.**

Distribution: As for the tribe.

Common name: Four-toed salamanders.

Content: Monotypic.

Commentary: Neill (1963b) provided a generic CAAR account.

Hemidactylium scutatum (Schlegel, 1838)

Salamandra scutata Schlegel *in* Temminck & Schlegel, 1838: 119.—**O: H**: RMNH 2301 (sd., Hoogmoed, 1978: 103-104).—**OT**: U.S., Tennessee, Davidson County, Nashville. Placed on official list of specific names (ICZN, 1997b).

Hemidactylium scutatum: Tschudi, 1838: 59, 94; Cope, 1889b: 130; Stejneger & Barbour, 1917: 14; 1943: 21; Dunn, 1926b: 196; Viosca, 1949: 9; Schmidt, 1953: 41; Cochran & Goin, 1970: 47; Gorham, 1974: 35; Conant, 1975: 282; Behler & King, 1979: 331; Frost, 1985: 590; Conant & Collins, 1998: 479; Petranka, 1998: 290; Raffaëlli, 2007: 190.

Desmodactylus Scutatus: Duméril *et al.*, 1854c: 118.

Batrachoseps scutatus: Boulenger, 1882b: 59.

Salamandra melanosticta Gibbes, 1845: 80.—**O**: none stated nor known to exist.—**OT**: U.S., South Carolina, Abbeville County, Abbeville.

Desmodactylus Melanostictus: Duméril *et al.*, 1854c: 119.

Distribution: As for the tribe. Neill (1963c) and Petranka (1998) have range maps. **Can.**: NS, ON, QC. **U.S.**: AL, AR, CT, DE, FL, GA, IL, IN, LA, KY, MA, MD, MI, MN, MO, NC, NH, NJ, NY, OH, PA, RI, SC, TN, VA, VT, WI, WV.

Common name: Four-toed Salamander. *French Canadian name:* Salamandre à quatre doigts.

Commentary: Authorship is commonly attributed to Temminck & Schlegel (*e.g.*, Frost, 2009-13), but Neill (1963c) pointed out that, while they were co-authors of the section,

Schlegel wrote the description in Part 3 of the "Reptilia" of the "Fauna Japonica." Neill (1963c) provided a CAAR account and included details of the authorship. Mueller *et al.* (2004) as well as Vieites *et al.* (2007) indicated that this taxon is apparently a sister to the Bolitoglossinae, which Wiens *et al.* (2005a) indicated to be a sister taxon to all other plethodontids; thus it seemed basal to the family. But the analysis of Vieites *et al.* (2011) and Wake (2012) found that *Hemidactylium* is a sister to *Batrachoseps* plus the neotropical plethodontids (Bolitoglossini), and that entire clade is a sister group to the spelerpines. Pyron & Wiens (2011) found *Hemidactylium* basal to both spelerpines and bolitoglossines, based on a more limited dataset. Listed as at risk in Quebec.

Tribe **Spelerpini** Cope, 1859

Spelerpinae Cope, 1859a: 123.—**NG**: *Spelerpes* Rafinesque, 1832 (= *Eurycea* Rafinesque, 1822)

Spelerpsidi: Acloque, 1900: 493.

Spelerpinae: Chippindale *et al.*, 2004.

Spelerpini: Dubois, 2005b: 20; Camp *et al.*, 2009: 92; Vieites *et al.*, 2011: 633; Wake, 2012: 78.

Typhlomolgidae Steneger & Barbour, 1917: 2.—**NG**: *Typhlomolge* Stejneger, 1896.

Distribution: Eastern to central North America.

Common name: None in general use. Cave and Stream Salamanders may be appropriate.

Content: 5 genera with 36 species, all of which are North American.

Commentary: Cope (1859a) erected the subfamily for species of today's *Eurycea*, but it was not in general use; Chippindale *et al.* (2004) resurrected it for the four genera now contained, and Macey (2005) supported this use, and that treatment is followed here. Vieites *et al.* (2007) presented evidence that this is the sister taxon to the Bolitoglossinae, which is supported by the study of Camp *et al.* (2009). Vieites *et al.* (2011) and Wake (2012) found this taxon to be a sister to a larger clade containing the rest of the subfamily, including bolitoglossines, *Batrachoseps,* and *Hemidactylium.* Pyron & Wiens (2011) used subfamily ranking for this taxon, and found it to be a sister to bolitoglossines, which they also ranked at subfamily level. Their bolitoglossines contain

Batrachoseps and the neotropical genera. Wake (2012) suggested an alternative treatment would be to recognize two hemidactyline infrafamilies, one with Spelerpini and one with the other tribes.

Subtribe **Pseudotritonina** Dubois & Raffaëlli, 2012

Pseudotritonina Dubois & Raffaëlli, 2012: 143.—**NG**: *Pseudotriton* Tschudi, 1838, od.

Distribution: Eastern North America.

Common name: None in general use. Spring and Mud Salamanders may be appropriate.

Content: Three genera, with 13 species.

Commentary: Members of this subtribe form a monophyletic clade that is sister to the other subtribe (Spelerpina). This is supported by Bonett *et al.* (2013a) and earlier studies.

Genus **Gyrinophilus** Cope, 1869

Gyrinophilus Cope, 1869b: 108-109.—**NS**: *Salamandra porphyritica* Green, 1827 (mo.).

Gyrinophilus: Stejneger & Barbour, 1917: 17; 1933: 15; Schmidt, 1953: 45.

Distribution: Northern Alabama and Georgia to New England and into southern Quebec. See Brandon's (1967a) range map.

Common name: Spring Salamanders.

Content: Three species are recognized here, two of which are polytypic.

Commentary: Brandon (1967a) provided a CAAR generic account. Mueller *et al.* (2004) found this genus to be the sister group of *Pseudotriton,* and those two together sister to *Stereochilus.* This relationship is supported by the study of Bonett *et al.* (2013a) Vieites *et al.* (2007) found this genus to be the sister group of *Stereochilus* plus *Pseudotriton.* Some workers have suggested placing this genus in the synonymy of *Pseudotriton* (*e.g.,* Grobman, 1959; Organ, 1961), but current studies suggest they are sufficiently differentiated to warrant separate generic status, although generic placement of some species is questionable. The currently recognized species, *G. palleucus,* can no longer be treated as separate from *G. porphoriticus,* due to the findings of Bonett *et al.* (2013a).

Gyrinophilus montanus Baird, 1850

Synonymy: See under subspecies, *G. m. montanus,* below.

Distribution: Gulf Coast from eastern Louisiana to northern Florida, and Atlantic coastal area north to Maryland. Petranka (1998) has a range map, in the genus *Pseudotriton.* **U.S.**: AL, DE, FL, GA, KY, LA, MD, MS, NC, NJ, OH, PA, SC, TN, VA, WV.

Common name: Mud Salamander (SECSN) or Eastern Mud Salamander (C&T).

Content: Currently three subspecies are recognized.

Commentary: Martof (1975b) provided a CAAR account in the genus *Pseudotriton.* He did not recognize any subspecies. The study of Bonett *et al.* (2013a) indicated a sister relationship between this species and a lineage identified as *G. porphoriticus,* which seems to indicate this species should be treated in the genus *Gyrinophilus,* rather than *Pseudotriton,* as it is currently treated.

Gyrinophilus montanus montanus Baird, 1850, **new comb.**

Pseudotriton montanus Baird, 1850b: 293.—**O**: **S**: USNM 3839 (3) *fide* Dunn (1926b: 291) and Cochran (1961a: 23), though Baird mentioned only two specimens.—**OT**: U.S., Pennsylvania, Cumberland County, South Mountain, near Carlisle; corrected/restricted to U.S., Pennsylvania, Franklin County, Caledonia State Park (South Mountain) (McCoy, 1992: 93), based on evidence of habitat and occurrence in Pennsylvania; accepted here as a valid restriction or correction.

Spelerpes montana: Gray, 1850: 45.

Pseudotriton montanum: Baird *in* Heck & Baird, 1851: 255.

Spelerpes ruber montana: Cope, 1869b: 107.

Spelerpes ruber var. *montanus*: Boulenger, 1882b: 63.

Geotriton rubra montanus: Garman, 1884a: 39.

Spelerpes montanus: Brimley, 1917: 87.

Eurycea montana: Stejneger & Barbour, 1917: 19.

Pseudotriton montanus: Dunn, 1920c: 132; Bishop, 1924: 91; Viosca, 1949: 9; Frost, 1985: 604.

Pseudotriton montanus montanus: Stejneger & Barbour, 1923: 14; 1943: 28; Dunn, 1926b: 286; Schmidt, 1953: 47; Brame, 1967: 14; Cochran & Goin, 1970: 61; Gorham, 1974: 39; Conant, 1975: 285; Behler & King, 1979: 353; Conant & Collins, 1998: 484; Petranka, 1998: 296; Raffaëlli, 2007: 174 ; Bonett *et al.*, 2013a: 7.

Gyrinophilus montanus montanus: ***new combination***

Spelerpes ruber sticticeps Baird *in* Cope, 1869b: 108.—**NU**.

Geotriton rubra sticticeps: Garman, 1884a: 39.

Spelerpes ruber sticticeps Baird *in* Cope, 1889b: 178.—**S**: USNM 11475 (2) *fide* Dunn, 1926b: 291, and Cochran, 1961a: 26.—**OT**: U.S., South Carolina. Corrected (Dunn: 1926b: 291) to U.S., Georgia, no locality.—**Comment**: Schmidt, 1953: 48, restricted the OT to U.S., Georgia, Rabun County; Neill (1957b: 141) disputed that action & suggested it may have come from Georgia, Richmond County, Augusta. Schmidt's action is clearly invalid; Neill did not actually make a restriction.—**SS**: Dunn, 1926b: 287, with *P. montanus*.

Distribution: Atlantic coast from eastern Georgia to Maryland and New Jersey. See range map in Petranka (1998). **U.S.**: DE, GA, MD, NC, NJ, PA, SC, VA.

Common name: Eastern Mud Salamander.

Gyrinophilus montanus diastictus Bishop, 1941, new comb.

"Triturus hypoxanthus" Rafinesque, 1820: 4. (a *nomen nudum* so unavailable [Gray, 1850]). Assigned as a synonym of this species by Frost (2009-13) based on geography.

Pseudotriton montanus diastictus Bishop, 1941: 14.—**O**: **H**: given as "in Bishop collection," now in MCZ.—**OT**: U.S., Kentucky, Carter County, Cascade Caverns.

Pseudotriton montanus diastictus: Stejneger & Barbour, 1943: 28; Schmidt, 1953: 48; Brame, 1967: 14; Cochran & Goin, 1970: 61; Gorham, 1974: 39; Conant, 1975: 286; Behler & King, 1979: 353; Conant & Collins, 1998: 485; Petranka, 1998: 296; Raffaëlli, 2007: 175; Bonett *et al.*, 2013a: 7.

Pseudotriton diastictus: Collins, 1991: 43.

Gyrinophilus montanus diasticus: ***new combination***

Distribution: Unglaciated plateau region west of the Appalachian divide, from Tennessee to Ohio and Indiana. Petranka (1998) has a range map. **U.S.**: IN, KY, OH, TN, VA, WV.

Common name: Midland Mud Salamander.

Commentary: Collins (1991) treated this as a distinct species, but with no justification, so we continue to treat it as a subspecies, pending further investigation. Bonett *et al.* (2013a) found no justification for treating it as a full species.

Gyrinophilus montanus flavissimus Hallowell, 1857, **new combination**

Pseudotriton flavissimus Hallowell, 1857c: 130.—**O**: **H:** ANSP 576 (Fowler & Dunn, 1917: 19, sd.); verified by Malnate, 1971:347.—**OT**: U.S., Georgia, Liberty County.—**SS**: Löding, 1922: 14, with *P. montanus*.

Spelerpes flavissimus: Strauch, 1870: 83.

Spelerpes ruber flavissimus: Cope, 1889b: 176.

Eurycea montana flavissima: Löding, 1922: 14.

Pseudotriton montanus flavissimus: Stejneger & Barbour, 1923: 48; 1943: 29; Dunn, 1926b: 291; Neill, 1948a: 136; Schmidt, 1953: 48; Cochran & Goin, 1970: 61; Gorham, 1974: 39; Conant, 1975: 286; Behler & King, 1979: 353; Conant & Collins, 1998: 485; Petranka, 1998: 296; Raffaëlli, 2007: 175; Bonett *et al.*, 2013a: 7.

Pseudotriton flavissimus flavissimus: Bishop, 1943: 378.

Gyrinophilus montanus flavissimus: ***new combination***

Distribution: Eastern Louisiana to eastern Alabama & southern Georgia & southern South Carolina. Petranka (1998) has a range map. **U.S.**: AL, FL, GA, LA, MS, SC.

Common name: Gulf Coast Mud Salamander.

Gyrinophilus montanus floridanus Netting & Goin, 1942, **new combination**

Pseudotriton montanus floridanus Netting & Goin, 1942: 175.—**O**: **H**: CM 16850.—**OT**: U.S., Florida, Gainesville, University of Florida campus, seepage area along C Creek.

Pseudotriton flavissimus floridanus: Bishop, 1943: 381.

Pseudotriton montanus floridanus: Stejneger & Barbour, 1943: 29; Neill, 1948a: 136; Schmidt, 1953: 48; Brame, 1967: 14; Cochran & Goin, 1970: 62; Gorham, 1974: 39; Conant, 1975: 286; Behler & King, 1979: 353; Conant & Collins, 1998: 485; Petranka, 1998: 295; Raffaëlli, 2007: 175.

Gyrinophilus montanus floridanus: ***new combination***

Distribution: Northern Florida and adjacent Georgia. **U.S.**: FL, GA.

Common name: Rusty Mud Salamander.

Gyrinophilus porphyriticus (Green, 1827)

Synonymy: See subspecies, *G. p. porphyriticus,* below.

Distribution: Northern Mississippi and Alabama to New England and adjacent Canada. Brandon (1967c) and Petranka (1998) have range maps, and Fisher *et al.* (2007) map the Canadian range. **Can.**: ON, QC. **U.S.**: AL, CT, GA, KY, MA, MD, ME, NC, NH, NJ, NY, OH, PA, RI, SC, TN, VA, VT, WV.

Common name: Spring Salamander. *French Canadian name:* Salamandre pourpre.

Content: Seven subspecies are here recognized.

Commentary: Brandon (1967c) provided a CAAR account. Listed in Ontario as Extirpated. There could be some unrecognized cryptic species composing this population (see Beachy, 2005). Populations currently treated as a separate species, *G. palleucus,* are here treated as members of *G. porphoriticus,* following the study of Bonett *et al.,* which showed no clear differentiation of the two as separate species.

Gyrinophilus porphyriticus porphyriticus (Green, 1827)

Salamandra porphyritica Green, 1827: 3.—**O**: none stated or known to exist.—**OT**: U.S., Pennsylvania, Crawford County, French Creek near Meadville.—**N**: MCZ 35778 (Brandon, 1966a: 32);—**OT** (of **N**): U.S., Pennsylvania, Crawford County, Meadville, a small spring-fed stream (flowing directly into French Creek) at Liberty and Linden streets (close equivalent of the original OT).

Triton porphyriticus: Holbrook, 1842: 83.

Spelerpes? porphyriticus: Gray, 1850: 46.

Amblystoma porphyriticum: Hallowell, 1855a: 8. Apparently in error; based on specimens of *Ambystoma texanum, fide* Cope, 1868a: 206.

Gyrinophilus porphyriticus: Cope, 1869b: 108; 1889b: 155; Stejneger & Barbour, 1917: 18; Dunn, 1926b: 259; Brandon, 1966a: 29.

Spelerpes porphyriticus: Smith, 1877: 91; Boulenger, 1882b: 64.

Gyrinophilus porphyriticus porphyriticus: Stejneger & Barbour, 1933: 15; 1943: 27; Schmidt, 1953: 45; Brame, 1967: 14: Cochran & Goin, 1970: 44; Gorham, 1974: 35; Conant, 1975: 284; Behler & King, 1979:

330; Frost, 1985: 589; Conant & Collins, 1998: 481; Petranka, 283; Raffaëlli, 2007: 175.

Pseudotriton porphyriticus: Organ, 1961: 53.

Salamandra salmonea Storer *in* Holbrook, 1838b: 101.—**O**: **H**: specimen in pl. 22.—**OT**: U.S., Massachusetts, neighborhood of Danvers, and Vermont, Green Mountains; corrected to U.S., Vermont (Storer, 1840: 58), stating description based on a specimen from Vermont, and implied the one illustrated may have come from Danvers.—**Comment:** Schmidt (1953: 45) restricted the OT to Vermont. Adler (1976: xli) rejected the restriction saying the figure is more likely based on Massachusetts material. As Schmidt provided no data to support that the H came from there, the action is invalid.—**SS**: Baird, 1850a: 287; Boulenger, 1882b: 64.

Salamandra salmonea: Holbrook, 1842: 5: 33.

Pseudotriton salmoneus: Baird, 1850a: 287.

Spelerpes? salmonea Gray, 1850: 46.

Pseudotriton salmoneum: Baird *in* Heck & Baird, 1851: 255.

Ambystoma salmoneum: Duméril, Bibron, & Duméril, 1854c: 110.

Pseudotriton salmonea: Hallowell, 1858b: 342.

Spelerpes salmoneus: Cope, 1866a: 98.

Gyrinophilus porphyriticus inagnoscus Mittleman, 1942a: 27.—**O**: **H**: USNM 115520.—**OT**: U.S., Ohio, Hocking County, Good Hope Township, 4 miles southwest of Bloomingville at Salt Creek. Corrected to U.S., Ohio, Hocking County, Salt Creek Township, 4 miles southwest of South Bloomingville at Salt Creek (Condit, 1958: 46-47).—**SS**: Brandon, 1963: 210-211; 1966a: 31, with *G. p. porphyriticus.*

Gyrinophilus porphyriticus inagnoscus: Stejneger & Barbour, 1943: 28; Schmidt, 1953: 46; Conant, 1958: 242.

Distribution: Central Virginia to New England and adjacent Canada. See range maps of Brandon (1967c) and Petranka (1998). There is some lack of agreement in the southern extent of the range of the subspecies. **Can.**: ON, QC. **U.S.**: AL, CT, GA, MA, MD, ME, MS, NC, NH, NJ, NY, OH, PA, RI, TN, VA, VT, WV.

Common name: Northern Spring Salamander. *French Canadian name:* Salamandre pourpre.

Commentary: Brandon (1967c) included this subspecies in his CAAR account.

Gyrinophilus porphyriticus danielsi (Blatchley, 1901)

Spelerpes danielsi Blatchley, 1901: 760.—**O: S:** MCZ 6638-6639 (Barbour & Loveridge, 1929: 339. sd.).—**L**: MCZ 6638 (Brandon, 1966a: 56).—**OT**: U.S., Tennessee, Sevier County; corrected, in error, to U.S., North Carolina, McDowell County (Barbour & Loveridge, 1929: 339); re-corrected to U.S., Tennessee, Sevier County, side of Mt. Collins or Indian Pass, elev. 3000-5000 feet (Brame, 1972: 181).—**SS**: Stejneger & Barbour, 1933: 15.

Gyrinophilus danielsi: Stejneger & Barbour, 1917: 17; 1943: 26; Fowler & Dunn, 1917: 19; Bishop, 1924: 87; 1928a: 161; Dunn, 1926b: 266.

Gyrinophilus porphyriticus danielsi: Stejneger & Barbour, 1933: 15; Brame, 1967: 14; Cochran & Goin, 1970: 46; Gorham, 1974: 35; Conant, 1975: 285; Behler & King, 1979: 330; Conant & Collins, 1998: 482; Petranka, 1998: 283; Raffaëlli, 2007: 176.

Gyrinophilus danielsi danielsi: Bishop, 1943: 361; Schmidt, 1953: 46.

Gyrinophilus danielsi polystictus Reese, 1950: 2.—**O**: **H**: FMNH 91108.—**OT**: U.S., North Carolina, Yancey County, Mt. Mitchell, elev. 6000 feet.—**SS**: Brandon, 1963: 210-211; 1966a: 56 with *G. porphyriticus danielsi*.

Gyrinophilus danielsi polystictus: Schmidt, 1953: 47.

Distribution: Eastern Tennessee and western North Carolina. See range map in Petranka (1998). **U.S.**: NC, TN.

Common name: Blue Ridge Spring Salamander.

Gyrinophilus porphyriticus dunni Mittleman & Jopson, 1941

Gyrinophilus dunni Mittleman & Jopson, 1941: 2.—**O**: **H**: USNM 113230.—**OT**: U.S., South Carolina, Pickens County, Clemson, campus of Clemson College (now Clemson University), elev. 700 feet.—**SS**: Bishop, 1943: 365, with *G. porphyriticus*.

Gyrinophilus dunni: Stejnger & Barbour, 1943: 27.

Gyrinophilus danielsi dunni: Bishop, 1943: 365; Schmidt, 1953: 47; Brame, 1967: 14.

Gyrinophilus porphyriticus dunni: Brandon, 1966a: 50; Cochran & Goin, 1970: 46; Gorham, 1974: 35; Conant, 1975: 285; Behler & King, 1979: 330; Conant & Collins, 1998: 482; Petranka, 1998: 283; Raffaëlli, 2007: 176.

Distribution: Northeastern Alabama and northern Georgia to western Carolinas. See range map of Petranka (1998). **U.S.**: AL, GA, NC, SC.

Common name: Carolina Spring Salamander.

Gyrinophilus porphyriticus duryi (Weller, 1930)

"Triturus lutescens" Rafinesque, 1832c: 121.—**O**: not stated or known to exist.—**OT**: U.S., west Kentucky.—**SS**: Newcomer, 1961: 21, Brandon, 1963: 210.—**Comment:** Placed on official index of rejected and invalid species names (ICZN, 1965: 167-168).

Gyrinophilus lutescens: Mittleman, 1942a: 33; Stejneger & Barbour, 1943: 28.

Pseudotriton duryi Weller, 1930: 6-9.—**O**: **S**: CSNM 499a-499g (7).—**L**: USNM 84300 (ex CSNM 499d), Walker & Weller, 1932: 81.—**OT**: U.S., Kentucky, Carter County, Cascade Caves, about 10 miles from Grayson.—**Comment:** Nomen placed on the official list of specific names (ICZN, 1965).

Gyrinophilus duryi: Weller: 1931a: 8.

Gyrinophilus porphyriticus duryi: Stejneger & Barbour, 1933: 15; 1943: 27; Schmidt, 1953: 46; Brame, 1967: 14; Cochran & Goin, 1970: 46; Gorham, 1974: 35; Conant, 1975: 284; Behler & King, 1979: 330; Conant & Collins: 481; Petranka, 1998: 283; Raffaëlli, 2007: 176.

Gyrinophilus danielsi duryi: King, 1939: 556.

Distribution: Mostly eastern Kentucky and western West Virginia. See range map in Petranka (1998). **U.S.**: KY, OH, WV.

Common name: Kentucky Spring Salamander.

Gyrinophilus porphoriticus gulolineatus Brandon, 1965

Gyrinophilus palleucus gulolineatus Brandon, 1965a: 347.—**O**: **H**: FMNH 142237.—**OT**: U.S., Tennessee, Roane County, Berry Cave.

Gyrinophilus palleucus gulolineatus: Cochran & Goin, 1970: 47; Gorham, 1974: 35; Conant, 1975: 285; Behler & King, 1979: 329; Conant & Collins, 1998: 484; Petranka, 280.

Gyrinophilus gulolineatus: Brandon, Jacobs, Wynn, and Sever, 1986: 2, by implication; Collins, 1991: 43; Raffaëlli, 2007: 178.

Distribution: Tennessee River Valley, eastern Tennessee. See range map of Petranka (1998), as *G. palleucus gulolineatus*. **U.S.**: TN.

Common name: Berry Cave Salamander.

Commentary: Niemiller & Miller (2010) provided a CAAR account as a full species. Bonett *et al.* (2013a) did not find sufficient differentiation of the form to afford full species status. They found the race to be a sister to one lineage of *G. porphoriticus porphoriticus,* making recognition of separate species status of *G. palleucus* untenable.

Gyrinophilus porphoriticus necturoides Lazell & Brandon, 1962, **new combination**

Gyrinophilus palleucus necturoides Lazell & Brandon, 1962: 301.—**O**: **H**: MCZ 34100.—**OT**: U.S., Tennessee, Grundy County, Big Mouth Cave, near Pelham.

Gyrinophilus palleucus necturoides: Brame, 1967: 14; Cochran & Goin, 1970: 47; Conant, 1975: 285; Behler.& King, 1979: 329; Conant & Collins, 1998: 483; Petranka, 1998: 280; Raffaëlli, 2007: 177.

Gyrinophilus porphoriticus necturoides: ***new combination***

Distribution: Northern Alabama into southern Tennessee. **U.S.**: AL, TN.

Common name: Big Mouth Cave Salamander.

Commentary: This race has also been treated under the species name *G. palleucus,* but again, the study of Bonett *et al.* (2013a) makes recognition of this as a separate species untenable.

Gyrinophilus porphoriticus palleucus McCrady, 1954

Gyrinophilus palleucus McCrady, 1954: 201.—**O**: **H**: CNHM 72585.—**OT**: U.S., Tennessee, Franklin County, hardwood forest at north end of Sinking Cove, 5 miles west of Sherwood across Burned Stand Ridge, and 15 miles southwest of Sewanee, elev. 900 feet.

Gyrinophilus palleucus: Conant, 1958: 243; Frost, 1985: 589.

Pseudotriton palleucus: A.P. Blair, 1961: 499.

Gyrinophilus palleucus palleucus: Lazell & Brandon, 1962: 301; Brame, 1967: 14; Cochran & Goin, 1970: 47; Gorham, 1974: 35; Conant, 1975: 285; Behler & King, 1979: 329; Conant & Collins, 1998: 483; Petranka, 1998: 280; Raffaëlli, 2007: 177.

Distribution: South central Tennessee, northeastern Alabama and adjacent Georgia, in subterranean waters.

See range map of Petranka (1998) as a full species; see also the more recent range map of Miller and Niemiller (2012), who recognize *gulolineatus* as a full species. The subspecies identity of the more northern populations in Tennessee is uncertain. **U.S.**: AL, GA, TN

Common name: Pale Salamander.

Commentary: This form was found to be part of a monophyletic cluster with two lineages identified as *G. p. nectroides,* and that cluster to be a sister of a lineage of *G. p. porphoriticus* plus *G. p. gulolineatus Bonett et al. (2013a).* Thus, we can no longer support separate species status for *G. palleucus* and *G. porphoriticus.* Accordingly, we treat the two under the older name, *G. porphoriticus.* Brandon (1967b) provided a CAAR account as a full species, and a more recent update was provided by Miller and Niemiller (2012).

Gyrinophilus subterraneus Besharse & Holsinger, 1977

Gyrinophilus subterraneus Besharse & Holsinger, 1977: 626—**O**: **H**: USNM 198533.—**OT**: U.S., West Virginia, Greenbrier County, a few km NE Alderson, at General Davis Cave.

Gyrinophilus subterraneus: Behler & King, 1979: 330; Collins, 1997: 7; Conant & Collins, 1998: 483; Petranka, 1998: 287; Raffaëlli, 2007: 177

Distribution: Known only from the cave system of the type locality. **U.S.**: WV.

Common name: West Virginia Spring Salamander.

Commentary: The study of Bonett *et al.* (2013a) found this to be sister to one lineage of *G. p. porphoriticus,* so for the present we will continue to recognize this as a full species.

Genus **Pseudotriton** Tschudi, 1838

Pseudotriton Tschudi, 1838: 60.—**NS**: *Salamandra subfusca* Green, 1818 (mo.).

Pseudotriton: Schmidt, 1953: 47; Montanucci, 2006; Frost, 2009-13.

Mycetoglossus Bonaparte, 1839e: 267. He attributes the name to Bibron, no year. Substitute name for *Pseudotriton* Tschudi, 1838. Suppressed for Priority (ICZN, 1997b: 140-141).—**SS**: Hallowell, 1858b: 347.

Batrachopsis Fitzinger, 1843: 34. Substitute name for *Pseudotriton* Tschudi 1838.

Pelodytes Gistel, 1848: xi. Substitute name for *Pseudotriton* Tschudi, 1838. Junior homonym of *Pelodytes* Fitzinger, 1843 (Anura).

Distribution: Most of the eastern US, east of the Mississippi River.

Common name: Red and Mud Salamanders.

Content: A single species is recognized here, with four subspecies.

Commentary: Wake (1966) considered this a distinct genus on morphological grounds, and molecular studies later have not suggested it should be placed in synonymy of another genus, although its species seem closely related to *Eurycea* and *Gyrinophilus*. Martof (1975a) provided a generic CAAR account. Some authors have suggested placing *Pseudotriton* into the synonymy of *Eurycea*. But most recent U.S. authors (*e.g.*, Montanucci, 2006; Frost, 2009-13), have continued to recognize *Pseudotriton* (though Stejneger & Barbour, 1917, also placed it in the synonymy of *Eurycea*, but they recognized it as distinct again in 1923). Bonett *et al.* (2013a) show this genus as a distinct clade, but with one species, *P. montanus*, appearing to be so closely related to species of *Gyrinophilus* that we have moved it there.

Pseudotriton ruber (Sonnini & Latreille, 1802)

Synonymy: See subspecies *P. r. ruber*, below.

Distribution: Much of US east of the Mississippi. Petranka (1998) has a range map, as does Martof (1975c). **U.S.**: AL, DC, DE, FL, GA, IN, KY, LA, MD, MS, NC, NJ, NY, OH, PA, SC, TN, VA, WV.

Common name: Red Salamander.

Content: Four subspecies are usually recognized, although Martof (1975c) did not recognize any.

Commentary: Schmidt (1953) restricted the type locality to vicinity of Philadelphia, Pennsylvania; although this is reasonable, based on the known range, no evidence was cited to indicate that onomatophores (type specimens) came from there, so that is apparently an invalid restriction. Martof (1975c) provided a CAAR account. The study of Bonett *et al.* (2013a) found more than one lineage in three of the subspecies, with differing interelations, indicating that the current treatment may need to be revised with further study.

Pseudotriton ruber ruber (Sonnini & Latreille, 1802)

Salamandra rubra Sonnini & Latreille, 1802: 305.—**O**: none stated or known to exist, but may have originally been in MNHN.—**OT**: U.S.

Salamandra rubra: Daudin, 1803: 227; Harlan, 1827: 332; Sonnini & Latreille, 1830d: 305; Holbrook, 1840: 123, pl. 27; 1842 (v5): 35.

Molge rubra: Merrem, 1820: 185.

Triton ruber: Fitzinger, 1826: 66.

Triton rubra: Wagler, 1830: 208.

Spelerpes rubra: Gray, 1850: 45.

Mycetoglossus ruber: Gray, 1850: 45.

Pseudotriton ruber: Baird, 1850a: 286.

Pseudotriton rubrum: Baird *in* Heck & Baird, 1851: 255.

Bolitoglossa rubra: Duméril, Bibron, & Duméril, 1854c: 89.

Spelerpes ruber: Cope, 1869b: 107; 1889b: 172; Knauer, 1878: 98.

Spelerpes ruber ruber: Davis & Rice, 1883: 27.

Geotriton rubra: Garman, 1884a: 39.

Eurycea rubra: Fowler, 1925a: 58.

Eurycea rubra rubra: Stejneger & Barbour, 1917: 20.

Pseudotriton ruber ruber: Dunn, 1918b: 467; 1926b: 272; Bishop, 1924: 91; Stejneger & Barbour, 1923: 15; 1943: 29; Schmidt, 1953: 48; Brame, 1967: 14; Cochran & Goin, 1970: 62; Gorham, 1974: 35; Conant, 1975: 286; Behler & King, 1979: 354; Conant & Collins, 1998: 485; Petranka, 1998: 299; Raffaëlli, 2007: 173.

Pseudotriton ruber: Frost, 1985: 604.

Salamandra subfusca Green, 1818: 351.—**O**: not stated or known to exist. **OT**: not stated but probably vicinity of U.S., New Jersey, Princeton.—**SS**: Baird, 1850a: 286, and Gray, 1850: 45.

Pseudotriton subfuscus: Tschudi, 1838: 60.

Mycetoglossus subfuscus: Bonaparte, 1839**e**: unnumbered (*fide* Frost, 2009-13). **Note**: we were unable to find this name in the volume.

Batrachopsis subfuscus: Fitzinger, 1843: 34.

Salamandra rubriventris Green, 1818: 353.—**O**: not stated or known to exist. **OT**: not stated but probably vicinity of U.S., New Jersey, Princeton. Homonym of *Salamandra rubriventris* Daudin, 1803.—**SS**: Say, 1818: 406.

Salamandra brevicauda Wied, 1865: 127.—**O**: not stated or known to exixt.—**OT**: U.S., Pennsylvania.—**SS**: Dunn, 1926b: 274.

Amblystoma brevicauda: Boulenger, 1882b: 38.

"Gyrinophilus warneri" Sinclair, 1955: 133.—**O**: **S**: 3 larvae, disposition not stated; apparently Sinclair's personal collection.—**OT**: U.S., Tennessee, but not stated. *Nomen nudum*.—**Comment**: Brandon, 1966a: 5, examined the "types" and identified them as *Pseudotriton*. Frost (2009-13) listed this nomen in the synonymy of *P. ruber* based on geography.

Distribution: Northern range of the species, south to northern Alabama and Georgia; see Petranka's (1998) range map. **U.S.**: AL, DC, DE, FL, GA, IN, KY, MD, NC, NJ, NY, OH, PA, SC, TN, VA, WV.

Common name: Northern Red Salamander.

Pseudotriton ruber nitidus Dunn, 1920

Pseudotriton ruber nitidus Dunn, 1920b: 133.—**O**: **H**: MCZ 5649.—**OT**: U.S., Virginia, White Top Mountain, elev. 4000 feet.

Pseudotriton ruber nitidus: Stejneger & Barbour, 1923: 15; 1943: 29; Dunn, 1926b: 281; Schmidt, 1953: 49; Brame, 1967: 14; Cochran & Goin, 1970: 63; Gorham, 1974: 39; Conant, 1975: 287; Behler & King, 1979: 354; Conant & Collins, 1998: 486; Petranka, 1998: 299; Raffaëlli, 2007: 174.

Distribution: Restricted range in southwestern Virginia and northwestern North Carolina. Petranka (1998) has a range map. **U.S.**: NC, VA.

Common name: Blue Ridge Red Salamander.

Pseudotriton ruber schencki (Brimley, 1912)

Spelerpes ruber schencki Brimley, 1912: 139.—**O**: **H**: USNM 49679 (Cochran, 1961a: 26, sd.).—**OT**: U.S., North Carolina, Haywood County, Sunburst (now Spruce), elev. 3200 feet.

Spelerpes schencki: Brimley, 1915: 198.

Eurycea rubra schencki: Stejneger & Barbour, 1917: 20.

Pseudotriton ruber schencki: Dunn, 1918b: 467; 1926b: 283; Stejneger & Barbour, 1923: 15; 1943: 30; Bishop, 1928a: 162; Schmidt, 1953: 49; Brame, 1967: 14; Cochran & Goin, 1970: 63; Gorham, 1974: 39; Conant, 1975: 287; Behler & King, 1979: 354; Conant & Collins, 1998: 486; Petranka, 1998: 299; Raffaëlli, 2007: 174.

Distribution: Northeastern Georgia, northwestern South Carolina, southeastern Tennessee, southwestern North Carolina. Petranka (1998) has a range map. **U.S.**: GA, NC, SC, TN.

Common name: Blackchinned Red Salamander.

Pseudotriton ruber vioscai Bishop, 1928

Pseudotriton ruber vioscai Bishop, 1928b: 247.—**O**: **H**: USNM 75057.—**OT**: U.S., Louisiana, Washington Parish, 10 miles west of Bogalusa, in a spring run.

Pseudotriton ruber vioscai: Stejneger & Barbour, 1943: 30; Viosca, 1949: 10; Schmidt, 1953: 49; Brame, 1967: 14; Cochran & Goin, 1970: 62; Gorham, 1974: 39; Conant, 1975: 287; Behler & King, 1979: 354; Conant & Collins, 1998: 486; Petranka, 1998: 299; Raffaëlli, 2007: 174.

Distribution: Southern range of the species, north to eastern Ten-nessee and Kentucky; see range map in Petranka (1998). **U.S.**: AL, FL, GA, KY, LA, MS, NC, SC, TN.

Common name: Southern Red Salamander.

Genus **Stereochilus** Cope, 1869

Stereochilus Cope, 1869b: 100.—**NS**: *Pseudotriton marginatus* Hallowell, 1856, **(**mo.**)**.

Stereochilus: Stejneger & Barbour, 1917: 17; 1933: 14; Schmidt, 1953: 44.

Distribution: Atlantic coast from southeastern Virginia to north-eastern Florida. Petranka (1998) has a range map.

Common name: Many-lined Salamanders.

Content: Monotypic.

Commentary: Studies of Mueller *et al.* (2004) and Vieites *et al.* (2007) indicated that this genus shares a sister-taxon relationship with *Pseudotriton* and *Gyrinophilus*. Bonett *et al.* (2013a) showed this genus has a basal sister relationship to the other two genera. Rabb (1966) provided a CAAR generic account.

Stereochilus marginatus (Hallowell, 1856)

Pseudotriton marginatus Hallowell, 1857c: 130.—**O**: **H**: ANSP 514 (Fowler & Dunn, 1917: 21, sd.); verified by Malnate, 1971: 348.—**OT**: U.S., Georgia, Liberty County; corrected to U.S., Georgia, Liberty County, Riceboro (Dunn, 1926a: 247).

Stereochilus marginatum: Cope, 1869b: 101; 1889b: 152; Dunn, 1926b: 244.
Spelerpes marginatus: Strauch, 1870: 83.
Geotriton marginatus: Garman, 1884a: 40.
Stereochilus marginatus: Cope, 1889b: 152; Stejneger & Barbour, 1917: 17; 1943: 25; Bishop, 1928a: 161; Schmidt, 1953: 44; Brame, 1967: 14; Cochran & Goin, 1970: 63; Gorham, 1974: 39; Conant, 1975: 282; Behler & King, 1979: 354; Frost, 1985: 604; Conant & Collins, 1998: 479; Petranka, 1998: 304; Raffaëlli, 2007: 189.

Distribution: As for the genus. **U.S.**: FL, GA, NC, SC, VA.

Common name: Many-lined Salamander.

Commentary: Rabb (1966) provided a CAAR species account.

Subtribe **Spelerpina** Cope, 1859

Spelerpeae Cope, 1859a: 124.
Spelerpini: Vieites *et al.*, 2011: 633; Wake, 2012: 78.
Spelerpina: Dubois & Raffaëlli, 2012: 143.

Distribution: Eastern North America.

Common name: None in general use.

Content: Two genera, with 26 species.

Commentary: This subtribe is a sister taxon to the other subtribe of the spelerines, Pseudotritonina. Each is supported as a monophyletic clade by Bonett *et al.* (2013a).

Genus **Eurycea** Rafinesque, 1822

Synonymy: See subgenus *Eurycea,* below.

Distribution: Eastern North America.

Common name: Brook Salamanders (SECSN) or Brook and Cave Stream Salamanders (C&T).

Content: 27 species in four subgenera, all North American.

Commentary: The names *Typhlotriton, Typhlomolge,* and *Haideotriton* were in general use for genera of some cave-adapted forms, as was *Manculus,* but all have recently been synonymized with *Eurycea.* Hillis *et al.* (2001) coined some other names for taxa within their subgenus *Paedomolge,* but some of those names are presumably unavailable under the International Code (see discussions and detailed synonymy, Frost, 2009-13).

Subgenus **Eurycea** Rafinesque, 1822

Eurycea Rafinesque, 1822a: 3.—**NS**: *Eurycea lucifuga* Rafinesque, 1822 (mo.).

Spelerpes Rafinesque, 1832a: 22.—**NS**: *Eurycea lucifuga* Rafinesque, 1822 (mo.).—**SS**: Dunn, 1926b: 294.

Cylindrosoma Tschudi, 1838: 95.—**NS**: *Salamandra longicauda* Green, 1818 (mo.).—**SS**: Baird, 1850a: 287, with *Spelerpes*; Dunn, 1926b: 294, with *Eurycea*.

Saurocercus Fitzinger, 1843: 34. Substitute name for *Cylindrosoma* Tschudi.

Haideotriton Carr, 1939: 334.—**NS** (od.): *Haideotriton wallacei* Carr, 1939.—**SS**: Dubois, 2005b: 20; Frost *et al.* (2006: 359), with *Eurycea*.

Distribution: Eastern North America.

Common name: Brook Salamanders (no name is in general use for the subgenus).

Content: Nine species are here recognized in this subgenus.

Commentary: Bonett *et al.* (2013a) supplied phylogentic information on relationships among members of this taxon. One species, *E. quadridigitata,* is distinct enough to have been placed in a separate genus (*Manculus*) for many years. *E. chamberlaini* was recently described and is apparently a sister species of that form. There are four subgenera recognized here; these are often treated as species-groups.

Eurycea (Eurycea) aquatica Rose & Bush, 1963

Eurycea aquatica Rose & Bush, 1963: 121.—**O**: **H** (od.), USNM 147138.—**OT**: U.S., Alabama, Jefferson County, small springs and permanent streams two miles west of Bessemer, along county highway 20.

Eurycea aquatica Brame, 1967: 13; Cochran & Goin, 1970: 40; Gorham, 1974: 34; Conant, 1975: 290; Behler & King, 1979: 321; Frost, 1985: 587; Conant & Collins, 1998: 490.

Eurycea (Manculus) aquatica: Dubois & Raffaëlli (2012: 143).

Misapplication of names:

Eurycea bislineata: Petranka, 1998: 241-242 (part).

Distribution: The Appalachians of northern Alabama, and confirmed into southern Tennessee and northwestern Georgia (Timpe *et al.,* 2009). Rose (1971) provided a CAAR account, which shows known localities; a map by Timpe *et al.* (2009) shows additional localities. **U.S.**: AL, GA, TN.

Common name: Dark-sided Salamander (SECSN) or Brown-backed Salamander (C&T). We prefer the common name "Brown-backed" because "Dark-sided" is also used for *E. longicauda melanopleura.*

Commentary: Mount (1975) and Petranka (1998: 241-242) argued that this is not a species distinct from *E. bislineata,* nor did Raffaëlli (2007) treat it as a distinct species; however, other studies seem to validate the species level analysis. Jacobs' (1987) phylogram indicates *E. aquatica* and *E. junaluska* are distinct, sister species of the *bislineata* group. The analysis of Bonett *et al.* (2013a) supports that relationship. Timpe *et al.* (2009) confirmed the distinctiveness of this species from *E. cirrigera,* and found three distinct clades which may represent additional full species. Dubois & Raffaëlli (2012) included this species in their Supraspecies *bislineata,* along with four other species.

Eurycea (Eurycea) bislineata (Green, 1818)

Salamandra bislineata Green, 1818: 352.—**O**: **S:** No type stated by Green. ANSP 695-698 and/or USNM 3738 (Fowler & Dunn, 1917: 20, sd.); Dunn (1926b: 297) doubted that there ever were types in existence. Fowler (1907: 65) mentioned 4 co-types, all from localities in New Jersey.—**OT:** given as New Jersey (probably Princeton?), but may be in error; Stejneger & Barbour (1917) listed **OT** as western Pennsylvania, based on the museum record for USNM 3738, which was claimed to be designated by Fowler (1907), but was not mentioned by him.

Salamandra bis-lineata: Harlan, 1827: 332.

Spelerpes bis-lineata: Harlan, 1835: 97.

Pseudotriton bilineata: Tschudi, 1838: 60.—**IS.**

Salamandra bilineata: Holbrook, 1838b: 127, pl. xxix, fig. 30.—**IS.**

Mycetoglossus bilineata: Bonaparte, 1839e: 267. (Frost, 2009-2013, cited the page as 131, in error; the pages are not numbered and there is "131" under the last printed line on the page, as on several nearby pages).—**IS.**

Spelerpes bilineata: Baird, 1850a: 287; Gray, 1850: 38.—**IS.**

Bolitoglossa bilineata: Duméril, Bibron, & Duméril, 1854c: 91.—**IS**.

Spelerpes (Cylindrosoma) bilineata: Hallowell, 1857e: 101.—**IS**.

Pseudotriton bilineatus: Jan, 1857: 55.—**IS.**

Spelerpes bilineatus: Verrill, 1865a: 199; Boulenger, 1882b: 66; Davis & Rice, 1883: 27; Werner, 1909: 25.—**IS**.

Geotriton bilineata: Garman, 1884a: 39.—**IS**.

Spelerpes bislineatus: Hay, 1892: 448; Rhoads, 1895: 401; Fowler, 1907: 63.

Eurycea bislineata: Stejneger & Barbour, 1917: 18; Frost, 1985: 587; Conant & Collins, 1998: 488.

Eurycea bislineata bislineata: Dunn, 1920c: 134; 1926b: 295; Stejneger & Barbour, 1923: 15; 1943: 30; Schmidt, 1953: 52; Brame, 1967: 13; Cochran & Goin, 1970: 40; Gorham, 1970: 4; 1974: 34; Conant, 1975: 288; Behler & King, 1979: 321; Petranka, 1998: 242.

Eurycea (Eurycea) bislineata: Raffaëlli, 2007: 185.

Eurycea (Manculus) bislineata: Dubois & Raffaëlli, 2012.

Salamandra flavissima Harlan, 1826b: 286.—**O**: **H**, in ANSP (Harlan, 1826b); not now known to exist.—**OT**: U.S., Pennsylvania.—**Comment**: Schmidt (1953: 52) restricted the type locality to "vicinity of Philadelphia" but provided no evidence the specimen came from there, so the restriction is invalid.—**SS**: Harlan, 1835: 97, with *Spelerpes bis-lineata*; Gray, 1850: 44, with *Spelerpes bilineata.*

Eurycea montana flavissima: Löding, 1922: 14

Salamandra haldemani Holbrook, 1840: 125.—**O**: **H**, specimen in pl. 28.—**OT**: U.S., Pennsylvania, Maryland, and Virginia.—**Comment**: Schmidt (1953: 52) restricted the type locality to U.S., Pennsylvania, Harrisburg, but provided no evidence that the **H** actually came from there, so the restriction is invalid.—**SS**: Cope, 1889b: 166; Dunn, 1926b: 297.

Ambystoma? haldemani: Gray, 1850: 38.

Spelerpes haldemani: Hallowell, 1858b: 347.

Desmognathus haldemani: Cope, 1859a: 124.

Desmognathus haldemanni: Strauch, 1870: 61.—**IS**.

Bolittoglossa dorsata Duméril, Bibron & Duméril, 1854c: 93 (attributed to Valenciennes by those authors and by Dunn, 1926b: 297).—**O**: **H** (sd., Mittleman, 1967b: 45.1) specimen in Fig. 1, Vélin 88?? du MNHN, Duméril *et al.*, 1854c: 93.—**OT**: none stated.—**SS**: Dunn, 1926b: 297, with *Eurycea bislineata* Green.—**Comment**: Schmidt (1953:52) restricted the type locality to Harrisburg, Pennsylvania, but did not cite any evidence that the **H** came from there, so that action is invalid.

Spelerpes bilineatus borealis Baird *in* Cope, 1889b: 165.—**O**: **S**: USNM 4735 (11: 9 ad., 2 larvae; Cochran, 1961a: 25, sd.).—**OT**: U.S. Maine, Franklin County, Lake Oquassa (= Kennebago Lake), near Oquossoc.—**SS**: Dunn, 1926b: 295, with *Eurycea bislineata bislineata.*

Eurycea bislineata major Trapido & Clausen, 1938: 119.—**O**: **H** (od.) USNM 104239.—**OT**: Can., Quebec, Lake St. John County, Val Jalbert, along Ouiatchouan River.—**SS**: Oliver & Bailey, 1939: 200; Mittleman, 1949: 90, with *E. b. bislineata.*

Eurycea bislineata major: Stejneger & Barbour, 1943: 31.

Eurycea bislineata rivicola Mittleman, 1949: 93.—**O**: **H**: USNM 129397.—**OT**: U.S., Indiana, Owen County, McCormick's Creek State Park, Echo Canyon.—**SS:** Sever, 1972: 323. with *E. b. bislineata.*

Eurycea bislineata rivicola: Schmidt, 1953: 52.

Misapplied names:

Desmognathus fuscus: Verrill, 1865b: 253, *fide* Dunn, 1926b: 297.

Desmognathus ochrophaea: Wilder, 1894: 216, *fide* Dunn, 1926b: 297.

Distribution: Northeastern North America. Petranka (1998) has a range map; only the portion indicated as *E. b. bislineata* should be regarded as this species. Mittleman (1967b) has a map showing the subspecies as recognized at that time; the populations designated as 2 and 3, plus possibly part of 4, are no longer regarded as this species. Sever (1999a) has a newer map. Raffaëlli (2007) also provided a map of the revised species range. **Can.**: NB, NL, ON, QC. **U.S.**: CT, DE, MA, MD, ME, NH, NJ, NY, OH, RI, VA, VT, WV.

Common name: Northern Two-lined Salamander. *French Canadian name:* Salamandre à deux lignes.

Content: Populations formerly considered subspecies of *E. bislineata,* (excluding the nominotypical one) are currently considered as other full species in this complex, except for some which are synonymized with *E. bislineata* (above).

Commentary: Mittleman (1967b) provided a CAAR account, and discussed the types. Sever (1999a), provided an updated CAAR account. Bonett *et al.* (2013a) indicated this species is a sister to one of three lineages of *E. cirrigera,* and that pair is sister to a clade including two other *cirrigera* lineages plus *E. wilderae.*

Eurycea (Eurycea) cirrigera (Green, 1831)

Salamandra cirrigera Green, 1831: 253.—**O**: **S**: USNM 4734 (2) (Cope, 1889b: 165, 168, sd.); Green stated there were 4 specimens, perhaps two were in ANSP: all apparently now lost (see Mittleman's, 1967b, discussion in the synonymy of *E. bislineata cirrigera*).—**OT**: U.S., Louisiana, near New Orleans.

Salamandra cirrigera: Holbrook, 1840: 119, pl. 30.

Spelerpes cirrigera: Baird, 1850a: 287; Gray, 1850: 44; Hallowell, 1858b: 346; Cope, 1870b: 401.

Spelerpes cirrigerus: Strauch, 1870: 82.

Spelerpes bilineatus cirrigera: Smith, 1877: 84; Cope, 1889b: 165.

Eurycea bilineata cirrigera: Dunn, 1920b: 135; 1926b: 307; Viosca, 1949: 10; Schmidt, 1953: 53; Cochran & Goin, 1970: 41; Brame, 1967: 13; Conant, 1975: 289;; Behler & King, 1979: 321; Petranka, 1998: 243.

Eurycea bislineata cirrigera: Stejneger & Barbour, 1923: 16; 1943: 30; Gorham, 1974: 34.

Eurycea cirrigera: Jacobs, 1987: 437; Conant & Collins, 1998: 488.

Eurycea (Eurycea) cirrigera: Raffaëlli, 2007: 185.

Misapplication of names:

Eurycea bislineata: Bishop, 1924: 91; Brimley and Mabee, 1925: 14.

Distribution: Part of the southeastern US. Petranka (1998) has a range map for the form as a subspecies of *Eurycea bislineata,* and Sever (1999b) as a full species. **U.S.**: AL, FL, GA, IL, IN, KY, LA, MS, NC, OH, SC, TN, VA, WV.

Common name: Southern Two-lined Salamander.

Commentary: This species was regarded as a subspecies of *E. bislineata* until Jacobs (1987) provided evidence that it is a separate species; many workers still treat it as a subspecies (*e.g.,* Petranka, 1998). Sever (1999b) provided a CAAR account. Bonett *et al.* (2013a) found three lineages, indicating that this species may represent at least two cryptic species; further study is needed.

Eurycea (Eurycea) guttolineata (Holbrook, 1838)

Salamandra gutto-lineata Holbrook, 1838a: 61.—**O**: **S**, based partly on Plate 12, and includes ANSP 716-717 (2) *fide* Fowler & Dunn (1917: 21).—**OT**: U.S., South Carolina middle country; Dunn (1926b: 327) listed the locality for the syntypes as Greenville, so a restriction or correction of OT to U.S., South Carolina, Greenville County, Greenville seems valid.

Salmandra guttolineata: Troost, 1840: 39.

Salamandra gutto-lineata: Holbrook, 1942: 29.

Spelerpes gutto-lineata: Baird, 1850a: 287; Gray, 1850: 45; Hallowell, 1857a: 11.

Cylindrosoma guttolineatum: Duméril, Bibron, & Duméril, 1854c: 80.

Spelerpes guttolineatus: Cope, 1869b: 167; 1889b: 170; Brimley, 1912: 139; Dunn, 1917b: 616.

Geotriton guttolineata: Garman, 1884a: 39.

Eurycea gutto-lineata: Stejneger & Barbour, 1917: 19; Bishop, 1924: 91; 1928a: 163; Dunn, 1926b: 327.

Eurycea longicauda guttolineata: Bailey, 1937: 8; Stejneger & Barbour, 1943: 32; Schmidt, 1953: 53; Brame, 1967: 13; Cochran & Goin, 1970: 42; Gorham, 1974: 34; Conant, 1975: 41; Behler & King, 1979: 323; Conant & Collins, 1998: 492; Petranka, 1998: 249.

Eurycea longicauda gutto-lineata: Viosca, 1949: 10.

Eurycea guttolineata: Carlin, 1997: 212.

Eurycea (Eurycea) guttolineata: Raffaëlli, 2007: 187.

Distribution: Southeastern US. Petranka (1998) has a range map. **U.S.**: AL, FL, GA, LA, MS, NC, SC, VA.

Common name: Three-lined Salamander.

Commentary: Ireland (1979) provided a brief CAAR account as a subspecies of *E. longicauda*. Carlin (1997) provided evidence of the species status for this form. Bonett *et al.* (2013a) found this species is a sister to *E. l. longicauda,* and those two taxa are sister to *E. l. melanopleura*.

Eurycea (Eurycea) junaluska Sever, Dundee, & Sullivan, 1976

Eurycea junaluska Sever, Dundee, & Sullivan, 1976: 26.—**O**: **H**: USNM 198421.—**OT**: U.S., North Carolina, Graham County, US Route 129, 3.2-11.2 km SE Tapoco.

Eurycea junaluska: Behler & King, 1979: 322; Frost, 1985: 587; Conant & Collins, 1998: 490; Petranka, 1998: 251.

Eurycea (Eurycea) junaluska: Raffaëlli, 2007: 186.

Distribution: Disjunct in North Carolina and Tennessee. Sever (1983) provided a CAAR account with a range map; Petranka (1998) also provided a range map. **U.S.**: NC, TN.

Common name: Junaluska Salamander.

Commentary: A sister relationship of this species to *E. aquatica* is supported by Bonett *et al.* (2013a).

Eurycea (Eurycea) longicauda (Green, 1818)

Synonymy: See the subspecies, *E. l. longicauda,* below.

Distribution: East-central US, from Oklahoma to New York. Petranka (1998) has a range map. **U.S.**: AR, IL, IN, KY, MD, MO, NJ, NY, OH, OK, PA, TN, VA, WV.

Common name: Long-tailed Salamander.

Content: Two subspecies are currently recognized.

Commentary: Schmidt (1953) restricted the type locality to vicinity of Princeton, New Jersey, apparently with inadequate justification. Ireland (1979) provided a CAAR account. One of the subspecies described and mapped, *E. l. guttolineata,* is now recognized as a full species. See commentary for *E. guttolineata* regarding relationships.

Eurycea (Eurycea) longicauda longicauda (Green, 1818)

Salamandra longicauda Green, 1818: 361.—**O**: none known to exist.—**OT**: U.S., New Jersey (probably near Princeton), as stated by Dunn (1926b: 321); invalidly restricted (Schmidt, 1953: 53).

Salamandra longicauda: Rafinesque, 1832d: 63; Holbrook, 1838b: 111

Salamandra longicaudata: Harlan, 1835: 96.—**IS.**

Cylindrosoma longicauda: Tschudi, 1838: 58, 95.

Salamandra longe-caudata: Troost, 1840: 40.—**IS.**

Plethodon (Saurocercus) longicauda: Fitzinger, 1843: 35.

Cylindrosoma longicaudatum: Duméril, Bibron, & Duméril, 1854c: 78.—**UE**.

Spelerpes longicauda: Baird, 1850a: 287; Gray, 1850: 43; Boulenger, 1882b: 64.

Spelerpes longicaudus: Cope, 1861b: 124; Davis & Rice, 1883: 27.

Saurocercus longicauda: Fitzinger, 1864: pl. 97, fig. 179.

Geotriton longicauda: Garman, 1884a: 39.

Eurycea longicauda: Stejneger & Barbour, 1917: 19; Dunn, 1926b: 320; Frost, 1985: 587.

Eurycea longicauda longicauda: Bailey, 1937: 8; Stejneger & Barbour, 1943: 31; Schmidt, 1953: 53; Brame, 1967: 13; Cochran & Goin, 1970: 42; Gorham, 1974: 34; Conant, 1975: 291; Behler & King, 1979: 323; Conant & Collins, 1998: 491; Petranka, 1998: 254.

Eurycea (Eurycea) longicauda longicauda: Raffaëlli, 2007: 187

Eurycea longicauda pernix Mittleman, 1942b: 101.—**O**: **H**: MCZ 25569.—**OT**: U.S., Indiana, Brown County, 2.5 miles southeast of Nashville, Brown County State Park, Jimmie Strahl Creek (tributary of Salt Creek).

Eurycea longicauda pernix: Stejneger & Barbour, 1943: 32; Schmidt, 1953: 54.

Distribution: As for the species, but generally east of the Mississippi. See Petranka's (1998) range map. **U.S.**: IL, IN, KY, MD, NJ, NY, OH, PA, TN, VA, WV.

Common name: As for the species.

Eurycea (Eurycea) longicauda melanopleura (Cope, 1894)

Spelerpes melanopleurus Cope, 1894a: 383.—**O**: **S** ANSP 10456-10460 (Fowler & Dunn, 1917: 21, sd.); confirmed by Malnate (1971: 348).—**OT**: U.S., Missouri, banks of Raley's Creek (= Riley's Creek, *fide* Dunn, 1926b: 316), one of the head tributaries of the White River.

Eurycea melanopleura: Stejneger & Barbour, 1917: 17; Dunn, 1926b: 316.

Eurycea longicauda melanopleura: Bishop, 1941: 20; Stejneger & Barbour, 1943: 32; Schmidt, 1953: 53; Brame, 1967: 13; Cochran & Goin, 1970: 43; Gorham, 1974: 34; Conant, 1975: Conant & Collins, 1998: 493: Petranka, 1998: 255.

Eurycea (Eurycea) longicauda melanopleura: Raffaëlli, 2007: 187

Spelerpes stejnegeri Eigenmann, 1901: 189.—**O**: **H** USNM 61259 (Brame, 1972: 172, sd.); not listed by Cochran, 1961a.—**OT**: U.S., Missouri, Barry County, Rock House Cave.—**SS**: Dunn (1926b: 316) with *Spelerpes melanopleura*.

Eurycea stejnegeri: Stejneger & Barbour, 1917: 20.

Distribution: Arkansas and Missouri, slightly into adjacent states. See Petranka's (1998) range map. **U.S.**: AR, IL, MO, OK.

Common name: Dark-sided Salamander.

Eurycea (Eurycea) lucifuga Rafinesque, 1822

Eurycea lucifuga Rafinesque, 1822a**:** 3.—**O:** none designated or known to exist.—**OT**: U.S., Kentucky, Fayette County, near Lexington in a cave.

Spelerpes lucifuga: Rafinesque, 1832a: 22.

Eurycea lucifuga: Stejneger & Barbour, 1917: 19; 1943: 33; Dunn, 1926b: 338; Bishop, 1928a: 163; 1943: 431; Schmidt, 1953: 54; Brame, 1967: 13; Cochran & Goin, 1970: 43; Gorham, 1974: 34; Conant, 1975: 292; Behler & King, 1979: 324; Frost, 1985: 587; Conant & Collins, 1998: 493; Petranka, 1998: 258.

Eurycea longicauda lucifuga: Mittleman, 1942b: 105.

Eurycea (Eurycea) lucifuga: Raffaëlli, 2007: 188.

Gyrinophilus maculicaudus Cope, 1890: 967.—**O:** unknown.—**OT:** Indiana, Franklin County, Brookville.

Spelerpes maculicaudus: Hay, 1891: 1133; Fowler & Dunn, 1917: 20.

Spelerpes maculicauda: Blatchley, 1897a: 125; 1897b: 181.

Misapplication of names:

Spelerpes longicaudus: Cope, 1870b: 401.

Distribution: East central US, Oklahoma to Virginia. See map in Hutchinson (1966) and Petranka (1998). **U.S.**: AL, AR, GA, IL, IN, KY, KS, MO, MS, OH, OK, TN, VA, WV.

Common name: Cave Salamander.

Commentary: Cope (1870b: 401; 1889b: 168) treated this as a synonym of *Spelerpes longicaudus*. Hutchinson (1966) provided a CAAR account. Bonett *et al.* (2013a) found this to be a sister to *E. guttolineata* + *E. longicauda*.

Eurycea (Eurycea) wallacei (Carr, 1939)

Haideotriton wallacei Carr, 1939: 335.—**O**: **H**: MCZ 19875.—**OT**: U.S., Georgia, Dougherty County, Albany, from a 200-foot artesian well.

Haideotriton wallacei: Stejneger & Barbour, 1943: 26; Schmidt, 1953: 44; Brame, 1967: 14; Cochran & Goin, 1970: 47; Gorham, 1974: 35; Conant, 1975: 296; Behler & King, 1979: 331; Conant & Collins, 1998: 499; Petranka, 1998: 289.

Eurycea wallacei: Dubois, 2005b: 20; Frost *et al.*, 2006: 359.

Eurycea (Eurycea) wallacei: Raffaëlli, 2007: 182.

Eurycea (Haideotriton) wallacei: Dubois & Raffaëlli, 2012: 143.

Distribution: A few caves and wells in southern Georgia and northern Florida. See map in Brandon (1967d) and Petranka (1998), as *Haideotriton*. **U.S.**: FL, GA.

Common name: Georgia Blind Salamander.

Commentary: Virtually universally maintained in its own genus for over 60 years, since its description; the genetic similarities to other *Eurycea* became apparent and its original genus was synonymized with *Eurycea*, along with others (Dubois, 2005b; supported by Frost *et al.*, 2006). Brandon (1967d) provided a CAAR account under the old generic name. Dubois & Raffaëlli (2012) revived the name *Haideotriton* as a subgenus containing only this species. The analysis of Bonett *et al.* (2013a) found this species is a sister to the pair *E. aquatica* + *E. junaluska,* so we treat it appropriately in the subgenus *Eurycea*.

Eurycea (Eurycea) wilderae Dunn, 1920

Eurycea bislineata wilderae Dunn, 1920c: 134.—**O**: **H**: MCZ 5848.—**OT**: U.S., Virginia, Grayson County, White Top Mountain, elev. 4000 feet (under log in woods).

Eurycea bislineata wilderae: Stejneger & Barbour, 1923: 16; 1943: 31; Bishop, 1924: 87; 1928a: 162; Schmidt, 1953: 52; Brame, 1967: 13; Cochran & Goin, 1970: 41; Gorham, 1974: 34; Conant, 1975: 289; Behler & King, 1979: 321; Petranka, 1998: 242.

Eurycea wilderae: Jacobs, 1987: 437; Conant & Collins, 1998: 489.

Eurycea (Eurycea) wilderae: Raffaëlli, 2007: 186.

Distribution: Eastern US in the Appalachian Mountains. Mittleman (1967b) and Petranka (1998) provided range maps as a subspecies of *Eurycea bislineata*. Sever (1999c, 2005) has a map as a full species. **U.S.**: GA, NC, SC, TN, VA.

Common name: Blue Ridge Two-lined Salamander.

Commentary: Formerly treated as a subspecies of *Eurycea bislineata,* until Jacobs (1987) provided evidence of its full species status; however, many workers (*e.g.,* Petranka, 1998) have still treated it as a subspecies. Sever (1999c) provided a CAAR account. Bonett *et al.* (2013a) showed this to be a sister to two of the three lineages of *E. cirrigera* which they identified.

Subgenus **Typhlomolge** Stejneger, 1896

Typhlomolge Stejneger, 1896: 620.—**NS**: *Typhlomolge rathbuni* Stejneger, 1896 (od.).—**SS**: Mitchell & Reddell, 1965: 23; Mitchell & Smith, 1972: 347-353; Chippindale *et al.,* 2000: 1, as *Eurycea.*

Typhlomolge: Hillis, Chamberlain, Wilcox, & Chippindale, 2001: 277 (as subgenus).

Notiomolge Hillis *et al.,* 2001: 277.—**NS (**od.**)**: *Eurycea neotenes* Bishop & Wright, 1937.—**Comment**: proposed as a genus-series "division;" under the *Code,* considered a subgenus (Dubois, 2007a: 395; Dubois & Raffaëlli, 2012: 99, 144).

Blepsimolge Hillis, *et al.,* 2001: 277.—**NS** (od.): *Eurycea nana* Bishop, 1941. Treated as a junior synonym of *Notionmolge* (see Dubois & Raffaëlli, 2012).

Paedomolge Hillis *et al.,* 2001: 277.—**NS**: *Eurycea tonkawae* Chippindale *et al.,* 2000 (od.). Proposed as a genus-series "section," contrary to the *Code*, but under the *Code* considered as a subgenus (Dubois, 2007a, Dubois & Raffaëlli, 2012).

Septentriomolge Hillis *et al.,* 2001: 277.—**NS**: *Eurycea chisholmensis* Chippindale, Price, Wiens, & Hillis, 2000 (od.). Treated as a junior synonym of *Paedomolge.*

Distribution: Springs and caves in central Texas.

Common name: None in general use. Spring & Cave Salamanders may be appropriate.

Content: 15 species are treated here as members of the subgenus.

Commentary:. Mitchell & Reddell (1965: 23) synonymized the genus *Typhlomolge* with *Eurycea*; this was supported by Chippendale *et al.* (2000); they treated the genus as including only two species: *rathbuni* and *robusta*. Hillis *et al.* (2001) treated *Typhlomolge* as a subgenus, and also proposed a name, *Notiomolge,* to contain the subgenera *Blepsimolge* and *Typhlomolge,* as a “division” of the genus *Eurycea;* however, as only genus and subgenus are available ranks in the genus-series, their nomen *Notiomolge* has since been treated as a subgenus. Dubois & Raffaëlli (2012) placed only three species in this subgenus, and most of the others listed here in the subgenus *Notiomolge; those along with the ones they treated in the subgenus Manculus are also included* here in a more compact monophyletic clade under the oldest name, based on phylogenetic relatonships determined by Bonett *et al.* (2013a).

Eurycea (Typhlomolge) chamberlaini Harrison & Guttman, 2003

Eurycea chamberlaini Harrison & Guttman, 2003: 163.—**O**: **H**: USNM 547846.—**OT**: U.S., South Carolina, Richland County, Sesquicentennial State Park.

Eurycea (Eurycea) chamberlaini: Raffaëlli, 2007: 189.

Eurycea (Manculus) chamberlaini: Dubois & Raffaëlli, 2012: 143.

Distribution: Coastal plain and piedmont of South and North Carolina, possibly in Georgia and Florida. Mostly parapatric with *E. quadridigitata*. See map in Harrison (2005), and remarks on references to other species names that probably refer to this one. **U.S.**: FL?, GA?, NC, SC.

Common name: Chamberlain’s Dwarf Salamander.

Commentary: Presumably a cryptic sister-species of *E. quadri-digitata,* with which it was confused for many years. Dubois & Raffaëlli, 2012) place the two together as the Supraspecies *quadradigitata*. Bonett *et al.* (2013a) showed the two have a sister relationship, and they in turn are a sister to the clade treated here as the subgenus *Notiomolge*. See Commentary for *E. quadradigitata*.

Eurycea (Typhlomolge) chisholmensis Chippindale, Price, Wiens & Hillis, 2000

Eurycea chisholmensis Chippindale, Price, Wiens & Hillis, 2000: 40.—**O**: **H**: TNHC 58859.—**OT**: U.S., Texas, Bell County, Salado, side spring immediately adjacent to Main (= Salado, Big Boiling or Siren) Springs, 30° 56' 37" N, 97° 32' 31" W.

Eurycea (Septentriomolge) chisholmensis: Raffaëlli, 2007: 182.

Eurycea (Paedomolge) chisholmensis: Dubois & Raffaëlli, 2012: 99, 144.

Distribution: Known only from Salado Springs and Robertson Springs, Bell County, Texas; see map in Chippindale (2005a). **U.S.**: TX.

Common name: Salado Salamander (SECSN) or Chisholm Trail Salamander (C&T).

Eurycea (Typhlomolge) latitans Smith & Potter, 1946

Eurycea latitans Smith & Potter, 1946: 106.—**O**: **H**: USNM 123594.—**OT**: U.S., Texas, Kendall County, 4.6 miles by road (3 1/2 miles airline) southeast of Boerne, the first large pool deep within the recesses of Cascade Cavern.

Eurycea neotenes latitans: Schmidt, 1953: 55; Brame, 1967: 13; Conant, 1958: 250; Cochran & Goin, 1970: 44;.

Eurycea latitans: Mitchell & Reddell, 1965: 13; Gorham, 1974: 34; Conant, 1975: 294; Behler & King, 1979: 322; Garrett & Barker, 1987: 59; Dixon, 2000: 55.

Eurycea (Blepsimolge) latitans: Hillis, Chamberlain, Wilcox & Chippindale, 2001 (by inference); Raffaëlli, 2007: 179.

Eurycea (Notiomolge) latitans: Dubois & Raffaëlli, 2012: 99, 144.

Missaplied name.

Eurycea neotenes: Wright & Wright, 1938: 31 (in part).

Eurycea tridentifera: Sweet, 1984: 438 (in part).

Distribution: Known only from Cascade Caverns, Texas (OT); see map in Chippindale (2005b). **U.S.**: TX.

Common name: Cascade Caverns Salamander.

Commentary: Brown (1967a) provided a CAAR account, with a range map for this and related species. Petranka (1998) treated this as a junior synonym of *E. neotenes*. Chippindale (2005b) commented on the status and relationships of this form. Listed in Texas as threatened (Dixon, 2000).

Eurycea (Typhlomolge) nana Bishop, 1941

Eurycea nana Bishop, 1941: 7.—**O**: **H**: UMMZ 89759.—**OT**: U.S., Texas, Hays County, San Marcos, lake at the head of the San Marcos River.

Eurycea nana: Stejneger & Barbour, 1943: 33; Cochran & Goin, 1970: 43; Brame, 1967: 13; Gorham, 1974: 34; Conant, 1975: 251; Behler & King, 1979: 325; Frost, 1985: 588; Garrett & Barker, 1987: 60; Conant & Collins, 1998: 495; Petranka, 1998: 264; Dixon, 2000: 55; Chippindale *et al.*, 2000:.

Eurycea neotenes nana: Schmidt, 1953: 55.

Eurycea (Blepsimolge) nana: Raffaëlli, 2007: 178.

Eurycea (Notiomolge) nana: Dubois & Raffaëlli, 2012: 99, 144.

Distribution: Known only from vicinity of the type locality. Brown's (1967a) account of *E. latitans* shows the locality for this species as well. **U.S.**: TX.

Common name: San Marcos Salamander.

Commentary: This species is federally listed as threatened. Brown (1967b) provided a CAAR account.

Eurycea (Typhlomolge) naufragia Chippindale, Price, Wiens, & Hillis, 2000

Eurycea naufragia Chippindale, Price, Wiens, & Hillis, 2000: 37.—**O**: **H**: TNHC 58860.—**OT**: U.S., Texas, Williamson County, headsprings of Buford Hollow, a small tributary of the South San Gabriel River below Lake Georgetown, 30° 39' 39" N, 97° 43' 36" W.

Eurycea (Septentriomolge) naufragia: Raffaëlli, 2007: 182.

Eurycea (Paedomolge) naufragia: Dubois & Raffaëlli, 2012: 99, 144.

Distribution: Known only from springs of San Gabriel River drainage, Williamson County, Texas. See map in Chippindale (2005c), who reviewed the known localities. **U.S.**: TX.

Common name: Georgetown Salamander (SECSN) *or* San Gabriel Springs Salamander (C&T).

Eurycea (Typhlomolge) neotenes Bishop & Wright, 1937

Eurycea neotenes Bishop & Wright, 1937: 142.—**O**: **H**: USNM 103161.—**OT**: U.S., Texas, Bexar County, 5 miles north of Helotes, Culebra Creek.

Eurycea neotenes: Stejneger & Barbour, 1943: 33; Behler & King, 1979: 326; Frost, 1985: 588; Garrett & Barker, 1987: 61 Conant & Collins, 1998: 495. Petranka, 1998: 266; Dixon, 2000: 56; Chippendale *et al.*, 2000: 25.

Eurycea neotenes neotenes: Schmidt, 1953: 55; Brame, 1967: 13; Cochran & Goin, 1970: 43; Gorham, 1974: 34; Conant, 1975: 293.

Eurycea (Blepsimolge) neotenes: Raffaëlli, 2007: 179.

Eurycea (Notiomolge) neotenes: Dubois & Raffaëlli, 2012: 99, 144.

Distribution: A portion of the Edwards Plateau in Texas. Brown's (1967a) account of *E. latitans* shows localities for this species as well, but many are now attributed to other species. See map and review of assignment of references of this species to others in Chippindale (2005d). **U.S.**: TX.

Common name: Texas Salamander.

Commentary: Brown (1967c) provided a CAAR account.

Eurycea (Typhlomolge) pterophila Burger, Smith, & Potter, 1950

Eurycea pterophila Burger, Smith, & Potter, 1950: 51.—**O**: **H**: FPC A993; current location unknown, perhaps TNHC.—**OT**: U.S., Texas, Hays County, 6.3 miles northeast of Wimberley on the Blanco River road, the shallow stream flowing from Fern Bank Spring.

Eurycea neotenes pterophila: Schmidt, 1953: 56; Brame, 1967: 13; Cochran & Goin, 1970: 43; Gorham, 1974: 34; Conant, 1975: 293.—**SS:** Sweet, 1978: 106, with *E. neotenes*.

Eurycea pterophila: Chippendale *et al.*, 2000: 25.

Eurycea (Blepsimolge) pterophila: Raffaëlli, 2007: 180.

Eurycea (Notiomolge) pterophila: Dubois & Raffaëlli, 2012: 99, 144.

Distribution: Springs and caves of Blanco River drainage, Hays County, Texas. Brown's (1967a) account of *E. latitans* shows the locality for this species as well. **U.S.**: TX.

Common name: Fern Bank Salamander.

Commentary: Petranka (1998) treated this as a junior synonym of *P. neotenes,* but the study by Chippindale *et al.* (2000) found it to be a full species.

Eurycea (Typhlomolge) quadridigitata (Holbrook, 1842)

Salamandra quadridigitata Holbrook, 1842: 5: 65, pl. 21.—**O**: **H**: the specimen in Plate 21, said to be ANSP 490 (Fowler & Dunn, 1917: 21), but Dunn (1926b: 332) seemed uncertain; Malnate, 1971 listed it as H.—**OT**: U.S., mid South Carolina through Georgia and Florida to the Gulf of Mexico.—**Comment:** Holbrook worked in South Carolina so it might be reasonable to assume he would tend to use specimens from that state over those he received from Georgia and Florida, as a basis for the plate and the name, but Schmidt's (1953: 56) restriction was not accompanied by evidence so it is invalid.

Batrachoseps quadridigitata: Baird, 1850a: 287; Gray, 1850: 287.—**UE.**

Batrachoseps quadridigitatus: Gray, 1850: 42; Hallowell, 1857a: 11; Knauer, 1878: 98.

Manculus quadridigitatus: Cope, 1869b: 101; 1889b: 159; Werner, 1909: 25; Stejneger & Barbour, 1917: 18; 1943: 34; Viosca, 1949: 10; Schmidt, 1953: 56; Mittleman, 1967a: 1; Cochran & Goin, 1970: 52; Conant, 1958: 251.

Manculus quadridigitata: Garman, 1884a: 40.

Manculus quadridigitatus quadridigitatus: Stejneger & Barbour, 1923: 14; Strecker & Williams, 1928: 8.

Eurycea quadridigitata quadridigitata: Dunn, 1923: 40; 1926b: 331; Carr, 1940: 48.

Eurycea quadridigitata: Carr, 1940: 48; Wake, 1966: 64; Brame, 1967: 13; Gorham, 1974: 34; Conant, 1975: 292; Behler & King, 1979: 326; Frost, 1985: 588; Garrett & Barker, 1987: 62; Conant & Collins, 1998: 494; Petranka, 1998: 269; Dixon, 2000: 56.

Eurycea (Eurycea) quadridigitata Raffaëlli, 2007: 189.

Manculus remifer Cope, 1871b: 84.—**O**: **H**: in ANSP, but lost *fide* Dunn, 1926b: 337.—**OT**: U.S., Florida, Jacksonville.—**SS**: Dunn, 1923: 40.

Manculus remifer: Boulenger, 1882b: 76; Cope, 1889b: 138; Stejneger & Barbour, 1917: 18.

Eurycea quadridigitata remifera: Dunn, 1923: 40; Bailey, 1937: 9.

Manculus quadridigitatus remifer: Stejneger & Barbour, 1923: 14. Subspecies rejected (Bishop, 1941: 20).

Eurycea quadridigitata remifer: Carr, 1940: 48.—**UE.**

Manculus quadridigitatus paludicolus Mittleman, 1947a: 320.—**O**: **H**: USNM 123979.—**OT**: U.S., Louisiana, Grant Parish, Pollock.

Manculus quadridigitatus uvidus Mittleman, 1947a: 221.—**O**: **H**: USNM 123980 (ex SM 2338), existence verified by Cochran (1961a: 16).—**OT**: U.S., Louisiana, Caddo Parish, Gayle.

Distribution: Texas to North Carolina along Gulf and Atlantic coasts, north to southern Arkansas and southeastern Missouri. Petranka (1998) and Bonett & Chippindale (2005) have range maps. **U.S.**: AL, AR, FL, GA, LA, MO, MS, NC, SC.

Common name: Dwarf Salamander (SECSN) or Coastal Plain Dwarf Salamander (C&T).

Content: No subspecies are currently recognized, but during part of the 20th century, *E. q. remifer* Cope was recognized (*e.g.*, Dunn, 1923, 1926b), as was *E. q. uvidus* Mittleman, 1947.

Commentary: Wake (1966) synonymized *Manculus* with *Eurycea*, after nearly a century of treatment as a monotypic genus, although Dunn never accepted *Manculus* as distinct from *Eurycea* (1926b: 335). Mittleman (1967a) provided a CAAR account, under the old generic name. Bonett & Chippindale (2005) reviewed the literature and also suggested that the species as treated here may actually consist of four separate cryptic species. Bonett *et al.* (2013a) found two populations identified as *E. quadridigitata* indicated a sister relationship to *E. chamberlaini*, in a complex way which casts doubt on the distinctness of the latter. Three other populations identified as *E. quadridigitata* have a sister relationship to most members of the clade here treated as subgenus *Typhlomolge*. Those populations and *Notiomolge* are in turn sister to *Manculus*. This species may consist of at least two distinct species based on the analysis of Bonett *et al.* (2013a), but as they did not make such a determination, we retain it as a single species.

Eurycea (Typhlomolge) rathbuni (Stejneger, 1896)

Typhlomolge rathbuni Stejneger, 1896: 620.—**O**: **H**: USNM 22686.—**OT**: U.S., Texas, Hays County, subterranean waters near San Marcos; corrected to Texas: San Marcos, U.S. Fish Commission Well 42 (Dunn, 1926b). Stated as Texas, Hays Co. San Marcos, artesian well 188 feet deep, at U.S. Fish Commission Station (Stejneger & Barbour, 1933).

Typhlomolge rathbuni: Werner, 1909: 21; Strecker, 1915: 56; Stejneger & Barbour, 1917: 6; 1943: 26; Schmidt, 1953: 45; Wake, 1966; Brame, 1967: 14; Cochran & Goin, 1970: 64; Gorham, 1974: 39; Conant, 1975: 295; Behler & King, 1979: 355; Frost, 1985: 606; Garrett & Barker, 1987: 68; Conant & Collins, 1998: 497; Dixon, 2000: 58.

Eurycea rathbuni: Mitchell & Reddell, 1965: 23; Petranka, 1998: 272; Chippindale *et al.*, 2000: 23.

Eurycea (Typhlomolge) rathbuni: Hillis *et al.*, 2001: 277; Raffaëlli, 2007: 181.

Distribution: Type locality, and Ezell's Cave in San Marcos, plus other artesian wells in the vicinity, probably all fed by the same underground stream. **U.S.**: TX.

Common name: Texas Blind Salamander.

Commentary: This species is federally listed as endangered.

Eurycea (Typhlomolge) robusta (Longley, 1978)

Typhlomolge robusta Longley, 1978: 10.—**O**: **H** TNHC 20255 (Potter & Sweet, 1981: 70, sd.).—**OT**: U.S., Texas, Hays County, San Marcos, beneath the Blanco River, 5 airline km NE of the Hays County courthouse, 178 m elevation.

Typhlomolge robusta: Potter & Sweet, 1981: 70; Frost, 1985: 606; Garrett & Barker, 1987: 69; Conant & Collins, 1998: 498; Dixon, 2000: 59.

Eurycea robusta: Petranka, 1998: 275; Chippindale, Price, Wiens, & Hillis, 2000: 24.

Eurycea (Typhlomolge) robusta: Raffaëlli, 2007: 181.

Distribution: Known only from the OT. **U.S.**: TX.

Common name: Blanco Blind Salamander.

Commentary: Discovery of this form was by Floyd E. Potter, who described it in his Master's thesis (Potter, 1963), and used the name, *Typhlomolge robusta.* Longley (1978) was aware of the description, and used the name in a Fish &

Wildlife publication, giving a short description and compared it to *Typhlomolge rathbuni.* Because Potter's thesis is not a publication under the *Code,* Longley's publication became the first to make the name available, and he is officially the author. Subsequently, Potter & Sweet (1981) identified the holotype and gave a more complete description. Citation of this subsequent article sometimes lists it as "Longley *in* Potter & Sweet," but none of the material presented in that work is from Longley. Chippindale (2005e) reviewed the history of this species, which may be extinct or perhaps difficult to find due to its subterranean habitat. As material is not available for molecular work, we assume this is a sister to *E. rathbuni* and/or *E. waterlooensis.*

Eurycea (Typhlomolge) sosorum Chippindale, Price, & Hillis, 1993

Eurycea sosorum Chippindale, Price, & Hillis, 1993: 249.—**O**: **H:** TNHC 51184.—**OT**: U.S., Texas, Travis County, Austin, Zilker Park, outflow of Parthenia (main) Springs in Barton Springs Pool, 30° 15' 49" N, 97° 46' 14" W.

Eurycea sosorum: Petranka, 1998: 276; Dixon, 2000: 56; Chippindale *et al.*, 2000: 26.

Eurycea (Blepsimolge) sosorum: Raffaëlli, 2007: 179.

Eurycea (Notiomolge) sosorum: Dubois & Raffaëlli, 2012: 99, 144.

Misapplied name.

Eurycea neotenes: Brown, 1950: 29 (in part).

Distribution: Known only from the type locality. **U.S.**: TX.

Common name: Barton Springs Salamander.

Commentary: Petranka (1998: 267) mentioned this species, but did not present an account of it, nor did he challenge its status. This species is federally listed as endangered. The study of Bonett *et al.* (2013a) indicated a sister relationship to a complex of *E. neotenes, E. latitans, E. pterophila,* and *E. tridentifera.*

Eurycea (Typhlomolge) tonkawae Chippindale, Price, Wiens, & Hillis, 2000

Eurycea tonkawae Chippindale, Price, Wiens, & Hillis, 2000: 32.—**O**: **H**: TNHC 50952.—**OT**: U.S., Texas, Travis County, primary outflows of Stillhouse Hollow Springs, 30° 22' 28" N, 97° 45' 55" W.

Eurycea (Septentriomolge) tonkawae: Raffaëlli, 2007: 182.

Eurycea (Paedomolge) tonkawae: Dubois & Raffaëlli, 2012: 99, 144.

Distribution: Localities in Travis and Williamson Counties, Texas. See review and map in Chippindale (2005f). **U.S.**: TX.

Common name: Jollyville Plateau Salamander.

Commentary: Following the relationships indicated by Hillis *et al.* (2001), Dubois & Räffelli (2012) included this species in a subgenus *Paedomolge*. Although that clade seems monophyletic, we include it in the broader clade represented by the subgenus *Typhlomolge*.

Eurycea (Typhlomolge) tridentifera Mitchell & Reddell, 1965

Eurycea tridentifera Mitchell & Reddell, 1965: 14.—**O**: **H**: USNM 153780.—**OT**: U.S. Texas, Comal County, waters of Honey Creek Cave.

Typhlomolge tridentifera: Wake, 1966: 65; Brame, 1967: 14; Gorham, 1974: 39.

Eurycea tridentifera: Conant, 1975: 295; Frost, 1985: 588; Garrett & Barker, 1987: 63; Petranka, 1998: 277; Dixon, 2000: 57; Chippindale *et al.*, 2000: 26.

Eurycea (Blepsimolge) tridentifera: Raffaëlli, 2007: 179.

Eurycea (Notiomolge) tridentifera: Dubois & Raffaëlli, 2012: 99, 144.

Distribution: Southeastern edge of the Edwards Plateau in central Texas. **U.S.**: TX.

Common name: Comal Blind Salamander.

Commentary: This species is included in a complex with *E. latitans, E. neotenes,* and *E. pterophila,* in the study of Bonett *et al.* (2013a).

Eurycea (Typhlomolge) troglodytes Baker, 1957

Eurycea troglodytes Baker, 1957: 329.—**O**: **H**: TNHC 21791.—**OT**: U.S., Texas, Medina County, Valdina Farms, a pool approximately 600 feet from the entrance of the Valdina Farms Sinkhole.

Eurycea troglodytes: Conant, 1958, 251; Brame, 1967: 13; Cochran & Goin, 1970: 44; Gorham, 1974: 34; Conant, 1975: 294; Garrett & Barker, 1987: 64; Dixon, 2000: 57.

Eurycea (Blepsimolge) troglodytes: Raffaëlli, 2007: 180.

Eurycea (Notiomolge) troglodytes: Dubois & Raffaëlli, 2012: 99, 144.

Distribution: Known only from the type locality. Brown's (1967a) CAAR account of *E. latitans* shows the locality for this species as well. **U.S.**: TX.

Common name: Valdina Farms Salamander.

Commentary: Petranka (1998) considered this as conspecific with *P. neotenes*. Hillis *et al.*(2001) found this to be a complex including other undescribed cryptic species, and forming a sister group to the rest of the subgenus. The study of Bonett *et al.* (2013a) supports that, indicating three lineages of this species, with a sister relationship to an undescribed species.

Eurycea (Typhlomolge) waterlooensis Hillis, Chamberlain, Wilcox, & Chippindale, 2001

Eurycea waterlooensis Hillis, Chamberlain, Wilcox, & Chippindale, 2001: 268.—**O**: **H**: TNHC 60201.—**OT**: U.S., Texas, Travis County, Austin, Zilker Park, Sunken Gardens Spring, an outlet of Barton Springs.

Eurycea (Typhlomolge) waterlooensis: Raffaëlli, 2007: 181.

Distribution: Known only from the type locality. **U.S.**: TX.

Common name: Austin Blind Salamander.

Commentary: Bonett *et al.* (2013a) indicate this is a sister to *E. rathbuni*. Material of *E. robusta* is not available, but we assume that species is a sister in this complex as well.

Subgenus **Typhlotriton** Stejneger, 1893

Typhlotriton Stejneger, 1893a: 115.—**NS**: *Typhlotriton spelaeus* Stejneger, 1892 (od.).—**SS**: Bonett & Chippindale, 2004: 1199

Distribution: Streams and caves in the central U.S.

Common name: No name in general use.

Content: Three species are recognized in this taxon.

Commentary: Bonett *et al* (2013a) demonstrated this as a distinct monophyletic clade, as a sister taxon to all other *Eurycea.*

Eurycea (Typhlotriton) multiplicata (Cope, 1869)

Spelerpes multiplicatus Cope, 1869b: 106.—**O**: **S**: USNM 4038 (5).—**OT**: U.S., Arkansas, Red River; corrected to Red River in eastern Oklahoma (Dunn, 1926b); restricted to near Fort Towson, Choctaw County, Oklahoma (Dundee, 1950: 28), based on inferences from museum data, so seems a valid restriction.

Spelerpes multiplicatus: Boulenger, 1882b: 67; Cope, 1889b: 162.

Geotriton multiplicatus: Garman, 1884a: 39.

Eurycea multiplicata: Stejneger & Barbour, 1917: 20; 1943: 33; Dunn, 1926b: 313; Frost, 1985: 587.

Eurycea multiplicata multiplicata: Schmidt, 1953: 54; Brame, 1967: 13; Cochran & Goin, 1970: 41; Gorham, 1974: 34; Conant, 1975: 290; Behler & King, 1979: 325; Conant & Collins, 1998: 491; Petranka, 1998: 262.

Eurycea (Eurycea) multiplicata: Raffaëlli, 2007: 183.

Eurycea (Typhlotriton) multiplicata: Dubois & Raffaëlli, 2012: 99, 144.

Distribution: Ouachita Mountains in Oklahoma and Arkansas. Dundee (1965a) provided a CAAR account and has a range map, as does Petranka (1998). The distribution of the species is probably best represented by the portion of the map indicated as *E. m. multiplicata,* as the other subspecies shown is now recognized as a different species, *E. tynerensis.*

Common name: Many-ribbed Salamander.

Content: The population now recognized as *E. tynerensis* includes what was considered a subspecies of *E. multiplicata* for about 50 years, but no subspecies are currently recognized. However, Bonett & Chippindale (2004) indicated that this population may represent a complex of more than one species. Bonett *et al.* (2013a) found two lineages among populations identified as this species. **U.S.**: AR, OK.

Eurycea (Typhlotriton) spelaea (Stejneger, 1893)

Typhlotriton spelaeus Stejneger, 1893a: 116.—**O**: **H**: USNM 17903 (Cochran, 1961a: 28, sd.).—**OT**: U.S., Missouri, Barry County, Rock House Cave.

Typhlotriton spelaeus: Stejneger & Barbour, 1917: 23; 1943: 26; Dunn, 1926b: 248; Schmidt, 1953: 44; Brame, 1967: 14; Cochran & Goin, 1970: 64; Gorham, 1974: 39; Conant, 1975: 296; Behler & King, 1979: 356; Frost, 1985: 606; Conant & Collins, 1998: 498; Petranka, 1998: 307.

Eurycea spelaeus: Bonett & Chippindale, 2004: 1199.

Eurycea (Eurycea) spelaea: Raffaëlli, 2007: 184.

Eurycea (Typhlotriton) spelaea: Dubois & Raffaëlli, 2012: 99-144.

Typhlotriton nereus Bishop, 1944: 1.—**O**: **H**: FMNH 93143,—**OT**: U.S., Arkansas, Lawrence County, Imboden, York Spring.—**SS**: Brandon, 1966b: 560-561, with *T. spelaeus*.

Typhlotriton nereus: Schmidt, 1953: 45.

Typhlotriton braggi Smith, CC, 1968: 156.—**O**: **H**: USNM 167146.—**OT**: U.S., Arkansas, Independence County, Cushman Cave, 3.5 miles SE of Cushman; corrected to 1.5 miles NW of Cushman, in a cave sometimes called Cushman Cave locally, but not the better known Cushman Cave, 3 miles W of Cushman.—**SS**: Brandon & Black, 1970: 390.

Distribution: Caves and springs of the Ozarks in Oklahoma, Arkansas, Missouri, and Kansas. See range maps of Brandon (1965b) and Petranka (1998), as genus *Typhlotriton*. **U.S.**: AR, KS, MO, OK.

Common name: Grotto Salamander.

Commentary: For many years this form was maintained in a separate genus, *Typhlotriton*; this was supported by Wake (1966). But Bonett & Chippindale (2004) found genetic similarities too great to justify the separate genus and synonymized it with *Eurycea*. Brandon (1965b) provided a CAAR generic (*Typhlotriton*) and species account, and tentatively recognized a second species, *T. nereus* Bishop, but later (1966b) included that in the synonymy of *T. spelaeus*. The study of Bonett *et al.* (2013a) found two lineages among populations identified as this species.

Eurycea (Typhlotriton) tynerensis Moore & Hughes, 1939

Eurycea tynerensis Moore & Hughes, 1939: 697.—**O**: **S** (32): UNNM 108548 (12), UMMZ 85534 (5), MCZ 25533 (1, formerly UMMZ), OKMNH 21325 (6, but only 3 exist), OSUS (6, apparently all lost), CM 18525 (1), CAS-SU 4778 (1).—**OT**: U.S., Oklahoma, Adair County, near Proctor, at Tyner Creek, a tributary of Barron Fork Creek.

Eurycea tynerensis: Stejneger & Barbour, 1943: 33; Schmidt, 1953: 55; Cochran & Goin, 1970: 44; Gorham, 1974: 35; Behler & King, 1979: 328; Frost, 1985: 589; Conant & Collins, 1998: 495; Petranka, 1998: 278. Bonett & Chippindale, 2004: 1198.

Eurycea (Eurycea) tynerensis: Raffaëlli, 2007: 184.

Eurycea (Typhlotriton) tynerensis: Dubois & Raffaëlli, 2012: 99: 144.

Eurycea griseogaster Moore & Hughes, 1941: 139.—**O**: **H:** CNHM 37832.—**OT**: U.S., Oklahoma, Sequoyah County, 10 miles south of Gore, Swimmer's Creek, near its junction with the Illinois River.—**SS**: Bonett & Chippindale, 2004: 1198.

Eurycea griseogaster: Bishop, 1943: 418; Stejneger & Barbour, 1943: 31.

Eurycea multiplicata griseogaster: Schmidt, 1953: 55; Conant, 1958: 249; Brame, 1967: 13; Cochran & Goin, 1970: 42; Behler & King, 1979: 325; Petranka, 1998: 262; Conant & Collins, 1998: 491.

Distribution: Northeastern Oklahoma, northern Arkansas, south-western Missouri, and adjacent Kansas. The range map for *E. multiplicata griseogaster* by Petranka (1998) reflects the distribution as now understood better than that of Dundee (1965b), or Petranka's map for *E. tynerensis*. **U.S.**: AR, KS, MO, OK.

Common name: Oklahoma Salamander.

Content: As noted in the synonymy, *E. m. griseogaster* is sometimes treated as a separate subspecies, but we follow Bonett & Chippindale (2004) in treating it within the synonymy of *E. tynerensis*.

Commentary: Dundee (1965b) provided a CAAR account, but some of it is now outdated by our current understanding of the systematics. Bonett & Chippindale (2004) showed that populations then recognized as *E. m. griseogaster* and *E. tynerensis* were essentially genetically indistinguishable so should all be referred to the older name. Bonett *et al.* (2013a) identified six lineages in populations identified as this species.

Genus **Urspelerpes** Camp, Peterman, Milanovich, Lamb, Maerz, & Wake, 2009

Urspelerpes Camp *et al.*, 2009: 87.—**NS**: *Urspelerpes brucei*, Camp *et al.*, 2009 (mo.).

Distribution: See species account.

Common name: Patchnose Salamanders.

Content: A single species.

Commentary: Camp *et al.* (2009) proposed this as a full genus, and, in their molecular phylogenetic study, found the species to be the sister taxon of *Eurycea,* basal to that genus, which seems to justify generic status; that is further supported by Bonett *et al.* (2013a). *Urspelerpes* and *Eurycea* together form a clade (Spelerpina) that is a sister to the clade (Pseudotritonina) containing the other three genera of the subfamily. Camp *et al.* (2012) provided a CAAR account.

Urspelerpes brucei Camp, Peterman, Milanovich, Lamb, Maerz, & Wake, 2009

Urspelerpes brucei Camp *et al.*, 2009: 87.—**O**: **H**: USNM 558253.—**OT**: U.S., Georgia, Stephens County, foot of Blue Ridge escarpment in first order stream; 34° 39' N, 83° 18' W.

Distribution: Northeastern Georgia and northwestern South Carolina. See range map in Camp *et al.* (2012). **U.S.**: GA, SC.

Common name: Patchnose Salamander.

Commentary: Camp *et al.* (2012) provided a CAAR account. Bonett *et al.* (2013a) demonstrated this to be a sister to all members of the genus *Eurycea.*

Subfamily **Plethodontinae** Gray, 1850

Synonymy: See Tribe Plethodontini, below.

Distribution: Most species are North American, but others occur in Mediterranean Europe and Korea.

Common name: None in general use. Terrestrial Lungless Salamanders seems appropriate.

Content: Five tribes, together containing 8 genera with 101 species, of which 5 genera and 89 species are North American.

Commentary: See the family account, under *Content,* for remarks on changes in subfamilies. Pyron & Wiens (2011) found two major clades, identified as Plethodoninae and Hemidactylinae. Within the former, they identified a clade containing several monophyletic lines (*Hydromantes, Ensatina, Aneides,* and desmognathines) and another containing the species of *Plethodon.* The studies of Vieites *et al.* (2011) and Wake (2012) showed similar relationships, with some differences. We follow Wake (2012) and utilize five tribe ranks to recognize those clades. North America contains all five tribes and 88.1% of the species diversity of the subfamily.

Tribe **Aneidini** Wake, 2012

"Aneidini" Dubois, 2008b: 72. *Nomen nudum* (see Dubois, 2012).

"Aneiditoi" Dubois, 2008b: 74. *Nomen nudum* (see Dubois, 2012).

"Aneidini" Vieites *et al.,* 2011: 11, 633.—**NG:** *Aneides* Baird, 1851, mo. *Nomen nudum* (see Dubois, 2012).

Aneidini Wake, 2012: 75.—**NG:** *Aneides* Baird, 1851, od.

Aneidini: Dubois & Raffaëlli, 2012: 117.—**NG:** *Aneides* Baird, 1851, **OS.**

Distribution: As for the genus.

Common name: Climbing Salamanders.

Content: A single genus with six species.

Commentary: Vieites *et al.* (2011) and Wake (2012) treated the genus *Aneides* in a separate tribe, which we follow here. Both authors suggested that an alternative would be to include *Aneides* in the Tribe Ensatinii, but noted the sister relationship between *Aneides* and *Ensatina* is not well supported by current evidence. Wake (2012) suggested another alternative treatment would be to recognize an infrafamily including only Plethodontini and a second one including Aneidini and the other Plethodontine tribes.

Genus **Aneides** Baird, 1851

Synonymy: see the subgenus *Aneides,* below.

Distribution: Part of the eastern US, plus localized areas of New Mexico, and coastal western North America, entering northern Baja California. See map in Wake (1974).

Common name: Climbing Salamanders.

Content: Six species, all in North America.

Commentary: Wake (1974) provided a generic CAAR account. Sessions & Larson (1987) and Mahoney (2001) found this genus to be nested within *Plethodon,* but Macey (2005) presented data that indicated *Aneides* may be a sister group to *Hydromantes,* or to a clade made up of *Hydromantes* plus *Ensatina*; that entire clade appeared as a sister taxon to the desmognathines. That study used data from Mueller *et al.* (2004) and a methodology considered invalid by most current workers, and reached somewhat different conclusions from the work by him and his co-authors, who earlier found *Aneides* a sister to *Hydromantes,* and that pair sister to a clade of *Ensatina* and desmognathines. So *Aneides* seemed rather far removed from *Plethodon.* Similarly, the works of Chippindale *et al.* (2004), Wiens *et al.* (2006) and Vieites *et al.* (2007) suggest *Aneides* is a sister taxon to the desmognathines, while *Ensatina* is a sister group to *Aneides* plus the desmognathines. They did not include *Hydromantes* in their studies. Kozak *et al.* (2006) found one species of *Aneides* had a sister relationship to one species of *Desmognathus,* but their small sample did not include other species of those genera, nor of any *Hydromantes* or *Ensatina*; however, the one *Aneides* they included was nested within *Plethodon.* The study of Vieites *et al.* (2011) found some support for a sister relationship between *Aneides* and *Ensatina.* The disagreement among these findings leads us to retain the species of *Aneides* in that genus, and follow Vieites *et al.* (2011) and Wake (2012) in treating the genus in a separate tribe within the plethodontines. Certainly on morphological grounds the genus seems most closely related to *Plethodon* and *Ensatina* (Wake, 1963, 1966, 1974). Dubois & Raffaëlli (2012) split the genus into two subgenera, and we follow that treatment here.

Subgenus **Aneides** Baird, 1851

"Anaides" Baird, *in* Heck & Baird, 1851: 256.—**NS**: *Salamandra lugubris* Hallowell, 1849 (mo.). Preoccupied by *Anaides* Westwood (Coleoptera), and placed on the index of rejected and invalid generic names (ICZN, 1956a).

Aneides Baird, *in* Heck & Baird, 1851: 257.—**NS** (mo.): *Salamandra lugubris* Hallowell, 1849. Alternative spelling of *Anaides.*

"Autodax" Boulenger, 1887a: 67.—**NN** for *Anaides* Baird, 1851. Placed on the index of rejected generic names (ICZN, 1956a).

Aneides (Aneides) ferreus (Cope, 1869)

Anaides ferreus Cope, 1869b: 109. (generic name now permanently rejected)—**O**: **H:** USNM 14451 (Dunn, 1926b: 208, sd.) (ex 6794) verified by Cochran, 1961a: 8.—**OT**: U.S., Oregon, Douglas County, Fort Umpqua.

Autodax ferreus: Cope, 1889b: 185 (generic name now permanently rejected).

Aneides ferreus: Grinnell & Camp, 1917: 135; Stejneger & Barbour, 1917: 21; 1943: 24; Dunn, 1926b: 208; Schmidt, 1953: 50; Cochran & Goin, 1970: 29; Gorham, 1974: 30; Behler & King, 1979: 303; Frost, 1985: 571; Stebbins, 1985: 51; 2003: 180; Petranka, 1998: 314; Raffaëlli, 2007: 320.

Aneides (Aneides) ferreus: Dubois & Raffaëlli, 2012: 117, 144.

Distribution: Coastal northern California and Oregon. Wake (1965a) and Petranka (1998) have range maps. **U.S.**: CA, OR.

Common name: Clouded Salamander.

Commentary: Wake (1965a) provided a CAAR account. Vieites *et al.* (2007) found this species to be a sister of *A. vagrans.* Jackman (1998) found two distinctive clades in the native populations of coastal Oregon and California, suggesting there may be two species involved.

Aneides (Aneides) flavipunctatus (Strauch, 1870)

Plethodon flavipunctatus Strauch, 1870: 71.—**O**: **S:** ZISP 155-157 (Borkin *in* Duellman, 1993: 302, sd.; apparently lost).—**OT**: U.S., California, Mendocino County, (New) Albion.

Plethodon flavipunctatus: Cope, 1889b: 145.

Aneides flavipunctatus: Storer, 1925: 119; Dunn, 1926b: 220; Stejneger & Barbour, 1933: 21; 1943: 24; Schmidt, 1953: 50; Frost, 1985: 571; Stebbins, 1985: 50; 2003: 178; Petranka, 1998: 318; Raffaëlli, 2007: 319; Dubois & Räffelli, 2012: 144.

Aneides flavipunctatus flavipunctatus: Myers & Maslin, 1948: 134; Brame, 1967: 15; Cochran & Goin, 1970: 28; Gorham, 1974: 30; Behler & King, 1979: 304.

Aneides (Aneides) flavipunctatus: Dubois & Raffaëlli, 2012: 117, 144.

Plethodon iëcanus Cope, 1884: 24.—**O**: **H**: ANSP 14061 (Fowler & Dunn, 1917: 23, sd.); verified by Malnate, 1971: 348.—**OT**: U.S., California,

Shasta County, U.S. fish hatchery on the McCloud River; modified to Shasta County, McCloud River, Baird, by Dunn, 1926b: 223.—**SS**: Storer, 1925: 119.

Anaides iëcanus: Cope 1886b: 526. (generic name now permanently rejected)

Autodax iecanus: Cope, 1889b: 187; Werner, 1909: 33. (generic name now permanently rejected)

Aneides iëcanus: Stejneger & Barbour, 1917: 21.

Aneides iecanus: Grinnell & Camp, 1917: 135; Dubois & Räffelli, 2012: 144.

Aneides flavipunctatus niger Myers & Maslin, 1948: 132.—**O**: **H**: CAS-SU 2938.—**OT**: U.S., California, Santa Cruz County, near the forks of Waddell Creek.

Aneides flavipunctatus niger: Brame, 1967: 15; Cochran & Goin, 1970: 28; Behler & King, 1979: 304; Reilly *et al.*, 2013: 668.

Aneides niger: Dubois & Räffelli, 2012:144.

"*Aneides flavipunctatus sequoiensis*" Lowe, 1950b, *nomen nudum.*

"*Aneides sequoiensis*:" Dubois & Räffelli, 2012: 144.

Distribution: Lowland northwestern California and adjacent Oregon. Lynch (1974) and Petranka (1998) have range maps. **U.S.**: CA, OR.

Common name: Black Salamander (SECSN) or Speckled Black Salamander (C&T).

Content: Two subspecies were recognized from 1948 to 1981, but are no longer recognized by most workers.

Commentary: Lynch (1974) provided a CAAR account. Vieites *et al.* (2007) found this to be a sister species of the *ferreus-vagrans* pair. Some apparent cryptic species have been identified within this species, and some names have been used, including *iecanus, niger,* and *sequoiensis* (see Rissler & Apodaca, 2009; results agree with Rielly *et al.*, 2013); however, we await formal taxonomic studies.

Aneides (Aneides) hardii (Taylor, 1941)

Plethodon hardii Taylor, 1941a: 77.—**O**: **H**: EHT-HMS 23656, now FMNH 100103.—**OT**: U.S., New Mexico, Otero County, Sacramento Mountains at Cloudcroft (9000 feet).

Plethodon hardii: Stejneger & Barbour, 1943: 18.

Aneides hardii: Lowe, 1950a: 93; Gorham, 1974: 30; Behler & King, 1979: 305; Frost, 1985: 571; Stebbins,

1985: 50; 2003: 178; Petranka, 1998: 320; Raffaëlli, 2007: 322.

Aneides hardyi: Schmidt, 1953: 51; Brame, 1967: 15; Cochran & Goin, 1970: 28—**UE** (incorrectly attributed to Lowe, 1950a, by Schmidt).

Aneides (Aneides) hardii: Dubois & Raffaëlli, 2012: 117, 144.

Distribution: Three disjunct areas in south-central New Mexico. Wake (1965b) and Petranka (1998) have range maps. **U.S.**: NM.

Common name: Sacramento Mountains Salamander.

Commentary: Wake (1965b) provided a CAAR account. Vieites *et al.* (2007) found this to be a sister species of all other western *Aneides*.

Aneides (Aneides) lugubris (Hallowell, 1850)

"Triton tereticauda" Eschscholtz, 1833: 14.—**O**: not stated nor known to exist.—**OT**: U.S., California, Fort Ross. Nomen suppressed for purposes of priority and placed on the list of rejected specific names (ICZN, 1956a). Provisional synonymy with *A. lugubris* by Storer, 1925: 127, and Dunn, 1926b: 211.

Salamandra lugubris Hallowell, 1850: 126.—**O**: **H:** ANSP 1257 (Fowler & Dunn, 1917: 23, sd.).—**OT**: U.S., California, Monterrey County, Monterrey.

Taricha? lugubris: Gray, 1850: 26.

Aneides lugubris: Baird *in* Heck & Baird, 1851: 257; Dunn, 1918b: 463; 1926b: 211; Noble, 1921: 5; Schmidt, 1953: 50; Brame, 1967: 15; Cochran & Goin, 1970: 28; Gorham, 1974: 30; Behler & King, 1979: 305; Stebbins, 1985: 52; 2003: 181; Liner, 1994: 9; Petranka, 1998: 322; Raffaëlli, 2007: 321. **Comment:** Placed on the official list of specific names (ICZN, 1956a).

Anaides lugubris: Baird, 1859d: 13; Garman, 1884a: 38. (generic name now permanently rejected)

Autodax lugubris: Boulenger, 1887a: 67; Werner, 1909: 32. (generic name now permanently rejected)

Aneides lugubris lugubris: Grinnell & Camp, 1917: 134; Stejneger & Barbour, 1917: 21; 1943: 25.

Aneides lugubris: Frost, 1985: 571.

Aneides (Aneides) lububris: Dubois & Raffaëlli, 2012: 117, 144.

Ambystoma punctulatum Gray, 1850: 37.—**O:** none indicated.—**OT:** U.S., California, Monterrey County, Monterrey.—**SS:** Dunn, 1926b: 211, with *A. lugubris*.

Plethodon crassulus Cope, 1886b: 521.—**O**: **H**: USNM 9447 (Dunn, 1926b: 220, sd.).—**OT**: U.S., California.—**SS**: Van Denburgh, (1916: 220) and Slevin (1928: 54) with *Plethodon intermedius* (now *P. vehiculum*); and Dunn, (1926b: 211) and Cochran (1961a: 19) with *A. lugubris*.

Plethodon crassulus: Cope, 1889b: 147; Fowler & Dunn, 1917: 25.

Autodax lugubris farallonensis Van Denburgh, 1905: 5, pl. 2 (generic name now permanently rejected).—**O**: **H:** CAS 3731 (destroyed).—**OT**: U.S., California, South Farallon Island.—**SS:** Dunn, 1926b: 211, with *A. lugubris*.

Aneides lugubris farallonensis: Grinnell & Camp, 1917: 135; Stejneger & Barbour, 1917: 21; 1943: 25.

Distribution: Western coastal California and adjacent Baja Cali-fornia. Lynch & Wake (1974) and Petranka (1998) have range maps. **U.S.**: CA.

Common name: Arboreal Salamander.

Content: No subspecies are currently recognized, although there were two in the past.

Commentary: Lynch & Wake (1974) provided a CAAR account. Vieites *et al.* (2007) found this to be a sister species of the clade including *ferreus-vagrans-flavipunctatus*.

Aneides (Aneides) vagrans Wake & Jackman, 1999

Aneides vagrans Wake & Jackman, 1999: 1579.—**O**: **H**: MVZ 124876,—**OT**: U.S., California, Humboldt County, a point about 10 km S Maple Creek; 40° 42' N, 123° 50' W, ca. 500 m elevation.

Aneides vagrans: Stebbins, 2003: 180; Raffaëlli, 2007: 321.

Aneides (Aneides) vagrans: Dubois & Raffaëlli, 2012: 117, 144.

Distribution: Coastal northern California; also on and around Vancouver Island, British Columbia (introduced). Staub & Wake (2005) have a map of the U.S. distribution. Fisher et al. (2007) map the Canadian distribution. **Can.**: *BC. **U.S.**: CA.

Common name: Wandering Salamander. *French Canadian name:* Salamandre errant.

Subgenus **Castaneides** Dubois & Raffaëlli, 2012.

Castaneides Dubois & Raffaëlli, 2012: 117, 144. **NS:** *Aneides aeneus* (Cope & Packard, 1881), od.

Aneides (Castaneides) aeneus (Cope & Packard, 1881)

Plethodon aeneus Cope & Packard, 1881: 878.—**O**: **H:** ANSP 10461 (Dunn, 1926b: 205, sd.).—**OT**: U.S., Tennessee, Marion County, Nickajack Cave.

Plethodon aeneus: Cope, 1889b: 143; Stejneger & Barbour, 1917: 14

Aneides aeneus: Dunn, 1923: 39; 1926b: 205; Stejneger & Barbour, 1923: 17; 1943: 24; Bishop, 1928a: 163; Schmidt, 1953: 50; Brame, 1967: 14; Cochran & Goin, 1970: 27; Gorham, 1974: 30; Conant, 1975: 283; Behler & King, 1979: 302; Frost, 1985: 570; Conant & Collins, 1998: 479; Petranka, 1998: 310; Raffaëlli, 2007: 322.

Aneides (Castaneides) aeneus: Dubois & Raffaëlli, 2012: 117, 144.

Distribution: Northern Alabama and adjacent Mississippi, north-eastward through eastern Tennessee, eastern Kentucky, and central West Virginia, and adjacent Virginia, Ohio and Pennsylvania, plus a disjunct population in the western Carolinas. See range maps in Gordon (1967) and Petranka (1998). **U.S.**: AL, KY, MS, NC, OH, PA. SC, TN, VA, WV.

Common name: Green Salamander.

Commentary: Gordon (1967) provided a CAAR account. Vieites *et al.* (2007) found this species to be sister to a clade containing all other *Aneides.*

Tribe **Desmognathini** Gray, 1850

Desmognathina Gray, 1850: 40.—**NG**: *Desmognathus* Baird, 1850.

Desmognathidae: Cope, 1866a: 107; 1869b: 112.

Desmognathinae: Boulenger, 1882b: 1; Hay, 1892: 439; Wake, 1966: 47.

Desmognathini: Dubois, 2005b: 20; Wake, 2012: 79.

Distribution: Most of the eastern U.S., from eastern Texas and Oklahoma, to New England and adjacent southeastern Canada.

Common name: Dusky Salamanders.

Content: Two genera, containing 22 species, all North American.

Commentary: Based on studies cited earlier, finding desmognathines closely related to *Aneides, Ensatina,* and *Hydromantes,* Dubois (2005b) included all those genera along with the traditional desmognathines in his tribe Desmognathini, leaving only the genus *Plethodon* in the Plethodontini. At this time we prefer to include only the traditionally included genera, *Desmognathus* and *Phaeognathus,* based on the unique jaw mechanism that sets them apart from all other plethodontids, and the sister relationship found by Vieites *et al.* (2011) of this tribe with *Plethodon.* The recommendations of that study and Wake (2012) to also treat *Ensatina* and hydromantines in two other tribes, pending better evidence of their relations, is also followed here.

Genus **Desmognathus** Baird, 1850

Synonymy: See the subgenus *Desmognathus,* below.

Distribution: Same as for the tribe.

Common name: Dusky Salamanders.

Content: 21 species.

Commentary: Dunn (1926b) recognized three species in this genus; here we recognize 21; he did recognize several of the taxa listed here but as subspecies. If we were to list species in an approximate phylogenetic sequence, from more derived toward more ancestral they would be as follows; this is based on several recent works, which are cited in comments in the species accounts, about specific relationships: *D. fuscus, D. planiceps, D. auriculatus, D. carolinensis, D. abditus, D. orestes, D. ochrophaeus, D. apalachicolae, D. monticola, D. ocoee, D. conanti, D. santeetlah, D. welteri, D. brimleyorum, D. marmoratus, D. folkertsi, D. quadramaculatus, D. imitator, D. aeneus, D. wrighti, D. organi.* Dubois & Raffaëlli (2012) organized these species into 4 subgenera; they commented that there are probably over 20 cryptic species awaiting description.

Subgenus **Desmognathus** Baird, 1850

Desmognathus Baird, 1850a: 282.—**NS** (sd., Brown, 1908: 126): *Triturus fuscus* Rafinesque, 1820.—**Comment**: Placed on the official list of generic names (ICZN, 1926).

Desmagnathus: Knauer, 1878: 98.—**IS.**

Desmognathus (Desmognathus) abditus Anderson & Tilley, 2003

Desmognathus abditus Anderson & Tilley, 2003: 107.—**O**: **H**: MCZ A-135817.—**OT**: U.S., Tennessee, Cumberland County, Staples Spring Branch, a small tributary of Daddy's Creek; 36° 3.609' N, 84° 47.674' W.

Desmognatus abditus: Raffaëlli, 2007: 330.

Desmognathus (Desmognathus) abditus: Dubois & Raffaëlli, 2012: 145.

Distribution: North-central Tennessee, Cumberland Plateau. The original description contains a map with localities. Lannoo (2005) provided a range map. **U.S.**: TN.

Common name: Cumberland Dusky Salamander.

Commentary: Anderson & Tilley (2003) indicated that the species is most closely related to *D. carolinensis*. Tilley (2010) provided a CAAR account.

Desmognathus (Desmognathus) apalachicolae Means & Karlin, 1989

Desmognathus apalachicolae Means & Karlin, 1989: 38.—**O**: **H**: USNM 269079.—**OT**: U.S., Florida, Liberty County, Liberty 7 W, Big Sweetwater Creek steephead; SE 1/4 Section 11, Township 1 N, Range 7 W, 60 m elevation.

Desmognathus apalachicolae: Conant & Collins, 1998: 457; Petranka, 1998: 162; Raffaëlli, 2007: 332.

Desmognathus (Desmognathus) apalachicolae: Dubois & Raffaëlli, 2012: 145.

Distribution: Area where Georgia, Florida, and Alabama meet, to northeastern Georgia. Means (1993) has a range map. **U.S.**: AL, FL, GA.

Common name: Apalachicola Dusky Salamander.

Commentary: Means (1993) provided a CAAR account. Means & Karlin (1989) suggested this species may have been derived from a *D. ochrophaeus*-like ancestor. Titus & Larson (1996) found it to be a sister species to *D. monticola*. Rissler & Taylor (2003) found it ancestral to a clade consisting of *D.*

brimleyorum, D. monticola, D. carolinensis, D. ochrophaeus, and *D. orestes,* but they did not sample *D. ocoee.* Kozak *et al.* (2005) and Beamer & Lamb (2008) found it to be closest to *D. ocoee* and *D. monticola.*

Desmognathus (Desmognathus) auriculatus (Holbrook, 1838)

Salamandra auriculata Holbrook, 1838b: 115.—**O**: **H**: Specimen in plate 28 (probably ANSP, assumed lost).—**OT**: U.S., Georgia, Liberty County, Riceborough.—**Comment**: Dunn, 1926b: 28, lists USNM 3901 as **H**, but not included by Cochran (1961a), and discounted by Adler (1976: xxxviii); Hallowell (1858b: 344) noted Holbrook deposited material in ANSP.

Salamandra auriculata: Holbrook, 1842, v. 5: 47.

Desmognathus auriculatus: Baird, 1850a: 286; Valentine, 1963: 130; Cochran & Goin, 1970: 34; Gorham, 1974: 33; Conant, 1975: 263; Behler & King, 1979: 312; Frost, 1985: 568; Conant & Collins, 1998: 450; Petranka, 1998: 164; Raffaëlli, 2007: 335.

Cylindrosoma auriculatum: Duméril, Bibron, & Duméril, 1854c: 81.

Plethodon auriculatum: Hallowell, 1858b: 344.

Desmognathus fusca var. *auriculata*: Cope, 1869b: 116; 1889b: 195.

Desmognathus fuscus auriculatus: Boulenger, 1882b: 78; Stejneger & Barbour, 1917: 22; 1943: 14; Bishop, 1924: 92; 1928a: 165; Dunn, 1926b: 97; Schmidt, 1953: 29; Conant, 1958: 218; Brame, 1967: 12.

Desmognathus auriculata: Lönnberg, 1895: 337.

Desmognathus (Desmognathus) auriculatus: Dubois & Raffaëlli, 2012: 145.

Desmognathus fuscus carri Neill, 1951b: 25.—**O**: **H** (od.), ERA-WTN 14188 (now FSM).—**OT**: U.S., Florida, Marion County, Ocala National Forest, Silver Glen Springs.—**SS:** Rossman (1959: 149) and Means (1974: 36) rejected the subspecies status and found it a synonym of *D. auriculatus.*

Desmognathus fuscus carri: Brame, 1967: 12.

Distribution: Coastal plain of Georgia and adjacent northern Florida, to eastern Texas, north along the Atlantic coast to Maryland. Map in Petranka (1998) includes range of other species which have since been defined. Means (1999a) has a map for the current species. **U.S.**: AL, FL, GA, LA, MD, MS, NC, SC, TX, VA.

Content: No subspecies recognized, but as recently as 1967, many workers considered this a subspecies of *D. fuscus*.

Common name: Southern Dusky Salamander.

Commentary: Titus & Larson (1996), Rissler & Taylor (2003), and Kozak *et al.* (2005) indicate this species is most closely related to *D. fuscus*. Means (1999a) provided a CAAR account.

Desmognathus (Desmognathus) carolinensis Dunn, 1916

Desmognathus ochrophaea carolinensis Dunn, 1916: 74.—**O**: **H**: USNM 31135.—**OT**: U.S., North Carolina, Buncombe County, spring near top of Mt. Mitchell, elevation over 6500 feet.

Desmognathus ochrophaeus carolinensis: Stejneger & Barbour, 1917: 23; 1943: 14; Bishop, 1924: 87; 1943: 203; Schmidt, 1953: 30; Conant, 1958: 224; Brame, 1967: 13; Cochran & Goin, 1970: 35; Gorham, 1974: 33.

Desmognathus fuscus carolinensis: Pope, 1924: 4; Dunn, 1926b: 105; Bishop, 1928a: 165.

Desmognathus carolinensis: Brimley, 1928a: 21; Tilley & Mahoney, 1996: 23. Petranka, 1998: 169; Raffaëlli, 2007: 331.

Desmognathus (Desmognathus) carolinensis: Dubois & Raffaëlli, 2012: 144.

Distribution: Western North Carolina, eastern Tennessee, in the Blue Ridge region. See map in Petranka (1998). **U.S.**: NC, TN.

Common name: Carolina Mountain Dusky Salamander (SECSN) or Carolina Dusky Salamander (C&T).

Content: No subspecies are recognized, but until about 15 years ago, it was considered by many workers to be a race of *D. ochrophaeus,* or merely a synonym.

Commentary: Most authors in the 1970s and 1980s treated this as a synonym of *D. ochrophaeus*. Titus & Larson (1996) treated it as a species closest to *D. orestes*. Rissler & Taylor (2003) found this species is closest to *D. fuscus* and *D. monticola*. Anderson & Tilley (2003) found this nearest to their new species, *D. abditus*. Kozak *et al.* (2005) found this to be a sister species to *D. fuscus*.

Desmognathus (Desmognathus) conanti Rossman, 1958

Desmognathus fuscus conanti Rossman, 1958: 158.—**H**: AMNH 62223.—**OT**: U.S., Kentucky, Livingston County, 2.1 miles South of Smithland, near US Highway 60, 400 feet elevation.

Desmognathus fuscus conanti: Brame, 1967: 12; Cochran & Goin, 1970: 33; Gorham, 1974: 33; Conant, 1975: 262; Behler & King, 1979: 313; Conant & Collins, 1998: 448; Petranka, 1998: 175.

Desmognathus conanti: Titus & Larson, 1996: 462; Raffaëlli, 2007: 334.

Desmognathus (Desmognathus) conanti: Dubois & Raffaëlli, 2012: 144.

Distribution: Much of the southeastern US. See range map in Means & Bonett (2005). **U.S.**: AL, AR, FL, GA, KY, IL, LA, MO?, MS, NC, SC, TN.

Common name: Spotted Dusky Salamander.

Commentary: Titus & Larson (1996) found this species to be a sister of *D. santeetlah*. Rissler & Taylor (2003) found it to be a sister of *D. monticola*. Kozak *et al.* (2005) found it to be most closely related to *D. auriculatus,* while Beamer & Lamb (2008) found some populations closest to *D. monticola,* while others seemed closer to *D. santeetlah,* suggesting that there may be at least two cryptic species under the name *conanti*. Petranka (1998) continued to treat this as a subspecies of *D. fuscus*.

Desmognathus (Desmognathus) fuscus (Rafinesque, 1820)

"Salamandra fusca" Green, 1818: 357.—**O**: not stated, none known to exist.—**OT**: not stated, probably U.S., New Jersey, vicinity of Princeton, *fide* Schmidt, 1953. Junior primary homonym of *Salamandra fusca* Laurenti, 1768 (= *Salamandra atra*), thus not available for this species.—**SS:** Gray, 1850: 31, with *Salamandra nigra*; Dunn, 1926b: 81 with *Triturus fuscus* Rafinesque, 1820.—**ND,** (Highton, Tilley, & Wake, *in* Crother *et al.,* 2003: 197).—**Comment:** Placed on the official list of rejected names (ICZN, 1956b).

?*"Salamandra nigra"* Green, 1818: 352.—**O**: not stated, none known to exist; however, Harlan (1827: 332) saw specimens in ANSP, and Baird (1850a: 285)examined an ANSP specimen that might be a type.—**OT**: not stated; presumably from near New Jersey, vicinity of Princeton;

given as U.S., Pennsylvania by Harlan, 1827: 332. Schmidt, 1953: 29, restricted the locality to U.S., New Jersey, probably vicinity of Princeton, presumably making the same assumption as Harlan.—**ND,** (Highton, Tilley & Wake, *in* Crother *et al.*, 2003: 197).—**SS**: Hallowell (1857a: 7), with *D. fuscus*.

"Salamandra sinciput-albida" Green, 1818: 352.—**O**: not stated, none known to exist.—**OT**: U.S., New Jersey.—**ND**, *fide* Highton *et al*, *in* Crother *et al.*, 2003: 197.—**SS**: Dunn (1926b: 81).

Triturus fuscus Rafinesque, 1820: 4.—**O**: not stated, none known to exist.—**OT**: U.S., New York, in the northern parts of the state, in small brooks. **Comment:** this name has been conserved on the official list of specific names (ICZN, 1956b).

Desmognathus fuscus: Baird, 1850a: 285; Viosca, 1949: 9; Schmidt, 1953: 28; Raffaëlli, 2007: 333.

Plethodon fuscum: Duméril, Bibron, & Duméril, 1854a or c?: 85.

Desmognathus fusca: Cope, 1859a: 124; 1889b: 194.

Desmognathus fusca var. *fusca*: Cope, 1889b: 195.

Spelerpes fuscus: Werner, 1909: 10.

Desmognathus fuscus fuscus: Stejneger & Barbour, 1917: 22; 1943: 13; Dunn, 1926b: 81; Bishop, 1928a: 164; Schmidt, 1953: 29; Brame, 1967: 12; Cochran & Goin, 1970: 33; Gorham, 1970: 3; 1974: 33; Conant, 1975: 261; Behler & King, 1979: 313; Conant & Collins, 1998: 447; Petranka, 1998: 175.

Desmognathus (Desmognathus) fuscus: Dubois & Raffaëlli, 2012: 144.

Triturus nebulosus Rafinesque, 1820: 5.—**O**: not stated nor known to exist.—**OT**: U.S., New York, near New York city, at Harlem and Long Island.—**SS**: Dunn (1926b: 81).—**Comment**: as first reviser, Dunn gave priority to *T. fuscus* Rafinesque.

Salamandra picta Harlan, 1825e: 136.—**O**: not stated or known to exist, but Harlan (1827: 333) saw specimens in ANSP.—**OT**: U.S, Pennsylvania, Philadelphia, shallow brooks in the vicinity.—**SS**: Gray, 1850: 40, with *S. nigra* Green; Boulenger, 1882b: 77.

Salamandra intermixta Green, 1825: 159.—**O**: not stated nor known to exist.—**OT**: U.S., Pennsylvania, Philadelphia; also found in the southern states. Schmidt, 1953: 29, invalidly restricted it to Jefferson College, Pennsylvania.—**SS**: Harlan, 1835: 98, with *S. picta* Harlan; Gray (1850: 40) with *S. nigra* Green; Baird (1850a: 285) and Boulenger (1882b: 77) with *T. fuscus* Rafinesque.

Salamandra frontalis Gray, 1831a: 107.—**NN** for *Salamandra sinciput-albida* Green, 1818.

Ambystoma frontale: Gray, 1850: 38.

Triton niger: Holbrook, 1842: 81.

Desmognathus niger: Baird, 1850a: 285; Hallowell, 1857a; 11.

Ambystoma nigrum: Duméril, Bibron, & Duméril, 1854c: 105.

Ambistoma nigrum: Jan, 1857: 55.

Plethodon niger: Hallowell, 1858b: 344.

Desmognathus nigra: Davis & Rice, 1883: 27; Blatchley, 1901: 759; Nash, 1905: 5.

"Molge brunnea" Duméril, Bibron, & Duméril, 1854c: 86.—**NU** attributed to Valenciennes.—**SS**: Dunn, 1926b: 82.

"Molge arenatus" Duméril, Bibron, & Duméril, 1854c: 86.—**NU** attributed to Valenciennes.—**SS**: Dunn, 1926b: 82.

Salamandra phoca Matthes, 1855: 273.—**O**: not stated or known to exist, perhaps originally NMW.—**OT**: U.S., Ohio, west bank of Miami River, Taylor's Creek, opposite the town of Miami (near Kentucky, Campbell County, about 14 miles northwest of Newport).—**SS**: Dunn, 1923: 39, as *D. monticola;* Grobman, 1945: 40, as *D. fuscus* (well supported). There is some suggestion that this may represent a synonym of *D. monticola*, so we list it there as well.

Desmognathus phoca: Dunn, 1923: 39; 1926b: 73; Bishop, 1924: 93; 1928a: 165.

Distribution: From the region of the western Carolinas, north-eastern Georgia and eastern Tennessee and Kentucky to New England and adjacent Canada. Petranka (1998) has a map showing the distribution as the subspecies *D. f. fuscus.* **Can.:** NB, ON, QC. **U.S.**: CT, DE, GA, IN, KY, MA, MD, ME, NC, NH, NJ, NY, OH, PA, RI, SC, TN, VA, VT, WV.

Common name: Northern Dusky Salamander. *French Canadian name:* Salamandre sombre du nord.

Content: Formerly, *D. conanti* and *D. santeetlah* were considered as subspecies, but they are now treated as full species, and generally not as close relatives of *D. fuscus*. No subspecies are currently recognized.

Commentary: Authorship is often given as Green, 1818 (*e.g.*, Cope, 1859a: 124; Schmidt, 1953: 28); but the three older names by Green are best regarded as *nomina dubia,* thus Rafinesque seems to be the valid author, and his nomen is now officially conserved. Several workers have noted that there may be more cryptic species currently identified under this name. Anderson & Tilley (2003) found this species to be a sister of *D. orestes,* while Kozak *et al.* (2005) and Beamer & Lamb (2008) found this species seemed to have a sister relation to either *D. carolinensis* or *D. auriculatus,* probably because some populations have diverged from those from which the species were described. Tilley *et al.* (2008) resurrected *D. planiceps* from the synonymy of *D. fuscus,* and found the two to have a sister-species relationship.

Desmognathus (Desmognathus) imitator Dunn, 1927

Desmognathus fuscus imitator Dunn, 1927a: 84.—**O**: **H:** USNM 75762.—**OT**: U.S., North Carolina, Great Smoky Mts., Indian Pass.

Desmognathus imitator: Tilley, Merritt, Wu & Highton, 1978: 100; Behler & King, 1979: 314; Frost, 1985: 569; Conant & Collins,1998: 456; Petranka, 1998: 182; Raffaëlli, 2007: 329.

Desmognathus (Desmognathus) imitator: Dubois & Raffaëlli, 2012: 144.

Desmognathus aureatagulus Weller, 1930: 8 (unnumbered).—**O**: **S**, includes USNM 93686, *fide* Cochran, 1961a: 11.—**OT**: U.S., Tennessee, Sevier County, Great Smoky Mountains National Park, Mount LeConte.—**SS**: Weller, 1931a: 8.

Distribution: Western North Carolina and eastern Tennessee, in the Great Smoky Mountains. See maps in Tilley (1985) and Petranka (1998). **U.S.**: NC, TN.

Common name: Imitator Salamander.

Commentary: In the past, some authors (e.g., Gorham, 1974: 33) treated this form as a synonym of D. ochrophaeus. Tilley (1985) provided a CAAR account. Titus & Larson (1996) found this species to be a basal sister to most *Desmognathus* species, except *D. aeneus, D. organi* and *D. wrighti.* Kozak *et al.* (2005) found similar results. Rissler & Taylor (2003) found it to be ancestral to many of those same species, but not to others.

Desmognathus (Desmognathus) monticola Dunn, 1916

?Salamandra phoca Matthes, 1855: 273.—**O:** not stated or known to exist, perhaps originally NHMW.—**OT:** U.S., Ohio, west bank of Miami River, Taylor's Creek, opposite the town of Miami (near Kentucky, Campbell County, abut 14 miles northwest of Newport). Most workers consider this a synonym of *D. fuscus,* but there is some suggestion that it may be a senior synonym of this species; if that should be determined to be correct, the nomen *D. phoca,* having priority, would replace the current name.

Desmognathus monticola Dunn, 1916: 73.—**H**: USNM 38313.—**OT**: U.S., North Carolina, Transylvania County, Elk Lodge Lake, near Brevard, elevation about 3000 feet.

Desmognathus monticola: Stejneger & Barbour, 1917: 22; Gorham, 1974: 33; Frost, 1985: 569; Conant & Collins, 1998: 453; Petranka, 1998: 187; Raffaëlli, 2007: 332.

Desmognathus monticola monticola: Hoffman, 1951: 251; Schmidt, 1953: 31; Brame, 1967: 12; Cochran & Goin, 1970: 35; Conant, 1975: 266; Behler & King, 1970: 315.

Desmognathus (Desmognathus) monticola: Dubois & Raffaëlli, 2012: 144.

Desmognathus monticola jeffersoni Hoffman, 1951: 250.—**H**: USNM 126891.—**OT**: U.S., Virginia, Albemarle County, 2 miles west of Crozet, Saddle Hollow on Jarman's Mountain, elev. 1600 ft.

Desmognathus monticola jeffersoni: Schmidt, 1953: 31; Brame, 1967: 12; Cochran & Goin, 1970: 35; Conant, 1975: 266; Behler & King, 1979: 315.

Distribution: An inland and montane range from southwestern Alabama and western Florida to southwestern Pennsylvania (see range map in Petranka, 1998). There is also an isolated population in Arkansas, believed to be introduced (Bonett *et al.*, 2007). **U.S.**: AL, *AR, FL, GA, KY, NC, PA, TN, VA, WV.

Common name: Seal Salamander.

Content: No subspecies are currently recognized, but in the past two were recognized, the nominotypical and *D. m. jeffersoni* Hoffman.

Commentary: Titus & Larson (1996) found this species to be a sister of *D. apalachicolae.* Rissler & Taylor (2003) found it to be a basal sister to a clade composed of *D. carolinensis, D. ochrophaeus,* and *D. orestes,* and a southern Alabama population a sister species to *D. conanti*; they suggested the

Alabama population may be a separate cryptic species. Kozak *et al.* (2005) found it to be a sister to *D. ocoee,* while Beamer & Lamb (2008) found it to be a sister to *D. apalachicolae* + *D. ocoee.*

Desmognathus (Desmognathus) ochrophaeus Cope, 1859

Desmognathus ochrophaea Cope, 1859a: 124.—**O**: **H**: Formerly ANSP, but destroyed (Dunn, 1917a: 415).—**OT**: U.S., Pennsylvania, Susquehanna County.

Plethodon ochrophaeus: Smith, 1877: 71.

Desmognathus ochrophaeus: Boulenger, 1882b: 77; Gorham, 1974: 33; Conant, 1975: 266; Behler & King, 1979: 316; Frost, 1985: 569; Tilley & Mahoney, 1996: 25; Conant & Collins, 1998: 454; Petranka, 1998: 192; Raffaëlli, 2007: 330.

Desmognathus ochrophaea: Cope, 1889b: 191.

Desmognathus ochrophaea ochrophaea: Dunn, 1917a: 415.

Desmognathus ochrophaeus ochrophaeus: Stejneger & Barbour, 1917: 22; 1943: 14; Bishop, 1943: 199; Schmidt, 1953: 30; Conant, 1958: 224; Brame, 1967: 12; Cochran & Goin, 1970: 34.

Desmognathus fuscus ochrophaeus: Dunn, 1926b: 114.

Desmognathus (Desmognathus) ochrophaeus: Dubois & Raffaëlli, 2012: 144.

Distribution: New York and adjacent Canada, southwestward through Pennsylvania and adjacent Ohio and New Jersey, West Virginia, western Virginia, eastern Kentucky and eastern Tennessee, in montane areas. Petranka (1998) has a range map. **Can.**: ON, QC. **U.S.**: KY, NJ, NY, OH, PA, TN, VA, WV.

Common name: Allegheny Mountain Dusky Salamander. *French Canadian name:* Salmandre sombre des montagnes.

Content: No subspecies are currently recognized, but species formerly treated as subspecies of this form have been elevated to full species status.

Commentary: Tilley (1973) provided a CAAR account, but included forms that have since been separated as distinct species. Tilley & Mahoney (1996) found this species to be ancestral to other members of their *ochrophaeus*-group, with *D. ocoee* the nearest relative. Rissler & Taylor (2003), Kozak *et al.* (2005), and Beamer & Lamb (2008) found *D. orestes* to be the closest relative.

Desmognathus (Desmognathus) ocoee Nicholls, 1949

Desmognathus ocoee Nicholls, 1949: 127.—**O**: **H**: USNM 128007.—**OT**: U.S., Tennessee, Polk County, 9 airline miles west of Ducktown, beside US Highway 64 in Ocoee Gorge, on the surface and in crevices of cliffs at Ship's Prow Rock.

Desmognathus ocoee: Schmidt, 1953: 31; Conant, 1958: 225; Valentine, 1961: 315; Gorham, 1974: 33; Tilley & Mahoney, 1996: 25; Petranka, 1998: 196; Raffaëlli, 2007: 331.

Desmognathus (Desmognathus) ocoee: Dubois & Raffaëlli, 2012: 144.

Desmognathus perlapsis Neill, 1950a: 1.—**O**: **H**: ERA-WTN 14150 (to be deposited FSM).—**OT**: U.S., Georgia, Rabun County, Tallulah Gorge, near Town of Tallulah Falls).—**SS**: Valentine, 1961: 315.

Desmognathus perlapsis: Schmidt, 1953: 31; Conant, 1958: 225.

Misapplied names:

Desmognathus ochrophaeus: Martof & Rose, 1963: 376.

Distribution: Disjunct populations, one mainly in northeastern Alabama, another in northern Georgia and adjacent Carolinas and Tennessee (see map in Petranka, 1998). AL, GA, NC, SC, TN.

Common name: Ocoee Salamander.

Commentary: Valentine (1964) provided a CAAR account. For many years, most authors treated this as a synonym of *D. ochrophaeus* (*e.g.*, Martof & Rose, 1963: 376). Possible relationships are covered in commentaries of other species; to that we can add that Anderson & Tilley (2003), who found this species to be a sister of *D. conanti*.

Desmognathus (Desmognathus) orestes Tilley & Mahoney, 1996

Desmognathus orestes Tilley & Mahoney, 1996: 27.—**O**: **H**: AMNH 146066.—**OT**: U.S., Virginia, Smyth County, a seepage area at 1329 m above sea level in the headwaters of Daves Branch, along the Elk Garden trail just north of Elk Garden on the divide between Mt. Rogers and Whitetop Mountain.

Desmognathus orestes: Petranka, 1998: 202; Raffaëlli, 2007: 330.

Desmognathus (Desmognathus) orestes: Dubois & Raffaëlli, 2012: 144.

Distribution: The Blue Ridge area of northeastern Tennessee, northwestern North Carolina, and southwestern Virginia. Tilley & Mahoney (1996) have a map with localities. **U.S.**: NC, TN, VA.

Common name: Blue Ridge Dusky Salamander.

Commentary: Tilley & Mahoney (1996) indicated that *D. caro-linensis* is the closest relative, and that there may be two cryptic species. Rissler & Taylor (2003), Kozak *et al.* (2005), and Beamer & Lamb (2008) found *D. ochrophaeus* to be the closest relative.

Desmognathus (Desmognathus) planiceps Newman, 1955

Desmognathus planiceps Newman, 1955: 83.—**O**: **H**: WBN 1316, (to be deposited USNM).—**OT**: U.S., Virginia, Patrick County, from a portion of the stream (approximate elevation 2800 feet) dropping down into the gorge below the Dan River Dam near Meadows of Dan.

Desmognathus planiceps: Conant, 1958: 219; Raffaëlli, 2007: 334; Tilley, Eriksen, & Katz, 2008: 124.

Desmognathus (Desmognathus) planiceps: Dubois & Raffaëlli, 2012: 144.

Distribution: Southwestern Virginia, maybe into North Carolina. **U.S.**: NC?, VA.

Common name: Flat-headed Salamander (SECSN).

Commentary: Resurrected from the synonymy of *D. fuscus* by Tilley *et al.* (2008). Often treated as a synonym of *D. fuscus,* following the assertion of Martof & Rose (1962: 216), that the paratypes consisted of specimens of *D. fuscus* and *D. ochrophaeus*, and Newman's photos of the *planiceps* holotype (holophorant) also appeared to be *D. fuscus*. Tilley *et al.* (2008) were not able to distinguish *planiceps* and *fuscus* morphologically, so there remains the possibility that the holophorant may be *D. fuscus*. Also see commentary under *D. fuscus*.

Desmognathus (Desmognathus) santeetlah Tilley, 1981

Desmognathus santeetlah Tilley, 1981: 3.—**O**: **H**: USNM 214218.—**OT**: U.S., Tennessee, Monroe County, Unicoi Mtns., below Cherry Log Gap, in a seepage area at in the headwaters of the N. Fork of Citico Creek, ca. 1219 m elevation.

Desmognathus santeetlah: Frost, 1985: 569; Conant & Collins, 1998: 449; Raffaëlli, 2007: 335.

Desmognathus fuscus santeetlah: Petranka, 1998: 175.

Desmognathus (Desmognathus) santeetlah: Dubois & Raffaëlli, 2012: 144.

Distribution: Restricted range in North Carolina and Tennessee, in Blue Ridge area. See map in Petranka (1998), as a subspecies of *D. fuscus,* and Tilley (2000) as a full species. **U.S.**: NC, TN.

Common name: Santeetlah Dusky Salamander.

Commentary: Petranka (1998) treated this as a subspecies of *D. fuscus.* Titus & Larson (1996) and Beamer & Lamb (2008) found it to be a sister to *D. conanti.* Rissler & Taylor (2003) found it to be a basal sister to a clade including *D. monticola, D. fuscus,* and several members of the "*ochrophaeus* complex." Kozak *et al.* (2005) found it to be a sister to the pair *D. auriculatus* + *D. conanti.* Tilley (2000) provided a CAAR species account.

Desmognathus (Desmognathus) welteri Barbour, 1950

Desmognathus fuscus welteri Barbour, 1950: 277.—**O**: **H**: USNM 129312.—**OT**: U.S., Kentucky, Harlan County, near Lynch, at Looney Creek, elevation 2300 feet.

Desmognathus fuscus welteri: Schmidt, 1953: 29; Conant, 1958: 218; Brame, 1967: 12; Cochran & Goin, 1970: 33; Gorham, 1974: 33.

Desmognathus welteri: Rubenstein, 1971: 329; Conant, 1975: 264; Behler & King, 1979: 317; Conant & Collins, 1998: 452; Petranka, 1998: 211; Raffaëlli, 2007: 335.

Desmognathus (Desmognathus) welteri: Dubois & Raffaëlli, 2012: 144.

Distribution: Northern Tennessee, eastern Kentucky, adjacent Virginia, possibly West Virginia. See map in Petranka (1998). **U.S.**: KY, TN, VA, WV?.

Common name: Black Mountain Salamander.

Commentary: Titus & Larson (1996) found this to be the sister species of *D. brimleyorum*. Rissler & Taylor (2003) found it to be in an ancestral position to all other *Desmognathus* except *D. aeneus* and *D. wrighti*. Kozak *et al.* (2005) found it to be closest to *D. fuscus* and some *D. auriculatus*. Beamer & Lamb (2008) found it to be a sister to some *D. auriculatus populations*. The relationships of this species remain somewhat unclear.

Subgenus **Geognathus** Dubois & Raffaëlli, 2012

Geognathus Dubois & Raffaëlli, 2012: 145.—**NS**: none indicated; here designated as *Desmognathus aeneus* Brown & Bishop, 1947.

Distribution: Mountainous areas of Alabama, Georgia, North and South Carolina, Tennesse, and Virginia.

Content: Three species, and more cryptic species may await description.

Desmognathus (Geognathus) aeneus Brown & Bishop, 1947

Desmognathus aeneus Brown & Bishop, 1947: 163.—**O**: **H**: USNM 123977.—**OT**: U.S., North Carolina, Cherokee County, 1/2 mile SSE of Peachtree, seepage branch 100 feet north of Peachtree Creek.

Desmognathus aeneus aeneus: Chermock, 1952: 29; Brame, 1967: 12; Cochran & Goin, 1970: 36; Conant, 1975: 269.

Desmognathus aeneus: Schmidt, 1953: 28; Gorham, 1974: 33; Behler & King, 1979: 311; Frost, 1985: 568; Conant & Collins, 1998: 349; Petranka, 1998: 159; Raffaëlli, 2007: 329.

Desmognathus (Geognathus) aeneus: Dubois & Raffaëlli, 2012: 145.

Desmognathus chermocki Bishop & Valentine, 1950: 39.—**O**: **H**: FMNH 59232.—**OT**: U.S., Alabama, Tuscaloosa County, Hurricane Creek, 1 1/8 mille ENE of bridge crossing creek on Alabama St. route 116.—**SS**: Chermock, 1952: 29.

Desmognathus aeneus chermocki: Chermock, 1952: 29; Conant, 1958: 227; Brame, 1967: 12; Cochran & Goin, 1970: 36.

Distribution: Northeastern Georgia and adjacent Tennessee and North Carolina, plus northern Alabama. See maps in Harrison (1992) and Petranka (1998). **U.S.**: AL, GA, NC, TN.

Common name: Seepage Salamander (also called Cherokee Salamander, which seems more appropriate).

Content: No subspecies are currently recognized, but for a while there were two, rejected by Mount (1975).

Commentary: Harrison (1992) provided a CAAR account. Virtually all phylogenetic studies of the genus have found this species to be ancestral to all other *Desmognathus* except *D. wrighti* and *D. organi.*

Desmognathus (Geognathus) organi Crespi, Browne & Rissler, 2010

Desmognathus organi Crespi, Browne & Rissler, 2010: 291.—**O: H**: MVZ-UA 16183.—**OT:** U.S., Virginia, Smyth County, Whitetop Mountain, north-facing slope, 36° 38.361' N, 81° 36.573' W.

Desmognathsus (Geognathus) organi: Dubois & Raffaëlli, 2012: 145.

Distribution: Mountainous areas generally above 1100 m, in western Virginia, Tennessee, and North Carolina. See map in Crespi *et al.* (2010). **U.S.:** NC, TN, VA.

Common name: Northern Pigmy Salamander (SECSN), or Southern Pigmy Salamander (C&T).

Commentary: Sister species to *D. wrighti,* the two together being a sister to *D. aeneus,* and the three forming a clade sister to all other species of the genus.

Desmognathus (Geognathus) wrighti King, 1936

Desmognathus wrighti King, 1936: 57.—**O**: **H** (od), USNM 101794.—**OT**: U.S., Tennessee, Sevier County, Great Smoky Mountains National Park, Mount LeConte.

Desmognathus wrighti: Stejneger & Barbour, 1943: 15; Schmidt, 1953: 32; Cochran & Goin, 1970: 35; Gorham, 1974: 33; Conant, 1975: 268; Behler & King, 1979: 318; Frost, 1985: 569; Conant & Collins, 1998: 458; Crespi, 1996; Petranka, 1998: 213; Raffaëlli, 2007: 328; Crespi, Browne & Rissler, 2010: 290.

Desmognathus (Geognathus) wrighti: Dubois & Raffaëlli, 2012: 145.

Distribution: Western North Carolina, plus adjacent South Carolina and Tennessee, in the Great Smoky Mountains.

Mostly above 1400 m, but as low as 950 m in parts of North Carolina. Petranka (1998) and Harrison (2000) have range maps, but include the more recently described *D. organi*; see map in Crespi *et al.* (2010). **U.S.**: NC, TN.

Common name: Pygmy Salamander.

Commentary: Harrison (2000) provided a CAAR account. All recent phylogenetic analyses of the genus agree that this species is ancestral to virtually all other *Desmognathus* species. Crespi *et al.* (2010) found evidence that distinguished this from a second species, *D. organi.* These two are apparently sisters, which together are sister to *D. aeneus,* and the three basal to all other species of the genus.

Subgenus **Hydrognathus** Dubois & Raffaëlli, 2012

Hydrognathus Dubois & Raffaëlli, 2012: 145.—**NS**: *Desmognathus brimleyorum* Stejneger, 1895, mo.

Distribution: As for the single species.

Content: Monotypic.

Desmognathus (Hydrognathus) brimleyorum Stejneger, 1895

Desmognathus brimleyorum Stejneger, 1895: 597.—**O**: **H** (sd., Cochran, 1961a: 11) USNM 22157.—**OT**: U.S., Arkansas, Garland County, Hot Springs.

Desmognathus brimleyorum: Stejneger & Barbour, 1917: 22; Valentine, 1963: 130; Cochran & Goin, 1970: 34; Gorham, 1974: 33; Conant, 1975: 264; Behler & King, 1979: 312; Frost, 1985: 568; Conant & Collins, 1998: 451; Petranka, 1998: 167; Raffaëlli, 2007: 336.

Desmognathus fuscus brimleyorum: Dunn, 1926b: 101; Stejneger & Barbour, 1943: 14; Schmidt, 1953: 29; Conant, 1958: 219; Brame, 1967: 12.

Desmognathus (Hydrognathus) brimleyorum: Dubois & Raffaëlli, 2012, 145.

Distribution: West-central Arkansas and southeastern Oklahoma, in the Ouachita Mountains. See maps in Petranka (1998) and Means (1999b). **U.S.**: AR, OK.

Common name: Ouachita Dusky Salamander.

Content: No subspecies recognized, but as recently as 1967, many workers considered this a subspecies of *D. fuscus.*

Commentary: Means (1999b) provided a CAAR account. Titus & Larson (1996) found this species a sister to *D. welteri*. Rissler & Taylor (2003) found it to be a basal sister to a clade that includes *D. auriculatus, D. carolinensis, D. conanti, D. fuscus, D. monticola, D. ochrophaeus, D. ocoee, D. planiceps,* and *D. santeetlah*. Kozak *et al.* (2005) found it a basal sister to a clade including all species listed above except some *D. ocoee,* plus *D. apalachicolae* and *D. welteri.*

Subgenus **Leurognathus** Moore, 1899

Leurognathus Moore, 1899: 316.—**NS (**mo.**)**: *Leurognathus marmorata* Moore, 1899.—**SS**: Bernardo, 1994: 15, with *Desmognathus.*

Desmognathus (Leurognathus) aureatus Martof, 1956

Leurognathus marmorata aureata Martof, 1956: 2.—**O**: **H** (od.), UMMZ 111566.—**OT**: U.S., Georgia, Lumpkin County, about 9 air miles north-northeast of Dahlonega, Jarrard's Creek, 0.2 mile below its crossing of U.S. Route 19, elev. 1550 feet.—**SS**: Status later rejected, Martof (1962: 30), followed by Gorham (1974) and others.

Leurognathus marmoratus aureatus: Cochran & Goin, 1970: 51; Conant, 1975: 228.

Desmognathus (Leurognathus) aureatus: Dubois & Raffaëlli, 2012: 145.

Distribution: Northwestern Georgia. **U.S.:** GA.

Commentary: Divergent lineages have been recognized in *D. marmoratus* (Voss *et al.,* 1995; Jackson, 2005; Kozak *et al.,* 2005; Jones *et al.,* 2006. The nomen *aureatus* was already available for one of these, and was revived by Dubois & Raffaëlli (2012).

Desmognathus (Leurognathus) folkertsi Camp, Tilley, Austin, & Marshall, 2002

Desmognathus folkertsi Camp, Tilley, Austin, & Marshall, 2002: 477.—**O**: **H**: USNM 536397.—**OT**: U.S., Georgia, Union County, south of Wolf Creek Road on an upper tributary of the West Fork of Wolf Creek; 34° 46' 05" N, 83° 56' 37" W, elevation 834 m.

Desmognathus folkertsi: Raffaëlli, 2007: 337.

Desmognathus (Leurognathus) folkertsi: Dubois & Raffaëlli, 2012: 145.

Distribution: Known from a limited area in northeastern Georgia and southwestern North Carolina. See map in Camp (2004). **U.S.**: GA, NC.

Common name: Dwarf Black-bellied Salamander.

Commentary: Camp (2004) provided a CAAR account. Kozak *et al.* (2005) found this species has a sister relationship to the pair *D. marmoratus* + *D. quadramaculatus,* while Beamer & Lamb (2008) found it to be a sister of *D. marmoratus,* with *D. quadramaculatus* as a sister to that pair.

Desmognathus (Leurognathus) marmortus (Moore, 1899)

Leurognathus marmorata Moore, 1899: 316.—**O**: **H:** ANSP 19610 (Fowler & Dunn, 1917: 21, sd.).—**OT**: U.S., North Carolina, Avery County, between Linville and Blowing Rock, south flank of Grandfather Mt., a large clear rocky pool, elevation about 3500 feet.

Leurognathus marmoratus: Brimley, 1907: 151; Stejneger & Barbour, 1917: 23; Martof, 1962: 30; Gorham, 1974: 35; Behler & King, 1979: 334; Frost, 1985: 570; Conant & Collins, 1998: 460.

Leurognathus marmorata: Bishop, 1924: 94; 1928a: 166; Dunn, 1926b: 120; Brame, 1967: 13.

Leurognathus marmorata marmorata: Pope, 1928: 14; Stejneger & Barbour, 1933: 23, by implication; 1943: 15; Schmidt, 1953: 32; Martof, 1956: 2.

Leurognathus marmoratus marmoratus: Cochran & Goin, 1970: 51; Conant, 1975: 227.

Desmognathus marmoratus: Bernardo, 1994: 15; Titus & Larson, 1996: 451; Petranka, 1998: 184; Raffaëlli, 2007: 337.

Desmognathus (Leurognathus) marmoratus: Dubois & Raffaëlli, 2012: 144.

Leurognathus marmorata intermedia Pope, 1928: 14.—**O**: **H** (od.), AMNH 25557.—**OT**: U.S., North Carolina, Haywood County, Waynesville, Davis Gap (modified by Bishop, 1943: 224, to Davis Farm, about 2 miles east and a little north of Waynesville on Highway 276). Subspecies rejected by Martof, 1962: 30.

Leurognathus marmorata intermedia: Stejneger & Barbour, 1933: 23.

Leurognathus intermedia: Pope and Hairston, 1947: 160.

Leurognathus marmoratus intermedius: Stejneger & Barbour, 1943: 16; Schmidt, 1953: 33; Cochran & Goin, 1970: 51; Conant, 1975: 228.

Leurognathus marmorata roborata Martof, 1956: 2.—**O**: **H** (od.), UMMZ 111568.—**OT**: U.S., Georgia, Rabun County, about 3.5 miles northwest of Pine Mountain community, at Reed Creek, along Burrell Ford Road, 0.5 mile from its junction with Glade School Road, elev. 2350 feet. Status later rejected (Martof, 1962: 30).

Leurognathus marmoratus roboratus: Cochran & Goin, 1970: 51; Conant, 1975: 228.

Distribution: In several river drainages of western North Carolina, plus slightly into adjacent Georgia, South Carolina, Tennessee, and Virginia. Petranka (1998) has a range map. **U.S.**: GA, NC, SC, TN, VA.

Common name: Shovel-nosed Salamander (SECSN), or Shovelnose Salamander (C&T).

Content: Five subspecies have been recognized, but have since been rejected (Martof, 1962), and none is currently recognized. However, Jones *et al.* (2006) suggested that this taxon may be composed of several cryptic species, and two have been revived by Dubois & Raffaëlli (2012).

Commentary: Martof (1963) provided a CAAR account, under the genus *Leurognathus*. This species was treated in that monotypic genus for over 60 years, but it is nested within *Desmognathus,* and recognition of *Leurognathus* as a genus or subgenus would require transferring several other species to avoid paraphyly. Nevertheless, Dubois & Raffaëlli (2012) have recognized *Leurognathus* as a subgenus, and we will follow that treatment pending further analysis. Titus & Larson (1996), Rissler & Taylor (2003), and Kozak *et al.* (2005) found this species to be a sister of *D. quadramaculatus*. Beamer & Lamb (2008) found *D. quadramaculatus* is a sister to the pair *D. marmoratus* + *D. folkertsi*.

Desmognathus (Leurognathus) melanius Martof, 1956

Leurognathus marmorata melania Martof, 1956: 6.—**O**: **H** (od.), UMMZ 111564.—**OT**: U.S., North Carolina, Macon County, 0.5 mile west of Tellico Gap, at Otter Creek (tributary of Nantahala River), elev. 3600 feet. Status later rejected (Martof, 1962: 30).

Leurognathus marmoratus melanius: Cochran & Goin, 1970: 51; Conant, 1975: 229.

Desmognathus (Leurognathus) melianus: Dubois & Raffaëlli, 2012: 117-118, 145.—**IS.**

Distribution: Parts of North Carolina and Tennessee. **U.S.**: NC, TN.

Common name: None in use. Otter Creek Dusky Salamander seems appropriate.

Commentary: This species nomen was revived by Dubois & Raffaëlli (2012) for one of the divergent cryptic lineages identified by Jones *et al.* (2006).

Desmognathus (Leurognathus) quadramaculatus (Holbrook, 1840)

Salamandra quadra-maculata Holbrook, 1840: 121.—**H**: specimen of plate numbered as 27 (error, 26th plate) in Holbrook's work; ANSP 490 *fide* Brame (1972: 160, unpublished manuscript, *fide* Frost, 2009-13).—**OT**: U.S., common in Georgia and Carolina and inhabits Pennsylvania.—**Comment:** This nomen is used on the plate, but "*Salamandra maculo-quadrata*" is listed in the table of contents and in the species account (see details, Adler, 1976: xl).

Salamandra quadrimaculata: Holbrook, 1842: 5: 49.—**IS.**

Ambystome quadrimaculatum: Duméril, Bibron, & Duméril, 1854c: 109.—**IS.**

Plethodon quadrimaculatus: Hallowell, 1857a: 11.—**IS.**

Desmognathus quadrimaculata: Stejneger, 1903: 557.—**IS.**

Desmognathus quadrimaculatus: Brimley, 1907: 154.—**IS.**

Desmognathus quadramaculata: Dunn, 1917a: 401.

Desmognathus quadra-maculatus: Stejneger & Barbour, 1917: 23; Bishop, 1924: 87; 1928a: 166; Dunn, 1926a: 63.

Desmognathus quadramaculatus: Bishop, 1941: 12, 1943: 210; Schmidt, 1953: 32; Cochran & Goin, 1970: 36; Gorham, 1974: 33; Conant, 1975: 266; Behler & King, 1979: 316; Frost, 1985: 569; Conant & Collins, 1998: 454; Petranka, 1998: 206; Raffaëlli, 2007: 336.

Desmognathus quadramaculatus quadramaculatus: Bishop, 1943: 210; Stejneger & Barbour, 1943: 15.

Desmognathus (Leurognathus) quadramaculatus: Dubois & Raffaëli, 2012: 145.

Triton niger Holbrook, 1842: 5: 81.—**O**: none stated nor known to exist.—**OT**: U.S., Atlantic states from Massachusetts and Pennsylvania, to Carolina, Georgia, and Louisiana.—**Comment**: Secondary junior homonym of *Salamandra niger* Green, 1818 (now *Desmognathus fuscus*).—**SS**: Dunn (1926b: 64).

Triton nigra: Holbrook, 1842: 27.—**UE.**

Desmognathus niger: Baird, 1850a: 285; Boulenger, 1882b: 79.

Desmognathus nigra: Cope, 1869b: 117.—**UE.**

Desmognathus quadramaculatus amphileucus Bishop, 1941: 12.—**O**: **H:** (od.), UMMZ 89767.—**OT**: U.S., Georgia, Habersham County, Demorest. Subspecies rejected (Pope, 1949: 1).

Desmognathus quadramaculatus amphileucus: Bishop, 1943: 214; Stejneger & Barbour, 1943: 15.

Distribution: Northern Georgia (introduced), western North Carolina, western Virginia, southern West Virginia, and adjacent South Carolina and Tennessee. Range maps are in Valentine (1974) and Petranka (1998). **U.S.**: *GA, NC, SC, TN, VA, WV.

Common name: Black-bellied Salamander (SECSN) or Blackbellied Salamander (C&T).

Content: No subspecies currently recognized, but Bishop (1941) described two races, redefined by Neill (1948c), and then rejected (Pope, 1949).

Commentary: Unjustified restriction of type locality to Great Smoky Mountains (Schmidt, 1953). Valentine (1974) provided a CAAR account, and has a more extensive synonymy, as does Frost (2009-13). See comments under *D. marmoratus* and *D. folkertsi* regarding relationships among these three species. Rissler & Taylor (2003) showed a southwestern North Carolina population is probably a separate cryptic species. Jones *et al.* (2006) also found this taxon may be composed of more than one species. Wake (*in litt*) says there is no evidence that Holbrook knew the populations currently referred to under this name; the name may be a junior synonym of *D. fuscus,* and the populations currently referred to the name *quadramaculatus* are without a name, unless the nomen *amphileucus*, in the synonymy above, is applicable.

Genus **Phaeognathus** Highton, 1961

Phaeognathus Highton, 1961: 66.—**NS** (od. and mo.): *Phaeognathus hubrichti* Highton, 1961.

Distribution: Southern Alabama. Petranka (1998) has a range map.

Common name: Red Hills Salamanders.

Content: The genus is monotypic.

Phaeognathus hubrichti Highton, 1961

Phaeognathus hubrichti Highton, 1961: 67.—**O**: **H:** USNM 142486.—**OT**: U.S., Alabama, Butler County, 3 miles northwest of McKenzie on US Route 31.

Phaeognathus hubrichti: Brame, 1967: 13; Cochran & Goin, 1970: 52; Gorham, 1974: 36; Conant, 1975: 271; Behler & King, 1979: 335; Frost, 1985: 570; Conant & Collins, 1998: 461; Petranka, 1998: 216; Raffaëlli, 2007: 338.

Distribution: As for the genus. **U.S.**: AL.

Common name: Red Hills Salamander.

Commentary: Brandon (1966c) provided a CAAR account. All recent phylogenetic analyses involving the Desmognathini have consistently found that this is a sister-taxon (ancestral) to all *Desmognathus*. This species is federally listed as threatened.

Tribe **Ensatinini** Gray, 1850

Ensatinina Gray, 1850: 48.—**NG**: *Ensatina* Gray, 1850.

Ensatinini: Vietites *et al.,* 2011: 633; Wake, 2012: 79; Dubois & Raffaëlli, 2012: 145.

Distribution: As for the single genus.

Common name: Ensatinas.

Content: One genus with one polytypic species.

Commentary: See comments on the possible relationships of the genus, and comments under *Aneides*. Pyron & Wiens (2011) analysis agreed that *Ensatina* is most closely related to *Aneides,* and including it in the Aneidini would be as reasonable as this separate tribe ranking. We are aware of a forthcoming article that rejects that and demonstrates a different relationship.

Genus **Ensatina** Gray, 1850

Ensatina Gray, 1850: 48.—**NS** (mo.): *Ensatina eschscholtzii* Gray, 1850.—**SS**: Cope, 1868a: 167, with *Plethodon*.

Ensatina: Dunn, 1923: 39

Heredia Girard, 1857a: 140.—**NS** (mo.): *Heredia oregonensis* Girard, 1857.—**SS**: Cope, 1868a: 167, with *Plethodon*; Dunn, 1923: 39, with *Ensatina*.

Urotropis Jiménez de la Espada, 1875a: 70.—**NS** (mo.): *Urotropis platensis* Jiménez de la Espada, 1875. Preoccupied by *Urotropis* Rafinesque, 1822 (now *Cryptobranchus* Leuckart, 1821).—**SS**: Dunn, 1926b: 181, with *Ensatina*.

Distribution: Forming a species-ring around the Sierra Nevada in California, with coastal populations ranging south into Baja California and north into British Columbia. See range map in Petranka (1998).

Common name: Ensatinas.

Content: There are several rather distinct populations which are usually treated as subspecies of a single species. In the past, one or two of these seemed to be reproductively isolated from others in sympatry, but then intergrade with intervening populations, and these have been suggested to be distinct species. There has been considerable research recently (*e.g.*, Moritz *et al.*, 1992; Kuchta *et al.*, 2009a; Kuchta, *et al.*, 2009b), and the study of Pereira & Wake (2009) demonstrates that none of the populations can be considered separate, reproductively isolated species. Thus, we continue to recognize a single species with seven named subspecies.

Commentary: Morphologically, the genus seems closest to *Plethodon* (Wake, 1966), but recent genetic studies seem to demonstrate a phylogenetic relationship with desmognathines or *Hydromantes*. Macey (2005) found the genus is a sister group to *Hydromantes*, and the two have a sister-group relation to *Aneides* and desmognathines, but see remarks about that study in Commentary for genus *Aneides*. Wiens *et al.* (2006) and Vieites *et al.* (2007) suggest *Ensatina* is a sister group to *Aneides* plus the desmognathines, but they did not include *Hydromantes* in their studies. The analysis of Pyron & Wiens (2011) showed *Ensatina* as a sister to *Aneides* and desmognathines, with *Hydromantes* a sister to that entire clade. The later study by Vieites *et al.* (2011) found a sister-taxon relationship between *Ensatina* and *Aneides,* and that clade being a sister to one containing desmognathines and *Plethodon.*

Ensatina eschscholtzii Gray, 1850

Synonymy: See the subspecies, *E. e. eschscholtzii,* below.

Distribution: As for the genus. **Can.**: *BC. **U.S.**: CA, OR, WA. See Jackman (1998) for evidence of introduction of the species into Vancouver Island, BC.

Common name: Ensatina (SECSN) or Common Ensatina (C&T).

Content: There are seven major populations, sometimes considered semispecies, most often as subspecies. As indicated above, under *Commentary* for the genus, we here recognize all these populations as subspecies of *E. eschscholtzii,* following the study of Pereira & Wake (2009), who found hybridization wherever the named populations meet, so that there is never reproductive isolation, but a level of genetic integrity persists in all these populations. See the range map in Petranka (1998) for all the subspecies.

Ensatina eschscholtzii eschscholtzii Gray, 1850

Ensatina eschscholtzii Gray, 1850: 48.—**O**: **S:** BM 1947.2.24.45-47 (Brame, 1972: 199, *fide* Frost, 2009-13, sd.).—**OT**: U.S., California; corrected to Monterey, California by Boulenger, 1882b: 55.

Plethodon eschscholtzii: Grinnell & Camp, 1917: 132; Stejneger & Barbour, 1917: 15; Dunn, 1918b: 459.

Ensatina eschscholtzii: Dunn, 1923: 39; 1926b: 188; Stejneger & Barbour, 1933; 13; Frost, 1985: 586.

Ensatina eschscholtzi eschscholtzi: Schmidt, 1953: 42; Cochran & Goin, 1970: 37; Gorham, 1974: 34; Behler & King, 1979: 319.—**UE.**

Ensatina eschscholtzii eschscholtzii: Wood, 1940: 426; Stejneger & Barbour, 1943: 21; Brame, 1967: 15; Liner, 1994: 12; Stebbins, 1985: 49; 2003: 175; Petranka, 1998: 326; Raffaëlli, 2007: 324.

Distribution: West coast, from northwestern Baja California northward about half the length of California. **U.S.**: CA.

Common name: Monterey Ensatina.

Commentary: Kuchta *et al.* (2009b) found this population to be a rather monophyletic unit. It interbreeds in a fairly broad zone with *E. e. xanthoptica* at the northern end of its range, but seems more isolated where it meets *E. e. klauberi* at the south end (Wake *et al.*, 1989; Alexandrino *et al.*, 2005); however, Devitt *et al.* (2011) reported extensive hybridization at one site.

Ensatina eschscholtzii croceater (Cope, 1868)

Plethodon croceater Cope, 1868a: 210.—**O**: **H**: USNM 4701, lost.—**OT**: U.S.: California, Kern County, Fort Tejon (questioned by Van Denburgh, 1916: 220).—**SS**: Wood, 1940: 426, and Stebbins, 1949: 457, with *E. eschscholtzii*.

Plethodon croceater: Boulenger, 1882b: 55; Cope, 1889b: 150; Stejneger & Barbour, 1917: 14.

Ensatina croceater: Dunn, 1923: 39; Stejneger & Barbour, 1923: 12; 1943: 20; Dunn, 1926b: 185.

Ensatina eschscholtzii croceater: Wood, 1940: 426; Stebbins, 1985: 49; 2003: 177; Petranka, 1998: 327, 328.

Ensatina eschscholtzii croceator: Stebbins, 1949: 457; Brame, 1967: 15; Raffaëlli, 2007: 326.—**UE.**

Ensatina eschscholtzi croceater: Schmidt, 1953: 43; Gorham, 1974: 34.—**UE** (sp.).

Ensatina eschscholtzi croceator: Cochran & Goin, 1970: 37; Behler & King, 1979: 319.—**UE.**

Distribution: Southern Sierra Nevada, south of Kern River, through the Tehachapi Mountains. It contacts *E. e. platensis,* but not *E. e. eschscholtzii.* **U.S.**: CA.

Common name: Yellow-blotched Ensatina.

Commentary: Kuchta *et al.* (2009b) found this population to be a rather monophyletic unit. This subspecies hybridizes with *E. e. platensis* in a 10 km overlap (Pereira & Wake, 2009) and with *E. e. klauberi* in at least one locality (Devitt *et al.*, 2013).

Ensatina eschscholtzii klauberi Dunn, 1929

Ensatina klauberi Dunn, 1929: 1.—**H**: USNM 75337.—**OT**: U.S., California, San Diego County, Descanso.—**SS**: Stebbins, 1949: 467, with *E. eschscholtzii*.

Ensatina eschscholtzii klauberi: Stebbins, 1949: 467; 1985: 49; 2003: 177; Brame, 1967: 15; Liner, 1994: 12; Petranka, 1998: 327, 328; Raffaëlli, 2007: 326; Devitt *et al.*, 2013: 1650.

Ensatina eschscholtzi klauberi: Schmidt, 1953: 42; Cochran & Goin, 1970: 39; Gorham, 1974: 34; Behler & King, 1979: 319.—**UE.**

Ensatina klauberi: Frost & Hillis, 1990: 97.

Distribution: Central southern California, where it contacts *E. e. eschscholtzii,* with some interbreeding. **U.S.**: CA.

Common name: Large-blotched Ensatina.

Commentary: Kuchta *et al.* (2009b) found this population to be a rather monophyletic unit. The case for recognizing this form as a distinct species was deemed the best of any of the forms for several years, but Devitt *et al.* (2011) found extensive hybridization where it contacts *E. e. eschscholtzii;* the study of Pereira & Wake (2009) makes it clear that none of these populations can be considered separate reproductively isolated species.

Ensatina eschscholtzii oregonensis (Girard, 1857)

Heredia oregonensis Girard, 1857a: 140.—**O**: **S:** USNM 7022, 15479-15480 (Dunn, 1926b: 194, sd.)—**OT**: U.S., Washington, Puget Sound.—**SS**: Dunn, 1926b: 189, and Stejneger & Barbour, 1933: 13, with *E. eschscholtzii.*

Plethodon oregonensis: Cope, 1869b: 100; 1889b: 148; Boulenger, 1882b: 54.

Ensatina eschscholtzii oregonensis: Stebbins, 1949: 393; 1985: 49; 2003: 176; Brame, 1967: 15; Petranka, 1998: 327; Raffaëlli, 2007: 325.

Ensatina eschscholtzi oregonensis: Schmidt, 1953: 43; Cochran & Goin, 1970: 39; Gorham, 1974: 34; Behler & King, 1979: 320.—**UE.**

Distribution: Northward from the mid-California coast, ranging inland around the north slopes to contact *E. e. platensis,* and continuing northward to central coastal British Columbia. **Can.**: BC. **U.S.:** CA, OR, WA.

Common name: Oregon Ensatina. *French Canadian name:* Salamandre variable.

Commentary: There is a zone of intergradation between this subspecies and *E. e. platensis* (Wake *et al.*, 1989). Kuchta *et al.* (2009b) found this subspecies to be genetically polyphyletic. Pereira & Wake (2009) found high genetic differentiation in the zone of overlap, even though dominated by hybrids.

Ensatina eschscholtzii picta Wood, 1940

Ensatina eschscholtzii picta Wood, 1940: 425.—**O**: **H** (od.), MVZ 27471.—**OT**: U.S., California, Del Norte County, Klamath.

Ensatina eschscholtzii picta: Stejneger & Barbour, 1943: 21; Brame, 1967: 15; Stebbins, 1985: 49; 2003: 176; Petranka, 1998: 327, 328; Raffaëlli, 2007: 325.

Ensatina eschscholtzi picta: Schmidt, 1953: 43; Cochran & Goin, 1970: 39; Gorham, 1974: 34; Behler & King, 1979: 320.—**UE.**

Distribution: A restricted coastal area of northwestern California and southwestern Oregon, in contact with *E. e. oregonensis*. **U.S.**: CA, OR.

Common name: Painted Ensatina.

Commentary: This subspecies is the least studied, probably because it is not as involved in the ring distribution.

Ensatina eschscholtzii platensis (Jiménez de la Espada, 1875)

Urotropis platensis Jiménez de la Espada, 1875a: 71.—**O**: **H**: MNCN [no number].—**OT**: Uruguay,?Montevideo (in error). Designated as U.S., California, Mariposa County, Yosemite Valley (Stebbins, 1949: 434); corrected to U.S. California, Calaveras County, North Grove Calaveras Big Trees State Park (Wake, 1993: 233), based on photos and documents.—**SS**: Myers & Carvalho, 1945: 2-5, with *E. sierrae*, a junior synonym; they maintained the Uruguay locality must be an error.

Plethodon platensis: Boulenger, 1882b: 55.

Plethodon platense: Gadow, 1901: 94.

Ensatina platensis: Dunn, 1923: 39.

Ensatina eschscholtzii platensis: Stebbins, 1949: 434; 1985: 49; 2003: 177; Brame, 1967: 15; Petranka, 1998: 327, 328; Raffaëlli, 2007: 325.

Ensatina eschscholtzi platensis: Schmidt, 1953: 43; Cochran & Goin, 1970: 39; Gorham, 1974: 34; Behler & King, 1979: 320.—**UE.**

Ensatina sierrae Storer, 1929: 448.—**O**: **H**: MVZ 10202.—**OT**: U.S., California, Mariposa County, Yosemite Valley, near Dewey Point on south rim, elev. 7300 ft.—**SS**: Myers & Carvalho, 1945: 2-5, with *E. eschscholtzii*; Schmidt, 1953: 43, with *E. e. platensis*.

Ensatina sierrae: Stejneger & Barbour, 1933: 14; 1943: 21.

Ensatina sierra: Bishop, 1943: 303.—**UE.**

Distribution: Inland California along the eastern slopes, meeting *E. e. oregonensis* at the north end of its range, the eastern population of *E. e. xanthoptica* in the middle of its range, and *E. e. croceater* at the southern end. **U.S.**: CA.

Common name: Sierra Nevada Ensatina.

Commentary: Intergrades with *E. e. oregonensis* and *E. e. croceater,* but does not interbreed extensively with *E. e. xanthoptica* (Wake *et al.*, 1989; Alexandrino *et al.*, 2005). Kuchta *et al.* (2009b) found this subspecies to be genetically polyphyletic. See also Pereira & Wake (2009) for further analysis of genetic differentiation.

Ensatina eschscholtzii xanthoptica Stebbins, 1949

Ensatina eschscholtzii xanthoptica Stebbins, 1949: 407.—**H**: MVZ 41726.—**OT**: U.S., California, Napa County, 4.5 miles east of Schellville.

Ensatina eschscholtzi xanthoptica: Schmidt, 1953: 44; Cochran & Goin, 1970: 39; Gorham, 1974: 34; Behler & King, 1979: 320.—**UE.**

Ensatina eschscholtzii xanthoptica: Brame, 1967: 15; Stebbins, 1985: 49; 2003: 176; Petranka, 1998: 327; Raffaëlli, 2007: 325.

Distribution: One population on the west slopes contacts *E. e. eschscholtzii* at the south end of its range, and *E. e. oregonensis* on the western edge of its range. Another population has apparently crossed the mountains and is on the east slopes, where it contacts *E. e. platensis*. **U.S.**: CA.

Common name: Yellow-eyed Ensatina.

Commentary: Alexandrino *et al.* (2005) provided evidence of strong selection against hybrids where *xanthoptica* and *platensis* meet in a parapatric situation; there seems to be well-developed habitat isolation and behavioral reproductive isolation, so there is no zone of intergradation. However, there are rather broad zones of intergradation where this subspecies meets *E. e. eschscholtzii* and *E. e. oregonensis,* Kuchta *et al.* (2009b) found this population to be a generally monophyletic unit, but Pereira & Wake (2009) found it to be diphyletic.

Tribe **Hydromantini** Wake, 2012

"Hydromantini" Vieites, *et al.,* 2011: 633.—**NG:** *Hydromantes* Gistel, 1848. *Nomen nudum* (see Dubois, 2012).

Hydromantini Wake, 2012: 80.—**NG:** *Hydromantes* Gistel, 1848, od.

Hydromantina: Dubois & Raffaëlli, 2012: 145.

Distribution: One genus in the U.S., and one in Europe and Asia.

Common name: Web-toed salamanders (as for our only genus).

Content: One genus in North America, with 3 North American species. Other species occur in Europe and Asia, sometimes placed in different genera. About 37.5% of the species diversity is in North America.

Genus **Hydromantes** Gistel, 1848

Hydromantes Gistel, 1848: xi.—**NS (**sd.**)**: *Salamandra genei* Schlegel, 1838, = *H platycephalus* Camp, 1916 (ICZN, 1997a).

Hydromantoides: Lanza & Vanni, 1981 (name now rejected and invalid; ICZN, 1997a).

Distribution: North American species with restricted distributions in the mountains of central, west-central, and northern California.

Common name: Web-toed Salamanders.

Content: Three species in North America.

Commentary: Gorman (1964a) provided a generic CAAR account. Wake (1966) placed this genus in his Bolitoglossinae, but some recent work suggested it is a member of the Plethodontini. Macey (2005) found the genus to be a sister-taxon of *Ensatina,* and those two share a sister group relationship to *Aneides* and desmognathines. Wake *et al.* (2005) presented a history of the nomenclature for this genus and the European ones, and proposed that all be treated as members of the genus *Hydromantes,* and the North American clade and the two European clades, be treated as subgenera, rather than full genera. The classification of Dubois and Raffaëlli (2012) treated the European species in a separate genus. That remains beyond the scope of this checklist, and is not considered here.

Hydromantes brunus Gorman, 1954

Hydromantes brunus Gorman, 1954: 153.—**O**: **H**: MVZ 59530.—**OT**: U.S., California, Mariposa County, 0.7 miles NNE Briceburg, base of low cliffs beside State Route 140, confluence of Bear Creek and Merced River, elevation 390 m.

Hydromantes brunus: Brame, 1967: 17; Cochran & Goin, 1970: 50; Gorham, 1974: 35; Behler & King, 1979: 332; Frost, 1985: 590; Stebbins, 1985: 62; 2003: 197; Petranka, 1998: 236; Dubois & Raffaëlli, 2012: 145.

Hydromantoides brunus: Lanza & Vanni, 1981: 120.

Hydromantoides (Hydromantoides) brunus: Dubois, 1984b: 108.

Hydromantes (Hydromantes) brunus: Wake, Salvador, & Alonso-Zarazaga, 2005: 543-548; Raffaëlli, 2007: 318.

Distribution: Vicinity of the type locality, along the Lower Merced River drainage. Gorman (1964b) and Petranka (1998) have maps. **U.S.**: CA.

Common name: Limestone Salamander.

Commentary: Gorman (1964b) provided a brief CAAR account. Rovito (2010) showed that this is a monophyletic taxon, but is included within a clade composed of some populations of *H. platycephalus,* suggesting to us that it may be best treated as a subspecies of the latter, but pending further study we will treat it as a separate species.

Hydromantes platycephalus (Camp, 1916)

Spelerpes platycephalus Camp, 1916a: 11.—**O**: **H**: MVZ 5693, but **S:** MVZ 5691-5694, *fide* Dunn, 1926b: 354, sd.; however, Camp stated the only material he used was the H and one paratype (topotype), MVZ 5694.—**OT**: U.S., California, Tuolumne County, Yosemite National Park, head of Lyell Canyon, 10,800 feet elevation.—**Comment**: Placed on the official list of specific names (International Commission on Zoological Nomenclature, 1997a).

Eurycea platycephala: Stejneger & Barbour, 1917: 20.

Hydromantes platycephalus: Dunn, 1923: 40; Stejneger & Barbour, 1943: 34; Schmidt, 1953: 56; Brame, 1967: 17; Cochran & Goin, 1970: 48; Gorham, 1974: 35; Behler & King, 1979: 333; Frost, 1985: 591; Stebbins, 1985: 61; 2003: 195; Petranka, 1998: 237; Dubois & Raffaëlli, 2012: 145.

Hydromantoides platycephalus: Lanza & Vanni, 1981: 120

Hydromantoides (Hydromantoides) platycephalus: Dubois, 1984b: 108.

Hydromantes (Hydromantes) platycephalus: Wake, Salvador, & Alonso-Zarazaga, 2005: 543-548; Raffaëlli, 2007: 318.

Distribution: West-central California, at high elevations in the Sierra Nevada. Gorman (1964b) and Petranka (1998) have range maps, although the range has been expanded by more recent records. **U.S.**: CA.

Common name: Mount Lyell Salamander.

Commentary: Gorman (1964b) provided a brief CAAR account. Rovito (2010) found this species to be composed of two divergent, monophyletic clades, so more than the single species may later be recognized.

Hydromantes shastae Gorman & Camp, 1953

Hydromantes shastae Gorman & Camp, 1953: 39.—**O**: **H**: MCZ 52314.—**OT**: U.S., California, Shasta County, 0.7 miles east of Squaw Creek Road, 18.4 miles north and 15.3 mi. east of Redding, entrance to limestone caves at the edge of Flat Creek Road in the narrows of Low Pass Creek, 460 m elevation.

Hydromantes shastae: Brame, 1967: 17; Cochran & Goin, 1970: 50; Gorham, 1974: 35; Behler & King, 1979: 333; Frost, 1985: 591; Stebbins, 1985: 61; 2003: 196; Petranka, 1998: 239; Dubois & Raffaëlli, 2012: 145.

Hydromantoides shastae: Lanza & Vanni, 1981: 120.

Hydromantoides (Hydromantoides) shastae: Dubois, 1984b: 108.

Hydromantes (Hydromantes) shastae: Wake, Salvador, & Alonso-Zarazaga, 2005: 543-548; Raffaëlli, 2007: 319.

Distribution: Isolated populations near Shasta Lake, Shasta County, California. Gorman (1964b) and Petranka (1998) have maps. **U.S.**: CA.

Common name: Shasta Salamander.

Commentary: Gorman (1964b) provided a brief CAAR account.

Tribe **Plethodontini** Gray, 1850

Plethodontidae Gray, 1850: 31.—**NG**: *Plethodon* Tschudi, 1838. Nomen placed on the official list of family-group names (ICZN, 1970).

Plethodontina: Gray, 1850: 38; Schultze, 1891: 169.

Plethodontinae: Boulenger, 1882b: 1.

Plethodontida: Knauer, 1878: 97.

Plethodontini: Wake, 1966: 51; Dubois, 2005b: 20; Vieites *et al.*, 2011: 633.

Distribution: As for the genus *Plethodon.*

Content: One genus, endemic to North America. Wake (1966) included three genera in this tribe, then *Hydromantes* (plus the European and Korean forms) were also moved here from Wake's Bolitoglossini. But Vieites *et al.* (2011) and Wake (2012) transferred the other genera to new tribes, as treated here.

Common name: Woodland Salamanders.

Genus **Plethodon** Tschudi, 1838

Synonymy: See the subgenus *Plethodon,* below.

Distribution: North America, but with a central gap in the Great Plains region.

Common name: Woodland Salamanders (also sometimes called Slimy Salamanders)

Content: 55 species, all North American. We recognize two subgenera.

Subgenus **Plethodon** Tschudi, 1838

Plethodon Tschudi, 1838: 58.—**NS**: *Salamandra glutinosa* Green, 1818 (Bibron *in* Bonaparte, 1839c; see ICZN, 1970, sd.).—**Comment**: Placed on the official list of generic names (ICZN, 1970). Vieites *et al.* (2011: 632) designated the subgenus.

Sauropsis Fitzinger, 1843: 33.—**NS (**od.**)**: *Salamandra erythronota* Rafinesque, 1818.—**SS**: Duméril *et al.* (1854c: 86), in synonymy of *Plethodon erythronotum*; Boulenger, 1882b: 53.

Saurophis Gray, 1850: 35.—**IS**.

Distribution: Most of eastern North America.

Common name: Eastern Woodland Salamanders.

Content: 46 species, generally arranged into four or five species-groups. The arrangement here is based on the initial arrangement into groups by Highton (1962), and subsequent modifications by various authors, as new species were described, others reduced to junior synonyms, or changes indicated by further studies. We include them in five species groups.

Plethodon glutinosus group. 22 species form a closely related group, and are arranged in approximate phylogenetic order, from most derived to most basal, according to the analysis of Wiens *et al.* (2006). The first 11 species compose Clade A of that study, the other 11 species make up a monophyletic Clade B of that study. Clade A: *P. ocmulge, P. savannah, P. grobmani, P. kisatchie, P. albagula, P. sequoyah, P. kiamichi, P. mississippi, P. glutinosus, P. aureolus, P. jordani.* Clade B: *P. chlorobrionis, P. variolatus, P. teyahalee, P. cylindraceus, P. shermani, P. chattahoochee, P. cheoah, P. amplus, P. meridianus, P. metcalfi, P. montanus.* The last 4 species of Clade B together are a sister clade to the other 7 species.

Plethodon ouachitae group. Three species of larger size, closely related to the *glutinosus* group: *P. ouachitae, P. fourchensis, P. caddoensis.*

Plethodon yonahlossee group. Four smaller species forming a clade basal to the *glutinosus* group: *P. petraeus, P. kentucki, P. yonahlossee, P. ainsworthi.*

Plethodon wehrlei-welleri group. Formerly treated as two separate species-groups, but the analysis of Wiens *et al.* (2006) found that the *wehrlei* group was nested within the *welleri* group. The analysis of Mahoney (2001) indicates that this group is a sister-taxon to the *glutinosus* group. It currently includes seven species: *P. wehrlei, P. punctatus, P. dorsalis, P. ventralis, P. angusticlavius, P. welleri, P. websteri.*

Plethodon cinereus group. The analysis of Mahoney (2001) indicates that this clade is a sister group to a clade containing all the other eastern *Plethodon* (*i.e.,* all other members of the subgenus *Plethodon*). The work of Wiens *et al.* (2006) agrees, and Kozak *et al.* (2006) partially agree. The group contains ten species: *P. richmondi, P. electromorphus, P. hubrichti, P. nettingi, P. hoffmani, P. virginia, P. cinereus, P. shenandoah, P. serratus, P. sherando.*

Commentary: In Dunn's (1926b) classic monograph of the family, he recognized 11 species of *Plethodon*. Currently that number has grown to the 55 that are recognized here, in two subgenera.

Plethodon (Plethodon) ainsworthi Lazell, 1998

Plethodon ainsworthi Lazell, 1998: 967.—**O**: **H**: MCZ 125869.—**OT**: U.S., Mississippi, Jasper County, 2 miles south of Bay Springs.

Plethodon ainsworthi: Raffaëlli, 2007: 293.

Plethodon (Plethodon) ainsworthi: Vieites *et al.*, 2011: 632, by inference; Wake, 2012: 80, by inference; Dubois & Raffaëlli, 2012: 146.

Distribution: Known only from the vicinity of the type locality. **U.S.**: MS.

Common name: Bay Springs Salamander.

Commentary: Lazell (1998) said this species does not seem related to any other *Plethodon*. Raffaëlli (2007) suggested it may be in the *P. glutinosus* group, but placement there is provisional; it has not been included in any phylogenetic analyses. We tentatively include it in the *Plethodon yonahlossee* group. Recent attempts at collection of the species (Lazell, 2005; Folt *et al.*, 2013) have been unsuccessful and it may be extinct.

Plethodon (Plethodon) albagula Grobman, 1944

Plethodon glutinosus albagula Grobman, 1944: 283.—**O**: **H**: CM 9652.—**OT**: U.S., Texas, Bexar County, 20 miles north of San Antonio.

Plethodon glutinosus albagula: Schmidt, 1953: 34; Brame, 1967: 15; Cochran & Goin, 1970: 59; Gorham, 1974: 37; Conant, 1975: 277; Behler & King, 1979: 340; Garrett & Barker, 1987: 65.

Plethodon albagula: Highton, 1989: 71; Conant & Collins, 1998: 471; Dixon, 2000: 57; Raffaëlli, 2007: 305.

Plethodon (Plethodon) albagula: Vieites *et al.*, 2011: 632, by inference; Dubois & Raffaëlli, 2012: 146.

Distribution: A disjunct population in central Texas, and a more extensive population in eastern Oklahoma, much of Arkansas, and southern Missouri. **U.S.**: AR, MO, OK, TX.

Common name: Western Slimy Salamander.

Commentary: In Highton's (1989) study, this species was found to be genetically a basal sister to the rest of the *glutinosus* group and seemed closest to *P. grobmani.* But both Wiens *et al.* (2006) and Kozak *et al.* (2006) found this species to be a sister to *P. sequoyah.* Petranka (1998) did not recognize this species as distinct from *P. glutinosus.* Frost & Hillis (1990) suggested there may be multiple cryptic species included in the present concept of a single species. Baird *et al.* (2006) found up to five separate lineages among central Texas populations, indicating that further study may show undescribed cryptic species. The more recent study of Davis & Pauly (2011) supports this view further, suggesting even more possible species may be living under this name.

Plethodon (Plethodon) amplus Highton & Peabody, 2000

Plethodon amplus Highton & Peabody, 2000: 60.—**O**: **H**: USNM 446296.—**OT**: U.S., North Carolina, Henderson County, south slope of Little Pisgah Mountain (35° 29' 42" N, 82° 20' 08" W), elevation 1109 m.

Plethodon amplus: Rafaelli, 2007: 306.

Plethodon (Plethodon) amplus: Vieites *et al.*, 2011: 632, by inference; Dubois & Raffaëlli, 2012: 146.

Distribution: Blue Ridge Mountains in North Carolina. **U.S.**: NC.

Common name: Blue Ridge Gray-cheeked Salamander.

Commentary: The analysis of Highton & Peabody (2000) indicated that this is a sister-species of *P. metcalfi.* Wiens *et al.* (2006) found it to be a sister species of *P. meridianus,* as did Kozak *et al.* (2006).

Plethodon (Plethodon) angusticlavius Grobman,1944

Plethodon cinereus angusticlavius Grobman, 1944: 302.—**O**: **H**: AMNH 40366.—**OT**: U.S., Missouri, Stone County, Mud Cave, near Fairy Cave.

Plethodon cinereus angusticlavius: Schmidt, 1953: 33.

Plethodon dorsalis angusticlavius: Thurow, 1956a: 177; 1966: 1; Brame, 1967: 15; Cochran & Goin, 1970: 55; Gorham, 1974: 37; Conant, 1975: 274; Behler & King,

1979: 337; Conant & Collins, 1998: 468; Petranka, 1998: 347.

Plethodon angusticlavius: Collins, 1991: 43; Highton, 1997: 354; Raffaëlli, 2007: 299.

Plethodon (Plethodon) angusticlavius: Vieites *et al.*, 2011: 632, by inference; Dubois & Raffaëlli, 2012: 146.

Distribution: Northeastern Arkansas and adjacent Missouri and Oklahoma. Thurow (1966) has a range map for the species as a subspecies of *P. dorsalis,* as does Petranka (1998). Meshaka & Trauth (2006) have a map as a full species. **U.S.**: AR, MO, OK.

Common name: Ozark Zigzag Salamander.

Commentary: Originally described as a subspecies of *P. cinereus,* then recognized as a subspecies of *P. dorsalis,* eventually Highton (1997) provided protein-electrophoretic support for full species status, though Petranka (1998) still retained the form as a subspecies of *P. dorsalis*. Meshaka & Trauth (2006) provided a CAAR account.

Plethodon (Plethodon) aureolus Highton, 1984

Plethodon aureolus Highton, 1984: 2.—**O**: **H**: USNM 238341.—**OT**: U.S., Tennessee, Monroe County, Unicoi Mountains, Farr Gap.

Plethodon aureolus: Conant & Collins, 1998: 472; Petranka, 1998: 332; Raffaëlli, 2007: 305.

Plethodon (Plethodon) aureolus: Vieites *et al.*, 2011: 632, by inference; Dubois & Raffaëlli, 2012: 146.

Misapplication of names:

Plethodon glutinosus: Highton, 1970; Peabody, 1978.

Distribution: Restricted range in southeastern Tennessee and southwestern North Carolina. See range maps in Highton (1986a) and Petranka (1998). **U.S.**: NC, TN.

Common name: Tellico Salamander.

Commentary: Highton (1986a) provided a CAAR account. In the study by Highton (1989), this species was near basal genetically to the rest of the *glutinosus* group, and genetically closest to *P. mississippi*. It was also one of the two most differentiated species of the *glutinosus* complex.

Plethodon (Plethodon) caddoensis Pope & Pope, 1951

Plethodon caddoensis Pope & Pope, 1951: 148.—**O**: **H**: CNHM 61959.—**OT**: U.S., Arkansas, southwestern Montgomery County, Caddo Mountains, on Polk Creek Mountain, elev. 1200 feet.

Plethodon caddoensis: Schmidt, 1953: 37; Cochran & Goin, 1970: 58; Gorham, 1974: 37; Conant, 1975: 280; Behler & King, 1979: 335; Frost, 1985: 595; Conant & Collins, 1998: 476; Petranka, 1998: 333; Raffaëlli, 2007: 309.

Plethodon (Plethodon) caddoensis: Vieites *et al.*, 2011: 632, by inference; Dubois & Raffaëlli, 2012: 146.

Misapplication of names:

Plethodon ouachitae Grobman, 1944 (part).

Distribution: Caddo Mountains in west central Arkansas. Pope (1964) and Petranka (1998) have range maps. **U.S.**: AR.

Common name: Caddo Mountain Salamander.

Commentary: Pope (1964) provided a CAAR account.

Plethodon (Plethodon) chattahoochee Highton, 1989

Plethodon chattahoochee Highton, 1989: 55.—**O**: **H** (od.), USNM 168527.—**OT**: U.S., Georgia, Towns County, 0.3 km east of the top of Brasstown Bald, 34° 52' 21" N, 83° 48' 31" W, elevation 1353 m (Locality 9).

Plethodon chattahoochee: Conant & Collins, 1998: 471; Raffaëlli, 2007: 302.

Plethodon (Plethodon) chattahoochee: Vieites *et al.*, 2011: 632, by inference; Dubois & Raffaëlli, 2012: 146.

Distribution: Blue Ridge area of northern Georgia and southwestern North Carolina. See map in Beamer & Lannoo (2005a). **U.S.**: GA, NC.

Common name: Chattahoochee Slimy Salamander.

Commentary: Petranka (1998) did not recognize this species. Some of the analyses of Highton (1989) indicated this as a sister-taxon to the pair of *chlorobryonis-variolatus*. Of the species they used, Highton & Peabody (2000) found this species to be a sister-species to *P. chlorobryonis,* and that pair a sister group to *glutinosus+cheoah*. Wiens, *et al.* (2006) found this species is a sister to *P. cheoah,* and that pair is a sister to a clade containing *chlorobryonis-variolatus* and several other species. Kozak, *et al.* (2006) found a somewhat similar relationship for this species. Beamer & Lannoo (2005a) presented an account.

Plethodon (Plethodon) cheoah Highton & Peabody, 2000

Plethodon cheoah Highton and Peabody, 2000: 62.—**O**: **H**: USNM 459012.—**OT**: U.S., North Carolina, Graham-Swain County line, 0.6 km southwest of the top of Cheoah Bald, Bellcollar Gap (35° 19' 20" N, 83° 41' 11" W), elev. 1445 m.

Plethodon cheoah: Raffaëlli, 2007: 308.

Plethodon (Plethodon) cheoah: Vieites *et al.*, 2011: 632, by inference; Dubois & Raffaëlli, 2012: 146.

Distribution: Vicinity of type locality. **U.S.**: NC.

Common name: Cheoah Bald Salamander.

Commentary: Sister-taxon to *P. glutinosus* in the study by Highton and Peabody (2000). Wiens *et al.* (2006) found this a sister species to *P. chattahoochee*. Kozak *et al.* (2006) found this to be a sister-species of *P. shermani*.

Plethodon (Plethodon) chlorobryonis Mittleman, 1951

Plethodon glutinosus chlorobryonis Mittleman, 1951: 108.—**O**: **H:** USNM 129933 (Cochran, 1961a: 20, sd.); verified by Highton (1989: 58).—**OT**: U.S., North Carolina, Craven County, 13 miles north of New Bern, along US Highway 17, in the dry bottomlands along a small creek.

Plethodon glutinosus chlorobryonis: Schmidt, 1953: 35.

Plethodon chlorobryonis: Highton, 1989: 58-59; Conant & Collins, 1998: 471; Raffaëlli, 2007: 303.

Plethodon (Plethodon) chlorobryonis: Vieites *et al.*, 2011: 632, by inference; Dubois & Raffaëlli, 2012: 146.

Distribution: Coastal plain, northeastern Georgia to southeastern Virginia. See map in Beamer and Lannoo (2005b). **U.S.**: GA, NC, SC, VA.

Common name: Atlantic Coast Slimy Salamander.

Commentary: The analysis of Highton (1989) suggests this is a sister species of *P. variolatus*; this relationship was supported by Wiens, *et al.* (2006), and partially supported by Kozak, *et al.* (2006). Gorham (1974) and Petranka (1998) considered this name in the synonymy of *P. glutinosus*. Beamer and Lannoo (2005b) presented an account.

Plethodon (Plethodon) cinereus (Green, 1818)

"Salamandra erythronota" Rafinesque, 1818a: 25 (March).—**O: S:** ANSP 1227-1238 (Fowler & Dunn, 1917: 24, sd.); verified by Malnate (1971: 347).—**OT**: U.S., New York, the highlands; revised to Hudson Highlands of New York (Highton, 1962: 285).—**Comment**:

suppressed for purposes of priority, placed on Official Index of Rejected and Invalid Specific Names (ICZN, 1963b: 199).

Salamandra erythronota: Harlan, 1827: 329; Holbrook, 1840: 113, 1842: 5: 43.

Sauropsis erythronotus: Fitzinger, 1843: 33.

Plethodon erythronota: Baird, 1850a: 285.

Ambystoma erythronota: Gray, 1850: 37.

Plethodon erythronotus: Baird, 1850a: 285; Hallowell, 1857a: 11; Boulenger, 1882b: 57.

Plethodon erythronotum: Duméril, Bibron, & Duméril, 1854c: 86.

Saurophis erythronotus: Fitzinger, 1864: pl. 98, fig. 176.

Plethodon erythronotus var. *erythronotus*: Cope, 1869b: 100.

Spelerpes erythronota: Kennicott *in* Anonymous, 1869: 144.

Plethodon cinereus erythronotus: Yarrow, 1883a: 154; Davis & Rice, 1883: 27; Cope, 1889b: 135.

Plethodon cinereus erthronota: Britcher, 1903: 120.

Salamandra cinerea Green, 1818 (Sept.): 356.—**O**: **S**: ANSP 1227-1238 (Fowler & Dunn, 1917: 24, sd.); only ANSP 1232-1234 and 1237 *fide* Highton (1962: 286).—**L**: ANSP 1232 (Highton, 1962: 286).—**OT**: U.S., New Jersey; revised to near Princeton (Fowler, 1907: 57).—**Comment**: Name conserved and placed on official list of specific names (ICZN, 1963b: 199).

Salamandra cinerea: Harlan, 1826b: 330; Haldeman, 1848: 315.

Plethodon cinereus: Tschudi, 1838: 92 (as first reviser, selected *S. cinerea* to have priority over *S. erythronota*); Werner, 1909: 33; Stejneger & Barbour, 1917: 15; Bishop, 1924: 89; 1928a; Dunn, 1926b: 163; 159; Muchmore, 1955: 170; Behler & King, 1979: 336; Frost, 1985: 595; Conant & Collins, 1998: 462; Petranka, 1998: 335; Raffaëlli, 2007: 293.

Plethodon erythronotus cinereus Cope, 1869b: 100; Brimley, 1907: 154.

Plethodon cinereus cinereus: Yarrow, 1883a: 154; Davis & Rice, 1883: 27; Cope, 1889b: 134; Stejneger & Barbour, 1943: 16; Schmidt, 1953: 33; Cochran & Goin, 1970: 56; Gorham, 1970: 3; 1974: 37; Conant, 1975: 272.

Plethodon (Plethodon) cinereus: Vieites *et al.*, 2011: 632, by inference; Dubois & Raffaëlli, 2012: 146.

Salamandra agilis Sager, 1839: 322.—**O**: **S**: USNM 3770 (15) (Dunn, 1926b: 179, sd.); Cochran, 1961a: 24, stated "possibly".—**OT**: not stated but presumably U.S., Michigan, Wayne County, Detroit.—**SS**: Cope, 1889b: 133; Highton, 1962: 285.

"Salamandra puncticulata" Duméril, Bibron, & Duméril, 1854c: 87. **NU**, attributed to Valenciennes, in description of *Plethodon erythronotum* Tschudi, 1838.—**SS**: Duméril *et al.* (1854c: 87).

Plethodon huldae Grobman, 1949: 136.—**O**: **H**: USNM 127955.—**OT**: U.S., Virginia, Madison County, 100 yards from Skyline Drive, along the foot trail to Hawksbill Mountain, elev. approx. 3500 ft.—**SS**: Muchmore (1955: 172).

Distribution: Northeastern quarter of the U.S., and southeastern Canada. See map in Petranka (1998). The map in Smith (1963) includes disjunct populations now considered to be a separate species, *P. serratus* (those labeled as subspecies *P. c. serratus* and *P. c. polycentratus,* plus the population in Missouri). **Can.**: NB, NS, ON, PE, QC. **U.S.**: CT, DE, IL, IN, MA, MD, ME, MI, MN, NC, NH, NJ, NY, OH, PA, RI, VA, VT, WI, WV.

Common name: Eastern Red-backed Salamander (SECSN) or Northern Redback Salamander (C&T). *French Canadian name:* Salamandre rayée.

Content: Currently there are no subspecies recognized, but some populations currently treated as full species were formerly considered subspecies of *P. cinereus.*

Commentary: Fowler (1907) revised the type locality to "near Princeton?", and Schmidt (1953) further revised it to vicinity of Princeton, New Jersey; we consider those restrictions invalid, as they provided no evidence the S came from there. Smith (1963) provided a CAAR account and discussed the history of the two oldest synonyms. Wiens *et al.* (2006) and Kozak *et al.* (2006) found this to be a sister species to *P. shenandoah.*

Plethodon (Plethodon) cylindraceus (Harlan, 1825)

Salamandra cylindracea Harlan, 1825b: 156.—**O**: none known to exist, though Harlan (1827) referenced specimens in ANSP.—**OT**: U.S., South Carolina.—**N**: USNM 257522 (Highton, 1989: 71);—**OT(**of **N)**: U.S., South Carolina, Chester County, locality 112 (34° 32' 10" N,

81° 95' 36" W), elevation 137 m.—**Comment**: Schmidt (1953: 34) restricted the original OT to vicinity of Charleston, South Carolina, but gave no evidence the O came from there, so the action is invalid.

Plethodon cylindraceus Highton, 1989: 70; Conant & Collins, 1998: 471; Raffaëlli, 2007: 304.

Plethodon (Plethodon) cylindraceus: Vieites *et al.*, 2011: 632, by inference; Dubois & Raffaëlli, 2012: 146.

Distribution: Much of western North Carolina plus adjacent Georgia and into Virginia and eastern West Virginia. Beamer & Lannoo (2005c) have a range map. **U.S.**: GA, NC, SC, VA, WV.

Common name: White-spotted Slimy Salamander.

Commentary: Cope (1889b: 139) treated this species as a synonym of *P. glutinosus,* as did Dunn (1926b: 137) and Gorham (1974: 37); Schmidt (1953: 34) treated it as a synonym of *P. g. glutinosus.* Petranka (1998) did not recognize this species as distinct from *P. glutinosus.* In contrast to the studies cited in the account above, Highton (1989) found this species to be closest to *P. grobmani,* while the analysis of Highton & Peabody (2000) indicated this is a sister-taxon to the *shermani-teyahalee* pair. The analysis of Pyron & Wiens (2011) showed this to be a sister of the clade including *P. teyahalee+aureolus+cheoah,* in the *P. glutinosus* complex.

Plethodon (Plethodon) dorsalis Cope, 1889

"Plethodon cinereus dorsalis" Cope, 1869b: 100.—**NU**.

"Plethodon erythronotus var. *dorsalis"* Baird *in* Cope, 1869b: 100.—**NU**.

Plethodon cinereus dorsalis Cope, 1889b: 138.—**O**: **S**: USNM 3776A-3776D (4);—**L**: USNM 3776A (Highton, 1962: 277).—**OT**: U.S., Kentucky, Jefferson County, Louisville.

Plethodon dorsalis: Stejneger & Barbour, 1917: 15; Schmidt, 1953: 34; Gorham, 1974: 37; Frost, 1985: 596; Raffaëlli, 2007: 298.

Plethodon cinereus dorsalis: Bishop, 1943: 236; Stejneger & Barbour, 1943: 16.

Plethodon dorsalis dorsalis: Thurow, 1956a: 179; Brame, 1967: 15; Cochran & Goin, 1970: 55; Conant, 1975: 273; Behler & King, 1979: 337; Conant & Collins, 1998: 468; Petranka, 1998: 347.

Plethodon (Plethodon) dorsalis: Vieites *et al.*, 2011: 632, by inference; Dubois & Raffaëlli, 2012: 146.

Distribution: Northern Alabama through Tennessee and Kentucky into Indiana, plus adjacent Illinois and North Carolina. See map of the nominotypical subspecies in Thurow (1966) and Petranka (1998). **U.S.**: AL, IL, IN, KY, NC, TN.

Common name: Northern Zigzag Salamander (SECSN) or Zigzag Salamander (C&T).

Commentary: Salamanders treated as *P. erythronotus* by Garman (1894: 38) are apparently this species (Dunn, 1926b: 158). Frost (2009-13) also listed two *nomina nuda* in his synonymy. Thurow (1966) provided a CAAR account, which included *P. angusticlavius* as a subspecies.

Plethodon (Plethodon) electromorphus Highton, 1999

Plethodon electromorphus Highton, 1999: 66.—**O**: **H**: USNM 507747.—**OT**: U.S., West Virginia, Gilmer County, Cedar Creek State Park, 274 m elevation (Locality 35).

Plethodon electromorphus: Raffaëlli, 2007: 295.

Plethodon (Plethodon) electromorphus: Vieites *et al.*, 2011: 632, by inference; Dubois & Raffaëlli, 2012: 146.

Distribution: Ohio and adjacent parts of Indiana, West Virginia, and Pennsylvania. See range map in Regester (2000a). **U.S.**: IA, OH, PA, WV.

Common name: Northern Ravine Salamander.

Commentary: Regester (2000a) provided a CAAR account.

Plethodon (Plethodon) fourchensis Duncan & Highton, 1979

Plethodon fourchensis Duncan & Highton, 1979: 109.—**O**: **H**: USNM 204835.—**OT**: U.S., Arkansas, Polk County, 1.5 km west, 0.3 km south of the top of Wolf Pinnacle Mountain.

Plethodon fourchensis: Frost, 1985: 596; Conant & Collins, 1998: 476; Raffaëlli, 2007: 309.

Plethodon (Plethodon) fourchensis: Vieites *et al.*, 2011: 632, by inference; Dubois & Raffaëlli, 2012: 146.

Distribution: Western Arkansas, Forche and Iron Fork Mountains. See range map of Highton (1986e); Petranka (1998) did not recognize this species. **U.S.**: AR.

Common name: Fourche Mountain Salamander.

Commentary: Highton (1986e) provided a CAAR account. Petranka (1998) considered this form to be conspecific with *P. ouachitae.* Other authors have referred to salamanders of this population as *P. ouachitae,* prior to its description as a separate species (*e.g.,* Blair & Lindsay, 1965).

Plethodon (Plethodon) glutinosus (Green, 1818)

Salamandra glutinosa Green, 1818: 357.—**O**: not stated, probably originally in ANSP, but not known to exist (Dunn, 1926b: 138).—**OT**: not stated but U.S., New Jersey, vicinity of Princeton, *fide* Stejneger & Barbour (1917: 10) and Dunn (1926b: 138).—**Comment:** Placed on the official list of specific names (ICZN, 1970).

Salamandra glutinosa: Harlan, 1827: 330; Holbrook, 1839b: 129; 1842: 5: 39.

Plethodon glutinosus: Tschudi, 1838: 92; Fitzinger, 1843: 34; Baird, 1850a: 285; Hallowell, 1857a: 11; Cope, 1869b: 100; 1889b: 139; Knauer, 1878: 97; Boulenger, 1882b: 56; Garman, 1894: 38; Stejneger & Barbour, 1917: 16; Dunn, 1926b: 136; Bishop, 1924: 89; 1928a: 159; Viosca, 1949: 9; Frost, 1985: 596; Conant & Collins, 1998: 471; Petranka, 1998: 354; Raffaëlli, 2007: 303.

Plethodon glutinosum: Gray, 1850: 39.

Cylindrosoma glutinosum: Duméril, Bibron, & Duméril, 1854c: 80.

Plethodon glutinosus glutinosus: Dunn, 1920c: 131; Stejneger & Barbour, 1943: 17; Bishop, 1943: pp; Grobman, 1944: 266; Schmidt, 1953: 34; Cochran & Goin, 1970: 59; Gorham, 1974: 37; Conant, 1975: 276; Behler & King, 1979: 340; Petranka, 1998: pp.

Plethodon (Plethodon) glutinosus: Vieites *et al.*, 2011: 632, by inference; Dubois & Raffaëlli, 2012: 146.

?*"Cylindrosoma albopunctata"* Duméril, Bibron, & Duméril, 1854c: 81.—**O**: MNHN.—**OT**: U.S., Georgia, "Savanah". The authors indicated this was a *nomen nudum* and referred it to the synonymy of *Cylindrosoma glutinosum*;—**SS:** Dunn, 1926b: 137. Frost (2009-13) suggested it may not be a synonym of *P. glutinosus* because the OT is beyond the known range of this species.

?*Plethodon variolosum* Duméril, Bibron, & Duméril, 1854c: 83.—**O**: **S** (sd., Thireau, 1986: 80) MNHN 4666 (2).—**OT**: Etats-Unis, l'Amérique sept., corrected to U.S., South Carolina by Thireau, 1986:

80.—**SS:** Hallowell, 1858b: 343. Frost (2009-13) suggests this may not be a synonym of *P. glutinosus* because the OT is beyond the known range of that species.

Plethodon melanoleuca Wied, 1865: 130.—**O**: none known to exist.—**OT**: U.S., Pennsylvania, Northhampton County, Nazareth.—**SS**: Dunn, 1926b: 137.

Amblystoma melanoleuca Boulenger, 1882b: 38.

Distribution: Eastern United States. Petranka (1998) has a range map, but it includes several forms that have been separated as distinct species by others, as noted in those accounts. Beamer & Lannoo (2005d) have an updated map. **U.S.:** AL, AR, CT, GA, IL, IN, KY, LA, MD, MO, MS, NC, NH, NJ, NY, OH, PA,?SC, TN, TX, VA, WV.

Common name: Northern Slimy Salamander.

Content: No subspecies are currently recognized, but some of the full species now recognized were formerly considered subspecies of this species.

Commentary: In contrast to Highton's (1989) analysis, which found this species closest to *P. ocmulgee,* Wiens *et al.* (2006) found this to be a sister species of *P. aureolus,* and that pair to have a sister-taxon relationship to a clade of several other species of the group. Kozak *et al.* (2006) found results congruent with that.

Plethodon (Plethodon) grobmani Allen & Neill, 1949

Plethodon glutinosus grobmani Allen & Neill, 1949: 112.—**O**: **H**: ERA-WTN 19220; present deposition not verified.—**OT**: U.S., Florida, Marion County, about one-half mile northeast of Silver Springs, about 1/4 mile west of the junction of the creek and Florida State Highway 40, Half-mile Creek Swamp.

Plethodon glutinosus grobmani: Schmidt, 1953: 35.

Plethodon grobmani: Highton, 1989: 69; Conant & Collins, 1998: 471; Raffaëlli, 2007: 304.

Plethodon (Plethodon) grobmani: Vieites *et al.*, 2011: 632, by inference; Dubois & Raffaëlli, 2012: 146.

Distribution: Central Florida along the Gulf coast to Alabama. See map in Beamer & Lannoo (2005e). **U.S.**: AL, FL.

Common name: Southeastern Slimy Salamander.

Commentary: Highton (1989) indicated a sister-species relationship to *P. cylindraceus*. That study also indicated that this species (together with *P. cylindraceus*) seemed to have a sister-taxon relation to *P. kiamichi* + *P. mississippi* + *P. kisatchie* + *P. sequoyah*. Wiens *et al.* (2006) found this is a sister species to *P kisatchie,* while Kozar *et al.* (2006) found this species to be a sister to *P. ocmulgee*. Carr (1996) provided evidence suggesting that the population may represent more than a single species. *Gorham (1974) and* Petranka (1998) did not recognize this species as distinct from *P. glutinosus.*

Plethodon (Plethodon) hoffmani Highton, 1972

Plethodon hoffmani Highton, 1972: 151.—**O**: **H**: USNM 135203.—**OT**: U.S., Virginia, Allegheny County, Clifton Forge.

Plethodon richmondi hoffmani: Smith, 1978: 118. Listing as subspecies may be a *lapsus,* as he stated this form is now regarded as a full species.

Plethodon hoffmani: Behler & King, 1979: 341; Frost, 1985: 596; Conant & Collins, 1998: 466; Petranka, 1998: 361; Raffaëlli, 2007: 294.

Plethodon (Plethodon) hoffmani: Vieites *et al.*, 2011: 632, by inference; Dubois & Raffaëlli, 2012: 146.

Misapplied names:

Plethodon richmondi Netting & Mittleman, 1938 (part).

Plethodon cinereus cinereus Hoffman, 1945: 200 (part).

Distribution: Western Virginia and eastern West Virginia, north into central Pennsylvania. Highton (1986f) and Petranka (1998) have range maps. **U.S.**: PA, VA, WV.

Common name: Valley and Ridge Salamander.

Commentary: Highton (1986f) provided a CAAR account. Mahoney (2001) indicated that *P. cinereus* and *P. hoffmani* are sister-species, but Highton (1999) placed *P. hoffmani* as a sister-species of *P. virginia* (though at a low level of support). Sites, *et al.* (2004) found agreement with Highton, that this species is closest to *P. virginia*. Wiens *et al.* (2006) and Kozak *et al.* (2006) also found this to be the sister species of *P. virginia,* and that pair to be sister to the clade formed by *P. electromorphus* + *P. hubrichti* + *P. nettingi* + *P. richmondi.*

Plethodon (Plethodon) hubrichti Thurow, 1957

Plethodon hubrichti Thurow, 1957: 59.—**O**: **H**: USNM 139087 (ex GRT 1221).—**OT**: U.S., Virginia, Bedford County near the Bedford-Botecourt County line, roughly 10 miles ESE of Buchanan, by the Blue Ridge Parkway, 0.9 miles south of cement milepost 80 (and sign reading "view of Black Rock Hill"), about 3100 feet elevation.

Plethodon richmondi hubrichti: Highton, 1962: 307; Brame, 1967: 16; Cochran & Goin, 1970: 57; Conant, 1975: 232.

Plethodon nettingi hubrichti: Thurow, 1968: 16; Behler & King, 1979: 345.

Plethodon hubrichti: *Gorham, 1974: 37;* Highton, 1986g: 393.1; Conant & Collins, 1998: 465; Petranka, 1998: 363; Raffaëlli, 2007: 297.

Plethodon (Plethodon) hubrichti: Vieites *et al*., 2011: 632, by inference; Dubois & Raffaëlli, 2012: 146.

Distribution: Blue Ridge Mountains of central Virginia. See map in Highton (1986g). **U.S.**: VA.

Common name: Peaks of Otter Salamander.

Commentary: Highton (1986g) provided a CAAR account. The analysis of Highton (1999) indicated a sister-species relationship with *P. nettingi,* or a remote relationship with *P. serratus* and the other species of the group (support levels were low for both). The analyses by Sites, *et al.* (2004) and Kozak *et al.* (2006) agreed with the former alternative, that this is a sister-species to *P. nettingi.* Wiens *et al.* (2006) found this is a sister species to the *richmondi-electromorphus* pair.

Plethodon (Plethodon) jordani Blatchley, 1901

Plethodon jordani Blatchley, 1901: 762.—**O**: **S**: (2) lost, originally in Blatchley's private collection (Dunn, 1926b: 145).—**OT**: U.S., Tennessee, Sevier County, on Mt. Collins and Indian Pass, 3000-5000 feet.

Plethodon jordani: Stejneger, 1906: 560; Stejneger & Barbour, 1917: 16; 1943: 18; Dunn, 1920c: 131; Bishop, 1928a: 159; Cochran & Goin, 1970: 59; Gorham, 1974: 37; Conant, 1975: 280; Behler & King, 1979: 341; Frost, 1985: 597; Conant & Collins, 1998: 477; Petranka, 1998: 367; Raffaëlli, 2007: 307.

Plethodon jordani jordani: Hairston, 1950: 271; Schmidt, 1953: 35; Conant, 1958: 235.

Plethodon (Plethodon) jordani: Vieites *et al.*, 2011: 632, by inference; Dubois & Raffaëlli, 2012: 146.

Distribution: Western North Carolina and adjacent Georgia, South Carolina, Tennessee, and Virginia. See maps in Highton (1973) and Petranka (1998). **U.S.**: GA, NC, SC, TN, *VA.

Common name: Red-cheeked Salamander (SECSN) or Jordan's Redcheek Salamander (C&T).

Content: No subspecies are currently recognized.

Commentary: Highton (1973) provided a CAAR account, in which he included several other forms in the synonymy of this species, which are now recognized as separate species or synonyms for others. There has been a history of confusion of this species with others (Highton & Peabody, 2000). In their study, this species was a basal sister to a clade composed of most of the *glutinosus*-group species of Clade B, while Wiens *et al.* (2006) found it to be basal to the entire Clade A (see genus *Content*, above).

Plethodon (Plethodon) kentucki Mittleman, 1951

Plethodon kentucki Mittleman, 1951: 105.—**O**: **H**: USNM 129937 (ex CSNM 1521A).—**OT**: U.S., Kentucky, Harlan County, Pine Mountain, ca. 2000 feet elevation.

Plethodon jordani kentucki: Schmidt, 1953: 37.

Plethodon kentucki: Highton & MacGregor, 1983: 199; Frost, 1985: 597; Conant & Collins, 1998: 473; Petranka, 1998: 374; Raffaëlli, 2007: 305.

Plethodon (Plethodon) kentucki: Vieites *et al.*, 2011: 632, by inference; Dubois & Raffaëlli, 2012: 146.

Distribution: Cumberland Plateau in western Virginia, southern West Virginia, eastern Kentucky and adjacent Tennessee. Known range indicated by map in Petranka (1998). **U.S.**: KY, TN, VA, WV.

Common name: Cumberland Plateau Salamander.

Commentary: Shortly after its description, Clay *et al.* (1955) suggested that this form was not distinct from *P. g. glutinosus; this was followed by Gorham (1974: 37).* Highton & MacGregor (1983) presented evidence that it is a valid species. Highton (1986b) provided a CAAR account. In the analysis by Highton (1989), this species was the most differentiated of the *glutinosus* complex of any of the species sampled, but closest to *P. savannah*. Wiens *et al.*

(2006) found the pair *petraeus-kentucki* to be a basal sister to all other species of the combined *glutinosus-ouachitae-yonahlossee* groups except *P. yonahlossee,* as did Kozak *et al.* (2006).

Plethodon (Plethodon) kiamichi Highton, 1989

Plethodon kiamichi Highton, 1989: 62.—**O**: **H**: USNM 257314.—**OT**: U.S., Oklahoma, LeFlore County, Round Mountain, 34° 36' 55" N, 94° 29' 50" W, elevation 640 m (Locality 76).

Plethodon kiamichi: Conant & Collins, 1998: 471; Raffaëlli, 2007: 304.

Plethodon (Plethodon) kiamichi: Vieites *et al.*, 2011: 632, by inference; Dubois & Raffaëlli, 2012: 146.

Distribution: Restricted range in the Kiamichi and Round Mountains in Oklahoma and Arkansas. **U.S.**: AR, OK.

Common name: Kiamichi Slimy Salamander.

Commentary: The analysis by Highton (1989) indicated this species seemed genetically closest to *P. mississippi* and *P. grobmani,* while Wiens *et al.* (2006) found its sister species to be *P. grobmani,* and this pair a sister to the *ocmulgee-savannah* pair; Kozak *et al.* (2006) found congruent results. Petranka (1998) did not recognize this species as distinct from *P. glutinosus.*

Plethodon (Plethodon) kisatchie Highton, 1989

Plethodon kisatchie Highton, 1989: 67.—**O**: **H**: USNM 257348.—**OT**: U.S., Louisiana, Grant Parish, along Indian Creek, Locality 90 (31° 43' 15" N, 92° 28' 02" W), elevation 30 m.

Plethodon kisatchie: Conant & Collins, 1998: 471; Raffaëlli, 2007: 304.

Plethodon (Plethodon) kisatchie: Vieites *et al.*, 2011: 632, by inference; Dubois & Raffaëlli, 2012: 146.

Distribution: Southern Arkansas into central Louisiana. **U.S.**: AR, LA.

Common name: Louisiana Slimy Salamander.

Commentary: The analysis by Highton (1989) indicated a closest relation of this species to *P. sequoyah,* and that pair seemed closest to *P. mississippi.* Petranka (1998) did not recognize this species as distinct from *P. glutinosus.* The analysis of Pyron & Wiens (2011) indicated this is a sister to *P. grobmani,* while that pair is in turn a sister clade to that containing *P. ocmulgee* and *P. savannah.*

Plethodon (Plethodon) meridianus Highton & Peabody, 2000

Plethodon meridianus Highton & Peabody, 2000: 61.—**O**: **H**: USNM 454653.—**OT**: U.S., North Carolina, Burke-Cleveland County line, South Mountains (35° 35' 08" N, 81° 41' 22" W, Locality 34), elevation 823 m.

Plethodon meridianus: Raffaëlli, 2007: 307.

Plethodon (Plethodon) meridianus: Vieites *et al.*, 2011: 632, by inference; Dubois & Raffaëlli, 2012: 146.

Distribution: Isolated populations in western North Carolina. See map in Beamer & Lannoo (2005f). **U.S.**: NC.

Common name: South Mountain Gray-cheeked Salamander (SECSN) or South Mountain Graycheek Salamander (C&T).

Commentary: The analysis of Highton & Peabody (2000) indicated a sister-taxon relationship between this species and the group *amplus-metcalfi-montanus.* The analysis of Pyron & Wiens (2011) showed results congruent with that study.

Plethodon (Plethodon) metcalfi Brimley, 1912

Plethodon metcalfi Brimley, 1912: 138.—**O**: **H**: USNM 49682 (Cochran, 1961a: 21, sd.).—**OT**: U.S., North Carolina, Haywood County, near Sunburst, and Avery County, Grandfather Mountain, elevation about 3500-4000 feet.—**Comment**: Schmidt, 1953: 35, restricted the locality to Sunburst, but did not provide any evidence that the H actually came from there, so that action is invalid.

Plethodon metcalfi: Stejneger & Barbour, 1917: 16; 1943: 18; Bishop, 1924: 96; 1928a: 160; Highton & Peabody, 2000: 59; Raffaëlli, 2007: 306.

Plethodon metcalfi metcalfi: Mittleman, 1948b: 418

Plethodon jordani metcalfi: Hairston, 1950: 271; Schmidt, 1953: 35; Conant, 1958: 236.

Plethodon (Plethodon) metcalfi: Vieites *et al.*, 2011: 632, by inference; Dubois & Raffaëlli, 2012: 146.

Plethodon clemsonae Brimley, 1927: 73.—**O**: **H** (od.), USNM 73849.—**OT**: U.S., South Carolina, Oconee County, Jocassee, elev. 1200-1500 ft.

Plethodon clemsonae: Stejneger & Barbour, 1933: 11; 1943: 17; Bishop, 1941: 20.

Plethodon shermani clemsonae: Hairston & Pope, 1948: 274.

Plethodon metcalfi clemsonae: Mittleman, 1948b: 418.

Plethodon jordani clemsonae: Hairston, 1950: 272; Conant, 1958: 236.

Plethodon shermani rabunensis Pope & Hairston, 1948: 106.—**O**: **H** (od.), FMNH 47697.—**OT**: U.S., Georgia, Rabun County, Rabun Bald Mountain, elev. 4200-4600 ft.—**SS**: Mittleman (1948b: 417) with *P. metcalfi clemsonae*.

Plethodon jordani rabunensis: Hairston, 1950: 272.

Plethodon shermani melaventris Pope & Hairston, 1948: 107.—**O**: **H** (od.), FMNH 47614.—**OT**: U.S., North Carolina, Macon County, Highlands. **SS**: Mittleman, 1948b: 417, with *P. metcalfi clemsonae*.

Plethodon jordani melaventris: Hairston, 1950: 272.

Distribution: Southwestern North Carolina, northwestern South Carolina, and adjacent Georgia. See map in Beamer & Lannoo (2005g). **U.S.**: GA, NC, SC.

Common name: Southern Gray-cheeked Salamander.

Commentary: As evidenced by the synonymy here, this species has had a confused taxonomic history. Highton & Peabody (2000) found this is a sister species to *P. amplus,* while Wiens *et al.* (2006) found it to be a sister to *P. montanus*. Kozak *et al.* (2006) found this to be a sister species to *P. jordani,* in a more basal position. Gorham (1974) considered this a synonym of *P. jordani*; Petranka (1998) apparently also treated it as a synonym of *P. jordani,* but did not mention the name.

Plethodon (Plethodon) mississippi Highton, 1989

Plethodon mississippi Highton, 1989: 65.—**O**: **H**: USNM 257388.—**OT**: U.S., Mississippi, Tishomingo County, Tishomingo State Park, Locality 79 (34° 36' 38" N, 88° 11' 56" W), elevation 177 m.

Plethodon mississippi: Conant & Collins, 1998: 471; Raffaëlli, 2007: 304.

Plethodon (Plethodon) mississippi: Vieites *et al.*, 2011: 632, by inference; Dubois & Raffaëlli, 2012: 146.

Distribution: Southeastern Louisiana to western Tennessee and Kentucky. See map in Beamer & Lannoo (2005h). **U.S.**: AL, KY, LA, MS, TN.

Common name: Mississippi Slimy Salamander.

Commentary: Petranka (1998) did not recognize this species as distinct from *P. glutinosus*. Highton (1989) had mixed results regarding relationships of this species. One analysis indicated it was closest to *P. kisachie*, another to *P. sequoyah,* and still another to *P. kiamichi.* The analysis of Pyron & Wiens (2011) showed this species as a sister to *P. kiamichi,* with that clade having a sister relationship with the clade including *P. sequoyah* and *P. albagula*. Beamer & Lanoo (2005h) presented an account.

Plethodon (Plethodon) montanus Highton & Peabody, 2000

Plethodon montanus Highton & Peabody, 2000: 58.—**O**: **H**: USNM 438400.—**OT**: U.S., Virginia, Grayson-Smyth County line, Deep Gap (36° 39' 28" N, 81° 33' 25" W), 1 km west of the top of Mt. Rogers, elevation 500 m.

Plethodon montanus: Raffaëlli, 2007: 306.

Plethodon (Plethodon) montanus: Vieites *et al.*, 2011: 632, by inference; Dubois & Raffaëlli, 2012: 146.

Distribution: Isolated montane populations in western North Carolina and adjacent Tennessee and Virginia. See map in Beamer & Lannoo (2005i). **U.S.**: NC, TN, VA.

Common name: Northern Gray-cheeked Salamander.

Commentary: The analysis of Highton & Peabody (2000) indicates that this species is a sister of the *amplus-metcalfi* pair, while Wiens *et al.* (2006) found it to be a sister to *P. metcalfi*; the latter study also found *amplus-meridianus* and *metcalfi-montanus* pairs to be sister clades.

Plethodon (Plethodon) nettingi Green, 1938

Plethodon nettingi Green, 1938a: 295.—**O**: **H**: CM 10279.—**OT**: U.S., West Virginia, Randolph County, near Cheat Bridge, above 4000 feet on Barton Knob.

Plethodon nettingi: Stejneger & Barbour, 1943: 19; Schmidt, 1953: 37; Highton, 1972: 144; Gorham, 1974: 37; Highton & Larson, 1979: 587; Frost, 1985: 598;

Conant & Collins, 1998: 467; Petranka, 1998: 381; Raffaëlli, 2007: 296.

Plethodon richmondi nettingi: Highton & Grobman, 1956: 187; Conant, 1958: 232; Brame, 1967: 16; Cochran & Goin, 1970: 57.

Plethodon nettingi nettingi: Thurow, 1968: 36; Conant, 1975: 275; Highton & Larson, 1979: 587; Behler & King, 1979: 345.

Plethodon (Plethodon) nettingi: Vieites *et al.*, 2011: 632, by inference; Dubois & Raffaëlli, 2012: 146.

Distribution: Cheat Mountains, eastern West Virginia. See maps of Highton (1986c) and Petranka (1998). **U.S.**: WV.

Common name: Cheat Mountain Salamander.

Commentary: Highton (1986c) provided a CAAR account. The analysis of Highton (1999) indicated a possible sister-species relationship with either *P. hubrichti* or *P. shenandoah,* but support levels for either were low. Wiens *et al.* (2006) found this a sister species to the clade formed by the *electromorphus+hubrichti+richmondi.* This species is federally listed as threatened.

Plethodon (Plethodon) ocmulgee Highton, 1989

Plethodon ocmulgee Highton, 1989: 60.—**O**: **H**: USNM 257426.—**OT**: U.S., Georgia, Wheeler County, Little Ocmulgee State Park, Locality 32 (32° 05' 38" N, 82° 53' 35"), elevation 49 m.

Plethodon ocmulgee: Conant & Collins, 1998: 471; Raffaëlli, 2007: 303.

Plethodon (Plethodon) ocmulgee: Vieites *et al.*, 2011: 632, by inference; Dubois & Raffaëlli, 2012: 146.

Distribution: Central Georgia, mainly Ocmulgee River drainage. See map in Beamer & Lannoo (2005j). **U.S.**: GA.

Common name: Ocmulgee Slimy Salamander.

Commentary: The analysis by Highton (1989) found this to be closest to *P. glutinosus,* but the study by Wiens *et al.* (2006) found it to be a sister species of *P. savannah,* as did Kozak *et al.* (2006). Petranka (1998) did not recognize this species as distinct from *P. glutinosus.*

Plethodon (Plethodon) ouachitae Dunn & Heinze, 1933

Plethodon ouachitae Dunn & Heinze, 1933: 121.—**O**: **H**: USNM 92484.—**OT**: U.S., Arkansas, Polk County, Ouachita National Forest, on Rich Mountain.

Plethodon ouachitae: Stejneger & Barbour, 1943: 19; Schmidt, 1953: 37; Brame, 1967: 16; Cochran & Goin, 1970: 58; Gorham, 1974: 37; Conant, 1975: 279; Behler & King, 1979: 345; Frost, 1985: 598; Conant & Collins, 1998: 475; Petranka, 1998: 386: Raffaëlli, 2007: 308.

Plethodon (Plethodon) ouachitae: Vieites *et al.*, 2011: 632, by inference; Dubois & Raffaëlli, 2012: 146.

Distribution: Ouachita Mountains in eastern Oklahoma and western Arkansas. Blair (1967) and Petranka (1998) have range maps, but they include what has been recognized as a separate species, *P. fourchensis*. Anthony (2005) has an updated map. **U.S.**: AR, OK.

Common name: Rich Mountain Salamander.

Commentary: Blair (1967) provided a CAAR account, which included references to a population later described as a separate species, *P. fourchensis*. This species, plus *P. fourchensis* and *P. caddoensis*, are often referred to as the "large" eastern *Plethodons,* while the other species of the subgenus are called the "small" eastern *Plethodons*. The three are close relatives.

Plethodon (Plethodon) petraeus Wynn, Highton, & Jacobs, 1988

Plethodon petraeus Wynn, Highton, & Jacobs, 1988: 135.—**O**: **H**: USNM 267205.—**OT**: U.S., Georgia, Walker County, eastern slope of Pigeon Mountain, at the mouth of Dickson Gulf; 34° 39' 50" N, 85° 22' 10" W, elevation 310 m.

Plethodon petraeus: Conant & Collins, 1998: 474. Petranka, 1998: 389; Raffaëlli, 2007: 301.

Plethodon (Plethodon) petraeus: Vieites *et al.*, 2011: 632, by inference; Dubois & Raffaëlli, 2012: 146.

Distribution: Mountain localities in northwestern Georgia. Jensen & Camp (2004) and Petranka (1998) have range maps. **U.S.**: GA.

Common name: Pigeon Mountain Salamander.

Commentary: Jensen & Camp (2004) provided a CAAR account. The study by Mahoney (2001) suggested that this species has a sister-group relationship to a clade composed of *P. jordani, P. ouachitae, P. fourchensis, P. teyahalee,* and *P. yonahlossee*. The work of Wiens *et al.* (2006) found this to be a sister species to *P. kentucki,* as did Kozak *et al.* (2006).

Plethodon (Plethodon) punctatus Highton, 1972

Plethodon punctatus Highton, 1972: 176.—**O**: **H**: USNM 190224.—**OT**: U.S., West Virginia state line, Pendleton County, between 0.1 and 0.2 mile north-northwest of the top of Cow Knob.

Plethodon punctatus: Conant, 1975: 278; Behler & King, 1979: 346; Frost, 1985: 598; Conant & Collins, 1998: 470; Petranka, 1998: 390; Raffaëlli, 2007: 300.

Plethodon (Plethodon) punctatus: Vieites *et al.*, 2011: 632, by inference; Dubois & Raffaëlli, 2012: 146.

Distribution: Mountains of northeastern West Virginia and north-western Virginia. See map in Highton (1988b) or Petranka (1998). **U.S.**: VA, WV.

Common name: Cow Knob Salamander (SECSN) or White-spotted Salamander (C&T).

Commentary: Highton (1988b) provided a CAAR account. Petranka (1998) tentatively retained separate species status, but suggested this form could eventually fall into junior synonymy of *P. wehrlei,* or be reduced to subspecies status.

Plethodon (Plethodon) richmondi Netting & Mittleman, 1938

Plethodon richmondi Netting & Mittleman, 1938: 288.—**O**: **H:** CM 14189.—**OT**: U.S., West Virginia, Cabell County, Huntington, Ritter Park, elevation 600-700 feet.

Plethodon richmondi: Stejneger & Barbour, 1943: 19; Schmidt, 1953: 37; Conant, 1975: 274; Behler & King, 1979: 346; Frost, 1985: 598; Conant & Collins, 1998: 465; Petranka, 1998: 392; Raffaëlli, 2007: 295.

Plethodon richmondi richmondi: Highton & Grobman, 1956: 187; Conant, 1958: 231; Brame, 1967: 16; Cochran & Goin, 1970: 57; Gorham, 1974: 37.

Plethodon (Plethodon) richmondi: Vieites *et al.*, 2011: 632, by inference; Dubois & Raffaëlli, 2012: 146.

Plethodon richmondi popei Highton & Grobman, 1956: 187.—**O**: **H**: FSM 8226.—**OT**: U.S., Virginia, Grayson-Wythe County line, Comers Rock.

Plethodon richmondi popei: Brame, 1967: 16.

Distribution: Western parts of North Carolina, Virginia, West Virginia, and Pennsylvania; eastern Kentucky and much of Ohio. Petranka (1998) and Regester (2000b) have range maps. **U.S.**: KY, NC, OH, PA, VA, WV.

Common name: Southern Ravine Salamander.

Content: The subspecies indicated in the synonymy, above, are not currently recognized.

Commentary: Regester (2000b) provided a CAAR account. The analysis of Mahoney (2001) indicates that this species is a sister-taxon to the *cinereus-hoffmani* pair. The analysis of Highton (1999) indicated it may be a sister-species of *P. electromorphus* (but with a low level of support). The analyses by Sites, *et al.* (2004), Wiens *et al.* (2006), and Kozak *et al.* (2006) all agree with Highton that this species is closest to *P. electromorphus.*

Plethodon (Plethodon) savannah Highton, 1989

Plethodon savannah Highton, 1989: 73.—**O**: **H:** USNM 257465.—**OT**: U.S., Georgia, Richmond County, Locality 128 (33° 19' 48" N, 82° 03' 49" W), elevation 101 m.

Plethodon savannah: Conant & Collins, 1998: 471; Raffaëlli, 2007: 305.

Plethodon (Plethodon) savannah: Vieites *et al.*, 2011: 632, by inference; Dubois & Raffaëlli, 2012: 146.

Distribution: In the Savannah area of eastern Georgia. See map in Beamer & Lannoo (2005k). **U.S.**: GA.

Common name: Savannah Slimy Salamander.

Commentary: Highton (1989) found this species was nearly basal to the rest of the *glutinosus* group. Highton (1989) found this species was highly differentiated and interpreted this as having a near basal sister relationship to other members of the *glutinosus* complex. More recent genetic studies by Wiens *et al.* (2006) and Kozak *et al.* (2006) found this species as a highly derived member of one clade of the *glutinosus*-group. Petranka (1998) did not recognize this species as distinct from *P. glutinosus.*

Plethodon (Plethodon) sequoyah Highton, 1989

Plethodon sequoyah Highton, 1989: 68.—**O**: **H**: USNM 257485.—**OT**: U.S., Oklahoma, McCurtain County, Locality 91 (34° 07' 29" N, 94° 40' 15" W), elevation 140 m.

Plethodon swquoyah: Conant & Collins, 1998: 471; Raffaëlli, 2007: 304.

Plethodon (Plethodon) sequoyah: Vieites *et al.*, 2011: 632, by inference; Dubois & Raffaëlli, 2012: 146.

Distribution: Known only from the type locality. See map, Hun-tington *et al.* (1993). **U.S.**: OK.

Common name: Sequoyah Slimy Salamander.

Commentary: Huntington *et al.* (1993) provided a CAAR account. In contrast to the more recent studies cited in the *P. albagula* account, the analysis of Highton (1989) indicated that this species is closest to *P. kisatchie,* and the pair, *P. kisatchie* + *P. sequoyah,* have a sister-taxon relation to *P. kiamichi* + *P. mississippi.* The study of Wiens *et al.* (2006) found the *albagula-sequoyah* pair to be a sister taxon to the clade *ocmulgee* + *savannah* + *kisatchi* + *grobmani.* Petranka (1998) did not recognize this species as distinct from *P. glutinosus.*

Plethodon (Plethodon) serratus Grobman, 1944

Plethodon cinereus serratus Grobman, 1944: 306.—**O**: **H**: FMNH 39464.—**OT**: U.S., Arkansas, Polk County, Rich Mountain, elev. 2500 feet.

Plethodon cinereus serratus: Schmidt, 1953: 33; Brame, 1967: 15; Cochran & Goin, 1970: 56; Gorham, 1974: 37; Conant, 1975: 273.

Plethodon serratus: Highton & Webster, 1976: 40; Behler & King, 1979: 347; Garrett & Barker, 1987: 67; Frost, 1985: 598; Conant & Collins, 1998: 464; Petranka, 1998: 395; Dixon, 2000: 58; Raffaëlli, 2007: 294.

Plethodon (Plethodon) serratus: Vieites *et al.*, 2011: 632, by inference; Dubois & Raffaëlli, 2012: 146.

Plethodon cinereus polycentratus Highton & Grobman, 1956: 185.—**O**: **H**: FSM 8376.—**OT**: U.S., Georgia, Fulton County, 2 miles northeast of Palmetto.

Plethodon cinereus polycentratus: Brame, 1967: 15; Conant, 1975: 273.—**SS:** Highton & Larson, 1979, by inference.

Distribution: Four widely disjunct populations: northern Georgia and adjacent Alabama, Tennessee, North Carolina; central Louisiana and eastern Texas; western Arkansas, adjacent Oklahoma; southeastern Missouri. Highton (1986h) has a partial range map; Petranka (1998) has a more current map. **U.S.**: AL, AR, GA, LA, NC, OK, TN, TX.

Common name: Southern Red-backed Salamander.

Commentary: Highton (1986h) provided a CAAR account. The analysis of Highton (1999) suggested this species is either related to the *cinereus-shenandoah* pair, or is remote to most other species of the group. The analysis of Mahoney (2001) indicated that this species is a sister-taxon to the *cinereus-hoffmani-richmondi* taxon. The analyses by Sites, *et al.* (2004), Wiens *et al.* (2006) and Kozak *et al.* (2006) all found this species is indeed remote to the rest of the *cinereus* group. The species has apparently been extirpated in Texas (Dixon, 2000).

Plethodon (Plethodon) shenandoah Highton & Worthington, 1967

Plethodon richmondi shenandoah Highton & Worthington, 1967: 617.—**O**: **H**: USNM 157379.—**OT**: U.S., Virginia, Page County, Shenandoah National Park, 0.4 air mile west of the top of Hawksbill Mountain, Appalachian Trail, 0.02 miles northeast of its junction with Naked Top Mountain Trail, 3650 feet elevation.

Plethodon nettingi shenandoah: Highton, 1972: 150-151; Gorham, 1974: 37; Conant, 1975: 276.

Plethodon shenandoah: Highton, 1977: 15; Highton & Larson, 1979: 587; Behler & King, 1979: 348; Frost, 1985: 599; Conant & Collins, 1998: 466; Peranka, 1998: 397; Raffaëlli, 2007: 296.

Plethodon (Plethodon) shenandoah: Vieites *et al.*, 2011: 632, by inference; Dubois & Raffaëlli, 2012: 146.

Misapplied names:

Plethodon cinereus Thurow, 1968: 32 (part).

Distribution: Northern Virginia, Shenandoah National Park. Highton (1988a) provided a map. Introduced in Illinois. **U.S.**: *IL, VA.

Common name: Shenandoah Salamander.

Commentary: Highton (1988a) provided a CAAR account. The analysis of Highton (1999) indicated the closest relationship of this species may be to *P. nettingi* or to *P. cinereus* and *P. serratus*. The analysis of Sites, *et al.* (2004) suggests this species is closest to *P. cinereus,* but remote from *P. serratus*. Wiens *et al.* (2006) found the *cinereus-shenandoah* pair to be a sister group to all other members of the group except *P. serratus,* while Kozak *et al.* (2006) found it to be a sister to the clade formed by *net tingi+hubrichti+richmondi+electro-morphus*. The species is federally listed as endangered.

Plethodon (Plethodon) sherando Highton, 2004

Plethodon sherando Highton, 2004: 10.—**O**: **H**: USNM 556159.—**OT**: U.S., Virginia, August County, Northwest slope of Bald Mountain; 37° 55' 09" N, 79° 04' 00" W, elev. 1055 m.

Plethodon sherando: Raffaëlli, 2007: 295.

Plethodon (Plethodon) sherando: Vieites *et al.*, 2011: 632, by inference; Dubois & Raffaëlli, 2012: 146.

Distribution: Blue Ridge Mountains of Virginia, vicinity of the type locality. **U.S.**: VA.

Common name: Big Levels Salamander.

Commentary: Highton (2004) found this form to be a sister-species of *P. serratus*. It has not been included in any other phylogenetic studies.

Plethodon (Plethodon) shermani Stejneger, 1906

Plethodon shermani Stejneger, 1906: 559.—**O**: **H**: USNM 36214 (Cochran, 1961a: 21, sd.).—**OT**: U.S., North Carolina, Nantahala Mountain, between Andrews and Aquone; corrected to east of Wayah Gap, Macon County, North Carolina (Brimley, 1912: 135-140), then to Wayah Bald Mountain, in Nantahala Range, between Franklin and Aquone, North Carolina (Dunn, 1926b: 146).

Plethodon shermani: Stejneger & Barbour, 1917: 16; Bishop, 1928a: 160;: Highton & Peabody, 2000: 62; Raffaëlli, 2007: 307.

Plethodon glutinosus shermani: Bishop, 1941: 18; Stejneger & Barbour, 1943: 17.

Plethodon shermani shermani: Pope & Hairston, 1948: 106.

Plethodon jordani shermani: Hairston, 1950: 271; Schmidt, 1953: 36; Conant, 1958: 236; Brame, 1967: 16.

Plethodon (Plethodon) shermani: Vieites *et al.*, 2011: 632, by inference; Dubois & Raffaëlli, 2012: 146.

Distribution: Restricted montane range in southwestern tip of North Carolina and adjacent Tennessee and Georgia. **U.S.**: GA, NC, TN.

Common name: Red-legged Salamander.

Commentary: The analysis of Highton & Peabody (2000) indicated this is a sister species of *P. teyahalee*, while Wiens *et al.* (2006) found it is a sister-taxon to the *tehahalee-cylindraceus* pair. The analysis of Kozak *et al.* (2006) found this species a more basal sister to a small clade including that pair. Gorham (1974: 37) treated this name as a synonym of *P. jordani*. Petranka (1998) did not mention this name, so we assume he may have considered it conspecific with *P. teyahalee* (*P. oconalufte*).

Plethodon (Plethodon) teyahalee Hairston, 1950

Plethodon jordani teyahalee Hairston, 1950: 269.—**O**:**H**: UMMZ 100807.—**OT**: U.S., North Carolina, boundary between Graham and Cherokee Counties, Snowbird Mountains, on Teyahalee Bald (= Johanna Bald), elev. 4525 feet.

Plethodon jordani teyahalee: Schmidt, 1953: 36; Conant, 1958: 237; Brame, 1967: 16.

Plethodon teyahalee: Highton, 1984: 7; Conant & Collins, 1998: 473; Highton, Tilley, & Wake, 2001: 28; Raffaëlli, 2007: 302.

Plethodon (Plethodon) teyahalee: Vieites *et al.*, 2011: 632, by inference; Dubois & Raffaëlli, 2012: 146.

Plethodon oconaluftee Hairston, 1993: 67.—**O**: **H**: GSMNP 33339.—**OT**: U.S., North Carolina, Transylvania County, Pisgah National Forest, by Forest Service Road 140, near North Fork of the French Broad River on south-facing slope of Balsam Mountains, elev. 930 m. (erroneous **NN** for *teyahalee*, thus a junior synonym; see Commentary).

Plethodon oconaluftee: Petranka, 1998: 383.

Distribution: Western North Carolina into South Carolina, plus adjacent Georgia and Tennessee. See map in Highton (1987a) and also the one in Petranka (1998) under the name *P. oconaluftee*. **U.S.**: GA, NC, SC, TN.

Common name: Southern Appalachian Salamander.

Commentary: Highton (1987a) provided a CAAR account. Gorham (1974: 37) treated this name as a synonym of *P. jordani*. Highton & Peabody (2000) listed it as a member of the *P. glutinosus* complex, but their analysis indicates it is a member of the *P. jordani* complex. In the analyses of Highton (1989), this species seemed one of the least closely related to other members of the *glutinosus* group, but always was indicated as closest to *P. cylindraceus*. This species apparently arose by hybridization, but members are not considered as hybrids under the *Code* (Highton, Tilley, & Wake, 2001), so it seems a correctly treated species. The analysis by Wiens *et al.* (2006) found that this is a sister species to *P. cylindraceus,* as did Kozak *et al.* (2006). Petranka (1998) disagreed with Highton's argument for recognizing this name instead of *P. oconaluftee,* and used that name instead.

Plethodon (Plethodon) variolatus (Gilliams, 1818)

Salamandra variolata Gilliams, 1818: 460.—**O**: in the cabinet of the Academy (ANSP); none currently known to exist.—**OT**: U.S., the southern states, not infrequently in small streams, not seen as far north as Maryland; restricted invalidly by Schmidt (1953: 34) to vicinity of Charleston, South Carolina.—**N**: USNM 167104 (Highton, 1989: 60);—**OT** (of **N**): U.S, South Carolina, Berkeley County, Beechtree Recreation Area, locality 27 (33° 08' 00" N, 79° 47' 06" W), elevation 6 m.

Plethodon variolosum: Duméril *et al.*, 1854c: pp.—**IS.**

Plethodon variolatus: Highton, 1989: 60; Conant & Collins, 1998: 741; Raffaëlli, 2007: 303.

Plethodon (Plethodon) variolatus: Vieites *et al.*, 2011: 632, by inference; Dubois & Raffaëlli, 2012: 146.

Distribution: Atlantic coastal plain in South Carolina and northeastern Georgia. **U.S.**: GA, SC.

Common name: South Carolina Slimy Salamander.

Commentary: Cope (1889b: 139) and Dunn (1926b: 137) treated this as a synonym of *P. glutinosus, and Gorham (1974: 37) followed that treatment.* Petranka (1998) also did not recognize this species as distinct from *P. glutinosus*. Some of Highton's (1989) analyses indicated this is probably closest to *P. chlorobryonis,* and the analysis of Pyron & Wiens (2011) showed these to be sister species.

Plethodon (Plethodon) ventralis Highton, 1997

Plethodon ventralis Highton, 1997: 351.—**O**: **H**: USNM 176841.—**OT**: U.S., Tennessee, Blunt County, Great Smoky Mountains National Park, near the entrance to a cave on the trail from Schoolhouse Gap to White Oak Sinks; 35° 38' 20" N, 83° 44' 52" W, elev. 549 m.

Plethodon ventralis: Raffaëlli, 2007: 298.

Plethodon (Plethodon) ventralis: Vieites *et al.*, 2011: 632, by inference; Dubois & Raffaëlli, 2012: 146.

Distribution: Western North Carolina, eastern Tennessee, south-western Virginia, and southeastern Kentucky, plus adjacent Georgia, Alabama, and Mississippi. **U.S.**: AL, GA, KY, MS, NC, TN, VA.

Common name: Southern Zigzag Salamander.

Commentary: Highton (1997) recognized this as a cryptic species that he split from *P. dorsalis.*

Plethodon (Plethodon) virginia Highton, 1999

Plethodon virginia Highton, 1999: 67.—**O**: **H**: USNM 507764.—**OT**: U.S., along the Pendleton County, West Virginia-Rockingham County, Virginia state line, along the jeep trail from 0.2-0.9 km southeast and south-southeast of the top of Cow Knob; 38° 41' 25" N, 79° 05' 17" W, elevation 1100-1200 m.

Plethodon virginia: Raffaëlli, 2007: 295.

Plethodon (Plethodon) virginia: Vieites *et al.*, 2011: 632, by inference; Dubois & Raffaëlli, 2012: 146.

Distribution: Mountains along the eastern West Virginia, western Virginia border. **U.S.**: VA, WV.

Common name: Shenandoah Mountain Salamander.

Commentary: Highton (1999) indicated that this species is most closely related to *P. hoffmani,* and is a member of the *cinereus* group.

Plethodon (Plethodon) websteri Highton, 1979

Plethodon websteri Highton, 1979: 32.—**O**: **H**: USNM 204814.—**OT**: U.S., Alabama, Etowah County, 0.6 km east, 0.9 km south of Howelton.

Plethodon websteri: Conant & Collins, 1998: 469; Frost, 1985: 599; Petranka, 1998: 407; Raffaëlli, 2007: 298.

Plethodon (Plethodon) websteri: Vieites *et al.*, 2011: 632, by inference; Dubois & Raffaëlli, 2012: 146.

Distribution: Eastern Alabama and western Georgia, with isolates in Alabama, Louisiana, and Mississippi. See maps in Highton (1986d), Petranka (1998) and Conant & Collins (1998). **U.S.**: AL, GA, LA, MS.

Common name: Webster's Salamander.

Commentary: Highton (1986d) provided a CAAR account.

Plethodon (Plethodon) wehrlei Fowler & Dunn, 1917

Plethodon wehrlei Fowler & Dunn, 1917: 23.—**O**: **H**: ANSP 19123.—**OT**: U.S., Pennsylvania, Indiana County, Two Lick Hills.

Plethodon wehrlei: Stejneger & Barbour, 1917: 17; 1943: 20; Dunn, 1926b: 133; Schmidt, 1953: 38; Highton, 1962; Cochran & Goin, 1970: 58; Gorham, 1974: 38; Behler & King, 1979: 350; Frost, 1985: 599; Conant & Collins, 1998: 469; Petranka, 1998: 409; Raffaëlli, 2007: 299.

Plethodon wehrlei wehrlei: Bogert, 1954: 1195; Conant, 1958: 234.

Plethodon (Plethodon) wehrlei: Vieites *et al.*, 2011: 632, by inference; Dubois & Raffaëlli, 2012: 146.

Plethodon dixi Pope & Fowler, 1949: 1.—**O**: **H**: FMNH 56510.—**OT**: U.S., Virginia, Roanoke County, Dixie Caverns. **SS**: Conant, 1958: 234.

Plethodon dixi: Schmidt, 1953: 35.

Plethodon wehrlei dixi: Bogert, 1954: 1195; Conant, 1958: 234; Brame, 1967: 16.

Plethodon jacksoni Newman, 1954: 9.—**O**: **H**: USNM 134498.—**OT**: U.S., Virginia, Montgomery County, approximately one mile east of Blacksburg, Trillium Vale, elev. 2100 ft.—**SS**: Highton, 1962: 318.

Plethodon wehrlei jacksoni: Cochran, 1961a: 20.

Distribution: Most of West Virginia and much of central Penn-sylvania and western Virginia, plus adjacent New York, Ohio, North Carolina, and Kentucky, plus one record from Tennessee. See map in Petranka (1998). **U.S.**: KY, NC, NY, OH, PA, TN, VA, WV.

Common name: Wehrle's Salamander.

Commentary: Populations described by Pope & Fowler (1949) and Newman (1954) are only variants of *P. wehrlei,* not deserving of species or subspecies status (Highton, 1962), although Bogert (1954) and Conant (1958) recognized *P. wehrlei dixi* as a subspecies. Highton (1987b) provided a CAAR account.

Plethodon (Plethodon) welleri Walker, 1931

Plethodon welleri Walker, 1931: 48.—**O**: **H**: USNM 84135.—**OT**: U.S., North Carolina, Avery County, Grandfather Mountain, near Linville, above 5000 feet.

Plethodon welleri: Stejneger & Barbour, 1933: 13; 1943: 20; Schmidt, 1953: 38; Highton, 1962: 274; Cochran & Goin, 1970: 55; Gorham, 1974: 38; Behler & King, 1979: 351; Frost, 1985: 600; Conant & Collins, 1998: 467; Petranka, 1998: 412; Raffaëlli, 2007: 297.

Plethodon welleri welleri: Thurow, 1956a: 343; 1964: 1; Brame, 1967: 16; Conant, 1975: 276.

Plethodon (Plethodon) welleri: Vieites *et al.*, 2011: 632, by inference; Dubois & Raffaëlli, 2012: 146.

Plethodon welleri ventromaculatum Thurow, 1956b: 344.—**O**: **H**: AMNH 54448.—**OT**: U.S., Virginia, Grayson County, Mt. Rogers, elev. 5500 ft.—**SS:** Highton, 1962: 274.

Plethodon welleri ventromaculatum: Conant, 1958: 233.

Plethodon welleri ventromaculatus: Thurow, 1964: 2; Brame, 1967: 16; Conant, 1975: 276.

Distribution: Montane localities in northwestern North Carolina and southwestern Virginia, and adjacent Tennessee. See maps in Thurow (1964) and Petranka (1998). **U.S.**: NC, TN, VA.

Common name: Weller's Salamander.

Content: Thurow (1956b) described a subspecies, but it was rejected by Highton (1962). Thurow (1964) provided a CAAR species account and recognized subspecies, but few workers have continued to recognize any subspecies for this species.

Plethodon (Plethodon) yonahlossee Dunn, 1917

Plethodon yonahlossee Dunn, 1917b: 598.—**O**: **H** (od.), AMNH 4634.—**OT**: U.S., North Carolina, Avery County, near the Yonahlossee Road about 1 1/2 miles from Linville, elevation 4200 feet.

Plethodon yonahlossee: Stejneger & Barbour, 1917: 17; 1943: 20; Dunn, 1926b: 129; Schmidt, 1953: 38; Cochran & Goin, 1970: 58; Gorham, 1974: 38; Conant, 1975: 278; Guttman, Karlin, & Labanick, 1978: 445; Behler & King, 1979: 352; Frost, 1985: 600; Conant & Collins, 1998: 474; Petranka, 1998: 414; Raffaëlli, 2007: 301.

Plethodon (Plethodon) yonahlossee: Vieites *et al.*, 2011: 632, by inference; Dubois & Raffaëlli, 2012: 146.

Plethodon longicrus Adler & Dennis, 1962: 1.—**O**: **H**: USNM 145658.—**OT**: U.S., North Carolina, Rutherford County, 0.8 miles ESE of Bat Cave (city), northeast slope of Bluerock Mountain, below the Bat Caves, elevation 1645 feet.—**SS**: Guttman *et al.* (1978: 445).

Plethodon longicrus: Cochran & Goin, 1970: 57.

Distribution: Blue Ridge Mountains in northeastern Tennessee, southwestern Virginia, and western North Carolina. Pope (1965) and Petranka (1998) have range maps. **U.S.**: NC, TN, VA.

Common name: Yonahlossee Salamander.

Commentary: Pope (1965) provided a CAAR account under this name. Additionally, Adler (1965) provided one for the junior synonym, *P. longicrus*. The studies of Wiens *et al.* (2006) and Kozak *et al.* (2006) found this species to be in the most primitive, or basal position, sister to all other species of the *glutinosus* group.

Subgenus **Hightonia** Vieites, Nieto-Román, Wake & Wake, 2011

Hightonia Vieites *et al.*, 2011: 632.—**NS**: *Ambystoma vehiculum* Cooper, 1869, od.

Distribution: West coastal North America, from northern California to British Columbia, plus a species in north-central New Mexico.

Content: Nine species, currently treated in four species-groups.

Common name: Western Woodland Salamanders (no name for this assemblage in current use).

Commentary: This group of species in western North America has repeatedly been shown to be monophyletic, its members forming a sister group to the subgenus *Plethodon* (*e.g.*, Highton & Larson 1979, Wiens *et al.* 2006, Kozak *et al.* 2006; but Mahoney, 2001, found the four species-groups were not strictly monophyletic), as has the eastern subgenus *Plethodon,* as recognized here (*e.g.*, Sessions & Larson, 1987; Mahoney, 2001; Wiens *et al.*, 2006; Kozak *et al.*, 2006), but the genus *Aneides* is nested within the genus *Plethodon,* according to Sessions & Larson (1987) and Mahoney (2001). Studies of others (*e.g.*, Macey, 2005; Wiens *et al.*, 2006;

Kozak *et al.*, 2006) indicated different relationships for *Aneides* (see that genus). Molecular studies in the family continue to improve resolution of relationships.

Four species-groups are generally recognized, and we list them here. The **Plethodon vandykei group:** Brodie (1970) formulated the species-groups as used here (except for the *P. neomexicanus* group). There are three species in this group: *P. idahoensis, P. larselli, P. vandykae.*

The **Plethodon neomexicanus group:** This was not included in the western group by Brodie (1970) but Highton & Larson (1979) found it is more closely related to members of this group than to any of the eastern forms. There is only a single species in this group *P. neomexicanus* (unless Highton & Larson, 1979, were correct in including *P. larselli* in this group), and *neomexicanus* seems to be morphologically one of the most primitive of the subgenus.

The **Plethodon vehiculum group:** This group is more derived, *fide* Brodie (1970), and contains two species, *P. dunni* and *P. vehiculum.*

The **Plethodon elongatus group:** This is the most derived group, *fide* Brodie (1970), and includes three species: *P. asupak, P. elongatus,* and *P. stormi.*

Plethodon (Hightonia) asupak Mead, Clayton, Nauman, Olson, & Pfrender, 2005

Plethodon asupak Mead, Clayton, Nauman, Olson, & Pfrender, 2005: 169.—**O**: **H**: UMMZ 262026.—**OT**: U.S., California, Siskiyou County, Muck-a-Muck Creek above Scott Bar, at the confluence of the Scott and Klamath Rivers; 41.774° N, 123.031° W.

Plethodon asupak: Raffaëlli, 2007: 311.

Plethodon (Hightonia) asupak: Vieites *et al.*, 2011: 632;; Dubois & Raffaëlli, 2012: 146.

Distribution: An apparently restricted range in northwestern California and adjacent Oregon. Mead *et al.* (2005) have a map of localities from which members of the species have been taken. **U.S.**: CA, OR.

Common name: Scott Bar Salamander.

Commentary: Assigned to the *elongatus* group by Mead *et al.* (2005); they compared members of the species genetically and morphologically with *P. elongatus* and *P. stormi* and found this species to be a sister to the *elongatus-stormi* pair.

Plethodon (Hightonia) dunni Bishop, 1934

Plethodon dunni Bishop, 1934: 169.—**O**: **H**: USNM 95196.—**OT**: U.S., Oregon, Clackamas County, just outside the city limits of Portland.

Plethodon dunni: Stejneger & Barbour, 1943: 17; Schmidt, 1953: 38; Cochran & Goin, 1970: 54; Gorham, 1974: 37; Feder, Wurst, & Wake, 1978: 69; Behler & King, 1979: 338; Frost, 1985: 596; Stebbins, 1985: 44; 2003: 169; Petranka, 1998: 349; Raffaëlli, 2007: 312.

Plethodon (Hightonia) dunni: Vieites *et al.*, 2011: 632; Dubois & Raffaëlli, 2012: 146.

Plethodon gordoni Brodie, 1970: 497.—**O**: **H** (od.), USNM 166687.—**OT**: U.S., Oregon, Benton County, Dinner Creek, T13S, R7W, Sec. 1, NW quarter.—**SS**: Feder *et al.*, 1978: 69.

Distribution: Western Oregon and adjacent California and Washington. Storm & Brodie (1970a) and Petranka (1998) have range maps. **U.S.**: CA, OR, WA.

Common name: Dunn's Salamander.

Commentary: Storm & Brodie (1970a) provided a CAAR account.

Plethodon (Hightonia) elongatus Van Denburgh, 1916

Plethodon elongatus Van Denburgh, 1916: 216.—**O**: **H**: CAS 29096.—**OT**: U.S., California, Del Norte County, Requa.

Plethodon elongatus: Stejneger & Barbour, 1917: 15; 1943: 17; Dunn, 1926b: 156; Schmidt, 1953: 39; Cochran & Goin, 1970: 54; Gorham, 1974: 37; Behler & King, 1979: 338; Frost, 1985: 596; Petranka, 1998: 352; Raffaëlli, 2007: 310.

Plethodon elongatus elongatus: Stebbins, 1985: 48; 2003: 174.

Plethodon (Hightonia) elongatus: Vieites *et al.*, 2011: 632; Dubois & Raffaëlli, 2012: 146.

Distribution: Northwestern California and adjacent Oregon. Brodie & Storm (1971) and Petranka (1998) have range maps. **U.S.**: CA, OR.

Common name: Del Norte Salamander.

Commentary: Brodie & Storm (1971) provided a CAAR account. Highton & Larson (1979) found this to be a sister species of *P. stormi,* and that pair a sister-taxon to the *vehiculum-dunni* pair. The studies of Wiens *et al.* (2006) and Kozak *et al.* (2006) did not sample all the species of the subgenus as treated here.

Plethodon (Hightonia) idahoensis Slater & Slipp, 1940

Plethodon idahoensis Slater & Slipp, 1940: 38.—**O**: **H**: CPS 2710 (now USNM 110504).—**OT**: U.S., Idaho, Kootenai County, northeast corner of Coeur d'Alene Lake, elevation about 2160 feet.

Plethodon idahoensis: Stejneger & Barbour, 1943: 18; Highton & Larson, 1979: 587; Frost, 1985: 597; Petranka, 1998: 365; Stebbins, 2003: 172; Raffaëlli, 2007: 313.

Plethodon vandykei idahoensis: Lowe, 1950a: 93; Schmidt, 1953: 39; Cochran & Goin, 1970: 53; Gorham, 1974: 37; Behler & King, 1979: 349; Stebbins, 1985: 46.

Plethodon (Hightonia) idahoensis: Vieites *et al.*, 2011: 632; Dubois & Raffaëlli, 2012: 146.

Distribution: Northern Idaho and northwestern Montana, plus adjacent British Columbia. See maps in Brodie & Storm (1970; as a subspecies), Petranka (1998), and Wilson & Ohanjanian (2002). **Can.**: BC. **U.S.**: ID, MT.

Common name: Coeur d'Alene Salamander. *French Canadian name:* Salamandre de Coeur d'Alene.

Commentary: Brodie & Storm (1970) provided a CAAR account of *P. vandykei,* which included this form as a subspecies. Wilson & Ohanjanian (2002) provided a CAAR account at the species level.

Plethodon (Hightonia) larselli Burns, 1954

Plethodon vandykei larselli Burns, 1954: 83.—**O**: **H**: USNM 134129.—**OT**: U.S., Oregon, Multnomah County, north slope of Larch Mountain, 3 miles from summit, on the Multnomah Falls Trail.

Plethodon larselli: Burns, 1962: 177; Brame, 1967: 16; Cochran & Goin, 1970: 53; Gorham, 1974: 37; Behler & King, 1979: 343; Frost, 1985: 597; Stebbins, 1985: 46; 2003: 172; Petranka, 1998: 377; Raffaëlli, 2007: 310.

Plethodon (Hightonia) larselli: Vieites *et al.*, 2011: 632; Dubois & Raffaëlli, 2012: 146.

Distribution: Isolated localities in northern Oregon and southern Washington. Burns (1964) and Petranka (1998) have range maps. **U.S.**: OR, WA.

Common name: Larch Mountain Salamander.

Commentary: Burns (1964) provided a CAAR species account. Highton & Larson (1979) found this species to be a sister-species of *P. neomexicanus,* and revised the species-groups to include this species in the *neomexicanus*-group. Wiens *et al.* (2006) found this species to be a sister to the *vandykei-idahoensis* pair, which was supported by Pyron & Wiens (2011). The clade of those three species form a sister clade to the other members of the subgenus.

Plethodon (Hightonia) neomexicanus Stebbins & Riemer, 1950

Plethodon neomexicanus Stebbins & Riemer, 1950: 73.—**O**: **H**: MVZ 49033.—**OT**: U.S., New Mexico, Sandoval County, 12 miles west and 4 miles south of Los Alamos, 8750 feet elevation.

Plethodon neomexicanus: Schmidt, 1953: 39; Brame, 1967: 16; Cochran & Goin, 1970: 55; Gorham, 1974: 37; Behler & King, 1979: 344; Frost, 1985: 595; Stebbins, 1985: 47; 2003: 173; Petranka, 1998: 380; Raffaëlli, 2007: 309.

Plethodon (Hightonia) neomexicanus: Vieites *et al.*, 2011: 632; Dubois & Raffaëlli, 2012: 146.

Distribution: Montane localities in north central New Mexico. See maps in Williams (1973) and Petranka (1998). **U.S.**: NM.

Common name: Jemez Mountains Salamander.

Commentary: Williams (1973) provided a CAAR account. As noted above, Highton & Larson (1979) found this to be a sister species of *P. larselli,* and included the two in their revised *neomexicanus* group. Wiens *et al.* (2006) found this form to be a sister to the *vandykei* group, as recognized here.

Plethodon (Hightonia) stormi Highton & Brame, 1965

Plethodon stormi Highton & Brame, 1965: 1-2.—**O**: **H**: USNM 149964.—**OT**: U.S., Oregon, Jackson County, 1.25 miles south of Copper.

Plethodon stormi: Brame, 1967: 16; Cochran & Goin, 1970: 54; Gorham, 1974: 37; Behler & King, 1979: 348; Frost, 1985: 599; Petranka, 1998: 399; Raffaëlli, 2007: 311.

Plethodon elongatus stormi: Stebbins, 1985: 48; 2003: 174.

Plethodon (Hightonia) stormi: Vieites *et al.*, 2011: 632; Dubois & Raffaëlli, 2012: 146.

Distribution: Restricted range in southwestern Oregon and adjacent California. Brodie (1971) and Petranka (1998) have range maps. **U.S.**: CA, OR.

Common name: Siskiyou Mountains Salamander.

Commentary: Brodie (1971) provided a CAAR account.

Plethodon (Hightonia) vandykei Van Denburgh, 1906

Plethodon vandykei Van Denburgh, 1906: 61.—**O**: **H**: CAS 6910 (destroyed).—**OT**: U.S., Washington, Mt. Ranier Park, Paradise Valley.—**N**: CAS 47495 (Slevin & Leviton, 1956: 535), rejected by Highton (1962: 257) as not meeting criteria for a neotype.—**OT** of **N**: U.S., Washington, Clallam County, Forks.

Plethodon vandykei: Stejneger & Barbour, 1917: 17; 1943: 19; Frost, 1985: 599; Petranka, 1998: 401; Stebbins, 2003: 170; Raffaëlli, 2007: 313.

Plethodon vandykei vandykei Lowe, 1950a: 93; Schmidt, 1953: 39; Brodie & Storm, 1970: 1; Cochran & Goin, 1970: 53; Gorham, 1974: 37; Behler & King, 1979: 349; Stebbins, 1985: 46.

Plethodon (Hightonia) vandykei: Vieites *et al.*, 2011: 632; Dubois & Raffaëlli, 2012: 146.

Distribution: Western Washington. See map of Brodie & Storm (1970); they included *P. idahoensis* as a subspecies. Petranka (1998) also has a range map. **U.S.**: WA.

Common name: Van Dyke's Salamander.

Commentary: Brodie & Storm (1970) provided a CAAR account. Brodie (1970) analyzed members of this subgenus morphologically, and found species of this group to be the most primitive. Highton & Larson (1979) established *P. idahoensis* as a separate species, removing it from the synonymy of this species. Wiens *et al.* (2006) found this species to be a sister of *P. idahoensis,* and this pair to be genetically the most derived.

Plethodon (Hightonia) vehiculum (Cooper, 1860)

Ambystoma vehiculum Cooper, 1860: viii, Pl. 31, fig. 4.—**O**: none known to exist, but based on a clear figure in the original publication.—**OT**: U.S., Oregon, Astoria.

Plethodon vehiculus: Bishop, 1934: 171.—**IS.**

Plethodon vehiculum: Slater, 1940: 43; Stejneger & Barbour, 1943: 19; Schmidt, 1953: 39; Brame, 1967: 16; Cochran & Goin, 1970: 54; Gorham, 1974: 37; Behler & King, 1979: 349; Frost, 1985: 599; Stebbins, 1985: 45; 2003: 170; Petranka, 1998: 403; Raffaëlli, 2007: 312.

Plethodon (Hightonia) vehiculum: Vieites *et al.*, 2011: 632; Dubois & Raffaëlli, 2012: 146.

Plethodon intermedius Baird *in* Cope, 1868a: 209-210.—**O**: **H** (sd., Cope, 1889b: 147) USNM 4732.—**S**: USNM 4732a *fide* Cochran, 1961a: 20.—**OT**: U.S., California, Kern County, Fort Tejon; in error *fide* Van Denburgh, 1916: 218.—**SS**: Highton, 1962: 265.

Distribution: Western Oregon and Washington, plus southeastern British Columbia including Vancouver Island. See maps in Storm & Brodie (1970b) and Petranka (1998). **Can.**: BC. **U.S.**: OR, WA.

Common name: Western Red-backed Salamander. *French Canadian name:* Salamandre à dos rayé.

Commentary: Storm & Brodie (1970b) provided a CAAR account. Highton & Larson (1979) found this to be a sister species of *P. dunni,* and that pair to be a sister-taxon to the *neomexicanus-larselli* pair.

Family **Proteidae** Gray, 1825

Proteina Gray, 1825: 215.—**NG**: *Proteus* Laurenti, 1768

Proteideae: Tschudi, 1838: 97.

Proteidae: Hogg, 1838: 152; Schmidt, 1953: 12.

Proteida: Jan, 1857: 55; Knauer, 1878: 95.

Proteoidea: Dubois, 2005b: 20.

Phanerobranchoidea Fitzinger, 1826: 43.—**NG**: *Phanerobranchus* Leuckart, 1821. Proposed as family.

Pododysmolgae Ritgen, 1828: 277. Unavailable family-series name; not based on generic nomen.

Hypochthonina Bonaparte, 1840a: 101.—**NG**: *Hypochthon* Merrem, 1820. As subfamily of Sirenidae.

Hypochthonidae: Bonaparte, 1850: [2].

Hyponchthonina: Bonaparte, 1850: [2].

Necturi Fitzinger, 1843: 35.—**NG**: *Necturus* Rafinesque, 1819

Necturidae: Bonaparte, 1850: [2]; Stejneger & Barbour, 1917: 5; Hecht, 1957: 290.

Necturina Bonaparte, 1845b: 6.—**NG**: *Necturus* Rafinesque, 1819; Bonaparte, 1850: [2].

Menobranchida Knauer, 1878: 96.—**NG**: *Menobranchus* Harlan, 1825. Proposed as family.

Hylaeobatrachidae Abel, 1919: 329.—**NG**: *Hylaeobatrachus* Dollo, 1884. Synonymy by Brame, 1972, *fide* Frost (2009-13).

Hylaeobatrachoidea Kuhn, 1965: 39. Nomen attributed to Huene, 1939, *fide* Frost (2009-13).

Distribution: Eastern North America, southern Europe.

Common name: Mudpuppies, waterdogs, olm.

Content: One genus, *Necturus*, is in North America. The genus *Proteus* (the olm) occurs in Europe, and that genus is treated in a separate subfamily, Proteinae.

Commentary: Dubois' (2005b) placed Proteidae and Sirenidae together but in separate epifamilies. Frost *et al.* (2006) also found those families to be closely related. However, Vieites *et al.* (2009) presented evidence that the sirenids are not closely related to any other extant family, while Pyron & Wiens (2011) found sirens basal to salamandroids and proteids basal to other amphiumoids. But the more recent and better resolved analysis of Shen *et al.* (2013) demonstrated that Proteids are a sister taxon to amphiumids + rhyacotritonids + plethodontids, and that sirenids are sister to all other salamanders except the basal imperfectibrachia.

Hecht (1957) argued that *Necturus* should be in a separate family, Necturidae, based on evidence that *Proteus* and *Necturus* evolved in parallel and were probably not closely related. Should this partition be supported, the correct name for the family would become Phanerobranchidae Fitzinger, 1826. We treat the North American and European genera in separate subfamilies.

North America contains 85.7% of the species diversity of this family.

Subfamily **Phanerobranchinae** Fitzinger, 1826

Phanerobranchoidea Fitzinger, 1826: 43.—**NG:** *Phanerobranchus* Leuckart, 1821 (now *Necturus*).

Distribution: Eastern North America.

Common name: Mudpuppies, waterdogs.

Content: One genus with seven species.

Commentary: The other subfamily is Proteinae, which is found in Europe.

Genus **Necturus** Rafinesque, 1819

Synonymy: see the subgenus *Necturus,* below.

Distribution: Eastern North America. Vitt & Caldwell (2009) provided a range map.

Common name: Waterdogs and Mudpuppies. *French Canadian name:* Nectures.

Content: Seven species in two subgenera, endemic to North America.

Commentary: Guttman, *et al.* (1990), in an electrophoretic study, found evidence that *N. maculosus* seems to have a sister relationship to the sister-pair, *N. beyeri* and *N. alabamensis. N. lewisi* and *N. punctatus* are more isolated and are sister-species. They did not include *N. louisianensis,* but we consider it a sister to *N. maculosus.* Results of the study by Guttman *et al.* were congruent with those of Sessions & Wiley (1985) and Maxson *et al.* (1988) using different evidence.

Subgenus **Necturus** Rafinesque, 1819

"Exobranchia" Rafinesque, 1815: 78.—**NU.**—**SS:** Gray, 1850: 65.

Necturus Rafinesque, 1819: 418.—**NS** (sd., Brown, 1908: 127): *Siren maculosa* Rafinesque, 1818 (now *Necturus maculosus*).

Necturus: Stejneger & Barbour, 1917: 5; Schmidt, 1953: 12.

Phanerobranchus Leuckart, 1821: 260.—**NS** (sd., Fitzinger, 1826): *Phanerobranchus tetradactylus* Leuckart, 1821.—**SS**: Baird, 1850a: 290.

Menobranchus Harlan, 1825c: 233.—**NS** (sd. Schmidt, 1953: 12): *Triton lateralis* Say, 1823.—**SS:** Wagler, 1830: 209.

Menobranchus: Gray, 1825: 216.

Hemitriton Van der Hoeven, 1833: 305.—**NN** for *Hypochthon* Merrem, *Menobranchus* Harlan, and *Siredon* Wagler.

Distribution: Wide-ranging in eastern North America.

Common name: Waterdogs or Mudpuppies.

Content: Four species.

Commentary: We tentatively recognize the species as treated by Dubois & Raffaëlli (2012), but more investigations are needed to validate the species status of some forms, and the relationships among them. Some of the evidence suggests closer relationships between two species in different subgenera than between one of them and members of the same subgenus.

Necturus (Necturus) alabamensis Viosca, 1937

Necturus alabamensis Viosca, 1937: 121.—**O**: **H**: USNM 102676,—**OT**: U.S., Alabama, Tuscaloosa County, Tuscaloosa, Black Warrior River.

Necturus alabamensis: Stejneger & Barbour, 1943: 1; Neill, 1963a: 166; Gorham, 1974: 20; Conant, 1975: 244; Behler & King, 1979: 281; Frost, 1985: 606; Petranka, 1998: 418; Conant & Collins, 1998: 423; Raffaëlli, 2007: 69.

Necturus beyeri alabamensis: Hecht, 1958: 17; Conant, 1958: 201; Cochran & Goin, 1970: 26.

Necturus (Necturus) alabamensis: Dubois & Raffaëlli, 2012: 147.

Misapplied names:

Siren maculosa Rafinesque, 1818b: xx. See *N. maculosus*.

Distribution: Alabama, western Georgia, the Florida panhandle, and eastern Mississippi. Petranka (1998) has a range map. **U.S.**: AL, FL, GA, MS.

Common name: Black Warrior River Waterdog (SECSN) or Black Warrior Waterdog (C&T); Alabama Waterdog (Petranka, 1998).

Commentary: Schmidt (1953) considered this a synonym of *N. beyeri,* which he treated as a subspecies of *N. maculosus.* Viosca (1937) originally described this as a distinct species, and also described another specimen from a different locality as another distinct species, *N. lodingi,* which Chermock (1952) placed as a subspecies of *N. punctatus,* then Gunter & Brode (1964) placed in the synonymy of *N. alabamensis.* Hecht (1958) placed *alabamensis* as a subspecies of *N. beyeri.* Neill (1963a) returned it to full species status, and Guttman *et al.* (1990) recommended recognition as a distinct species, and a sister to *N. beyeri,* but indicated more intensive study of these taxa is needed. Bailey & Moler (2003) provided a CAAR account.

Necturus (Necturus) lewisi Brimley, 1924

Necturus maculosus lewisi Brimley, 1924: 167.—**O**: **H**: USNM 73848.—**OT**: U.S., North Carolina, Wake County, Neuse River, near Raleigh.

Necturus maculosus lewisi: Stejneger & Barbour, 1933: 1; Schmidt, 1953: 13.

Necturus lewisi: Bishop, 1943: 32; Stejneger & Barbour, 1943: 1; Gorham, 1974: 20; Conant, 1975: 244; Behler & King, 1979: 282; Frost, 1985: 607; Ashton, 1990: 1; Conant & Collins, 1998: 422; Petranka, 1998: 422; Raffaëlli, 2007: 69.

Necturus beyeri lewisi: Cochran & Goin, 1970: 26.

Necturus (Necturus) lewisi: Dubois & Raffaëlli, 2012: 147.

Distribution: Neuse and Tar drainages in North Carolina. Ashton (1990) has a range map. **U.S.**: NC.

Common name: Neuse River Waterdog.

Commentary: Ashton (1990) provided a CAAR account. The study by Guttman *et al.* (1990) found that this is a sister species to *N. punctatus,* and those two form a clade distinct from the other gulf coast species.

Necturus (Necturus) louisianensis Viosca, 1938

Necturus louisianensis Viosca, 1938: 143.—**O**: **H**: USNM 104238.—**OT**: U.S., Louisiana, Grant Parish, Big Creek, near Pollock.

Necturus louisianensis: Stejneger & Barbour, 1943: 2; Viosca, 1949: 9; Collins, 1991: 43.

Necturus maculosus louisianensis: Schmidt, 1953: 13; Hecht, 1958: 23; Brame, 1967: 6; Gorham, 1974: 20; Conant, 1975: 242; Behler & King, 1979: 273; Petranka, 1998: 425; Conant & Collins, 1998: 422; Raffaëlli, 2007: 71.

Necturus (Necturus) louisianensis: Dubois & Raffaëlli, 2012: 147.

Misapplication of names:

Siren maculosa Rafinesque, 1818b. See *N. maculosus*

Necturus maculosus: Frost, 2009-13.

Distribution: Northern Louisiana, north to Arkansas, eastern Oklahoma, southern Missouri, and southeastern Kansas. Petranka (1998) has a range map, but treats this form as a subspecies of *N. maculosus.* **U.S.**: AR, LA, KS, MO, OK.

Common name: Red River Mudpuppy.

Commentary: Viosca (1938) described this as a distinct species; Schmidt (1953) listed it as a subspecies of *N. maculosus,* and most authors have followed that treatment. Tilley *et al.* (2012) and Frost (2009-13) do not treat this population as a separate species, although the CNAH site does. We tentatively recognize this as a full species, but it needs further study.

Necturus (Necturus) maculosus (Rafinesque, 1818)

Siren maculosa Rafinesque, 1818b: 41.—**O**: not stated nor known to exist.—**OT**: U.S., Ohio River.

Necturus maculatus: Rafinesque, 1819: 417; Baird *in* Heck & Baird, 1851: 252; Boulenger, 1882b: 84; Cope, 1889b: 23; Werner, 1909: pl. 3.—**IS**

Necturus maculosus: Rafinesque, 1820; Stejneger & Barbour, 1917: 5; Cochran & Goin, 1970: 25; Frost, 1985: 607.

Salamandra maculosa: Fitzinger, 1826: 66; Wagler, 1830: 208.

Menobranchus maculatus: Holbrook, 1842: 111.—**IS**

Necturus maculosus maculosus: Brimley, 1924: 167; Stejneger & Barbour, 1943: 2; Schmidt, 1953: 12; Hecht, 1958: 18; Brame, 1967: 6; Gorham, 1974: 20; Conant, 1975: 242; Behler & King, 1979: 283; Conant & Collins, 1998: 422; Raffaëlli, 2007: 70.

Necturus (Necturus) maculosus: Dubois & Raffaëlli, 2012: 147.

Proteus tetradactylus Green, 1818: 358.—**NU.**—**SS:** Wagler, 1830: 209, with *Triton lateralis,* and Tschudi, 1838: 97, with *Menobranchus lateralis.*

Menobranchus tetradactylus: Harlan, 1825c: 233.

Necturus tetradactylus: Wagler, 1830: 210.

Hemitriton (Menobranchus) tetradactylus: Van der Hoeven, 1833: 305.

Triton lateralis Say, *in* James, 1823 (Philadelphia edition): 5, James, 1823 (London ed.): 301.—**O**: not stated, likely deposited in the Philadelphia Museum (not ANSP), perhaps now in MCZ (*fide* Frost, 2009-13).—**OT**: U.S., Pennsylvania, neighborhood of Pittsburgh.—**SS:** Gray, 1850: 84.

Menobranchus lateralis: Harlan, 1825c: 233; Leunis, 1844: 148; Knauer, 1878: 96.

Necturus lateralis: Wagler, 1830: 210; Baird *in* Heck & Baird, 1851: 252; Baird, 1859c: 45; Jordan, 1878: 198; Davis & Rice, 1883: 26.

Menobranchus sayii Gray, 1825: 216.—**O**: not stated, likely deposited in BM.—**OT**: not stated.—**SS**: Gray, 1850: 67.

Phanerobranchus cepedii Fitzinger, 1826: 66.—**O**: not stated, probably deposited in NMW, but not listed in recent type listings.—**OT**: North America.—**SS**: Baird, 1850a: 291.

Menobranchus lacepedii Gray, 1831: 108.—**NN** for *P. cepedii* Fitzinger, 1826.

Proteus canadensis Hodgkins, 1856: 22.—**O**: **H:** deposited in "Provincial Museum in the Normal School Building," probably now in ROM.—**OT**: Can., Ontario, Lake Ontario, Bay of Toronto.

Siredon hyemalis Kneeland, 1857: 152.—**O**: Not stated nor known to exist.—**OT**: uncertain; presumed U.S., Michigan upper peninsula, "Lake Superior . . . Portage Lake."—**SS**: Schmidt, 1953: 13.

Menobranchus latastei Garnier, 1888: 218.—**O**: **S:** include USNM 22331-22332 (Cochran, 1961a: 16, sd.).—**OT**: Can., Ontario, Maitland River.—**SS**: Cope *in* Garnier, 1888: 218.

Necturus lateralis var. *latastei*: Cope *in* Garnier, 1888: 218.

Necturus maculosus stictus Bishop, 1941: 9.—**O**: **H**: UMMZ 89765.—**OT**: U.S., Wisconsin, Lake Winnebago.

Necturus maculosus stictus: Stejneger & Barbour, 1943: 2; Schmidt, 1953: 13; Hecht, 1958: 23; Brame, 1967: 6; Behler & King, 1979: 283.

Distribution: Eastern North America, Atlantic and Gulf drainages. See map in Matson (2005). **Can.:** MA, ON, QC. **U.S.:** AL, AR, *CT, GA, IA, IL, IN, KS, KY, LA, *MA, *ME, MI, MN, MO, MS, NC, ND, *NH, NY, OH, OK, PA, *RI, SD, *VT, WI, WV.

Common name: Mudpuppy (SECSN), or Common Mudpuppy (C&T). *French Canadian name:* Necture tacheté.

Content: A subspecies, *N. m. sticta,* described by Bishop (1941), was listed by Schmidt (1953), but is no longer recognized. Two taxa (*N. louisianensis* and *N. beyeri*), treated here as full species, were earlier considered as subspecies of this species; Tilley *et al* (2012) still treat *louisianensis* as a subspecies of this species.

Subgenus **Parvurus** Dubois & Raffaëlli, 2012

Parvurus Dubois & Raffaëlli, 2012: 147, 156.—**NS**: *Menobranchus punctatus* Gibbes, 1850, od. Described as a subgenus.

Distribution: Somewhat isolated populations from the Carolinas and Georgia to Louisiana and eastern Texas.

Common name: No name in general use. Southeastern Waterdogs may be appropriate.

Content: Three species are included in this subgenus.

Necturus (Parvurus) beyeri Viosca, 1937

Necturus beyeri Viosca, 1937: 123.—**O**: **H** (od.), USNM 1020674.—**OT**: U.S., Louisiana, Allen Parish, Upper Calcasiew River, near Oakdale.

Necturus beyeri: Stejneger & Barbour, 1943: 1; Viosca, 1949: 9; Brame, 1967: 6; Gorham, 1974: 20; Conant, 1975: 244; Behler & King, 1979: 282; Frost, 1985: 607; Garrett & Barker, 1987: 57; Guttman, *et al.*, 1990; Petranka, 1998: 419; Conant & Collins, 1998: 423; Dixon, 2000: 59; Raffaëlli, 2007: 70.

Necturus maculosus beyeri: Schmidt, 1953: 13.

Necturus beyeri beyeri: Hecht, 1958: 16; Conant, 1958: 200; Cochran & Goin, 1970: 26.

Necturus (Parvurus) beyeri: Dubois & Raffaëlli, 2012.

Distribution: Isolated ranges in eastern Texas, western Louisiana and central Mississippi; see map in Petranka (1998). **U.S.**: LA, MS, TX.

Common name: Gulf Coast Waterdog.

Commentary: A sister taxon to *N. alabamensis,* but a distinct species according to evidence of Guttman *et al.* (1990: 172-173), who recommended full species status for both. Evidence that this species and the others here included in the separate subgenus, is a sister to the clade containing the other species, subgenus *Necturus,* is weak; but we tentatively follow the subgeneric treatment.

Necturus (Parvurus) lodingi Viosca, 1937

Necturus lödingi Viosca, 1937: 126.—**O**: **H**: USNM 61752;—**OT**: U.S., Alabama, Eslava Creek near Mobile;—**SS**: Gunter & Brode (1964: 122).—**Comment**: some authors have treated this as a synonym of *N. punctatus* (*e.g.*, Chermock, 1952, and Neill, 1954), while some suggest it may be a distinct species.

Necturus lödingi: Stejneger & Barbour, 1943: 2; Schmidt, 1953: 14.

Necturus punctatus lodingi: Chermock, 1952: 23; Neill, 1954: 75 (justified emendation); Hecht, 1958: 14; Conant, 1958: 201; Brame, 1967: 6; Gorham, 1974: 20.

Necturus (Parvurus) lodingi: Dubois & Raffaëlli, 2012: 147.

Distribution: Restricted portion of the lower Black Warrior River in Alabama. **U.S.:** AL.

Common name: None in use. Perhaps Loding's Waterdog would be appropriate.

Content: No subspecies have been described, but this species has been treated as a subspecies or synonym of *N. alabamensis* or *N. punctatus* in the past.

Commentary: Recognition of this species is very tentative, and probably most workers do not recognize it. More investigation of the relationships among the populations comprising this genus is badly needed.

Necturus (Parvurus) punctatus (Gibbes, 1850)

Menobranchus punctatus Gibbes, 1850: 159.—**O**: **S** (10) one apparently lost; **L:** USNM 11813 (Yarrow, 1883a: 145, by implication).—**OT**: U.S., South Carolina, Dr. Schoolbred's plantation on the South Santee River, a few miles from its mouth.

Necturus punctatus: Baird *in* Heck & Baird, 1851: 252; Garman, 1884a: 36. [*fide* Frost (2009-13), by implication]; Cope, 1889b: 27; Stejneger & Barbour, 1917: 5; 1943: 3; Schmidt, 1953: 14; Cochran & Goin, 1970: 26; Conant, 1975: 245; Behler & King, 1979: 284; Frost, 1985: 607; Conant & Collins, 1998: 424; Dundee, 1998: 1; Petranka, 1998: 429; Raffaëlli, 2007: 69.

Necturus punctatus punctatus: Neill, 1954: 75, by implication; Hecht, 1958: 13; Conant, 1958: 201; Brame, 1967: 6; Gorham, 1974: 20.

Necturus (Parvurus) punctatus: Dubois & Raffaëlli, 2012, by implication.

Distribution: Atlantic coastal drainages in the Carolinas and Georgia. See map in Dundee (1998). **U.S.**: GA, NC, SC.

Common name: Dwarf Waterdog.

Content: Subspecies have been described but are not well documented and not generally recognized.

Commentary: Dundee (1998) provided a CAAR species account. Guttman *et al.* (1990) found this to be a sister species to *N. lewisi,* and the two form a clade distinct from other gulf coast species.

Family **Rhyacotritonidae** Tihen, 1958

Rhyacotritoninae Tihen, 1958: 3. (as subfamily of Ambystomatidae)—**NG**: *Rhyacotriton* Dunn, 1920.

Rhyacotritonidae: Good & Wake, 1992.

Distribution: Northwestern US.

Common name: Torrent Salamanders.

Content: One genus with 4 species, all endemic to North America.

Commentary: This group of salamanders has been treated in the Ambystomatidae, the Dicamptodontidae, and in its own family, as followed here. The recent study of plethodontids by Vieites *et al.*(2011), indicates that rhyacotritonids are a basal sister to the clade of amphiumatoids, while salamandrids and ambystomatids are more basal, which supports several earlier studies. Pyron & Wiens (2011) found rhyacotritonids a basal sister to amphiumoids, but not closely related to salamandroids. The recent study of Shen *et al.* (2013) demonstrates that this family is a sister to amphiumoids, and we follow that analysis in our treament. Good & Wake (1992) summarized the nomenclatural and taxonomic history, and provided rationale for recognizing the family. Frost *et al.* (2006) included this family and others in a suprafamilial taxon Plethosalamandroiei, which is not valid for family-series nomina as it is not based on a generic nomen.

Genus **Rhyacotriton** Dunn, 1920

Rhyacotriton Dunn, 1920b: 56.—**NS**: *Ranodon olympicus* Gaige, 1917, od.

Distribution: As for the family.

Common name: Torrent Salamanders.

Content: Four species.

Commentary: Anderson (1968) provided a CAAR generic account.

Rhyacotriton cascadae Good & Wake, 1992

Rhyacotriton cascadae Good & Wake, 1992: 15.—**O**: **H**: MVZ 90795.—**OT**: U.S., Oregon, Multnomah County, base of Wahkeena Falls.

Rhyacotriton cascadae: Petranka, 1998: 434; Stebbins, 2003: 163; Raffaëlli, 2007: 168.

Distribution: Cascade Mountain slopes in Oregon and Washington. Petranka (1998) has a range map. **U.S.**: OR, WA.

Common name: Cascade Torrent Salamander.

Commentary: Members of this population were treated as montane intergrades between two subspecies of *R. olympicus* (*e.g.*, Anderson, 1968), until Good & Wake (1992) presented biochemical evidence that this and other populations represent distinct species.

Rhyacotriton kezeri Good & Wake, 1992

Rhyacotriton kezeri Good & Wake, 1992: 16.—**O**: **H**: MVZ 197300.—**OT**: U.S., Oregon, Clatsop County, Junction of highway 26 & Luukinen Road (at the Nehalim River Bridge).

Rhyacotriton kezeri: Petranka, 1998: 437; Stebbins, 2003: 162; Raffaëlli, 2007: 167.

Distribution: Northwestern Oregon and southwestern Washington. Petranka (1998) has a range map. **U.S.**: OR, WA.

Common name: Columbia Torrent Salamander.

Commentary: Members of this population have been treated as coastal intergrades between two subspecies of *R. olympicus* (*e.g.*, Anderson, 1968), until Good & Wake (1992) presented biochemical evidence that this and other populations represent distinct species.

Rhyacotriton olympicus (Gaige, 1917)

Ranodon olympicus Gaige, 1917: 2.—**O**: **H**: UMMZ 48608.—**OT**: U.S., Washington, Mason County, Lake Cushman.

Ranodon olympicus: Stejneger & Barbour, 1917: 13.

Rhyacotriton olympicus: Dunn, 1920b: 56; Stejneger & Barbour, 1923: 8; 1943: 8; Frost, 1985: 562; Petranka, 1998: 439; Stebbins, 2003: 162; Raffaëlli, 2007: 167.

Rhyacotriton olympicus olympicus: Stebbins & Lowe, 1951: 465; Schmidt, 1953: 17; Brame, 1967: 10; Cochran & Goin, 1970: 17; Gorham, 1974: 30; Behler & King, 1979: 301; Stebbins, 1985: 40.

Distribution: Northwestern Washington. **U.S.**: WA.

Common name: Olympic Torrent Salamander.

Commentary: Treated as a wider-ranging species for many years, and included *R. variegatus* as a subspecies. Anderson (1968) provided a CAAR account which used that treatment.

Rhyacotriton variegatus Stebbins & Lowe, 1951

Rhyacotriton olympicus variegatus Stebbins & Lowe, 1951: 471.—**O**: **H**: MVZ 45868.—**OT**: U.S., California, Trinity County, 1.3 miles west of Burnt Ranch Post Office.

Rhyacotriton olympicus variegatus: Schmidt, 1953: 17; Brame, 1967: 10; Cochran & Goin, 1970: 17; Gorham, 1974: 30; Behler & King, 1979: 301; Stebbins, 1985: 40.

Rhyacotriton variegatus: Good & Wake, 1992: 18; Petranka, 1998: 441; Stebbins, 2003: 163; Raffaëlli, 2007: 167.

Distribution: Mostly coastal Oregon and northwestern California. Anderson (1968) provided an account and range map as a subspecies of *R. olympicus;* Petranka (1998) has a range map. **U.S.**: CA, OR.

Common name: Southern Torrent Salamander.

Family **Salamandridae** Goldfuss, 1820

Salamandrae Goldfuss, 1820: 129.—**NG**: *Salamandra* Laurenti, 1768.

Salamandridae: Gray, 1825: 215; Bonaparte, 1840a: 100; 1850: [2]; Bronn, 1849: 683.

Salamandroidea: Fitzinger, 1826: 41. (explicit family); Schinz, 1833: 202; Wied, 1865: 124; Dunn, 1922: 423, explicit superfamily.

Salamandrina: Hemprich, 1820: xix; Wiegmann, 1832: 203; Bonaparte, 1838b: 393; 1839e: unnumb.; 1840a: 101; 1850: [2].

Salamandrinae: Tschudi, 1838: 26, 91.

Salamandrae: Tschudi, 1838: 26.

Salamandrinae: Fitzinger, 1843: 33.—**NG**: *Salamandrina* Fitzinger, 1826.

Salamandrini: Massalongo, 1853: 11.

Salamandrida: Strauch, 1870: 14; Knauer, 1878: 98. Explicit family.

Salamandroideae: Garman, 1884a: 37. Implied superfamily. Stejneger, 1907: 3. Explicit superfamily.

Salamandridi: Acloque, 1900: 494.

Salamandroidae: Dubois, 2005b: 19; Dubois & Raffaëlli, 2012: 148. Explicit epifamily.

"Cercopi" Wagler, 1828: 859. Explicit family to contain "*Salamandra etc.";* nomen unavailable.

Tritones Tschudi, 1838: 26.—**NG**: *Triton* Laurenti, 1768.

Tritonides: Tschudi, 1838: 26.—**NG**: *Triton* Laurenti, 1768.

Tritonines: Bronn, 1849: 683.

Tritonina: Bonaparte, 1850: 1.—**NG**: *Triton* Laurenti, 1768.

Tritonidae: Hallowell, 1857a: 10; 1858b: 337.

Tritonidi: Acloque, 1900: 495.

Seiranotina Gray, 1850: 29.—**NG**: *Seiranota* Barnes, 1826 (= *Salamandrina* Fitzinger, 1826).—**SS**: Cope, 1875a: 11 with Pleurodelidae; Boulenger, 1882b: 2, with Salamandrinae.

Seiranotidae: Hallowell, 1857a: 10.

Bradybatina Bonaparte, 1850: 1.—**NG**: *Bradybates* Tschudi, 1838.

Geotritonidae Bonaparte, 1850: 1.—**NG**: *Geotriton* Bonaparte, 1832 (= *Triturus* Rafinesque, 1815).

Geotritonina: Bonaparte, 1850: 1.

"Triturinae" Brame, 1957: 9 *fide* Frost (2009-13), not seen.—**NG**: *Triturus* Rafinesque, 1815. Unavailable nomen.

Triturinae Kuhn, 1965: 37. attributed to Brame, **NG** same.

Distribution: Much of Europe and Asia, northern Africa, and North America to northern Mexico.

Content: 21 genera with 86 species, of which two genera and seven species occur in North America (8.1% of the family species diversity). Three subfamilies are recognized. One extralimital subfamily, Salamandrinae, includes 2 genera and 21 species; the other contains one genus with two species.

Common name: Newts. *French Canadian name:* Tritones.

Commentary: Gray (1825) is often assigned authorship, but Goldfuss (1820) used the name earlier (with a different suffix) so it should be attributed to him. Shen *et al.* (2013) found this family to be sister to the clade formed by Ambystomatidae + Dicamptodontidae, and those three families should probably be assigned to a separate infraorder.

Subfamily **Pleurodelinae** Tschudi, 1838

Pleurodelinae Tschudi, 1838: pp.—**NG**: *Pleurodeles* Michahelles, 1830 (etymologically implied).

Pleurodelinae: Bonaparte, 1839e: unnumbered; Dubois & Raffaëlli, 2009: 42; Dubois & Raffaëlli, 2012: 148.

Pleurodelina: Bonaparte, 1840a: 11; 1850: [2].

Pleurodelae: Fitzinger, 1843: 33.

Pleurodelidae: Bonaparte, 1850: [2].

Pleurodelini: Massalongo, 1853: 11.

Pleurodelidae: Hallowell, 1857a: 10.

Distribution: Asia, Europe, the Middle East, northern Africa, and North America.

Content: Eighteen genera and 67 species, of which 2 genera and 7 species are in North America (10.4% of subfamily species diversity).

Tribe **Molgini** Gray, 1950

Molgini Gray, 1850: pp.—**NG:** *Molge* Merrem, 1820: 166 (od.).

Molgina: Bonaparte, 1850: [2].

Molgini: Dubois & Raffaëlli, 2009: 44; Dubois & Raffaëlli, 2012: 148.

Distribution: Asia, Europe and the Middle East, plus North Ame-rica.

Content: Fourteen genera with 53 species, of which 2 genera and 7 species are in North America (13.2% of tribal species diversity).

Commentary: Treated as a tribe by Dubois & Raffaëlli, 2009, 2012.

Subtribe **Tarichina** Dubois & Raffaëlli, 2009

Tarichina Dubois & Raffaëlli, 2009: 57.—**NG**: *Taricha* Gray, 1850 (od.).

Tarichina: Dubois & Raffaëlli, 2012: 150.

Distribution: Eastern and western North America, and south into northern Mexico, but absent from a central portion of North America.

Content: Two genera with seven species, all of which occur in North America.

Commentary: Recognized as a subtribe for North American sala-mandrids, separate from the Old World subtribe Molgina (Dubois & Raffaëlli, 2009). The study of Zhang *et al.* (2008) supports this action, finding this North American taxon monophyletic, derived from European ancestors in middle Eocene or late Cretaceous.

Genus **Notophthalmus** (Rafinesque, 1820)

Synonymy: See below under subgenus *Notophthalmus.*

Distribution: Eastern US and adjacent Canada, to coastal northeastern Mexico.

Common name: Eastern Newts.

Content: Two subgenera, with three species, all in North America.

Subgenus **Notophthalmus** (Rafinesque, 1820)

Notophthalmus Rafinesque, 1820: 5.—**NS** (mo.): *Triturus (Diemictylus) miniatus* Rafinesque, 1820 (= *viridescens*). On official list of generic names (ICZN, 1962).

"*Notopthalmus*:" Baird, 1950a: 284.—**IS**, on official list of rejected/invalid names (ICZN, 1962).

"*Notophthalmia*:" Gray, 1850: 22.—**UE**, on official list of rejected/invalid names (ICZN, 1962).

"*Notophthalma*:" Gray, 1858a: 138.—**UE**, on official list of rejected/invalid names (ICZN, 1962).

Notophthalmus: Gill, 1907: 256; Stejneger & Barbour, 1917: 7; Dubois & Raffaëlli, 2009: 57 (as genus).

Notophthalmus: Dubois & Raffaëlli, 2009: 57 (as subgenus).

Diemictylus Rafinesque, 1820: 5.—**NS** (od.): *Triturus (Diemictylus) viridescens* (as subgenus of *Triturus*).

"*Diemyctylus*:" Hallowell, 1857a: 11.—**IS**, on official list of rejected/invalid names (ICZN, 1962).

"*Diemichylus*:" Cope, 1859a: 128.—**IS**, on official list of rejected/invalid names (ICZN, 1962).

"*Diemyctelus*:" Günther, 1901c: 294.—**IS**, on official list of rejected/invalid names (ICZN, 1962).

Diemictylus: Schmidt, 1953: 23. (as a genus)

Tristella Gray, 1850: 23.—**NS**: *Salamandra symmetrica* Harlan, 1825, mo.

Distribution: Eastern U.S. and adjacent Canada.

Common name: Eastern Newts. *French Canadian name:* Tritones orientales.

Content: Two species, both in North America.

Commentary: Dubois and Rafaelli (2009) retained two species in this subgenus, but split *N. meridionalis* into a separate subgenus, *Rafinus*.

Notophthalmus (Notophthalmus) perstriatus (Bishop, 1941)

Triturus perstriatus Bishop, 1941: 3.—**O**: **H**: UMMZ 89761.—**OT**: Georgia, Charlton County, Dodge pond, 2 miles east of Chesser's Island.

Triturus perstriatus: Stejneger & Barbour, 1943: 6.

Diemyctylus perstriatus: Neill, 1952: 195; 1954: 79.

Diemictylus viridescens perstriatus: Schmidt, 1953: 25.

Diemictylus perstriatus: Conant, 1958: 215.

Notophthalmus perstriatus: H.M. Smith, 1953a: 98; Mecham, 1967a: 1; Brame, 1967: 7; Cochran & Goin, 1970: 22; Gorham, 1974: 22; Conant, 1975: 259; Behler & King, 1979: 276; Frost, 1985: 611; Conant & Collins, 1998: 444; Petranka, 1998: 448; Raffaëlli, 2007: 139.

Notophthalmus (Notophthalmus) perstriatus: Dubois & Raffaëlli, 2009: Table 5.

Distribution: Northern peninsular Florida and southern Georgia. Petranka (1998) has a range map. **U.S.**: FL, GA.

Common name: Striped Newt.

Commentary: Mecham (1967a) provided a CAAR account.

Notophthalmus (Notophthalmus) viridescens (Rafinesque, 1820)

Synonymy: See the subspecies *N. v. viridescens,* below.

Distribution: Eastern US and adjacent Canada. Petranka (1998) has a range map. **Can.**: NB, NS, ON, PE, QC. **U.S.**: AL, AR, CT, DC, DE, FL, GA, IA, IL, IN, KS, KY, LA, MA, MD, ME, MI, MN, MO, MS, NC, NH, NJ, NY, OH, OK, PA, RI, SC, TN, TX, VA, VT, WI, WV.

Common name: Eastern Newt. *French Canadian name:* Triton vert.

Content: Four subspecies are usually recognized.

Commentary: The species has a long and complex nomenclatural and taxonomic history. Mecham (1967b) provided a CAAR account of this species and its subspecies.

Notophthalmus (Notophthalmus) viridescens viridescens (Rafinesque, 1820)

"Salamandra americana" Houttuyn, 1782: 330.—**O**: **H:** specimen figured by Seba, 1734: pl. 89, fig 4, 5.—**OT**; not stated, but implied North America.—**NO:** *fide* Frost, 2009-13.—**SS**: Frost, 2009-13.

Triturus (Diemictylus) viridescens Rafinesque, 1820: 5.—**O**: none designated.—**OT**: in Lake George, Lake Champlain, the springs and brooks of the neighborhood; this could be in the U.S., New York, Lake George, or the U.S., Vermont and New York, Lake Champlain, or Can., Quebec, Lake Champlain.—**Comment:** The specific nomen was placed on the official list of specific names (ICZN, 1962), as it was selected by Cope (1859a) in preference to *miniatus,* as first reviser.

Notophthalmus viridescens: Baird, 1850a: 284; Gray, 1850: 23; Gill, 1907: 256. Stejneger & Barbour, 1917: 8; H.M. Smith, 1953a: 96; Mecham, 1967b: 1; Frost, 1985: 611.

Notophthalma viridescens: Gray, 1858a: 138.—**UE** (generic name).

Triton viridescens: Strauch, 1870: 50.

Triturus viridescens: Boulenger, 1878: 307; Viosca, 1949: 9.

Molge viridescens: Boulenger, 1882b: 21.

Diemyctylus miniatus viridescens: Yarrow, 1883a: 22; Davis & Rice, 1883: 27.

Diemyctylus viridescens viridescens: Cope, 1889b: 207; Strecker, 1915: 54; Schmidt, 1953: 24.

Notophthalmus viridescens viridescens: Stejneger & Barbour, 1917: 8; Smith, 1953a: 96; Mecham, 1967b: 2; Brame, 1967: 7; Gorham, 1974: 22; Conant, 1975: 257; Behler & King, 1979: 277; Conant & Collins, 1998: 442; Petranka, 1998: 452.

Triturus viridescens: Dunn, 1918b: 451.

Triturus viridescens viridescens: Stejneger & Barbour, 1923: 4; 1943: 7; Bishop, 1924: 88; 1928a: 156.

Diemictylus viridescens viridescens: Conant, 1958: 214.

Notophthalmus (Notophthalmus) viridescens viridescens: Dubois & Raffaëlli, 2009: Table 5.

Triturus (Notophthalmus) miniatus Rafinesque, 1820: 5.—**O**: none stated.—**OT**: Eastern U.S.—**SS:** Cope, 1859a.

Notophthalmus miniatus: Gray, 1850: 22; Baird *in* Heck & Baird, 1851; 254.

Diemyctylus miniatus miniatus: Yarrow, 1883a: 22; Davis & Rice, 1883: 27.

Salamandra symmetrica Harlan, 1825b: 157.—**O**: **S** not stated; ANSP 1582 and 1584, *fide* Malnate, 1971: 348.—**OT**: U.S., South Carolina.—**SS**: Gray, 1850: 23 and Baird, 1850a: 284, with *N. miniatus*; Bishop, 1941: 1, with *T. v. viridescens*.

Triton symetricus: Holbrook, 1842: 57; Duméril *et al.*, 1854c: 154.—**IS**.

Tristella symmetrica: Gray, 1850: 23.

Triturus viridescens symmetrica: Schmidt, 1924: 67; Myers, 1927: 337; Stejneger & Barbour, 1933: 4; Carr, 1940: 45.

Salamandra punctatissima Wood, 1825: 306.—**O**: not stated.—**OT**: unknown.—**SS**: Duméril, Bibron, and Duméril, 1854c: 154, with *N. viridescens*.

Triton punctatissimus: Duméril *et al.*, 1854c: 154.

"Salamandra greeni" Gray, 1831a: 107.—**NU**.—**SS**: Boulenger, 1882b: 22.

Salamandra coccinea DeKay, 1842: 81.—**O**: based partly on NYSM specimen in pl. 21, fig 54b.—**OT**: U.S., New York, Hamilton County, Lake Pleasant, in a forest.—**SS**: Gage, 1891: 1084.

"Molge ocellata" Gray, 1850: 23.—**O**: not stated, probably in BM.—**OT**: North America.—**NU**. Name attributed to Bell.—**SS**: Gray, 1850: 23 with *N. miniatus*.

Salamandra ventralis Provancher, 1875: 251.—**O**: not stated.—**OT**: Can., Quebec, "le lac sur la montagne d'Yamaska."—**SS**: Schmidt, 1953: 24, with *Diemictylus v. viridescens*.

Diemyctylus viridescens var. *vittatus* Garman, 1896: 49.—**O**: **S**, 15 deposited MCZ. Now includes USNM 23797-23799, *fide* Cochran, 1961a: 13, and MCZ 1992 (2) *fide* Barbour & Loveridge, 1929: 252.—**OT**: U.S., North Carolina, New Hanover County, near Wilmington in shallow pools.

Notophthalmus viridescens vittatus: Stejneger & Barbour, 1917: 8.

Distribution: Upland southeastern US north to southeastern Canada. Range maps in Mecham (1967b) and Petranka (1998). **Can.**: NB, NS, ON, PE, QC. **U.S.**: AL, CT, DC, DE, GA, IN, KY, MA, MD, ME, MI, NC, NH, NJ, NY, OH, PA, SC, TN, VA, VT, WV.

Common name: Red-spotted Newt; eft stage, Red Eft. *French Canadian name:* Triton vert.

Notophthalmus (Notophthalmus) viridescens dorsalis (Harlan, 1829)

Salamandra dorsalis Harlan, 1829: 101.—**O**: **S:** ANSP 1583, 1585, 1586 (Malnate, 1971: 347, sd.).—**OT**: South Carolina.—**SS**: Baird, 1850a: 284.

Triton dorsalis: Holbrook, 1842: 77.

Triturus dorsalis: Dunn, 1918b: 452.

Diemictylus viridescens dorsalis: Wolterstorff, 1925: 293; Conant, 1958: 215.

Triturus viridescens dorsalis: Bishop, 1943: 103; Stejneger & Barbour, 1923: 4; 1943: 7.

Notophthalmus viridescens dorsalis: Smith, 1953: 96; Cochran & Goin, 1970: 21; Gorham, 1974: 22; Conant, 1975: 258; Behler & King, 1979: 277; Conant & Collins, 1998: 444; Petranka, 1998: 452.

Diemictylus viridescens dorsalis: Wolterstorff, 1925; Schmidt, 1953: 24; Mecham, 1967b: 2; Brame, 1967.

Notophthalmus (Notophthalmus) viridescens dorsalis: Dubois & Raffaëlli, 2009: Table 5.

Salamandra millepunctata Storer, 1840: 60.—**O**: **S:** numerous indicated, but deposition unknown.—**OT**: U.S., Massachusetts, from Roxbury and Amherst. Coined provisionally as synonym of *S. dorsalis*.—**SS**: Baird, 1850a: 284.

Triton millepunctatus: DeKay, 1842: 81.

Distribution: Northeastern South Carolina and southeastern North Carolina. **U.S.**: NC, SC.

Common name: Broken-striped Newt.

Commentary: Schmidt (1953) restricted the type locality to north-eastern South Carolina, but with no explanation, so that action is invalid.

Notophthalmus (Notophthalmus) viridescens louisianensis (Wolterstorff, 1914)

Diemyctylus viridescens louisianensis Wolterstorff, 1914: 383.—**O**: none designated, but probably MM, likely now destroyed.—**OT**: U.S., Louisiana, New Orleans.

Diemictylus viridescens louisianensis: Wolterstorff, 1925: 293.

Triturus viridescens louisianae: Strecker & Williams, 1928: 5.—**IS**

Triturus viridescens louisianensis: Schmidt & Necker, 1935: 62 (*fide* Frost, 2009-13, not seen); Bishop, 1943: 106; Stejneger & Barbour, 1943: 7.

Notophthalmus viridescens louisianensis: Herre, 1936: 196 (*fide* Frost 2009-13, not seen); H.M. Smith, 1953: 97; Mecham, 1967b: 2; Brame, 1967: 7; Cochran & Goin, 1970: 21; Gorham, 1974: 22; Conant, 1975: 258; Behler & King, 1979: 277; Garrett & Barker, 1987: 71; Conant & Collins, 1998: 443; Petranka, 1998: 452; Dixon, 2000: 60.

Triturus louisianensis: Carr, 1940: 45.

Diemictylus viridescens louisianensis: Schmidt, 1953: 24; Conant, 1958: 214.

Notophthalmus (Notophthalmus) viridescens louisianensis: Dubois & Raffaëlli, 2009: Table 5.

Distribution: Gulf coast from eastern Texas to Florida, to Atlantic coast to South Carolina; Mississippi valley north to Great Lakes region and adjacent Canada. **Can.**: ON. **U.S.**: AL, AR, FL, GA, IA, IL, IN, KS, LA, MI, MN, MO, MS, OK, SC, TX, WI.

Common name: Central Newt. *French Canadian name:* Triton central.

Notophthalmus (Notophthalmus) viridescens piaropicola (Schwartz & Duellman, 1952)

Diemictylus viridescens piaropicola Schwartz & Duellman, 1952: 219.—**O**: **H**: UMMZ 106333.—**OT**: U.S., Florida, Collier County, 5.2 miles east of Monroe Station.

Notophthalmus viridescens piaropicola: H.M. Smith, 1953: 97; Mecham, 1967b: 3; Brame, 1967: 7; Cochran & Goin, 1970: 22; Gorham, 1974: 22; Conant, 1975: 258; Behler & King, 1979: 277; Conant & Collins, 1998: 444; Petranka, 1998: 452.

Diemictylus viridescens piaropicola: Conant, 1958: 215.

Notophthalmus (Notophthalmus) viridescens piaropicola: Dubois & Raffaëlli, 2009: Table 5.

Diemictylus viridescens evergladensis Peterson, 1952: 103.—**O**: **H**: UIMNH 28740.—**OT**: U.S., Florida, Monroe County, 60 mi. W Miami, U.S. Highway 94.

Distribution: Peninsular Florida. **U.S.**: FL.

Common name: Peninsula Newt.

Subgenus **Rafinus** Dubois & Raffaëlli, 2009

Rafinus Dubois & Raffaëlli, 2009: 57 (as subgenus).—**NS:** *Diemyctylus miniatus meridionalis* Cope, 1880 (od.).

Distribution: As for the only species.

Common name: No name in use. Southwestern Newts may be appropriate.

Content: A single species.

Notophthalmus (Rafinus) meridionalis (Cope, 1880)

Diemyctylus miniatus meridionalis Cope, 1880: 30.—**O**: none designated but the only specimen mentioned sent to USNM, apparently lost; ANSP 1104-1106, 15977 catalogued as types (Mecham, 1968a).—**OT**: Mexico, Matamoros, ... tributaries of the Medina River and southward. Restricted to Mexico, Tamaulipas, Matamoros, by Smith & Taylor, 1948: 15.

Molge meridionalis: Boulenger, 1888a: 24; 1890a: 326.

Diemyctylus viridescens meridionalis: Cope, 1889b: 211; Strecker, 1915: 54.

Diemyctylus meridionalis: Strecker, 1908: 48.

Notophthalmus meridionalis: Stejneger & Barbour, 1917: 7; Cochran & Goin, 1970: 22; Frost, 1985: 611; Garrett & Barker, 1987: 70;.

Triturus meridionalis: Dunn, 1918b: 452; Stejneger & Barbour, 1923: 3; 1943: 5.

Diemictylus meridionalis: Wolterstorff, 1925: 293; Schmidt, 1953: 25; Conant, 1958: 216.

Notophthalmus meridionalis meridionalis: Mecham, 1968b: 124; Gorham, 1974: 22; Conant, 1975: 259; Behler & King, 1979: 276; Conant & Collins,1998: 445; Petranka, 1998: 446; Dixon, 2000: 60.

Notophthalmus (Rafinus) meridionalis meridionalis: Dubois & Raffaëlli, 2009: Table 5.

Diemyctylus kallerti Wolterstorff, 1930: 147.—**O: H**: MM N.V. 44/29 ex. nr. 1, destroyed (Brame, 1972: 58, *fide* Frost, 2009-13).—**OT**: Mexico, Veracruz, Tampico.—**SS** with *N. viridescens* by Herre, 1936: 196; with *N. meridionalis* by Mecham, 1968b: 124.

Triturus kallerti: Smith, 1934a: 407.

Notophthalmus viridescens kallerti: Herre, 1936: 196.

Notophthalmus kallerti: H.M. Smith, 1953a: 98; Brame, 1967: 7.

Notophthalmus meridionalis kallerti: Mecham, 1968b: 124; Gorham, 1974: 22.

Distribution: Coastal southern Texas and northeastern Mexico. Mecham (1968a) and Petranka (1998) have range maps; that of Mecham shows the two subspecies. **U.S.**: TX.

Common name: Black-spotted Newt.

Content: Two subspecies are recognized; only the nominotypical one occurs north of Mexico (Texas Black-spotted Newt), so it is not listed separately here. *N. m. kallerti* occurs in Mexico.

Commentary: Mecham (1968a) provided a CAAR account.

Genus **Taricha** Gray, 1850

Taricha Gray, 1850: 25.—**NS**: *Triton torosa* Rathke, 1833 (mo.).

Taricha: Dubois & Raffaëlli, 2009: 58 (as subgenus).

Distribution: Coastal California to southwestern Alaska, plus northeastern California. Petranka (1998) has range maps of the species.

Common name: Pacific Newts.

Content: Four species, all endemic to North America.

Commentary: Nussbaum & Brodie (1981a) provided a CAAR generic account. Tihen (1974) reduced the fossil genus *Paleotaricha* van Frank to a subgenus of *Taricha.* This checklist deals only with extant species, so we do not list that subgenus. However, Dubois & Raffaëlli (2009) split one extant species into a separate subgenus, *Twittya,* leaving the others in a subgenus *Taricha*; we here abandon the subgenera for extant species.

Taricha granulosa (Skilton, 1849)

Salamandra (Triton) granulosa Skilton, 1849: 202.—**O**: **S**: 2 or more in Troy Lyceum of Natural History, Cabinet of A.J. Skilton, not now known to exist.—**OT**: U.S., Oregon. See commentary for the species.

Triturus granulosus: Twitty, 1935: 76.

Triturus granulosus granulosus: Bishop, 1941: 18; Stejneger & Barbour, 1943: 4.

Taricha granulosa: Wahlert, 1952: 30; Frost, 1985: 613; Stebbins, 1985: 41; 2003: 165.

Taricha granulosa granulosa: Schmidt, 1953: 26; Brame, 1967: 8; Cochran & Goin, 1970: 23; Gorham, 1974: 24; Behler & King, 1979: 278; Petranka, 1998: 464.

Taricha torosa granulosa: Pimentel, 1958: 168.

Taricha (Taricha) granulosa granulosa: Dubois & Raffaëlli, 2009: Table 5.

Triturus similans Twitty, 1935: 76.—**O**: **H**: MVZ 18149.—**OT**: U.S., California, Mendocino County, Ukiah, Robinson Creek.—**SS**: Fitch, 1938a: 149.

Triturus granulosus similans: Myers, 1942a: 79; Stejneger & Barbour, 1943: 5.

Taricha granulosa similans: Schmidt, 1953: 27.

Taricha torosa simulans: Pimentel, 1958: 168.—**IS.**

Triturus granulosus twittyi Bishop, 1941: 18.—**O**: **H**: UMMZ 89763.—**OT**: U.S., California, Santa Clara County, Saratoga. Subspecies rejected by Riemer, 1958: 385.

Triturus granulosus twittyi: Stejneger & Barbour, 1943: 5.

Taricha granulosa twittyi: Schmidt, 1953: 26.

Taricha torosa twittyi: Pimentel, 1958: 168.

Triturus granulosus mazamae Myers, 1942a: 80.—**O**: **H**: CAS-SU 7465.—**OT**: U.S., Oregon, Klamath County, Crater Lake National Park, near Crater Lake on the higher slopes of Mount Mazama.

Triturus granulosus mazamae: Bishop, 1943, 75; Stejneger & Barbour, 1943: 5.

Taricha granulosa mazamae: Schmidt, 1953: 26; Brame, 1967: 8; Gorham, 1974: 24; Petranka, 1998: 464.

Taricha torosa mazamae: Pimentel, 1958: 168.

Taricha (Taricha) granulosa mazamae: Dubois & Raffaëlli, 2009: Table 5.

Distribution: Coastal northern California to Alaska; introduced in northern Idaho and northwestern Montana. Petranka (1998) has range maps, including subspecies. **Can.**: BC. **U.S.**: AK, CA, *ID, *MT, OR, WA.

Common name: Roughskin Newt. *French Canadian name:* Triton rugueux.

Content: Two subspecies are sometimes recognized. One of these, *T. g. mazamae* seems to be an ecotype and is not recognized by workers in the area, so we do not recognize it here. Two others have been listed in the past but are no longer recognized.

Commentary: Fitch (1938a) restricted the type locality to U.S., Oregon, Clackamas County, Oregon City, based on the collector's home and travels, as well as the lack of suitable habitat in the OT indicated by the original description; we accept that restriction as valid. Nussbaum & Brodie (1981b) provided a CAAR account, including former subspecies rejected here.

Taricha rivularis (Twitty, 1935)

Triturus rivularis Twitty, 1935: 73.—**O: H**: MVZ 18131.—**OT**: U.S., California, Mendocino County, Gibson Creek, about one mile west of Ukiah.

Triturus rivularis: Stejneger & Barbour, 1943: 6.

Taricha rivularis: Wahlert, 1952: 30; Schmidt, 1953: 27; Twitty, 1964b: 1; Gorham, 1974: 24; Frost, 1985: 614; Stebbins, 1985: 43; 2003: 167; Petranka, 1998: 167.

Taricha (Twittya) rivularis: Dubois & Raffaëlli, 2009: Table 5.

Distribution: Coastal northern California. Petranka (1998) has a range map. **U.S.**: CA.

Common name: Red-bellied Newt (SECSN) or Redbelly Newt (C&T).

Commentary: Twitty (1964b) provided a CAAR account.

Taricha sierrae (Twitty, 1942)

Triturus sierrae Twitty, 1942: 65.-—**O: H**: CAS-SU 2425.—**OT**: U.S., California, Butte County, Cherokee Creek, in the hills above Chico.

Triturus sierrae: Stejneger & Barbour, 1943: 6.

Triturus torosus sierrae: Stebbins, 1951: 28.

Taricha sierrae: Wahlert, 1952: 30; Collins, 1991: 43.

Taricha torosa sierrae: Schmidt, 1953: 26; Brame, 1967: 8; Cochran & Goin, 1970: 24; Gorham, 1974: 24; Behler & King, 1979: 280; Stebbins, 1985: 43; 2003: 167; Petranka, 1998: 474.

Taricha (Taricha) sierrae: Dubois & Raffaëlli, 2009: Table 5.

Distribution: Northeastern California in the Sierra Nevada. Petranka (1998) shows the range on the map of *T. torosa,* as a subspecies, isolated to the east of *T. torosa* range. **U.S.**: CA.

Common name: Sierra Newt.

Taricha torosa (Rathke, 1833)

Triton torosa Rathke, *in* Eschscholtz, 1833: 12.—**O**: none designated nor known to exist.—**OT**: U.S., California, vicinity of San Francisco Bay.

Taricha torosa: Herre, 1934: 250; Gray, 1850: 25; Frost, 1985: 614; Stebbins, 2003: 167.

Notophthalmus torosus: Baird, 1850a: 284; Stejneger & Barbour, 1917: 7; Grinnell & Storer, 1924: 651.

Diemyctylus torosus: Cope, 1875b: 28; 1889b: 204.

Triton torosus: Wiedersheim, 1877: 465.

Molge torosa: Boulenger, 1882b: 20.
Cynops torosus: Cope, 1884: 25.
Diemictylus torosus: Fowler & Dunn, 1917: 28.; Wolterstorff, 1925: 294.
Triturus torosus: Dunn, 1918b: 450; Stejneger & Barbour, 1923: 3; Bishop, 1943: 95.
Taricha torosa torosa: Wolterstorff, 1935: 179, by implication; Schmidt, 1953: 25; Cochran & Goin, 1970: 24; Gorham, 1974: 24; Behler & King, 1979: 280; Stebbins, 1985: 43; Liner, 1994: 15.
Triturus torosus torosus: Stejneger & Barbour, 1943: 6.
Taricha (Taricha) torosa: Dubois & Raffaëlli, 2009: Table 5.

Triton ermani Wiegmann *in* Erman, 1835: 24.—**O**: **H**: ZMB 3674 (Bauer *et al.*, 1993: 300, sd.), lost.—**OT**: U.S., California, San Francisco.—**SS**: Baird, 1850a: 284; Gray, 1850: 25; note: Riemer (1958) expressed doubt about the synonymy, indicating it might instead be a synonym of *T. granulosa*.

Salamandra beecheyi Gray, 1839: 98.—**O**: **S**: 2 in "Museum of the College of Surgeons," probably lost.—**OT**: U.S., California, Monterey.—**SS**: Gray, 1850: 25.; Boulenger, 1882b: 20.

"Pleurodeles californiae" Gray, 1850: 25.—**NU**, attributed to Bibron.

Taricha laevis Baird & Girard, 1853a: 302.—**O**: **S**: USNM 4014 (5; Cochran, 1961a: 26, sd.).—**OT**: U.S., California, San Francisco County, San Francisco.—**SS**: Boulenger, 1882b: 20.
Triton laevis: Strauch, 1870: 50.

Amblystoma rubrum Reid, 1895: 600.—**O**: not stated nor known to exist.—**OT**: U.S., California, Los Angeles County, vicinity of Pasadena, pools in mountain canyons.—**SS**: Storer, 1925: 44.—**Comment**: This name was presented with no indication there is any type material, just a book about Pasadena, mentions that water puppies occur in the mountain ponds & are called "*amblystoma rubrum*". Certainly a *nomen nudum* unless it refers to an existing name, which seems not to be true.

Taricha torosa klauberi Wolterstorff, 1935: 179.—**O**: **H**: MM 108.22, destroyed (Brame, 1972: 83, *fide* Frost, 2009-13).—**OT**: U.S., California, San Diego County, Boulder Creek.—**SS**: Schmidt, 1953: 25.
Triturus torosus klauberi: Twitty, 1942: 65; Myers, 1942a: 77; Stejneger & Barbour, 1943: 7.
Triturus klauberi: Bishop, 1943: 80.
Taricha klauberi: Smith & Taylor, 1948: 15.

Taricha torosa forma *monstrosa* Scharlinski, 1939: 50 (*fide* Frost, 2009-13, not seen).—**O**: **S**: MM 4897 (many), destroyed (Brame, 1972: 83, *fide* Frost, 2009-13).—**NU.**

Distribution: Most of coastal California. Petranka (1998) has a range map, but the range illustrated in northeastern California is now recognized as for a separate species, *T. sierrae*. **U.S.**: CA.

Common name: California Newt.

Content: No subspecies are currently recognized. Formerly, *T. sierrae* was considered a subspecies of *T. torosa* but it has been elevated to full species status.

Commentary: The species was described by Rathke in the work by Eschscholtz (1833), so see the latter reference for the description. Prior to 1935, this was the only species of *Taricha* generally recognized, so some references to that name may actually be to populations now recognized as other species. Nussbaum & Brodie (1981c) provided a CAAR account.

Order **Anura** Duméril, 1805

Salientia Laurenti, 1768: 24 (this nomen is often used for the Superorder containing anurans and fossil taxa).

Salientia: Merrem, 1820: 164; Stejneger & Barbour, 1917: 25; Schmidt, 1953: 57.

Anoures Duméril, 1805: 91

Anuri: Fischer, 1813: 58

Anuria: Rafinesque, 1815: 58

Anoura: Gray, 1825: 213.

Anura: Hogg, 1839: 270; Frost. 2009-13.

Batrachia Merrem, 1820: 163.

Batrachia: Tschudi, 1838: 25.

Ranae Spix, 1824: 23.

Ranae: Wagler, 1830: 199; Bonaparte, 1850: [2].

Distribution: Nearly cosmopolitan, but absent from Antarctica, the arctic, and most oceanic islands.

Common name: Frogs (or, Frogs and Toads).

Content: There are 54 families, containing over 400 genera, with about 6200 species (Frost, 2009-13). Of those there are 12 families, containing 20 genera, with 109 species in North America (including 5 introduced species). Thus North America contains only about 1.7% of the anuran species diversity. This is in contrast to the much higher proportion of salamanders found here; the reason is that salamanders have a mostly holarctic distribution, whereas anurans have a more tropical distribution.

Commentary: Anura and Salientia have been used more or less interchangeably for the Order of living frogs (*sensu lato*). In recent usage, since Romer (1945) used Salientia as a superorder to include the Order Anura and an order of fossil ancestors to frogs, most workers seem to have followed that usage. Again, the *Code* does not currently apply to taxa above the family-series, so we are free to use whichever names we prefer, but many of us tend to extend the priority principle to the higher ranks. In this checklist, we follow the rules of Dubois (2006a; see also 2009a) in assigning authorship of this taxon to Duméril (see *Commentary* for Order Urodela). Laurenti's Salientia included the salamander *Proteus*, as well as frogs. Merrem (1820) seems to be the first to use it for the order including only frogs. Duméril (1805) was the first to use Anoures to apply only to frogs, and that name was later latinized by Hogg (1839) providing the necessary treatment to satisfy assigning authorship to Duméril. It is also the name most often used for the Order in the last 50 years. Ford & Cannatella (1993) defined the Order Anura cladistically, and that was conceptually adopted as the definition for Anura by Frost *et al.* (2006a). Frost (2009-13) has a more extensive synonymy for the order.

The families of anurans are arranged here in four superfamilies, which are listed in alphabetical order. Within superfamilies, where there are more than one family included, they are listed in alphabetical order. However, in the largest superfamily, Ranoidea, we treat the families in three epifamilies, which are listed in alphabetical order, then the families are alphabetized within epifamily. The genera are listed alphabetically within family, or in some families there are subfamilies, in which those are listed alphabetically, then the genera are alphabetized within subfamily. If there are subgenera, the nominotypical subgenus is listed first, then others follow alphabetically. Species are arranged alphabetically within genus or subgenus. If there are subspecies, the nominotypical one is listed first, then the others follow alphabetically.

North America contains only 1.7% of the species diversity of this Order.

Superfamily **Leiopelmatoidia** Mivart, 1869

Synonymy: See the family Leiopelmatidae, below.

Content: We recognize a single family, containing two genera.

Family **Leiopelmatidae** Mivart, 1869

Liopelmatina Mivart, 1869: 291.—**NG**: *Liopelma* Günther, 1869 (now *Leiopelma* Fitzinger, 1861). Incorrect original spelling of Leiopelmatidae, placed on official list of rejected and invalid names (ICZN, 1977).

Liopelmidae: Noble, 1924: 9.—**OS,** by ICZN, 1977.
Leiopelmidae: Turbott, 1942: 247.—**OS**, by ICZN, 1977.
Leiopelmatidae: Stephenson, 1951: 18.
Leiopelmatinae: Kuhn, 1965: 86; Dubois, 2005b: 8.
Liopelmoidea: Laurent, 1967: 207.
Leiopelmatoidia: Dubois, 2005b: 8.
Leiopelmatoidea: Dubois, 2005b: 8.

Ascaphidae Fejévåry, 1923: 178.—**NG**: *Ascaphus* Stejneger, 1899.

Ascaphidae: Schmidt, 1953: 57; Adams, 2005: 382; Pyron & Wiens, 2011: 554; Blackburn & Wake, 2011: 40.
Ascaphoidea: Lynch, 1973: 162.

Distribution: A North American genus in northwestern U.S. and adjacent Canada; another genus (*Leiopelma*) in New Zealand.

Common name: Tailed Frogs and New Zealand Frogs.

Content: Two genera, with six species; one genus with two species in North America.

Commentary: The species of this family are apparently the most primitive living frogs. The two genera are sometimes placed in separate families, especially on geographic grounds, but Frost *et al.* (2006a) and Roelants *et al.* (2007) found they are sister-taxa, phylogenetically ancestral to all other frogs, and Frost *et al.* (2006a) placed them together in one family, following Stephenson (1951), and we follow that treatment. This family name has been placed on the official list of family-series names (ICZN, 1977). The analysis of Pyron & Wiens (2011) supports the sister family relationship of the North American *Ascaphus* and the New Zealand *Leiopelma,* but they preferred to treat each in a different family. Their analysis also supports these two taxa as a basal sister to all other anurans.

North America contains 33.3% of the species diversity of this family.

Genus **Ascaphus** Stejneger, 1899

Ascaphus Stejneger, 1899a: 899.—**NS**: *Ascaphus truei* Stejneger, 1899 (od. and mo.).

Distribution: Northwestern U.S. and adjacent Canada.

Common name: Tailed Frogs.

Content: Two species, both endemic to North America.

Commentary: Metter (1968) provided a CAAR generic account.

Ascaphus montanus Mittleman & Myers, 1949

Ascaphus truei montanus Mittleman & Myers, 1949: 64.—**O: H**: USNM 102505.—**OT**: U.S., Montana, Flathead County, Glacier National Park, tributary of Lincoln Creek.

Ascaphus truei montanus: Schmidt, 1953: 57; Gorham, 1974: 40.

Ascaphus montanus: Nielson, Lohman, & Sullivan, 2001: 156; Elliott *et al*, 2009: 302; Dodd, 2013: 1.

Misapplication of name:

Ascaphus truei: Metter, 1964: 189; Frost, 1985: 234.

Distribution: Western Montana through northern Idaho into northeastern Oregon and southeastern Washington, plus adjacent British Columbia. See map in Dodd (2013). **Can.**: BC. **U.S.:** ID, MT, OR, WA.

Common name: Rocky Mountain Tailed Frog (SECSN) or Eastern Tailed Frog (C&T). *French Canadian name:* Grenouille-à-queue des Rocheuses.

Content: No subspecies are recognized. See *Content* for *A. truei.*

Commentary: Nielson *et al.* (2001) recognized the former subspecies as distinct species. Adams (2005) provided a species account. Listed in British Columbia as Threatened. Stebbins (2003) listed only *A. truei,* and showed both species' ranges on his map, apparently including this species in the synonymy of *A. truei.*

Ascaphus truei Stejneger, 1899

Ascaphus truei Stejneger, 1899a: 900.—**O**: **H**: USNM 25979.—**OT**: U.S., Washington, Grays Harbor County, Humptulips.

Ascaphus truei: Stejneger & Barbour, 1917: 25; 1943: 36; Schmidt, 1953: 57; Cochran & Goin, 1970: 69; Behler & King, 1979: 359; Frost, 1985: 234; Stebbins, 1985:

63; 2003: 199; Nielson, Lohman, & Sullivan, 2001: 156; Elliott *et al.*, 2009: 302; Dodd, 2013: 7.

Ascaphus truei truei: Mittleman & Myers, 1949: 62; Schmidt, 1953: 57; Gorham, 1974: 40.

Ascaphus truei californicus Mittleman & Myers, 1949: 63.—**O**: **H** (od.), MVZ 19142.—**OT**: U.S., California, Del Norte County, near Klamath.—**SS:** Metter (1964: 189), subspecies rejected.

Ascaphus truei californicus: Schmidt, 1953: 57; Gorham, 1974: 40.

Distribution: Coastal to montane northwestern California, through western Oregon and Washington to southwestern British Columbia. See map in Dodd (2013). **Can.**: BC. **U.S.**: CA, OR, WA.

Common name: Coastal Tailed Frog (SECSN) or Western Tailed Frog (C&T). *French Canadian name:* Grenouille-à-queue côtière.

Content: No subspecies are currently recognized, but Mittleman & Myers (1949) described the California populations as a separate subspecies (*A. t. californicus*) from the more northern ones (*A. t. truei*), and the eastern populations (now considered a separate species) as another subspecies (*A. t. montanus*).

Commentary: Metter (1968) provided a CAAR species account, but it includes what is currently considered the separate species, *A. montanus*. *Adams & Pearl (2005) provided a recent account.* Nielson *et al.* (2001) presented genetic evidence to recognize the eastern populations as a separate species (see account of *A. montanus*, above). Listed as Special Concern in British Columbia.

Superfamily **Pelobatoidea** Bonaparte, 1850

Pelobatidae Bonaparte, 1850: [2].

Content: Four families are recognized (Frost *et al.*, 2006), only one of which is North American. The extralimital families are Megophrynidae, Pelobatidae, and Pelodytidae.

Commentary: Dubois (2005b) recognized these four taxa as subfamilies of a single family Pelobatidae, but Frost *et al.* (2006) and Pyron & Wiens (2011) treated them as separate families, and we also follow that treatment here.

Family **Scaphiopodidae** Cope, 1865

Alytidae Cope, 1863: 51.—**NG**: *Scaphiopus* Holbrook, 1836, by implication; immediately below the family name was printed *Scaphiopus,* which we interpret to mean that genus was specified as NG, even though *Alytes* was also included within the family.

Scaphiopodidae Cope, 1865a: 104, 107.—**NG**: *Scaphiopus* Holbrook, 1836, by implication.

Scaphiopodidae: Cope, 1866a: 80; Yarrow, 1875: 525; Günther, 1901b: 258; Storer, 1925: 43; Stejneger & Barbour, 1943: 37; Rocek, 1981: 151; Frost, 1985: 422; García-Paris *et al*, 2003: 12; Blackburn & Wake, 2011: 45.

Scaphiopodina: Mivart, 1869: 291.

Scaphiopidae: Cope, 1887a: 12; 1889b: 248.—**IS.**

Scaphiopinae: Spinar, 1972: 234.—**IS.**

Scaphiopodinae: Dubois, 1983: 271.

Distribution: Southeastern U.S. to western U.S. and northern Mexico, north to southwestern Canada.

Common name: North American Spadefoot Toads.

Content: Two genera with seven species are currently recognized, all in North America.

Commentary: The North American genera were formerly placed with their European allies in the family Pelobatidae. Dubois (2005b) included them as a subfamily (Scaphiopodinae) of the Pelobatidae (along with subfamilies Pelobatinae, Megophryinae, and Pelodytinae), under the superfamily Pelobatoidea Bonaparte, 1850. Frost *et al.* (2006) separated this taxon as a family Scaphiopodidae Cope and placed it and the families Pelobatidae, Megophryidae and Pelodytidae in the superfamily Pelobatoidea; they found Scaphiopodidae to be a sister taxon to Pelodytidae, rather than to the Pelobatidae. Roelants *et al.* (2007) found this family to be a sister-taxon to all other Pelobatoids, and Pyron & Wiens (2011) found this family to be a basal sister to the other Pelobatoids.

North America contains 100% of the species diversity of this family.

Genus **Scaphiopus** Holbrook, 1836

Scaphiopus Holbrook, 1836: 85.—**NS** (mo.): *Scaphiopus solitarius* Holbrook, 1836 (synonym *Rana holbrookii* Harlan, 1835a),

Distribution: Northeastern U.S. to southwestern U.S. and northern Mexico.

Common name: North American Spadefoots (SECSN) or Southern Spadefoots (or Southern Spadefoot Toads) (C&T).

Content: Three species are currently recognized.

Commentary: Western spadefoot toads (*Spea*) were formerly also included in this genus (*e.g.,* Schmidt, 1953: 59), sometimes as a subgenus or merely a species-group, but others regarded the two as distinct genera; see *Commentary* under the genus *Spea.*

Scaphiopus couchii Baird, 1856

Scaphiopus couchii Baird, 1856*:* 62.—**O**: **S**: USNM 3713-3715 (Wasserman, 1970, sd.); lost *fide* Kellogg (1932: 21).—**OT**: Mexico, Coahuila and Tamaulipas.

Scaphiopus couchii: Baird, 1859e: 28; Cope, 1863: 52; 1887a: 12; Yarrow, 1875: 526; Boulenger, 1882a: 434; Strecker, 1915: 52; Stejneger & Barbour, 1917: 25; 1943: 37; Wasserman, 1970: 1; Gorham, 1974: 44; Behler & King, 1979: 363; Frost, 1985: 422; Stebbins, 1985: 65; 2003: 202; Garrett & Barker, 1987: 38; Liner, 1994: 28; Conant & Collins, 1998: 505; Brennan & Holycross, 2006: 32; Lemos-Espinal & Smith, 2007a: 55, 249; Elliott *et al.,* 2009: 262; Dodd, 2013: 753.

Scaphiopus couchi: Yarrow, 1883a: 177; Schmidt, 1953: 58; Conant, 1958: 253; Chrapliwy & Malnate, 1961: 160; Dixon, 2000: 61.—**UE.**

Scaphiopus (Scaphiopus) couchii: Tanner, 1939: 6; Dubois, 1987b: 130.

Scaphiopus couchii couchii: Smith & Sanders, 1952: 209.

Scaphiopus varius Cope, 1863: 52.—**O**: **S:** USNM 5893 (3) (Cochran, 1961a, sd.).—**OT**: Mexico, Baja California, Cabo San Lucas.—**SS**: Boulenger (1882a: 434).

Scaphiopus couchi var. *varius*: Cope, 1867a: 313; 1887b: 12; Yarrow, 1875: 526.

Scaphiopus varius varius: Cope, 1875b: 31.

Scaphiopus rectifrenis Cope, 1863: 53.—**O**: **S**: USNM 3714 (Coahuila, lost), 3715 (Tamaulipas).—**OT**: Mexico, Coahuila and Tamaulipas.—**SS**: Cope, 1889b: 301.

Scaphiopus varius rectifrenis: Cope, 1875b: 31

Scaphiopus rectifrenis: Boulenger, 1882a: 435.

Scaphiopus couchii rectifrenis: Smith & Sanders, 1952: 209.

Spea laticeps Cope, 1893a: 155.—**O**: **H:** ANSP 13610 (Malnate, 1971: 351, sd.).—**OT**: U.S., Texas, between Seymour (in northwest Texas south of the Red River) and Austin.—**SS**: Chrapliwy & Malnate (1961: 162).

Scaphiopus laticeps: Nieden, 1923: 49.

Distribution: Most of Texas, southwestern Oklahoma, eastern and southern New Mexico and Arizona, to southeastern California and northern Mexico, with an isolated population in southeastern Colorado. Wasserman (1970) has a range map, as do Elliott *et al.* (2009) and Dodd (2013). **U.S.**: AZ, CA, CO, NM, OK, TX.

Common name: Couch's Spadefoot (Toad).

Content: No subspecies are currently recognized, but some have been proposed (see synonymy and Wasserman, 1970).

Commentary: Unjustified restriction of type locality to Mexico, Tamaulipas, Matamoros (Smith & Taylor, 1950a). Wasserman (1970) provided a CAAR account.

Scaphiopus holbrookii (Harlan, 1835)

Rana holbrookii Harlan, 1835: 105.—**O**: **H**: given as ANSP but not known to exist.—**OT**: U.S., South Carolina. Invalid restriction by Schmidt, 1953: 58.

Scaphiopus holbrookii: Baird, 1859d: 12; Cope, 1889a: 298; Gorham, 1974: 44; Frost, 1985: 423; Elliott *et al.*, 2009: 256; Dodd, 2013: 761.

Scaphiopus holbrookii holbrookii: Stejneger & Barbour, 1917: 26; 1943: 38; Viosca, 1949: 10; Conant & Collins, 1998: 503.

Scaphiopus (Scaphiopus) holbrookii: Tanner, 1939: 6; Dubois, 1987b: 130.

Scaphiopus holbrooki: Schmidt, 1953: 58; Conant, 1958: 253; Cochran & Goin, 1970: 72.—**UE.**

Scaphiopus holbrooki holbrooki: Conant, 1975: 299.—**UE.**

Scaphiopus solitarius Holbrook, 1836: 85, pl. 12.—**O**: figured specimen; not known to exist.—**OT**: U.S., Carolina, Georgia, and Tennessee; invalidly restricted by Schmidt, 1953: 58.—**SS**: Cope, 1863: 54.

Scaphiopus albus Garman, 1876: 194.—**O**: **S:** MCZ 1453 (12) (Barbour & Loveridge, 1929: 334, sd.).—**OT**: U.S., Florida, Monroe County, Key West.—**Comment:** While Garman used the binomen, he referred to the taxon as a "variety" of *S. solitarius.*—**SS**: Stejneger & Barbour (1917: 26).

Scaphiopus holbrookii albus: Stejneger & Barbour, 1917: 26; 1943: 38; Schmidt, 1953: 58.

Distribution: East of the Mississippi River from southeastern Louisiana, north to southern parts of Missouri, Illinois, Indiana, and Ohio; eastward across the Gulf states to Florida and the Keys, and north along the coast to Massachusetts. Wasserman (1968) has a map, but also includes the range of *S. hurterii,* as a subspecies. There are also maps in Elliott *et al.* (2009) and Dodd (2013). **U.S.**: AL, AR, CT, DE, FL, GA, IL, IN, KY, LA, MD, MA, MO, MS, NC, NH, NJ, NY, OH, PA, RI, SC, TN, VA, VT, WV.

Common name: Eastern Spadefoot (or Eastern Spadefoot Toad).

Content: No subspecies are currently recognized, but formerly *S. hurterii* was often treated as a subspecies of *S. holbrookii.*

Commentary: Unjustified restriction of type locality of *S. holbrookii* and *S. solitarius* to Charleston, South Carolina (Schmidt, 1953). Wasserman (1968) provided a CAAR account.

Scaphiopus hurterii Strecker, 1910

Scaphiopus hurterii Strecker, 1910: 116.—**O**: **H**: SM 4179 (lost).—**OT**: U.S., Texas, McLennan County, 3 1/2 miles east of Waco.

Scaphiopus hurterii: Strecker, 1915: 52; Stejneger & Barbour, 1917: 26; 1943: 38; Gorham, 1974: 44; Collins, 1991: 43; 1997: 14; Dixon, 2000: 61; Elliott *et al.,* 2009: 260; Dodd, 2013: 772.

Scaphiopus holbrookii hurterii: Wright & Wright, 1933: 116; Wasserman, 1968: 2; Garrett & Barker, 1987: 39; Conant & Collins, 1998: 504.

Scaphiopus holbrooki hurteri: Schmidt, 1953: 58; Conant, 1975: 299.—**UE**

Scaphiopus (Scaphiopus) hurterii: Tanner, 1939: 6.

Scaphiopus hurteri: Conant, 1958: 253; Cochran & Goin, 1970: 72.—**UE**

Distribution: Southern Texas, south into Mexico and north to eastern Oklahoma and east to western Arkansas and Louisiana. Wasserman (1968) has a map, as a subspecies of *S. holbrookii*. Elliott *et al.* (2009) have a map of the separate species, as does Dodd (2013). **U.S.**: AR, LA, OK, TX.

Common name: Hurter's Spadefoot (Toad).

Content: No subspecies are described, but the species has been treated as a subspecies of *S. holbrookii*.

Commentary: Wasserman (1968) provided a CAAR account, as a subspecies of *S. holbrookii*.

Genus **Spea** Cope, 1866

Spea Cope, 1866a: 81.—**NS**: *Scaphiopus bombifrons* Cope, 1863.

Scaphiopus (Spea): Tanner, 1939: 6; Bragg, 1944: 522; Stebbins, 1951; Zweifel, 1956a: 35; Kluge, 1966: 21.

Spea: Brown, 1950: 40; Tihen, 1960a: 91; Tanner, 1989a: 55.

Neoscaphiopus Taylor, 1942b: 203.—**NS**: *Neoscaphiopus noblei* Taylor, 1942; fossil genus.—**SS**: Tihen, 1960a: 93. Zweifel (1956a) suggested it should be treated as another subgenus of *Scaphiopus*, along with *Spea*.

Misapplication of names:

Scaphiopus: Schmidt, 1953: 58-59.

Distribution: Much of western North America.

Common name: Western Spadefoots (or Western Spadefoot Toads).

Content: There are four species currently recognized, but a fifth awaits description.

Commentary: After Cope (1866a) described the genus *Spea*, most workers continued to treat this group of species as members of the genus *Scaphiopus* (*e.g.*, Schmidt, 1953), or to treat *Spea* as a subgenus (*e.g.*, Tanner, 1939; Bragg, 1944, 1945; Stebbins, 1951; Zweifel, 1956a; Kluge, 1966). But some (*e.g.*, Brown, 1950; Smith, 1950) treated *Spea* as a distinct genus. Tihen (1960a) made strong arguments for splitting *Spea* from *Scaphiopus*, and most workers since have accepted *Spea* as a separate genus.

Spea bombifrons (Cope, 1863)

Scaphiopus bombifrons Cope, 1863: 53.—**O**: **S**: USNM 3520 (Platte River), USNM 3703 (Llano Estacado), and USNM 3704 (Fort Union).—**OT**: U.S.: North Dakota, Fort Union, on Missouri River, lat. 48°; on Platte River, 200 miles west of Fort Kearney; and Llano Estacado, Texas; restricted to Fort Union, North Dakota (Schmidt, 1953). We here designate USNM 3704 as lectotype/lectophorant, thereby validating Schmidt's restriction.

Spea bombifrons: Cope, 1866a: 81; Firschein, 1950: 75; Gorham, 1974: 44; Tanner, 1989a: 56; Liner, 1994: 28; Dixon, 2000: 62; Stebbins, 2003: 205; Lemos-Espinal *et al.,* 2004: 6, 79; Brennan & Holycross, 2006: 32; Elliott *et al.,* 2009: 266; Dodd, 2013: 777.

Scaphiopus bombifrons: Boulenger, 1882a: 435; Stejneger & Barbour, 1943: 37; Schmidt, 1953: 59; Conant, 1958: 256; Cochran & Goin, 1970: 70; Behler & King, 1979: 362; Frost, 1985: 422; Stebbins, 1985: 67; Garrett & Barker, 1987: 37; Conant & Collins, 1998: 505.

Scaphiopus hammondii bombifrons: Cope, 1889a: 303; Strecker, 1915: 54.

Scaphiopus (Spea) bombifrons: Tanner, 1939: 11; Brown, 1976: 2.

Distribution: Western Texas east to eastern Arizona, south into Tamaulipas and Chihuahua (Mexico), and north through southeastern Utah, eastern Colorado, Nebraska, the Dakotas, and Montana, into southwestern Canada. See map in Elliott *et al.* (2009) or Dodd (2013). **Can.**: AB, MB, SK. **U.S.**: AR, AZ, CO, IA, KS, MO, MT, ND, NE, NM, OK, SD, TX, UT, WY.

Common name: Plains Spadefoot (Toad). *French Canadian name:* Crapaud pied-bêche des plaines.

Commentary: There is apparently another undescribed cryptic species, possibly morphologically indistinguishable from *S. bombifrons,* but reproductively isolated by a strong difference in advertisement call (Blair, 1955a; Bogert, 1958).

Wiens & Titus (1991) found this to be the most derived species of *Spea,* and a sister to *S. intermontana* from Colorado.

Spea hammondii (Baird, 1859)

Scaphiopus hammondii Baird, 1859d: 12.—**O**: **H**: USNM 3695.—**OT**: U.S., California, Fort Reading.

Spea hammondii: Cope, 1866a: 81; Tanner, 1989a: 56; Wiens & Titus, 1991; Liner, 1994: 29; Stebbins, 2003: 203; Elliott *et al.*, 2009: 272; Dodd, 2013: 786.

Spea hammondii hammondii: Cope, 1889a: 303; Tanner, 1989a: 57.

Scaphiopus (Spea) hammondii: Tanner, 1939: 12; Brown, 1976: 14; Dubois, 1987b: 130.

Scaphiopus hammondii hammondii: Grinnell & Camp, 1917: 140; Gorham, 1974: 44.

Scaphiopus hammondii: Stejneger & Barbour, 1917: 25; 1943: 37; Frost, 1985: 422; Stebbins, 1985: 66.

Scaphiopus hammondi hammondi: Schmidt, 1953: 59.—**UE.**

Scaphiopus hammondi: Cochran & Goin, 1970: 72; Behler & King, 1979: 364.—**UE.**

Distribution: Western California into Baja California. See map in Elliott *et al.* (2009) or Dodd (2013). **U.S.**: CA.

Common name: Western Spadefoot (Toad).

Commentary: Wiens & Titus (1991) found that this is a sister taxon to *S. bombifrons* + all *S. intermontana* populations. References to populations by this name ranging east of California are actually *S. multiplicata*.

Spea intermontana (Cope, 1884)

Scaphiopus intermontanus Cope, 1884: 15.—**O**: **S** (sd., Malnate, 1971: 351): ANSP 13787 (Salt Lake City); ANSP 13788-13789 (Pyramid Lake), all lost.—**OT**: U.S., Utah, Salt Lake County, Salt Lake City, and U.S., Nevada, Storey County, Pyramid Lake; restricted to Salt Lake City, Salt Lake County, Utah (Schmidt, 1953).—**L:** We designate ANSP 13787 (apparently the only existing specimen) as lectoype/ lectophorant, thus validating Schmidt's OT restriction.

Spea hammondii intermontana: Cope, 1889a: 303.

Scaphiopus (Spea) intermontanus: Tanner, 1939: 12; Brown, 1976: 2; Dubois, 1987b: 130.

Scaphiopus hammondi intermontanus: Schmidt, 1953: 59.—**UE.**

Scaphiopus intermontanus: Stejneger & Barbour, 1943: 39; Cochran & Goin, 1970: 73; Gorham, 1974: 44;

Behler & King, 1979: 366; Frost, 1985: 423; Stebbins, 1985: 67.

Spea intermontana: Tanner, 1989b: 508; Wiens & Titus, 1991: 25; Stebbins, 2003: 204; Brennan & Holycross, 2006: 32; Elliott *et al.*, 2009: 268; Dodd, 2013: 791.

Distribution: The Great Basin, northeastern New Mexico and northern Arizona, most of Nevada and adjacent California, to western Colorado and southwestern Wyoming, southern Idaho, through Oregon and Washington and into British Columbia. See map in Hall (1998) or Dodd (2013). **Can.**: BC. **U.S.:** AZ, CA, CO, ID, NV, OR, UT, WA, WY.

Common name: Great Basin Spadefoot (Toad). *French Canadian name:* Crapaud pied-bêche du Grand Bassin.

Commentary: Wiens & Titus (1991) found that samples from Colorado and Oregon were phylogenetically different, so that two species may be involved. They also reported that the Colorado population is a sister species to *S. bombifrons,* while the Oregon population is a sister taxon to *S. bombifrons* + *S. intermontana* Colorado population. Hall (1998) provided a CAAR account. Listed in British Columbia as Special Concern.

Spea multiplicata (Cope, 1863)

Synonymy: See the subspecies, *S. m. multiplicata,* below.

Distribution: Mexican Plateau northward through western Texas, New Mexico, and Arizona, to western Oklahoma, southern Colorado and southeastern Utah. See map in Elliott *et al.* (2009) or Dodd (2013). **U.S.**: AZ, CO, NM, OK, TX, UT.

Common name: Mexican Spadefoot (SECSN) or New Mexico Spadefoot (Toad)(C&T).

Content: Tanner (1989b), recognized two subspecies of *S. hammondii,* now treated as *S. multiplicata,* as well as a nominotypical subspecies now treated as a monotypic species *S. hammondii.* These subspecies have not been widely recognized, but because some of the populations they represent may be reproductively isolated by advertisement call differences (Fouquette, unpublished data), thus may not be conspecific, we tentatively recognize two subspecies.

Commentary: The phylogenetic study of *Spea* by Wiens & Titus (1991) found this to be the sister species to all other *Spea, i.e.,* occupying the basal or ancestral position. Smith &

Taylor (1950a) made an invalid restriction of the type locality to Mexico, Distrito Federal, Coyoacán.

Spea multiplicata multiplicata (Cope, 1863)

Scaphiopus multiplicatus Cope, 1863: 52.—**O**: **H**: USNM 3694 (Kellogg, 1932: 22, sd.).—**OT**: Valley of Mexico.

Spea multiplicata: Cope, 1866a: 81; Taylor, 1952a: 794; Tanner, 1989a: 55; Dixon, 2000: 62; Stebbins, 2003: 206; Lemos-Espinal & Smith, 2007a: 56, 252; Dodd, 2013: 798.

Scaphiopus multiplicatus: Boulenger, 1882a: 436; Cope, 1887a: 12; Frost, 1985: 423; Garrett & Barker, 1987: 40 (part); Conant & Collins, 1998: 506 (part).

Scaphiopus hammondii multiplicatus: Kellogg, 1932: 19; Gorham, 1974: 44.

Spea hammondii multiplicata: Firschein, 1950: 76; Tihen, 1960a: 91; Tanner, 1989a: 59.

Scaphiopus (Spea) multiplicatus: Brown, 1976: 14; Sattler, 1980: 608; Dubois, 1987b: 130.

Spea hammondii multiplicata: Tanner, 1989b: 507.

Spea multiplicata multiplicata Liner, 1994: 29; Frost *et al.*, 2008: 12, by implication.

Scaphiopus dugesi Brocchi, 1879: 23.—**O**: **S**: MNHN 1886.287-1886.288 (2).—**OT**: Mexico, Guanajuato.

Misapplication of names:

Scaphiopus hammondii: Coues, 1875: 630; Cope, 1887a: 12.

Scaphiopus hammondi hammondi: Schmidt, 1953: 59 (part).

Distribution: South of the range of the other subspecies, in Mexico, so extralimital; listed here only because it is the nominotypical subspecies.

Common name: None in use; Mexican Spadefoot (Toad), as for the species, is suggested.

Spea multiplicata stagnalis Cope, 1875

Spea stagnalis Cope *in* Yarrow, 1875: 525.—**O**: **S**: USNM 8558 *fide* Yarrow, 1883a; or USNM 8188, 8563, 25335 *fide* Cochran, 1961a.—**OT**: U.S., northwestern New Mexico; Yarrow (1883a) stated Alto dos Utas, New Mexico, perhaps a valid restriction.

Spea hammondii stagnalis: Tanner, 1989a: 58; 1989b: 507.

Spea stagnalis: Lemos-Espinal *et al.*, 2004: 6, 79.

Spea multiplicata stagnalis: Liner, 1994: 29; Frost *et al.*, 2008: 12.

Misapplication of names:

Scaphiopus hammondii: Strecker, 1915: 54.

Distribution: Northern Mexico northward through western Texas, New Mexico, and Arizona, to western Oklahoma, southern Colorado and southeastern Utah. See map in Elliott, *et al.* (2009), as *Spea multiplicata*. **U.S.:** AZ, CO, NM, OK, TX, UT.

Common name: Chihuahuan Desert Spadefoot (Toad) (not currently recognized by SECSN).

Commentary: Wiens & Titus (1991) rejected recognition of subspecies. Fouquette (unpublished) has call data that may indicate reproductive isolation between the two taxa, so they may eventually be separated as full species, but we retain subspecies rank for now.

*Superfamily **Pipoidea** Gray, 1825

Synonymy: See family Pipidae below.

Content: There are two families with six genera. One family (including one genus) is extralimital but there is an introduced species in North America.

*Family **Pipidae** Gray, 1825

Piprina Gray, 1825: 214.—**NG**: *Pipra* Laurent (junior synonym of *Pipa* Laurenti, 1768)

Pipoidea: Fitzinger, 1826: 37; Dubois, 2005b: 8.

Pipina: Gray, 1829: 203; Bonaparte, 1840a: 100, as a subfamily of Ranidae; 1850: [2], as a subfamily of Pipidae.

Pipae: Tschudi, 1838: 26.

Pipidae: Swainson, 1839: 88; Bolkay, 1919: 349, 354; Bonaparte, 1850: [2]; Blackburn & Wake, 2011: 44.

Pipaeformes: Duméril & Bibron, 1841: 761.

Pipridae: Gray, 1842a: 112

Pipadae: Hallowell, 1858c: 65.

Pipoides: Bruch, 1862: 221.

Pipida: Knauer, 1878: 103.

Pipinae: Metcalf, 1923: 391.

"Pipinomorpha": Baez & Pugener, 2003: 454. Unranked family series nomen, unavailable.

Pipoidia: Dubois, 2005b: 8.
"Aglossa" Wagler, 1830: 131.—nomen nudum.
Aglossa: Wiegmann, 1832: 200, as family.
Aglossidae: Sollas, 1906: 37.
Dactylethridae Hogg, 1838: 152.—**NG**: *Dactylethra* Cuvier, 1829, mo., inferred.
Dactylethridae: Hogg, 1839: 274; Günther, 1859: 1.
Dactylethrina: Bonaparte, 1850: [2].
Dactyletridae: Hoffmann, 1878: 598.—**IS**
Dactylethrae: Peters, 1882c: 179.
Dactylethrida: Knauer, 1878: 103; 1883: 103.
Dactylethrinae: Metcalf, 1923: 391.
Astrodactylidae Hogg, 1839: 152.—**NG**: *Astrodactylus* Hogg, 1838 (synonym *Asterodactylus* Wagler, 1827).—**SS**: Boulenger (1882a: 458).
Astrodactylae: Duméril, 1863: 300.
Xenopoda Fitzinger, 1843: 33.—**NG**: *Xenopus* Wagler, 1827.
Xenopodes: Fitzinger, 1861: 416.
Xenopidae: Cope, 1889b: 248.
Xenopodidae: Bolkay, 1919: 349, 356.
Xenopodinae: Metcalf, 1923: 391.
Xenopinae: Noble, 1931:489; Van Dijk, 1966: 248.
"Xenopodinomorpha": Baez & Pugener, 2003: 454. Unranked family series nomen, unavailable.
Hymenochiridae Bolkay, 1919: 348.—**NG**: *Hymenochirus* Boulenger, 1896, mo., inferred.

Distribution: Africa and South America. Non-native introductions in North America.

Common name: Tongueless Frogs.

Content: Two subfamilies containing 5 genera with 32 species.

Commentary: *The taxon is non-native (introduced). It is often included, with the Rhinophrynidae, in a Superfamily Pipoidea Gray, 1825. Because this is a non-native family, we essentially adopt much of the synonymy of Frost (2009-13), with little evaluation.

*Subfamily **Xenopodinae** Fitzinger, 1843

Xenopoda Fitzinger, 1843: 33.—**NG**: *Xenopus* Wagler, 1827.

Xenopodes: Fitzinger, 1861: 416.

Xenopidae: Cope, 1889b: 248.

Xenopodidae: Bolkay, 1919: 349, 356.

Xenopodinae: Metcalf, 1923: 391.

Xenopinae: Noble, 1931:489; Van Dijk, 1966: 248.

"Xenopodinomorpha": Baez & Pugener, 2003: 454. Unranked family series nomen, unavailable.

Distribution: Native to Africa, non-native introductions in North America.

Common name: Clawed Frogs.

Content: Four genera with 25 species. *None are native, but one genus and species introduced to North America.

*Genus **Xenopus** Wagler, 1827

Xenopus Wagler, 1827: 726 (footnote).—**NS**: *Xenopus boiei* Wagler, 1827 (as synonym *Bufo laevis* Daudin, 1802).

Dactylethra Cuvier, 1829b: 107.—**NS**: not stated; *Pipa laevis* Merrem, sd., Frost (2009-13).

Tremeropugus Smith, 1831: 18.—**NS** (mo.): *Tremeropugus typicus* Smith, 1831.—**SS**: Branch & Bauer, 2005: 4.

Distribution: Africa, but introduced to California and Arizona (see *Distribution* for *X. laevis,* below).

Common name: Clawed Frogs.

Content: 18 species. Kobel *et al.* (1998) treated the genus as composed of two subgenera. The introduced species is in the subgenus *Xenopus.*

***Xenopus (Xenopus) laevis** (Daudin, 1802)

Bufo laevis Daudin, 1802a: 85; 82, pl. 30, fig. 1.—**O**: included figured specimen; stated to be in MNHN but now lost (Poynton, 1964: 31).—**OT**: not stated, unknown.

Pipa laevis: Merrem, 1820: 180; Fitzinger, 1826: 40; Duvernoy *in* Cuvier, 1849: 155.

Dactylethra laevis: Cuvier, 1829: 107, by implication; Günther, 1859a: 2; Blanford, 1870: 459.

Xenopus laevis: Wagler, 1830: 200; Steindachner, 1867b: 4 (*fide* Frost, 2009-13, not seen); Knauer, 1878: 103; Frost, 1985: 428; Stebbins, 1985: 96; 2003: 244;

Liner, 1994: 29; Brennan & Holycross, 2006: 46; Elliott *et al.*, 2009: 308; Dodd, 2013: 828.

Dactylethra levis: Duméril & Bibron, 1841: 765.—**IS**

Xenopus laevis laevis: Parker, 1936: 597; Gorham, 1974: 41.

Xenopus (Xenopus) laevis: Kobel, Barandun, & Thiebaud, 1998: 13.

Pipa bufonia Merrem, 1820: 180.—**O**: based on "*Pipa* male. Pl. 21. f. 2", author unknown.—**OT**: not stated, unknown.

Xenopus boiei Wagler, 1827: 726.—**O**: specimens in Boie's collection, now possibly RMNH 2267 (see Gassó Miracle *et al.*, 2007).—**OT**: not stated.—**SS**: Tschudi, 1838: 90; Duméril & Bibron, 1841: 765.

Xenopus bojei: Van der Hoeven, 1833: 308; Leunis, 1844: 145; 1860: 145.—**IS.**

Leptopus boiei: Mayer, 1835: 35. Alternate name for *L. oxydactylus* Mayer, 1835.

Dactylethra bojei: Tschudi, 1838: 90.

Xenopus (Dactylethra) boiei: Schlegel, 1858: 59.

Dactylethra capensis Cuvier, 1830: pl. 7.—**O**: not stated, probably in MNHN.—**OT**: not stated, but Cape of Good Hope by implication of name. Synonymy with *laevis* by Duméril & Bibron (1841: 765). Also see discussion (Frost, 2009-13).

Tremeropugus typicus Smith, 1831: 19.—**O**: not stated nor known to exist.—**OT**: South Africa, most parts, in lakes and slow-running rivers. Synonymy with *X. laevis* by Branch & Bauer, 2005: 4.

Leptopus oxydactylus Mayer, 1835: 34, pl. 2, fig. 5.—**O**: **H**, specimen figured.—**OT**: not stated; Africa, by implication of alternate name.—**SS**: Tschudi, 1838: 90, and Duméril & Bibron, 1841: 765.

Pipa africana Mayer, 1835: 35. Alternate name for *Leptopus boiei* Mayer, 1835.

Dactylethra delalandii Cuvier, 1849: pl. 38, fig. 2 and 2a.—**O**: specimen figured, status unknown, said to be in Musée de Strasbourg, France.—**OT**: Midi de l'Afrique.

Xenopus laevis bunyoniensis Loveridge, 1932: 114.—**H**: MCZ 14616, od.—**OT**: southwestern Uganda, Kigezu District, west shore of Lake Bunyoni, Bufundi.

Xenopus (laevis) bunyoniensis: Tymowska & Fischberg, 1973: 335.

Xenopus laevis sudanensis Perret, 1966: 301.—**O**: **H** (od.), MHNG 1017.74.—**OT**: Sudan, Adamaoua, Ngaoundéré.

Distribution: *Introductions are reported established at sites in southwestern California and southeastern Arizona

(Lever, 2003; Crayon, 2005—see Crayon's map, or also maps in Elliott *et al.,* 2009, and Dodd, 2013). There are scattered reports from other states, but reproductive status has not been established. Indigenous to sub-Saharan Africa. **U.S.**: *AZ, *CA.

Common name: African Clawed Frog.

Content: There are three described subspecies; it is believed that all or most of the introduced frogs are of the nominotypical subspecies, *X. l. laevis* (Crayon, 2005).

Commentary: This is a tertraploid species, as are most of the species of this subgenus. There are reports of introduced Clawed Frogs in several other states, but they have not established breeding populations outside of California and Arizona (Crayon, 2005); these frogs may pose a threat to native aquatic animals, especially larvae of native anurans, though this is disputed by some biologists. Crayon (2005) recommends using all available means to extirpate the established populations.

Again, as this is a non-native species, we essentially follow the synonymy of Frost (2009-13) with little critical evaluation.

Family **Rhinophrynidae** Günther, 1859

Rhinophrynina Günther, 1859a: xiv.—**NG** (mo.): *Rhinophrynus* Duméril & Bibron, 1841.

Rhinophrynidae: Mivart, 1869: 286; Hoffman, 1878: 607; Walker, 1938: 10; Blackburn & Wake, 2011L 45.

Rhinophrynida: Knauer, 1878: 108.

Rhinophryninae: Noble, 1931: 500.

Distribution: Southern Texas and both coasts of Middle America.

Common name: Burrowing Toads.

Content: A single genus and species.

Commentary: Fouquette (1969) provided a CAAR family account. Frost *et al.* (2006) and Roelants *et al.* (2007) found this to be the sister taxon of the Pipidae. The latter authors included the two in a suprafamilial taxon Xenoanura Savage, which is unavailable in the family-series as it is not based on a generic nomen.

Genus **Rhinophrynus** Duméril & Bibron, 1841

Rhinophrynus Duméril & Bibron, 1841: 757.—**NS**: *Rhinophrynus dorsalis* Duméril & Bibron, 1841, mo.

Rhinophryne: O'Shaughnessy, 1879: 13.—**UE.**

Distribution: As for the family.

Common name: Burrowing Toad.

Content: A single species.

Commentary: Fouquette (1969) provided a generic CAAR account.

Rhinophrynus dorsalis Duméril & Bibron, 1841

Rhinophrynus dorsalis Duméril & Bibron, 1841: 758.—**O**: **H:** MNHN 693 (Guibé, 1950: 17, sd.).—**OT**: Mexico, Veracruz.

Rhinophrynus dorsalis: Günther, 1858a: 348; Knauer, 1878: 108; Fouquette, 1969: 1; 2005: 599; Gorham, 1974: 73; Conant, 1975: 297; Behler & King, 1979: 361; Frost, 1985: 551; Garrett & Barker, 1987: 49; Liner, 1994: 28; Conant & Collins, 1998: 501; Dixon, 2000: 60; Elliott *et al.*, 2009: 306; Dodd, 2013: 458.

Rhinophrynus rostratus Brocchi, 1877b: 196. **O**: **H:** MNHN 6335 (Guibé, 1950: 17, sd.).—**OT**: Mexico, Tehuantepec.—**SS**: Boulenger, 1882a: 329.

Distribution: Extreme southern Texas to Honduras on the Caribbean coast, and southern Mexico to Costa Rica on the Pacific coast. Fouquette (1969) has a range map, as does Dodd (2013). James (1966) first reported the species from Texas, where it may now be extinct; however, Dixon (2000) reported that a chorus had been heard in 1998. See a summary of U.S. records by Fouquette (2005). **U.S.:** TX.

Common name: Burrowing Toad (SECSN) or Mexican Burrowing Toad (C&T).

Commentary: Fouquette (1969) provided a CAAR species account.

Superfamily **Ranoidea** Rafinesque, 1814

Ranaridia Rafinesque, 1814b: 102. See family Ranidae, below.

Content: Three epifamilies (loosely following Dubois, 2005b) with 25 families, following Frost *et al.* (2006). All epifamilies have species in North America, as do six of the families.

Epifamily **Eleutherodactyloidia new epifamily**

"Terrarana" Hedges, Duellman, & Heinicke, 2008 (unranked taxon). Nomen unavailable for the family series, not based on a generic name. We feel this is unfortunate as it is an ideal name for the taxon, but we adhere to the *Code*. It could be used as an infraordinal name, however.

Typification: NF: Eleutherodactylidae.

Diagnosis: The same as for the unranked taxon Terrarana (Hedges *et al.*, 2008): terrestrial breeding, direct development, embryonic egg-tooth, arciferal (or sometimes pseudofirmisternal) pectoral girdle, partially fused calcanea and astragali, lacking intercalary elements in digits, and lacking Bidder's organ.

Distribution: From Texas, New Mexico and Arizona, southward through Middle America and most of tropical and sub-tropical South America, and many Caribbean islands of the West Indies, plus introductions in the southeastern U.S. and Hawaii.

Common name: We propose New World Direct-developing Frogs as appropriate.

Content: 883 species in 3 families, 25 genera; 2 families, 2 genera, and 6 species are in North America. Thus, 0.68% of the species diversity of this taxon is North American.

Commentary: Hedges *et al.* (2008) used an unranked taxon with an informal name (Terrarana) for these families of direct-developing anurans, stating that it would be better to avoid a formal superfamilial name. We believe the epifamily/ superfamily framework is the proper one for containing such related families, and is easily modified as necessary later on. Unfortunately, the name proposed by Hedges *et al.* is not based on a generic nomen thus is not available for the family-series, even though it seems an appropriate name for the taxon. Dubois (2009b) noted that if used above the family-series, the Hedges *et al.* name should be modified to match the proper Latin grammar for that rank, suggesting either Terraranae or Terranae. We consider this clade to be distinctive from the other superfamilies in the New World, with synapomorphies development of terrestrial eggs and no larval stage.

Family **Craugastoridae** Hedges, Duellman & Heinicke, 2008

Eleutherodactylinae Lutz, 1954: 175 (part)
Eleutherodactylini: Lynch, 1971: 142 (part)
Brachycephalidae Günther, 1858a: 347 (part)
Brachycephalinae: Dubois, 2005b: 11 (part)
Brachycephalidae: Frost *et al.*, 2006a: 63 (part)
Craugastoridae Hedges, Duellman, & Heinicke, 2008: 1.—**NG**: *Craugastor* Cope, 1862.
Craugastoridae: Blackburn & Wake, 2011: 41

Distribution: Central Texas to southern Arizona, south into northwestern South America, plus coastal Atlantic Brazil. See map, Hedges *et al.* (2008: 30).

Common name: Fleshbelly Frogs.

Content: Two genera with 114 species, of which one genus and species reaches North America.

Commentary: This was formerly a part of the family Leptodactylidae, and more recently the family Brachycephalidae, and the North American species was formerly a member of the large genus *Eleutherodactylus*. For further explanation of the synonymy and nomenclatural history, see Hedges *et al.* (2008).

North America contains only 0.88% of the species diversity of this family.

Genus **Craugastor** Cope, 1862

Craugastor Cope, 1862a: 153.—**NS**: *Hylodes fitzingeri* Schmidt, 1857 (Dunn & Dunn, 1940: 71, sd.). Described as a subgeneric "group" of *Hylodes*.

Leiyla Keferstein, 1868a: 296.—**NS:** *Leiyla güntheri* Keferstein, 1868, mo.—**SS:** Boulenger, 1882a: 198, with *Hylodes*; Savage, 1974: 290, with *Eleutherodactylus*.

Microbatrachylus Taylor, 1940b: 499.—**NS** (od.): *Eleutherodactylus hobartsmithi* Taylor, 1936.—**SS**: Lynch, 1965a: 3; with *Eleutherodactylus*.

Hylactophryne Lynch 1968: 511 (see the subgenus).

Campbellius Hedges, Duellman, & Heinicke, 2008: 33.—**NS** (od.): *Eleutherodactylus stadelmani* Schmidt, 1936, as a subgenus of *Craugastor*.

Distribution: Central Texas to southern Arizona, south to northwestern South America.

Common name: Northern Rainfrogs (SECSN) or Fleshbelly Frogs (C&T).

Content: 112 species in three subgenera; one subgenus with one species reaches North America.

Commentary: See Hedges *et al.* (2008) for more extensive synonymy.

Subgenus **Hylactophryne** Lynch, 1968

Hylactophryne Lynch, 1968: 511.—**NS**: *Hylodes augusti* Dugès, 1879, od. Proposed as a genus.

Hylactophryne: Hedges *et al.*, 2008: 42. Treated as a subgenus.

Distribution: Central Mexico, with a western arm extending to Arizona and possibly New Mexico, and an eastern arm extending into western and central Texas and southeastern New Mexico. See map (Hedges *et al.*, 2008: 43).

Common name: Barking Frogs.

Content: 21 species, one reaching North America.

Craugastor (Hylactophryne) augusti (Dugès, 1879)

Hylodes augusti Dugès *in* Brocchi, 1879: 21.—**O**: **S:** MDUG (3), lost;—**N**: MDUG, labeled "type," (Smith & Necker, 1943: 201).—**OT**: Mexico, Guanajuato, Guanajuato.

Eleutherodactylus augusti: Stejneger, 1904: 582, by implication; Stejneger & Barbour, 1943: 52; Schmidt, 1953: 60; Cochran & Goin, 1970: 118; Dixon, 2000: 63; Stebbins, 2003: 201; Lemos-Espinal & Smith, 2007b: 48, 245.

Eleutherodactylus augusti augusti: Zweifel, 1956b: 14; Liner, 1994: 18.

Hylactophryne augusti: Lynch, 1968: 511; Gorham, 1974: 61; Behler & King, 1979: 420; Frost, 1985: 332; Stebbins, 1985: 64; Garrett & Barker, 1987: 29.

Eleutherodactylus (Craugastor) augusti: Hedges, 1989: 318.

Craugastor augusti: Crawford & Smith, 2005: 551, by implication; Brennan & Holycross, 2006: 38; Frost *et al.*, 2006a: 360; Elliott *et al.*, 2009: 296; Dodd, 2013: 192.

Craugastor (Hylactophryne) augusti: Hedges, Duellman, & Heinicke, 2008: 44.

Craugastor augusti augusti: Frost *et al.*, 2008: 5.

Eleutherodactylus augusti fuscofemora Zweifel, 1956b: 24.—**O**: **H**: FMNH 48132, od.—**OT**: Mexico, Coahuila, 5 miles south of Cuatro Ciénegas, Sacaton.

Eleutherodactylus (Craugastor) fuscofemora: Lynch & Duellman, 1997: 224.

Distribution: Mainly central inland Mexico, with a western extension northward to southeastern Arizona, with a disjunct population in central Arizona, and an eastern extension northward into central and western Texas, and southern New Mexico. See map in Zweifel (1967) and updated maps in Schwalbe and Goldberg (2005) and Dodd (2013). **U.S.**: AZ, NM, TX.

Common name: Barking Frog.

Content: Four subspecies are recognized. The two listed below reach North America.

Commentary: Zweifel (1967) provided a CAAR account, as *Eleutherodactylus augusti*. Goldberg *et al.* (2004b) analyzed advertisement calls and genetic sequences, revealing differences between the Arizona (*cactorum*) and Texas-New Mexico (*latrans*) populations great enough to provide reproductive isolation, and distinctiveness at the species level. But to establish the level of speciation among these *augusti*-complex populations, more call and genetic data are needed from connecting Mexican populations. So for now, we retain the two as subspecies.

Craugastor (Hylactophryne) augusti cactorum (Taylor, 1939)

Eleutherodactylus cactorum Taylor, 1939b: 391.—**O**: **H** (od.), EHT-HMS 6383, now FMNH 100021 (Marx, 1976: 47).—**OT**: Mexico, Puebla, 20 miles northwest of Tehuacán, km 226, near Cacaloapam.—**SS**: Zweifel, 1956b: 17, with *E. augusti*.

Eleutherodactylus augusti cactorum: Zweifel, 1956b: 17; Cochran & Goin, 1970: 119; Liner, 1994: 18; Lemos-Espinal & Smith, 2007b: 49, 245.

Eleutherodactylus cactorum: Goldberg *et al.* (2004b: 312), conditional.

Hylactophryne augusti cactorum: Gorham, 1974: 61; Behler & King, 1979: 420; Stebbins, 1985: 65; Dodd, 2013: 192.

Craugastor (Hylactophryne) augusti cactorum: Frost *et al.* 2008: 5 (by implication).

Eleutherodactylus bolivari Taylor, 1942a: 298.—**O**: **H** (od.) UIMNH 15968 (*ex* EHT-HMS 29564).—**OT**: Mexico, Mexico, Ixtapan del Oro.—**SS**: Zweifel (1956b: 14).

Distribution: Southeastern Arizona, south into Mexico. **U.S.**: AZ.

Common name: Western Barking Frog.

Commentary: Based on the results of Goldberg *et al.* (2004b), this may be a distinct species.

Craugastor (Hylactophryne) augusti latrans (Cope, 1880)

Lithodytes latrans Cope, 1878a: 186. *Nomen nudum*; an editorial note attributing the name to Cope, but no description.

Lithodytes latrans Cope, 1880: 25.—**O**: **S** (sd.), USNM 10058 (2), 10529 (2), 10751-10753 (Cochran, 1961a: 65), and ANSP 10757-10758 (Piatt, 1934: 90).—**OT**: U.S., Texas, cliffs of Cretaceous limestone found in every direction along the borders and river valleys of the first plateau region of southwestern Texas; listed as U.S., Texas, Bexar County, Helotes, by Cochran (1961a: 65).

Hylodes latrans: Garman, 1884a: 42; Günther, 1901a: 241.

Lithodytes latrans: Cope, 1889b: 316; Strecker, 1915: 47.

Eleutherodactylus latrans: Stejneger, 1904: 582, by implication; Stejneger & Barbour, 1917: 34; 1943: 52; Schmidt, 1953: 60; Goldberg *et al.* (2004b: 312), conditional.

Eleutherodactylus augusti latrans: Zweifel, 1956b: 17; Conant, 1958: 260; Cochran & Goin, 1970: 118; Conant & Collins, 1998: 508; Lemos-Espinal & Smith, 2007a: 49, 245.

Hylactophryne augusti latrans: Conant, 1975: 303; Lynch, 1968: 511; Gorham, 1974: 61; Behler & King, 1979: 420; Stebbins, 1985: 65; Garrett & Barker, 1987: 29.

Eleutherodactylus (Craugastor) latrans: Lynch & Duellman, 1997: 227. Unintended combination.

Craugastor augusti latrans: Frost *et al.*, 2008: 5.

Distribution: Central Texas west to the Pecos River, northward along that drainage into southeastern New Mexico. **U.S.**: NM, TX.

Common name: Balcones Barking Frog (SECSN) or Eastern Barking Frog (C&T).

Commentary: This probably is a species distinct from *C. a. cactorum*, based on results of Goldberg *et al.* (2004b), but whether it is distinct from *C. augusti* requires more evidence, so for the present we retain subspecies status.

Family **Eleutherodactylidae** Lutz, 1954

Cornuferinae Noble, 1931: 521.—**NG**: *Cornufer* Tschudi, 1838. Synonymy by implication of synonymy of *Cornufer* with *Eleuthero-dactylus* by Zweifel, 1966: 167.—**Comment:** The generic nomen *Eleutherodactylus* was ruled to have priority over *Cornufer* (ICZN, 1978), so the latter is not available for formation of family-series names.

Eleutherodactylinae (part) Lutz, 1954: 157.—**NG**: *Eleutherodactylus* Duméril & Bibron, 1841.

Eleutherodactylini: Lynch, 1971: 142.

Eleutherodactylidae: Hedges, Duellman, & Heinicke, 2008: 47; Blackburn & Wake, 2011: 42.

Brachycephalinae Dubois, 2005b: 11. As a subfamily of Leptodactylidae.

Brachycephalidae: Frost *et al.*, 2006a: 63.

Distribution: Tropical South America northward through Middle America, into southern Texas, plus throughout the Caribbean islands of the West Indies, and introduced populations from Florida to Louisiana, and Hawaii, as well as other Pacific islands.

Common name: Free-toed Frogs.

Content: 728 species, in 21 genera of two subfamilies, including five species in one genus found in North America.

Commentary: Long a part of the family Leptodactylidae, and more recently in the family Brachycephalidae, this taxon was recognized as a distinct family by Hedges *et al.* (2008).

North America contains only 0.4% of the species diversity of this family.

Genus **Eleutherodactylus** Duméril & Bibron, 1841

Synonymy: See the subgenus *Eleutherodactylus,* below:

Distribution: Southern Texas through Mexico to Belize and Guatemala, plus introduced into Florida, possibly Louisiana, and Hawaii.

Common name: Rainfrogs (SECSN) or Robber Frogs (C&T).

Content: Five subgenera with 185 species, of which three subgenera and five species are found in North America. Thus, 2.7% of the species diversity of the genus is North American.

Commentary: Hedges, *et al.* (2008) included this in a subfamily Eleutherodactylinae. Note that the generic synonymy includes only the part of the genus which does not occur in North America. Remainder of synonymy is found under the three subgenera below. Frost *et al.* (2006) found the species composing the taxon *Eleutherodactylus* did not consist of a monophyletic unit, but for the present we retain this as a distinct genus, divisible into distinctive subgenera. The analysis of Pyron & Wiens (2011) indicated two major sister clades, which might be named as subgenera, but none are native to North America. Much systematic work remains to be done to achieve a better understanding of this assemblage, and our treatment is admittedly a tentative one.

*Subgenus **Eleutherodactylus** Duméril & Bibron, 1841

Eleutherodactylus Duméril & Bibron, 1841: 620.—**NS**: *Hylodes martinicensis* Tschudi, 1838.

Eleutherodactylus: ICZN, 1978.

Ladailadne Dubois, 1987a: 23. **NS** (od.): *Eleutherodactylus jasperi* Drewry & Jones, 1976.—**SS**: Hedges, 1989: 327.

Pelorius Hedges, 1989: 329.—**NS** (od.): *Leptodactylus inoptatus* Barbour, 1914. Proposed as subgenus of *Eleutherodactylus.*

Pelorius: Frost *et al.*, 2006a: 197. Proposed generic status.

Schwartzius Hedges, Duellman & Heinicke, 1008: 87. **NS** (od.): *Eleutherodactylus counouspeus* Schwartz, 1964. Proposed as a subgenus of *Eleutherodactylus.*

Distribution: Most islands of the West Indies, except Jamaica, and introduced into parts of the U.S. (see species).

Common name: None in general use; West Indies Rainfrogs seems appropriate.

Content: 54 species.

Commentary: *Members of this subgenus in North America are non-native.

***Eleutherodactylus (Eleutherodactylus) coqui** Thomas, 1966

Eleutherodactylus coqui Thomas, 1966: 376.—**O**: **H** (od.), MCZ 43208.—**OT**: Puerto Rico, Area Recreo La Mina, 22.8 km S Palmer.

Eleutherodactylus coqui: Gorham, 1974: 51; Behler & King, 1979: 418; Frost, 1985: 281; Schwartz & Henderson, 1991: 41; Conant & Collins, 1998: 509; Elliott *et al.*, 2009: 288; Dodd, 2013: 812.

Eleutherodactylus (Eleutherodactylus) coqui: Hedges, 1989: 327; Hedges *et al.*, 2008: 59.

Distribution: Puerto Rico, plus *introduced and established in the state of Hawaii (islands of Hawaii, Oahu, Kauai and Maui), and perhaps in Florida and Louisiana, as well as other Caribbean and Pacific islands (see Lever, 2003; Meshaka *et al.*, 2004; and Stewart & Lannoo, 2005, for details). Not included by McKeown (1996). Also see maps of North American distribution in Elliott *et al.* (2009) or Dodd (2013). **U.S.**: *FL?, *HI, *LA?.

Common name: Coquí (SECSN) or Puerto Rican Coqui (C&T).

Commentary: *This is a non-native species.

***Eleutherodactylus (Eleutherodactylus) martinicensis** (Tschudi, 1838)

Hylodes martinicensis Tschudi, 1838: 77.—**O**: **S** (sd., Guibé, 1950) MNHN 4881-4883 (6).—**OT**: Insel Martinique, probably an error; more likely Guadeloupe (Schwartz, 1967: 34-35).—**OS**: *Hyla martinicensis* Tschudi, 1838: 77.

Eleutherodactylus martinicensis: Duméril & Bibron, 1841: 620; Gorham, 1974: 54; Frost, 1985: 302; Schwartz & Henderson, 1991: 85.

Hylodes martinicensis: Werner, 1909: 10.

Eleutherodactylus (Eleutherodactylus) martinicensis: Hedges, 1989: 327.

Distribution: West Indies, *reported introduced to the island of Maui, Hawaii (Kraus *et al.*, 1999), but later found to be misidentified *E. coqui* (Kraus & Campbell, 2002. **This species is not introduced in North America**; we include it merely because it was previously listed as invasive in Hawaii.

Common name: Martinique Greenhouse Frog.

Commentary: *This species is not known in North America. Kaiser & Hardy (1994) provided a CAAR account.

*Subgenus **Euhyas** Fitzinger, 1843

Euhyas Fitzinger, 1843: 31.—**NS (**od.**)**: *Hylodes ricordii* Duméril & Bibron, 1841,—**SS**: Boulenger, 1882a: 198, with *Hylodes*; Barbour, 1910: 100, with *Eleutherodactylus.*

Euhyas: Hedges, 1989; Hedges *et al.*, 2008, as a subgenus; Frost *et al.* (2006a:197), as a genus.

Sminthillus Barbour & Noble, 1920: 402.—**NS** (od.): *Phyllobates limbatus* Cope, 1862.—**SS**: Hedges, 1989: 318.

Distribution: Cuba and most of the Greater Antilles and other Caribbean islands, but not Puerto Rico; *established introductions of one species to the gulf coast, from Louisiana to Florida, as well as Hawaii.

Common name: None in general use. Greater Antilles Rainfrogs may be appropriate.

Content: 95 species, including *one species introduced to North America.

Commentary: *Members of this taxon in North America are probably non-native.

***Eleutherodactylus (Euhyas) planirostris** (Cope, 1862)

Hylodes planirostris Cope, 1862a: 153.—**O**: not stated, but deposited in Museum of Salem (now Peabody Essex Museum), apparently lost.—**OT**: Bahamas, New Providence Island.

Hylodes (Lithodytes) planirostris: Cope, 1862a: 153 (implied).

Eleutherodactylus planirostris: Stejneger, 1904: 582, by implication; Behler & King, 1979: 419; Frost, 1985: 312; Schwartz & Henderson, 1991: 103; Meshaka, 2005: 499; Elliott *et al.*, 2009: 286; Dodd, 2013: 815.

Eleutherodactylus ricordii planirostris: Shreve, 1945: 117; Gorham, 1974: 56.

Eleutherodactylus ricordi planirostris: Schmidt, 1953: 236; Conant, 1958: 261; Cochran & Goin, 1970: 119.—**UE.**

Eleutherodactylus planirostris planirostris: Conant, 1975: 303; Liner, 1994: 20; Conant & Collins, 1998: 509.

Eleutherodactylus (Euhyas) planirostris: Hedges, 1989: 325; Hedges *et al.*, 2008

Euhyas planirostris: Frost, *et al.*, 2006a: 361.

Distribution: Cuba, Cayman Islands, Caicos Islands; *introduced or immigrated to Florida at least 130-150 years ago, now widespread in the state; also introduced to Louisiana, Mississippi, Alabama, Georgia, and Hawaii (islands of Hawaii and Oahu) in North America. See Lever (2003) for details of North American introductions. See map of North American distribution in Meshaka (2005), Elliott *et al.* (2009) or Dodd (2013). **U.S.**: *AL, *? FL, *GA, *HI, *LA, *MS.

Common name: Greenhouse Frog.

Commentary: *This is presumably a non-native species (again, it may possibly have immigrated naturally to Florida). In Florida the species was long known, in error, as *Eleutherodactylus ricordii.* Cope (1863: 48) first reported the species from Florida as *Hylodes planirostris*, and again (1889b: 314) reported it in southern Florida, as *Lithodytes ricordii,* and seemed to assume it was native to Florida. Barbour (1910) reported a specimen from further north, as *Eleutherodactylus recordii,* saying there was no doubt it was introduced into Florida. Virtually all species listed as *E. recordii* from North America are misapplications of the name, and should be corrected to *E. planirostris.* Schwartz (1974) provided a CAAR account.

Subgenus **Syrrhophus** Cope, 1878

Epirhexis Cope: 1866b: 96.—**NS**: *Batrachyla longipes* Baird, 1859, od.—Senior **SS** of *Syrrhophus* (Lynch, 1967). Nomen suppressed for priority but not homonymy, and placed on the official list of rejected and invalid generic names (ICZN, 1974).

Syrrhophus Cope, 1878b: 253.—**NS**: *Syrrhophus marnocki* Cope, 1878, mo.—**SS**: Myers (1962: 198) with *Eleutherodactylus*. Placed on official list of generic names (ICZN, 1974).

Syrrhopus: Boulenger, 1888b: 206.—**IS.**

Syrrhaphus: Günther, 1900: 215.—**IS.**

Syrrhophus: Hedges, 1989: 318, as a subgenus; Frost *et al.*, 2006a: 199, as a genus.

Malachylodes Cope, 1879b: 264. **NS** (mo.): *Malachylodes guttilatus* Cope, 1879.—**SS**: Boulenger, 1888a: 206.

Tomodactylus Günther, 1900: 219. **NS** (mo.): *Tomodactylus amulae* Günther, 1900.—**SS**: Hedges, 1989: 318.

Distribution: Southern Texas south through Mexico to Belize and Guatemala, plus Cuba.

Common name: Chirping Frogs.

Content: 26 species, of which three occur in North America.

Commentary: The work of Hedges *et al.* (2008) seems to show that this subgenus is the sister taxon of the subgenus *Euhyas,* and those two are a sister taxon of the subgenus *Eleutherodactylus.* Frost *et al.* (2006a) found this taxon to represent a distinct genus, but for the present we retain subgeneric rank. The analysis of Pyron & Wiens (2011) shows species of this taxon as basal to the subgenus *Euhyas,* as treated here. We have seen no studies to determine phylogenetic relationships among the species of *Syrrhophus.*

Eleutherodactylus (Syrrhophus) cystignathoides (Cope, 1878)

Phyllobates cystignathoides Cope, 1878c: 89.—**O**: **S** (sd.), USNM 32402-32404, 32406-32409 (Cochran, 1961a) and MCZ 25732 (ex USNM 32405) (Barbour and Loveridge, 1946: 170).—**OT**: Mexico, Veracruz, Potrero near Cordova (= Córdoba), under decayed trunks of trees.

Syrrhophus cystignathoides: Cope, 1879b: 268; Gorham, 1974: 70; Behler & King, 1979: 421; Frost, 1985: 340; Frost *et al.*, 2006: 362.

Eleutherodactylus cystignathoides: Myers, 1962: 198, by implication; Elliott *et al.*, 2009: 95; Dodd, 2013: 197.

Eleutherodactylus cystignathoides cystignathoides: Liner, 1994: 19.

Eleutherodactylus (Syrrhophus) cystignathoides: Hedges, 1989: 318; Hedges *et al.*, 2008.

See more synonymy under the subspecies, *E. c. campi,* below.

Distribution: Southern Texas and northeastern Mexico; introduced farther north in Texas but perhaps not established

(Wallace, 2005; see map) and in Louisiana. Also see map in Dodd (2013). **U.S.**: *LA, TX.

Common name: Rio Grande Chirping Frog.

Content: Of the described subspecies, only one is found within North America.

Eleutherodactylus (Syrrhophus) cystignathoides campi (Stejneger, 1915)

Syrrhophus campi Stejneger, 1915a: 131.—**O**: **H**: USNM 52290.—**OT**: U.S., Texas, Cameron County, Brownsville.—**SS**: Lynch, 1970: 15, with *S. cystignathoides.*

Syrrhophus campi: Stejneger & Barbour, 1917: 35; 1943: 53; Schmidt, 1953: 60; Conant, 1958: 261; Cochran & Goin, 1970: 120; Gorham, 1974: 70.

Syrrhophus cystignathoides campi: Conant, 1975: 304; Lynch, 1970: 15; Garrett & Barker, 1987: 31; Conant & Collins, 1998: 511; Dixon, 2000: 64.

Eleutherodactylus campi: Myers, 1962: 198, by implication.

Eleutherodactylus cystignathoides campi: Liner, 1994: 19.

Distribution: Southern Texas and northern Mexico. It has also been introduced in scattered localities in the eastern half of Texas. See map in Elliott *et al.* (2009). **U.S.**: TX.

Common name: Rio Grande Chirping Frog.

Commentary: Validity of this subspecies is questionable; a molecular analysis of the subgenus might help clarify the taxonomy. Martin (1958:50) was uncertain of the identity of frogs he listed as *S. cystignathoides* versus *S. campi,* but did not suggest they might be populations of the same species.

Eleutherodactylus (Syrrhophus) guttilatus (Cope, 1879)

Malachylodes guttilatus Cope, 1879b: 264.—**O**: **H**: USNM 9888 (Cochran, 1961a, sd.).—**OT**: Mexico, Guanajuato.

Syrrhopus guttulatus: Boulenger, 1888b: 206.—**IS.**

Syrrhophus guttilatus: Nieden, 1923: 399; Smith & Taylor, 1948: 51; Gorham, 1974: 70; Conant, 1975: 304; Behler & King, 1979: 421; Frost, 1985: 341; Garrett & Barker, 1987: 32; Conant & Collins, 1998: 511; Dixon, 2000: 64; Frost *et al.,* 2006a: 362; Lemos-Espinal & Smith, 2007a: 49, 246.

Eleutherodactylus guttilatus: Myers, 1962: 198, by implication; Liner, 1994: 19; Elliott *et al.*, 2009: 292; Dodd, 2013: 197.

Eleutherodactylus (Syrrhophus) guttilatus: Hedges, 1989: 318; Hedges *et al.*, 2008.

Syrrhophus smithi Taylor, 1940a: 43.—**O**: **H:** USNM 108594.—**OT**: Mexico, Nuevo León, 15 miles west of Galeana.—**SS**: Lynch, 1970: 22.

Eleutherodactylus smithi: Myers, 1962: 198, by implication.

Syrrhophus gaigeae Schmidt & Smith, 1944: 80.—**O**: **H**: FMNH 27361.—**OT**: U.S., Texas, Brewster County, Chisos Mountains, the Basin.—**SS:** Lynch, 1970, with *S. guttilatus*.

Syrrhophus petrophilus Firschein, 1954: 50.—**O**: **H**: UIMNH 7807.—**OT**: Mexico, San Luis Potosí, 3 miles southwest of San Luis Potosí. Synonymy with *S. guttilatus* by Lynch, 1970: 22. Also considered a **NN** for *S. gaigei* (Lynch, 1996: 279).

Eleutherodactylus petrophilus: Myers, 1962: 198, by implication.

Distribution: Davis Mountains area of western Texas, south into Mexico. See map in Elliott *et al.* (2009) and Dodd (2013). **U.S.**: TX.

Common name: Spotted Chirping Frog.

Commentary: There is some controversy regarding the synonymy of the name *Syrrhophus gaigeae;* Milstead *et al.* (1950) placed it in the synonymy of *S. marnockii,* while Lynch (1970) placed it in the synonymy of *S. guttilatus*; Lynch is probably correct.

Eleutherodactylus (Syrrhophus) marnockii (Cope, 1878)

Syrrhophus marnockii Cope, 1878b: 253.—**O**: **S**: ANSP 10765-10768 (Malnate, 1971, sd.).—**OT**: U.S., Texas, Bexar County, near San Antonio. Placed on the official list of specific names (ICZN, 1974: 130-132).

Syrrhophus marnochii: Cope, 1880: 26; Strecker, 1915: 48.—**IS.**

Hylodes marnockii: Garman, 1884a: 42.

Syrrhophus marnockii: Boulenger, 1888a: 205; Cope, 1889b: 318; Stejneger & Barbour, 1917: 35; 1943: 53; Gorham, 1974: 70; Behler & King, 1979: 422; Frost, 1985: 341; Garrett & Barker, 1987: 33; Conant

& Collins, 1998: 510; Frost *et al.*, 2006a: 362; Lemos-Espinal & Smith, 2007a: 52, 249.

Syrrhophus marnocki: Schmidt, 1953: 60; Conant, 1975: 304; Cochran & Goin, 1970: 119; Dixon, 2000: 64.—**UE**.

Eleutherodactylus marnockii: Myers, 1962: 196, by implication; Lemos-Espinal *et al.*, 2004: 6, 79; Elliott *et al.*, 2009: 290; Dodd, 2013: 201.

Eleutherodactylus (Syrrhophus) marnockii: Hedges, 1989: 318; Hedges *et al.*, 2008.

?Syrrhophus gaigeae Schmidt & Smith, 1944: 80.—**O**: **H**: FMNH 27361.—**OT**: U.S., Texas, Brewster County, Chisos Mountains, the Basin.—**SS:** Milstead *et al.*, 1950, with *S. marnocki*, followed by Schmidt, 1953: 60; see *Commentary* for *E. guttilatus*.

Distribution: Edwards Plateau region of central Texas. See map in Elliott *et al.* (2009) or Dodd (2013). **U.S.**: TX.

Common name: Cliff Chirping Frog.

Commentary: On the official list of specific names.

Epifamily **Hyloidia** Rafinesque, 1815

Synonymy: see below under Hylini.

Content: Ten families, of which four are in North America.

Family **Bufonidae** Gray, 1825

Bufonina Gray, 1825: 214.—**NG**: *Bufo* Laurenti, 1768

Bufonoidea: Fitzinger, 1826: 37.

Bufonidea: Fitzinger, 1827: 264.

Bufones: Wiegmann, 1832: 202.

Bufonina: Bonaparte, 1938b: 393; 1850: [2].

Bufonini: Bonaparte, 1839e: unnumbered.

Bufonidae: Bell, 1839: 105, first use of current spelling; Bonaparte, 1850: [2]; Bolkay, 1919: 349, 356; Miranda-Ribeiro, 1926: 19; Blackburn & Wake, 2011: 40.

Bufoniformes: Duméril & Bibron, 1841: 640.

Bufonia: Gravenhorst, 1845: 43.

Bufonides: Bruch, 1862: 221.

Bufonida: Haeckel, 1866: cxxxii; Knauer, 1878: 109; Bayer, 1885: 19.

Bufonidi: Acloque, 1900: 492.

Bufoninae: Fejérváry, 1917: 148.

Bufavidae: Fejérváry, 1920: 30.

Atelopoda Fitzinger, 1843: 32.—**NG**: *Atelopus* Duméril & Bibron, 1841.—**SS**: Duellman & Lynch, 1969: 239.

Atelopodes: Fitzinger, 1861: 414.

Atelopodidae: Parker, 1934: 8; Davis, 1935: 91.

Atelopodinae: Davis, 1935: 91.

Phryniscidae Günther, 1858a: 346.—**NG**: *Phryniscus* Wiegmann, 1834.—**SS**: Boulenger, 1894b: 374, by implication.

Phryniscina: Mivart, 1869: 288 (as subfamily).

Adenomidae Cope, 1861g: 371.—**NG**: *Adenomus* Cope, 1861.—**SS**: Cope, 1862c: 358, by implication; Boulenger, 1882a: 281.

Adenominae: Dubois, 1983: 273.

Dendrophryniscina de la Espada, 1871: 65.—**NG**: *Dendrophryniscus* Jiménez de la Espada, 1871.

Dendrophryniscidae: de la Espada, 1871: 65; Miranda-Ribeiro, 1926: 19.

Dendrophryniscinae: Gadow, 1901: 139.

Platosphinae Fejérváry, 1917: 147.—**NG**: *Platosphus* d'Isle, 1877; proposed as a fossil genus.

Tornieriobatidae Miranda-Ribeiro, 1926: 19.—**NG**: *Tornierobates* Miranda-Ribeiro, 1926.

Tornieriobatinae: Dubois, 1983: 273.

Nectophrynidae Laurent, 1942: 6.—**NG**: *Nectophryne* Buchholz & Peters, 1875.—**SS**: Dubois, 1987c: 27.

Nectophrynoidini: Dubois, 1982: 50, 1987c: 27.

Stephopaedini Dubois, 1987c: 27.—**NG**: *Stephopaedes* Channing, 1979.

Distribution: Essentially cosmopolitan, but absent from the Australian and Madagascar regions, most oceanic islands, and polar regions, but one species widely introduced in Australia, Philippines, Hawaii, and other islands.

Common name: True Toads.

Content: Currently about 36 genera with 558 species. We recognize one genus with three subgenera and 22 species found in North America.

Commentary: Frost *et al.* (2006a) partitioned the genus *Bufo,* by far the largest genus of the family and the only one in North America, into several genera matching the clades resulting from their phylogenetic study. Partly because some North American species have been well known to biologists whose work is outside of systematics or herpetology, while we retain the generic names of that study, or modifications that have followed, we reduce them in rank to subgenera of *Bufo.*

At least within our area, all the other clades branch below *Bufo,* so this arrangement still results in monophyletic groups. Pyron & Wiens (2011) also retained *Bufo* as the generic name for all North American bufonids, and did not suggest any subdivisions. Dodd (2013) elected to follow Frost *et al.* (2006a).

North America contains only 3.9% of the species diversity of this family.

Genus **Bufo** Garsault, 1764

The synonymy here includes only the subgenus *Bufo* and other **extralimital** subgenera. The remainder of the synonymy for subgenera occurring in North America is with those subgenera that follow.

Bufo Garsault, 1764: pl 672 (*nec Bufo* Laurenti, 1768).—**NS** (sd. Dubois & Bour, 2010b: 24): *Rana bufo* Linnaeus, 1758.

Bufo Laurenti, 1768: 25 (*nec Bufo* Garsault, 1764).—**NS** (sd. Fitzinger, 1843: 32): *Bufo viridis* Laurenti, 1768. Invalid senior OS of *Bufotes* Rafinesque, 1815.

"Buffo" de la Cepède, 1788b: 460.—**UE.**—Published in a work on the Official Index of Rejected and Invalid Works; an invalid senior OS of *Bufo.*

"Batrachus" Rafinesque, 1814b: 103 (*nec Batrachus* Schaeffer, 1760; *nec Batrachus* Walbaum, 1792; *nec Batrachus* Schneider, 1801).—**NN** of *Bufo* Laurenti, 1768, invalid SS.

Bufotes Rafinesque, 1815: 78.—**NN.** Senior nomen available to replace *Bufo* Laurenti as a valid subgeneric name for the subgenus containing *Bufo viridis* Laurenti.

"Phryne" Oken, 1816: 210 (*nec Phryne* Meigen, 1800; *nec Phryne* Herrich-Schaffer, 1843).—**NS:**: *Bufo vulgaris* Laurenti, 1768 (Fitzinger, 1843: 32, sd.).—**SS**: Boulenger, 1882a: 281. Junior synonym of *Bufo.* Unavailable in this genus as a junior homynym of Meigen's name.

"Calamita" Oken, 1816: Abth. 2: v, 209 (*nec Calamita* Schneider, 1799).—**NS:** *Bufo calamita* (Oken, 1816) (absolute tautonomy). Published in a work on the Official Index of Rejected and Invalid Works; an invalid senior OS of *Epidalea* Cope, 1864.

Chascax Ritgen, 1828: 278.—**NS**: *Bombinator strumosus* Merrem, 1820 (synonym *Bufo strumosus* Daudin, 1802). Invalid senior OS of *Peltophryne.*

Peltophryne Fitzinger, 1843: 344.—**NS**: *Bufo peltocephalus* Fitzinger, 1843.—**SS**: Günther, 1859a; removed by Pregill, 1981; replaced by Hedges, 1996 and Pramuk, 2000; removed by Frost *et al.*, 2006a. Valid as a subgenus.

Peltaphryne: Cope, 1862b: 344; 1869a: 312.—**IS**.

Peltophryne: Cope, 1862b: 344.

Epidalea Cope, 1864a: 181.—**NS**: *Bufo calamita* (Oken, 1816), mo. Valid as a subgenus.

Otaspis Cope, 1869a: 312.—**NS**: *Peltaphryne empusa* Cope, 1862, mo.—**SS**: Boulenger, 1882a: 281. Listed as a synonym of *Peltophryne* by Frost (2009-13).

Pegaeus Gistel & Tilesius, 1868: 161.—**NS**: *Rana bufo* Linnaeus, 1758, mo.—**SS**: Mertens, 1936. Invalid junior synonym of *Bufo*.

Nannophryne Günther, 1870: 402.—**NS**: *Nannophryne variegata* Günther, 1870, mo. Valid as a subgenus.

Platosphus de l'Isle, 1877: 472.—**NS**: *Platosphus gervaisii* de l'Isle, 1877, mo. (fossil species; synonym of *Rana bufo* Linnaeus, 1758, *fide* Sanchíz, 1998: 125).

Mertensophryne Tihen, 1960b: 226.—**NS**: *Bufo rondoensis* Loveridge, 1942, od. Valid as a subgenus.

Stephopaedes Channing, 1979: 394.—**NS**: *Bufo anotis* Boulenger, 1907. Frost *et al.*, 2006a, treated this as a subgenus of *Mertensophryne*.

Bufavus Portis, 1885: 1182.—**NS**: *Bufavus meneghinii* Portis, 1885, mo. (junior synonym of *Rana bufo* Linnaeus, 1758, *fide* Sanchiz (1998: 125).

Capensibufo Grandison, 1980: 294.—**NS**: *Bufo tradouwi* Hewitt, 1926, od. Valid as a subgenus.

"Torrentophryne" Rao & Yang, 1994: 142.—**NS**: *Torrentophryne aspinia* Rao & Yang, 1994, sd. Invalid as proposed.

Torrentophryne Yang *in* Yang, Liu, & Rao, 1996: 353.—**NS**: *Torrentophryne aspinia* Yang, Liu, & Rao, 1996 (Dubois & Bour, 2010b: 24, sd.).—**SS:** Liu *et al.*, 2000.

"Pseudepidalea" Frost *et al.*, 2006a: 219.—**NS**: *Bufo viridis* Laurenti, 1768, od.—**OS** (junior) of *Bufotes,* so invalid.

Vandijkophrynus Frost *et al.*, 2006a: 220.—**NS**: *Bufo angusticeps* Smith, 1848, od. Valid as a subgenus.

Amietophrynus Frost *et al.*, 2006a: 221.—**NS**: *Bufo regularis* Reuss, 1833, od. Valid as a subgenus.

Distribution: Generally the same as for the family, essentially cosmopolitan.

Common name: True Toads.

Content: Under our treatment, about 226 species, in 14 subgenera; 22 species (9.73%), representing three subgenera, occur in North America. The totally extralimital subgenera of *Bufo* we would tentatively recognize are *Amietophrynus, Bufo, Bufotes, Capensibufo, Epidalea, Mertensophryne, Nannophryne, Peltophryne, Stephopaedes, Torrentophryne,* and *Vandijkophrynus* (refer to the cladogram of Frost *et al.*, 2006a: 218, but note that the nomina *Cranopsis* and *Chaunus* used there are no longer valid).

Commentary: The author of *Bufo* has been given as Laurenti (1768) for most of its history. Recently, it was rediscovered (Welter-Schultes *et al.*, 2008) that some nomina attributed to Laurenti were actually attributable to Garsault (1764), who proposed them earlier. One of those is the genus *Bufo,* and Dubois & Bour (2010b) discussed nomenclatural changes necessitated by this discovery. Some of the names we list as junior synonyms were proposed as full genera by Frost *et al.* (2006a), but we propose that those and some others be treated as subgenera of *Bufo.* Those taxa are in clades composing the sister taxon of their restricted genus *Bufo* (Fig. 70, p. 218; our subgenus *Bufo*), or later-branching clades below *Bufo* in the strict sense of Frost *et al.* (2006a). It is a taxonomic choice to regard them as genera or subgenera without upsetting monophyly, and we elect the latter. In so doing, we are agreeing with Hedges *et al.* (2008) that this allows non-taxonomists to continue correctly using the familiar binomen, without having to learn a new generic name, while having a subgeneric name to further identify relationships.

Subgenus **Anaxyrus** Tschudi, 1845

Anaxyrus Tschudi, 1845: 170.—**NS**: *Anaxyrus melancholicus* Tschudi, 1845, mo.—Removed from synonymy of *Bufo* (Frost *et al.*, 2006a: 363) as a full genus.

Dromoplectrus Camerano, 1879: 822.—**NS**: *Bufo anomalus* Günther, 1859, mo.—**SS**: Boulenger, 1882a: 282 with *Bufo*; Frost (2009-13) lists it as a synonym of *Anaxyrus*.

Distribution: The entire continental United States, north into southern Canada, and into Alaska; southward through most of Mexico. **Can:** all. **U.S.:** all **except**: HI.

Common name: North American Toads (SECSN) or Nearctic Toads (C&T).

Content: 22 species, of which 19 are in North America (86.4%). The other three are endemic to Mexico.

Commentary: Frost *et al.* (2006a) recognized this taxon as a full genus. We must agree with Smith & Chiszar (2006) that it is more appropriately ranked as a subgenus. Several generations of North American biologists have learned the toad species of our continent by the generic name *Bufo*; that has been a stable name here for over 100 years. If an older synonym were discovered, we would expect immediate action petitioning the ICZN to suppress that older name. Now, when we can as easily express the better understood phylogenetic relationships among toads by either using more names for separate genera to recognize the monophyletic clades, or express equally well those relationships by recognizing the same clades at a subgeneric level, preserving the generic nomen to continue to be used as well, this latter solution seems preferable. Pauly *et al.* (2009) criticized Frost's (2006a) recognition of unnecessary genera as destabilizing, and suggested that the species of *Anaxyrus*, would be better treated as a subgenus or an unranked clade. Frost *et al.* (2009a) responded negatively to that criticism. Crother (2012) and Dodd (2013) retain generic status for the nomen. The analysis of the North American bufonids by Van Bocxlaer *et al.* (2010) suggested *Anaxyrus* may not be monophyletic, which could lead to additional genera being added, where subgenera would better serve stability. We disagree with the concept of unranked subgeneric clades if they are assigned names, as that is contrary to the rules of the *Code*; and at present, only genera and subgenera can be named in the genus-series.

Species in this subgenus have been organized into several distinct species-groups. Those generally recognized (and their North American species) are as follows: **americanus group** (Blair, 1959): *B. americanus, B. baxteri, B. californicus, B. fowleri, B. hemiophrys, B. houstonensis, B. microscaphus, B. terrestris, B. woodhousii*; **boreas group** (Blair, 1972): *B. boreas, B. canorus, B. exsul, B. nelsoni*; **cognatus group** (Blair, 1959, 1972): *B. cognatus, B. speciosus*; **debilis group**

(Ferguson & Lowe, 1969, as *punctatus* group): *B. debilis, B. punctatus, B. retiformis*; **quercicus group** (Blair, 1959): *B. quercicus* only. Some of these groups currently have a composition different from that when the group was originally cited.

Bufo (Anaxyrus) americanus Holbrook, 1836 **new combination**

Synonymy: See the subspecies *B. a. americanus,* below.

Distribution: Eastern United States through much of the Great Plains, as far south as central Louisiana, and north through central Manitoba to Hudson's Bay and southern Newfoundland. See map in Elliott *et al.* (2009). **Can.**: MB, NB, NL, NS, ON, PE, QC. **U.S.**: All **except:** AK, AZ, CA, CO, FL, HI, ID, LA, MT, NM, NV, OR, WA, WY. Introduced in MA (Kraus, 2009: 142).

Common name: American Toad. *French Canadian name:* Crapaud d'Amerique.

Content: Two subspecies are recognized. A third, *copei,* is sometimes recognized, but we do not find it different enough from the nominotyical population for such treatment.

Bufo (Anaxyrus) americanus americanus Holbrook, 1836, **new combination**

"Bufo americanus" Cuvier *in* McMurtrie, 1834: 83.—**NU.**

Bufo americanus Holbrook, 1836: 75.—**O**: **H:** ANSP 2474 (Adler, 1976: xxxvi, sd.).—**OT**: mountains of Maine through all the Atlantic states, common in the upper districts of South Carolina, along the western side of the Alleghanies, and in the Valley of the Mississippi. Invalidly restricted to vicinity of Philadelphia (Schmidt, 1953: 65).

Bufo lentiginosus var. *americanus*: Günther, 1859a: 63.

Chilophryne americana: Cope, 1862c: 358.

Incilius americanus: Cope, 1863: 50.

Bufo lentiginosus americanus: Cope, 1886b: 516; 1889b: 278.

Bufo americanus: Baird, 1859a: 25; 1859c: 44; 1859e: 25; Dickerson, 1907: 63; Stejneger & Barbour, 1917: 27; Bishop, 1928a: 166; A.P. Blair, 1957: 106; Gorham, 1970: 5; Frost, 1985: 35; Elliott *et al.*, 2009: 128.

Bufo americanus americanus: Gaige, 1932: 134; Stejneger & Barbour, 1933: 27; 1943: 39; Wright & Wright,

1949: 143; Smith, 1961: 74; Cochran & Goin, 1970: 95; Gorham, 1974: 76; Conant, 1975: 307; Behler & King, 1979: 387; Conant & Collins, 1998: 514; Elliott *et al.*, 2009: 128.

Bufo terrestris americanus: Netting & Goin, 1946: 107; Schmidt, 1953: 65.

Anaxyrus americanus: Frost *et al.*, 2006a; Elliott *et al.*, 2009: 128; Dodd, 2013: 17.

Anaxyrus americanus americanus: Frost *et al.*, 2008: 2.

Bufo (Anaxyrus) americanus americanus: **new combination**

Bufo americanus var. *quadripunctatus* Jan, 1857: 54.—**O**: not stated nor known to exist.—**OT**: U.S., Georgia. *Nomen nudum.*

Bufo copei Yarrow & Henshaw, 1878: 207.—**O**: **S:** USNM 5376 (Yarrow, 1883a: 163, sd.); and USNM 5388 (17) and 35925-35971 (ex 5937) Cochran (1961a: 32, sd.).—**OT**: Can., Ontario, Hudson's Bay and James Bay.—**SS**: Cope (1886b: 516, 1889b: 34) of *Bufo lentiginosus americanus* Günther, 1859.

Bufo americanus copei: Gaige, 1932: 134; Stejneger & Barbour, 1933: 27; 1943: 39; Conant, 1975: 307; Behler & King, 1979: 388.

Bufo terrestris copei: Netting & Goin, 1946: 107; Schmidt, 1953: 66.

Bufo copei: Sanders, 1987: 11. Species status rejected, Collins (1989: 19).

Bufo americanus var. *alani* Long, 1982: 16.—**O**: not stated, perhaps in WMNH.—**OT**: U.S., Michigan, Lake Michigan, Rock Island.

Misapplication of names:

Bufo lentiginosus: Davis & Rice, 1883: 27. A synonym of *B. terrestris.*

Distribution: All but the southwestern part of the species range. **Can.**: MB, NB, NL, NS, ON, PE, QC. **U.S.**: All **except:** AK, AZ, CA, CO, FL, HI, ID, KS, LA, MT, NM, NV, OK, OR, TX, WA, WY.

Common name: Eastern American Toad. *French Canadian name:* Crapaud d'Amerique de l'Est.

Bufo (Anaxyrus) americanus charlesmithi Bragg, 1954 **new combination**

Bufo terrestris charlesmithi Bragg, 1954: 247.—**O**: **H**: OKMNH 26359.—**OT**: U.S., Oklahoma, Cleveland County, 1.8 miles south, 7 miles east of Norman.

Bufo americanus charlesmithi: Conant, 1958: 264; Smith, 1961: 74; Cochran & Goin, 1970: 95; Gorham, 1974: 76; Behler & King, 1979: 388; Garrett & Barker, 1987: 4; Conant & Collins, 1998: 515; Dixon, 2000: 70; Elliott *et al.*, 2009: 128.

Anaxyrus americanus charlesmithi: Frost *et al.*, 2008: 2; Elliott *et al.*, 2009: 128.

Bufo (Anaxyrus) americanus charlesmithi: **new combination.**

Distribution: The southwestern part of the species range. **U.S.**: AR, KS, MO, OK, TX.

Common name: Dwarf American Toad.

Commentary: The molecular analysis by Masta *et al.* (2002) only weakly supported recognition of this subspecies. We tentatively retain it, pending further study.

Bufo (Anaxyrus) baxteri Porter, 1968, **new combination**

Bufo hemiophrys baxteri Porter, 1968: 593.—**O**: **H**: K.R. Porter collection 5-164, now USNM 166434.—**OT**: U.S., Wyoming, Albany County, 0.5 mile NW Laramie.

Bufo baxteri: Packard, 1971: 191; Collins, 1991: 43; Elliott *et al.*, 2009: 158.

Bufo hemiophrys baxteri: Sanders, 1987: 17.

Anaxyrus baxteri: Frost *et al.*, 2006a: 363; Elliott *et al.*, 2009: 158; Dodd, 2013: 43.

Bufo (Anaxyrus) baxteri: **new combination**

Misapplied names:

Bufo americanus hemiophrys: Cook, 1983: 1.

Distribution: Isolated populations in the Laramie Basin of southeastern Wyoming. See map in Elliott *et al.* (2009) and Dodd (2013). **U.S.**: WY.

Common name: Wyoming Toad.

Commentary: This species is federally listed as endangered. Odum & Corn (2005b) reported that populations had declined dramatically, and were considered extinct in 1983, and then were completely extirpated in the wild by

1994. Captive populations yielded individuals which were reintroduced where they had been extirpated, but with limited success. Apparently there is only one extant population remaining, in the area of Mortenson Lake, and the species is again on the verge of extinction.

Bufo (Anaxyrus) boreas Baird & Girard, 1852, **new combination**

Synonymy: See under the subspecies *B. b. boreas,* below.

Distribution: Southwestern California, into Baja California, and northward through central and northern California, Oregon, Washington, most of British Columbia, into southern Yukon and southeastern Alaska; eastward from the coast it is found in most of Nevada, western Wyoming, western Montana, all of Idaho, and much of western Alberta. Formerly it was also present in central Utah, west central Colorado into southern Wyoming, with a disjunct area in south central Colorado and adjacent north central New Mexico, but those populations are now mostly or perhaps completely extirpated (Muths & Nan-jappa, 2005). Also see map in Elliott *et al.* (2009) and for map of the range in Canada, see Fisher et al. (2007). Dodd (2013) maps the entire range. **Can.**: AB, BC, YT. **U.S.**: AK, CA, CO, ID, MT, NM, NV, OR, UT, WA, WY.

Common name: Western Toad. *French Canadian name:* Crapaud de l'Ouest.

Content: Two subspecies are currently recognized, but Goebel *et al.* (2009) have found that the species as treated here is apparently a complex of cryptic species which seem not to conform to the current subspecies arrangement, so there may soon be changes in taxonomy. Two other populations formerly considered subspecies of this species are now elevated to species-level, *B. exsul* and *B. nelsoni.*

Commentary: In British Columbia, Northwest Territories, and Yukon, the species is listed as of Special Concern.

Bufo (Anaxyrus) boreas boreas Baird & Girard, 1852, **new combination**

"Rana canagica" Pallas, 1831: 12.—**O**: not stated nor known to exist.—**OT**: Aleutian region, between Camtschatcam and America, corrected to U.S., Alaska, Kanaga Island (Kuzmin, 1996: 51).—**SS**: Lindholm, 1924: 46, with *Rana sylvatica*; Myers, 1930a: 62, with *Bufo boreas*. Said to be **ND** (Stejneger & Barbour, 1933: vi), but Kuzmin (1996: 51) treated this as a **NO**. Either is grounds to reject the name as unavailable.

Bufo canagicus canagicus: Myers, 1930a: 62.

Bufo boreas Baird & Girard, 1852a: 174.—**O**: **S** USNM 15467-15470 (Cochran, 1961a: 32, sd.).—**OT**: U.S., Washington, Columbia River and Puget Sound. Restricted to vicinity of Puget Sound by Schmidt (1953: 61), but with no evidence the **S** were all from there, hence that action is invalid.

Bufo boreas: Cooper, 1860: 303; Cook, 1983: 1; Frost, 1985: 37.

Bufo boreas boreas: Stejneger, 1893b: 220; Camp, 1917: 116; Stejneger & Barbour, 1917: 27; 1943: 40; Schmidt, 1953: 61: Cochran & Goin, 1970: 92; Gorham, 1974: 77; Behler & King, 1979: 388; Stebbins, 1985: 70; 2003: 209; Elliott *et al.*, 2009: 172.

Anaxyrus boreas: Frost *et al.*, 2006a: 363; Elliott *et al.*, 2009: 172; Dodd, 2013: 47.

Anaxyrus boreas boreas: Frost *et al.*, 2008: 3; Elliott *et al.*, 2009: 172.

Bufo (Anaxyrus) boreas boreas: **new combination**

Bufo columbiensis Baird & Girard, 1853b: 378.—**S**: including USNM 4182 (7) (Cochran, 1961a: 32, sd.).—**OT**: U.S., Oregon, on the Columbia River. Restricted to vicinity of the mouth of the Columbia River, by Schmidt (1953: 61), but as he presented no evidence that any of the **S** came from that specific area, the restriction is not valid.—**SS**: Boulenger (1882a: 296), Cope (1886b: 514; recognized the synonymy but used this name anyway), Camp (1917: 116).

Bufo columbiensis: Baird, 1859d: 12; Cooper, 1860: 304.

Bufo columbiensis columbiensis: Cope, 1889b: 269; again Cope recognized the synonymy with the older name (267) but used the junior name.

Bufo halophilus columbiensis: Stejneger, 1893b: 220.

Bufo lamentor Girard, 1860: 169.—**O: S:** including USNM 4194 (4) (Cochran, 1961a: 35, sd.).—**OT**: U.S., Utah Territory (now Wyoming), about Fort Bridger.—**SS**: Tanner, 1933: 42.

Bufo politus Cope, 1862a: 158.—**O**: **H**: USNM 5600.—**OT**: near Greytown, Nicaragua, in error *fide* Savage (1967: 226), probably U.S. Pacific coast.—**SS**: Savage (1967: 226).

Bufo pictus Cope *in* Yarrow, 1875: 522.—**O**: **H** USNM 8655 (Yarrow, 1883a: 164, sd.). Cochran (1961a: 36) also listed USNM 25332-25334 as additional **S**, apparently in error.—**OT**: U.S, Utah, neighborhood of Provo and Utah Lake. Schmidt (1953: 61) restricted it to Provo, but presented no evidence, thus invalid.—**SS**: Cope (1886b: 514; 1889b: 267).

Bufo lentiginosus pictus: Garman, 1884a: 43, by implication.

Distribution: As for the species, excluding the portion in southern California and Baja California, reaching southward only into northern California. **Can.**: AB, BC, YT. **U.S.**: AK, CA, CO, ID, MT, NM, NV, OR, UT, WA, WY.

Common name: Boreal Toad. *French Canadian name:* Crapaud de l'Ouest.

Commentary: Goebel *et al.* (2009) identified lineages within the *B. boreas* group, which may not correspond to current subspecies treatment, but did not suggest taxonomic changes.

Bufo (Anaxyrus) boreas halophilus Baird & Girard, 1853, new combination

Bufo halophila Baird & Girard, 1853a: 301.—**O**: **S** USNM 2589 (2) (Cochran, 1961a: 34, sd.).—**OT**: U.S., California, Solano County, Benicia.—**SS**: Cope (1886b: 514; 1889b: 267), Camp (1917: 116) with *B. boreas*.

Bufo halophila: Baird, 1859e: 26.

Bufo halophilus: Boulenger, 1882a: 295.

Bufo columbiensis halophilus: Cope, 1889b: 269.

Bufo boreas halophilus: Camp, 1917: 116; Stejneger & Barbour, 1917: 27; 1943: 40; Schmidt, 1953: 62; Cochran & Goin, 1970: 92; Gorham, 1974: 77; Behler & King, 1979: 389; Stebbins, 1985: 70; 2003: 209; Liner, 1994: 16; Elliott *et al.*, 2009: 172.

Bufo canagicus halophilus: Myers, 1930a: 62.

Bufo halophilus halophilus: Frank & Ramus, 1995: 40.

Anaxyrus boreas halophilus: Frost *et al.*, 2008: 3; Elliott *et al.*, 2009: 172; Dodd, 2013: 50.

Bufo (Anaxyrus) boreas halophilus: **new combination.**

Distribution: Southwestern California, into Baja California, and northward through most of central and northern California. **U.S.**: CA.

Common name: California Toad (C&T); not listed by SECSN.

Commentary: See the analysis of Goebel *et al.* (2009).

Bufo (Anaxyrus) californicus Camp, 1915, **new combination**

Bufo cognatus californicus Camp, 1915: 331.—**O**: **H**: MVZ 4364.—**OT**: U.S., California, Ventura County, Santa Paula, elev. 800 feet.

Bufo cognatus californicus: Stejneger & Barbour, 1917: 27.

Bufo californicus: Myers, 1930b: 73; Stejneger & Barbour, 1933: 27; 1943: 40; Frost & Hillis, 1990; Liner, 1994: 16; Stebbins, 2003: 212; Sweet & Sullivan, 2005: 396; Elliott *et al.*, 2009: 180.

Bufo compactilis californicus: Linsdale, 1940: 206.

Bufo woodhousii californicus: Shannon, 1949: 301.

Bufo microscaphus californicus: Stebbins, 1951: 274; 1985: 72; Schmidt, 1953: 64; Cochran & Goin, 1970: 94; Gorham, 1974: 82; Behler & King, 1979: 394.

Anaxyrus californicus: Frost *et al.*, 2006a: 363; Elliott *et al.*, 2009: 180; Dodd, 2013: 65.

Bufo (Anaxyrus) californicus: **new combination.**

Distribution: Southern California west of the deserts, south into Baja California, Mexico. Jennings & Hayes (1994) estimated this species has been extirpated from about 76% of its range, and the distribution is highly fragmented. Sweet & Sullivan (2005) discussed the conservation status, and have a general map. Also see map in Elliott *et al.* (2009) and Dodd (2013). **U.S.**: CA.

Common name: Arroyo Toad.

Commentary: This species is federally listed as endangered.

Bufo (Anaxyrus) canorus Camp, 1916, **new combination**

Bufo canorus Camp, 1916c: 59.—**O**: **H**: MVZ 5744 [ex Camp 2129].—**OT**: U.S., California, Mariposa County, Yosemite National Park, Porcupine Flat, 8100 feet.

Bufo canorus: Stejneger & Barbour, 1917: 27; 1943: 41; Schmidt, 1953: 62; Cochran & Goin, 1970: 92;

Gorham, 1974: 78; Behler & King, 1979: 389; Frost, 1985: 39; Stebbins, 1985: 71; 2003: 210; Elliott *et al.*, 2009: 182.

Anaxyrus canorus: Frost *et al.*, 2006a: 363; Elliott *et al.*, 2009: 182; Dodd, 2013: 70.

Bufo (Anaxyrus) canorus: **new combination.**

Distribution: Restricted to upper elevations of the Sierra Nevada in west central California. See map in Karlstrom (1973), Elliott *et al.* (2009), or Dodd (2013). **U.S.**: CA.

Common name: Yosemite Toad.

Commentary: Karlstrom (1973) provided a CAAR account. Stuart *et al.* (2008) indicated this form may be conspecific with *B. exsul,* and discussed the conservation status. Goebel *et al.* (2009) found that the species as currently treated is polyphyletic, and the name-bearing clade is imbedded in a clade that includes most of the *B. boreas* range. More study is needed to sort out the taxonomy of these forms.

Bufo (Anaxyrus) cognatus Say, 1822, **new combination**

Bufo cognatus Say, *in* James, 1822: 55; not found in James, 1823.—**O** (sd., Kellogg, 1932: 43), "Philadelphia Museum" (not ANSP), probably destroyed.—**OT**: U.S., Colorado, Prowers County, alluvial margins of the Arkansas River. Corrected to U.S., Colorado, Prowers County, approx. 3 miles west of Holly (Dundee, 1996: 83).

Bufo cognatus: Baird, 1859c: 44; 1859e: 27; Cope, 1889b: 275; Strecker, 1915: 52; Stejneger & Barbour, 1933: 28; 1943: 41; Schmidt, 1953: 62; Cochran & Goin, 1970: 93; Gorham, 1974: 78; Conant, 1975: 312; Behler & King, 1979: 389; Cook, 1983: 1; Frost, 1985: 41; Stebbins, 1985: 74; 2003: 215; Liner, 1994: 17; Conant & Collins, 1998: 523; Lemos-Espinal & Smith, 2007a: 37, 236; Elliott *et al.*, 2009: 152.

Chilophryne cognata: Cope, 1862b: 358.

Incilius cognatus: Cope, 1863: 50.

Bufo lentiginosus cognatus: Cope, 1875b: 29; Yarrow, 1875: 521.

Bufo cognatus cognatus: Camp, 1915: 331, by implication; Stejneger & Barbour, 1917: 27.

Anaxyrus cognatus: Frost *et al.*, 2006a: 363; Elliott *et al.*, 2009: 152; Dodd: 2013: 78.

Bufo (Anaxyrus) cognatus: **new combination**

Bufo dipternus Cope, 1879a: 437.—**O**: **S** (sd., Malnate, 1971: 350) ANSP 19769-19771.—**OT**: U.S., Montana, Chouteau County, north of the Missouri River east of Fort Benton.—**SS**: Cope, 1886b: 516, 1889b: 275.

Distribution: Western Texas north through the Great Plains, east of the Rocky Mountains, to southern Saskatchewan and Manitoba, west of Texas through New Mexico and Arizona into southeastern California, and south of Texas into Mexico. See map in Krupa (1990). **Can.**: MB, SK. **U.S.**: AZ, CA, CO, IA, KS, MN, MO, MT, ND, NE, NM, NV, OK, SD, TX, UT, WI, WY.

Common name: Great Plains Toad. *French Canadian name:* Crapaud des Grandes Plaines.

Commentary: Krupa (1990) provided a CAAR account. In Canada the species is listed as Special Concern. Say described the species in a volume compiled by James (1822); a second volume was also published that year. Another edition was published in London (James, 1823), differing in use of end notes rather than footnotes, and other details. The date has often been cited as 1823; that would be based on the London edition, but the information was published first in Philadelphia, in 1822, as verified by Woodman (2010), so new names in the Philadelphia edition take priority as 1822. The species is listed in Manitoba as Threatened.

Bufo (Anaxyrus) debilis Girard, 1856

Bufo debilis Girard, 1856a: 87.—**O**: **S** (sd., Kellogg, 1932: 52) USNM 2620 (lost) and 2621 (8). Cochran (1961a: 33) listed only 7 specimens, and Girard mentioned only 6. Bogert (1962) suggested that there may have been errors transferring specimens prior to attaching tags, and questioned the reliability of considering these as the type material.—**OT**: Lower Rio Grande valley and Mexican state of Tamaulipas. Kellogg (1932: 51-52) reported that data associated with USNM 2620 place it as from between Guerrero and Camargo in Tamaulipas, and for the 8 toadlets of USNM 2621, from Matamoros, Tamaulipas. Several authors incorrectly listed the **OT** as Brownsville, Texas (*e.g.*, Schmidt, 1953: 63; Frost, 1985: 43).

Bufo debilis: Baird, 1859e: 27; Cope, 1889b: 264; Stejneger & Barbour, 1917: 28; 1943: 41; Frost, 1985: 43; Stebbins, 1985: 76; 2003: 217.

Bufo debilis debilis: Smith, 1950: 75; Schmidt, 1953: 63; Cochran & Goin, 1970: 99; Gorham, 1974: 79; Conant,

1975: 314; Behler & King, 1979: 391; Garrett & Barker, 1987: 6; Liner, 1994: 17; Conant & Collins: 1998: 525; Dixon, 2000: 70-71; Lemos-Espinal & Smith, 2007a: 39, 237; Elliott *et al.*, 2009: 164.

Anaxyrus debilis: Frost *et al.*, 2006a: 363; Elliott *et al.*, 2009: 164; Dodd, 2013: 88.

Bufo (Anaxyrus) debilis: Pough, 2007: 207. Apparently the first treatment of *Anaxyrus* as a subgenus.

Anaxyrus debilis debilis: Frost *et al.*, 2008: 3.

Bufo insidior Girard, 1856a: 87.—**O**: **S** (sd., Kellogg, 1932: 52) USNM 2622 (2).—**OT**: Mexico, Chihuahua. Bogert (1962) doubted that the specimens 2622 actually were from the city of Chihuahua, but the form does occur there.

Bufo insidior: Baird, 1859e: 26; Taylor, 1938b: 513; Stejneger & Barbour, 1943: 42.

Bufo debilis insidior: Smith, 1950: 75; Schmidt, 1953: 63; Cochran & Goin, 1970: 101; Gorham, 1974: 79; Conant, 1975: 314; Behler & King, 1979: 391; Garrett & Barker, 1987: 7; Liner, 1994: 17; Conant & Collins, 1998: 525; Dixon, 2000: 71; Lemos-Espinal & Smith, 2007b: 39, 237; Elliott *et al.*, 2009: 164.

Distribution: Central and western Texas, western Oklahoma, and southwestern Kansas, with a disjunct population in southeastern Colorado, westward through eastern and southern New Mexico to southeastern Arizona, south into Mexico. See map in Elliott *et al.* (2009) or Dodd (2013). **U.S.**: AZ, CO, KS, NM, OK, TX.

Common name: Chihuahuan Green Toad (SECSN) or Green Toad (C&T).

Content: Two subspecies are sometimes recognized, but they are poorly differentiated and probably not validly recognized. We here reject them.

Bufo (Anaxyrus) exsul Myers, 1942, **new combination**

Bufo exsul Myers, 1942b: 3.—**H**: UMMZ 83357, od.—**OT**: U.S., California, Inyo County, Deep Springs Valley, Deep Springs.

Bufo exsul: Stejneger & Barbour, 1943: 42; Savage, 1959: 252; Gorham, 1974: 79; Behler & King, 1979: 391; Stebbins, 1985: 71; 2003: 210; Elliott *et al.*, 2009: 178.

Bufo boreas exsul: Schmidt, 1953: 61; Cochran & Goin, 1970: 92.

Anaxyrus exsul: Frost *et al.*, 2006a: 363; Elliott *et al.*, 2009: 178; Dodd, 2013: 92.

Bufo (Anaxyrus) exsul: **new combination.**

Distribution: Known only from around the **OT,** Deep Springs, in California. See map in Elliott *et al.* (2009) and Dodd (2013). **U.S.**: CA.

Common name: Black Toad.

Commentary: Often treated as a subspecies of *B. boreas.* Goebel *et al.* (2009), in a molecular study, found this form to be a sister taxon of a clade composed of some populations of *B. boreas halophilus, B. canorus,* and *B. nelsoni.* More study is needed to elucidate the relationships among these forms.

Bufo (Anaxyrus) fowleri Hinckley, 1882, **new combination**

"Bufo lentiginosus var. *fowlerii"* Putnam *in* Cope, 1875a: 29.—**NU**.

Bufo lentiginosus fowlerii: Boulenger, 1882a: 310.

Bufo fowleri Hinckley, 1882: 310,—**O**: **S** (sd., Barbour & Loveridge, 1929: 231) includes MCZ 518 (4). Frost (2009-13) pointed out that Peters (1952: 11) considered UMMZ 50246 a cotype, adding confusion. But Kluge (1983a: 24) indicated that none of the specimens previously considered syntypes were ever actually seen by Hinckley, so he designated a **N**: UMMZ 50246.—**OT** (original): U.S., Massachusetts, Norfolk County, Milton;—**OT** of **N**: U.S., Massachusetts, Essex County, Danvers.

Bufo lentiginosus fowleri: Garman, 1884a: 42, by implication; Cope, 1889b: 279.

Bufo fowleri: Dickerson, 1907: 93; Stejneger & Barbour, 1917: 28; Strecker & Williams, 1928: 8; Bragg & Sanders, 1951; Gorham, 1974: 79; Sanders, 1986: 23; 1987: 27. Elliott *et al.*, 2009: 132.

Bufo woodhousii fowleri: Stejneger & Barbour, 1943: 44; Viosca, 1949: 10; Conant & Collins, 1998: 520.

Bufo woodhousei fowleri: Schmidt, 1953: 67; Cochran & Goin, 1970: 97; Conant, 1975: 310; Behler & King, 1979: 399; Cook, 1983: 1.

Bufo fowleri fowleri: Sanders, 1987: 31.

Anaxyras fowleri: Frost *et al.*, 2006a: 363; Elliott *et al.*, 2009: 132; Dodd, 2013: 96.

Bufo (Anaxyrus) fowleri: **new combination**

Bufo woodhousii velatus Bragg & Sanders, 1951: 366.—**O**: **H** (od.), USNM 131869 (*ex* OS 1891) (Cochran, 1961a: 38.).—**OT**: U.S., Texas,

Anderson County.—**Comment**: This population may represent hybrids between *B. woodhousii* and *B. fowleri* (Sullivan *et al.,* 1996b), and if so cannot be a valid taxon. Sullivan *et al.,*1996b*,* referred the name to synonymy of *Bufo fowleri*.

Bufo woodhousei velatus: Schmidt, 1953: 67; Conant, 1958: 269.

Bufo woodhousii velatus: Cochran & Goin, 1970: 97; Gorham, 1974: 86; Collins, 1997: 11.

Bufo velatus: Garrett & Barker, 1987: 13; Sanders, 1986: 19; 1987: 52; Dixon, 2000: 73. Species status rejected, Collins, 1989: 19.

Bufo hobarti Sanders, 1987: 35.—**O**: **H** (od.), UIMNH 4574.—**OT**: U.S., Indiana, Montgomery County, Shades State Park.—**SS**: Collins, 1989: 19, of *B. woodhousii, sensu lato*. Frost (2009-13) lists it as a synonym of *B. fowleri*.

Distribution: Eastern Texas and Oklahoma, northeastward through Missouri to southwestern Michigan, eastward as far as New Hampshire, barely entering Ontario. Elliott *et al.* (2009) and Dodd (2013) have maps. **Can.**: ON. **U.S.**: AL, AR, CT, DE, FL, GA, IA, IL, IN, KY, LA, MA, MD, MI, MO, MS, NC, NH, NJ, NY, OH, OK, PA, RI, SC, TN, TX, VA, VT, WV.

Common name: Fowler's Toad. *French Canadian name:* Crapaud de Fowler.

Commentary: Sanders (1986) reviewed the species *B. woodhousii* in Texas, and doubted the presence of *B. fowleri* in the state; his treatment of the relationships among *B. woodhousii, B. fowleri* (as *B. velatus*), and *B. nebulifer* (as *B. valliceps*) differs from most others. In a molecular phylogenetic study, Masta *et al.* (2002) found this species most closely related to *Bufo terrestris* (rather than to *B. woodhousii* as previously believed), and those two species form a clade that is sister to a clade of *B. americanus* + *B. woodhousii.* They also found *B. fowleri* to contain three distinctive lines which might represent cryptic species, but further study is needed to elucidate that. Listed in Ontario as Endangered.

Bufo (Anaxyrus) hemiophrys Cope, 1886, new combination

Bufo hemiophys Cope, 1886b: 515.—**O**: **S** (od), USNM 11927 (6); Cochran (1961a: 34) noted seven, of which one had been exchanged to MCZ, renumbered 3728. Cope noted only 6 specimens in the description.—**OT**: U.S., northern boundary of Montana; corrected to U.S., North Dakota, Pembina County, Pembina (Cochran, 1961a: 34). Stejneger & Barbour (1917: 28) and Barbour & Loveridge (1929: 232) gave the **OT** as Pembina and Turtle Mountains. Schmidt's (1953: 67) further restriction was invalid as he presented no evidence to support it.

Bufo hemiophrys: Stejneger & Barbour, 1917: 28; 1943: 42; W.F. Blair, 1957: 99; Conant, 1958: 265; Cochran & Goin, 1970: 98; Gorham, 1974: 80; Frost, 1985: 47; Stebbins, 1985: 75; Sanders, 1987: 17; Elliott *et al.*, 2009: 156.

Bufo woodhousei hemiophrys: Schmidt, 1953: 67.

Bufo hemiophrys hemiophrys: Porter, 1968: 593, by implication; Conant, 1975: 309; Behler & King, 1979: 392; Conant & Collins, 1998: 517.

Bufo americanus hemiophrys: Cook, 1983: 1.

Anaxyrus hemiophrys: Frost *et al.*, 2006a: 363; Elliott *et al.*, 2009: 156; Dodd, 2013: 113.

Bufo (Anaxyrus) hemiophrys: **new combination.**

Distribution: Mostly Canadian, in southern Manitoba, most of Saskatchewan, eastern Alberta and into southern Northwest Territories and Nunavut; south into northern Montana, northern North Dakota, northeastern South Dakota and western Minnesota. See map in Elliott *et al.* (2009), and for Canadian map see Fisher et al. (2007); Dodd (2013) maps the entire range. **Can.**: AB, MB, NT, NU, SK. **U.S.**: MN, MT, ND, SD, WI.

Common name: Canadian Toad. *French Canadian name:* Crapaud du Canada.

Bufo (Anaxyrus) houstonensis Sanders, 1953 new combination

Bufo houstonensis Sanders, 1953: 27.—**O**: **H** (od.), UIMNH 33687 (*ex* OS 3006).—**OT**: U.S., Texas, Harris County, Fairbanks; corrected to U.S., Texas, Harris County, northwest Houston, of Tanner Road, 1-2 mi. W of its junction with Campbell Road (Brown, 1971: 195).

Bufo americanus houstonensis: A.P. Blair, 1957: 250.

Bufo houstonensis: Conant, 1958: 265; Cochran & Goin, 1970: 97; Brown, 1971: 185; Gorham, 1974: 80; Behler & King, 1979: 392; Frost, 1985: 48; Garrett & Barker, 1987: 8; Conant & Collins, 1998: 516; Dixon, 2000: 71; Elliott *et al.*, 2009: 148.

Anaxyrus houstonensis: Frost *et al.*, 2006a: 363; Elliott *et al.*, 2009: 148; Dodd, 2013: 120.

Bufo (Anaxyrus) houstonensis: **new combination**

Distribution: Isolated population in southeastern Texas. See map in Brown (1973). Shepard & Brown (2005) also have a map, and they reported on the conservation status of the species. Populations have been extirpated from much of the former range, including the area around Houston. **U.S.**: TX.

Common name: Houston Toad.

Commentary: A rare and endangered species with a restricted range; it is federally and state listed as endangered. Brown (1973) provided a CAAR account.

Bufo (Anaxyrus) microscaphus Cope, 1867, **new comb.**

Bufo microscaphus Cope, 1867a: 301.—**O**: **S** (sd., Cochran, 1961a: 35) USNM 4106 (lost), 4184, 132901.—**L**: USNM 4184, Shannon (1949: 307).—**OT**: U.S., Arizona, near parallel of 35°, along the valley of the Colorado from Fort Mojave to Fort Yuma; **OT** of **L** (Shannon, 1949), U.S., Arizona, Mohave County, vicinity of Fort Mohave.

Bufo microscaphus: Yarrow, 1875: 522; Frost, 1985: 54; Liner, 1994: 17; Stebbins, 2003: 213; Brennan & Holycross, 2006: 36; Elliott *et al.*, 2009: 170.

Bufo lentiginosus microscaphus: Garman, 1884a: 43, by implication.

Bufo woodhousii microscaphus: Shannon, 1949: 301.

Bufo microscaphus microscaphus: Stebbins, 1951: 274; 1985: 72; Schmidt, 1953: 64; Cochran & Goin, 1970: 94; Gorham, 1974: 82; Behler & King, 1979: 394.

Anaxyrus microscaphus: Frost *et al.*, 2006a: 363; Elliott *et al.*, 2009: 170; Dodd, 2013: 127.

Bufo (Anaxyrus) microscaphus: **new combination.**

Distribution: Southwestern Utah and southern Nevada and adjacent California, across central Arizona into New Mexico. See map in Price & Sullivan (1988), Elliott *et al.* (2009), or Dodd (2013). **U.S.**: AZ, CA, NM, NV, UT.

Common name: Arizona Toad.

Commentary: Cope (1889b: 267) synonymized this species with *Bufo columbiensis* Baird & Girard, 1853 (now *Bufo boreas* Baird & Girard, 1852a). Price and Sullivan (1988) provided a CAAR account.

Bufo (Anaxyrus) nelsoni Stejneger, 1893, **new combination**

Bufo boreas nelsoni Stejneger, 1893b: 220.—**O**: **H** (od.), USNM 18742.—**OT**: U.S., Nevada, Nye County, Oasis Valley.

Bufo boreas nelsoni: Stejneger & Barbour, 1943: 40; Schmidt, 1953: 61; Cochran & Goin, 1970: 93; Behler & King, 1979: 389.

Bufo nelsoni: Savage, 1959: 252; Gorham, 1974: 82; Blair, 1972: 350; Stebbins, 1985: 70; 2003: 209; Elliott *et al.*, 2009: 176.

Bufo halophilus nelsoni: Frank & Ramus, 1995: 40.

Anaxyrus nelsoni: Frost *et al.*, 2006a: 363; Elliott *et al.*, 2009: 176; Dodd, 2013: 132.

Bufo (Anaxyrus) nelsoni: **new combination.**

Distribution: A restricted range in the Amargosa River valley, Nye County, Nevada. See map in Elliott *et al.* (2009) or Dodd (2013). **U.S.** NV.

Common name: Amargosa Toad.

Commentary: Formerly treated as a subspecies of *B. boreas,* but distinctive morphologically, in adults as well as larvae, and larval habitat (Altig *et al.*, 1998), and allopatric with all populations related to *B. boreas*. Goebel *et al.* (2009), in their molecular analysis, found this population to have relationships with *B. boreas halophilus*. Further study is needed.

Bufo (Anaxyrus) punctatus Baird & Girard, 1852, **new combination**

Bufo punctatus Baird & Girard, 1852d: 173.—**O**: **S** sd., Kellogg, 1932:62) USNM 2618 (3); verified by Cochran (1961a: 37).—**OT**: U.S., Texas, Val Verde County, Rio San Pedro (= Devil's River).

Bufo punctatus: Baird, 1859e: 25; Yarrow, 1875: 523; Cope, 1889b: 262; Stejneger & Barbour, 1917: 29; 1943: 43; Schmidt, 1953: 64; Cochran & Goin, 1970: 99; Gorham, 1974: 83; Conant, 1975: 314; Behler & King, 1979: 394; Frost, 1985: 57; Stebbins, 1985: 73; 2003: 214; Garrett & Barker, 1987: 10; Liner, 1994:

17; Conant & Collins, 1998: 524; Dixon, 2000: 72; Lemos-Espinal & Smith, 2007b: 41, 249; Elliott *et al.*, 2009: 160.

Anaxyrus punctatus: Frost *et al.*, 2006a: 363; Elliott *et al.*, 2009: 160; Dodd, 2013: 136.

Bufo (Anaxyrus) punctatus: **new combination**

Bufo beldingi Yarrow, 1883b: 441.—**S**: USNM 12660 (6); also USNM 1267 (4) were listed by Yarrow in the description, but not as "types" but he did elsewhere identify them as such (Yarrow, 1883a: 163).—**OT**: Mexico, Baja California del Sur, La Paz.—**SS**: Boulenger (1883a: 23, 1883b: 19).

Distribution: Western two-thirds of Texas and western Oklahoma, to southern California, north into southern parts of Kansas, Colorado, and Utah, and south into Mexico. See map in Korky (1999) or Dodd (2013). **U.S.**: AZ, CA, CO, KS, NM, NV, OK, TX, UT.

Common name: Red-spotted Toad.

Commentary: Korky (1999) provided a CAAR account.

Bufo (Anaxyrus) quercicus Holbrook, 1840, **new combination**

Bufo quercicus Holbrook, 1840: 109, pl. 22.—**O**: including specimen shown in pl. 22, not known to now exist.—**OT**: U.S., South Carolina, near Charleston, and North Carolina, Smithville. Restricted to Charleston, S.C. by Schmidt (1953: 65), but with no evidence **O** is from there the restriction is not valid.

Bufo quercicus: Cope, 1886b: 516; 1889a: 291; Stejneger & Barbour, 1917: 29; 1943: 43; Viosca, 1949: 10; Schmidt, 1953: 65; Cochran & Goin, 1970: 94; Gorham, 1974: 83; Conant, 1975: 312; Behler & King, 1979: 395; Frost, 1985: 58; Conant & Collins, 1998: 522; Elliott *et al.*, 2009: 142.

Anaxyrus quercicus: Frost *et al.*, 2006: 363; Elliott *et al.*, 2009: 142; Dodd, 2013: 144.

Bufo (Anaxyrus) quercicus: **new combination**

Chilophryne dialopha Cope, 1862b: 341.—**O**: ANSP, apparently lost, not listed by Malnate, 1971.—**OT**: "Sandwich Islands" (= Hawaii), error *fide* Cope (1889b: 291-292), but not corrected. Restricted invalidly to U.S., South Carolina, Charleston, by Schmidt, 1953: 65.—**SS**: Cope, 1886b: 516, 1889a: 291.

Incilius dialophus: Cope, 1863: 50.

Bufo dialophus: Boulenger, 1882a: 319.

Distribution: Coastal plain from southeastern Louisiana east to peninsular Florida, and northward to southeastern Virginia. See map in Ashton & Franz, 1979 or Dodd (2013). **U.S.**: AL, FL, GA, LA, MS, NC, SC, VA.

Common name: Oak Toad.

Commentary: Ashton & Franz (1979) provided a CAAR account.

Bufo (Anaxyrus) retiformis Sanders & Smith, 1951, new combination

Bufo debilis retiformis Sanders & Smith, 1951: 153.—**O**: **H** (od.), UIMNH 5847 (*ex* EHT-HMS 449).—**OT**: U.S., Arizona, Pima County, 14.4 miles south of Ajo.

Bufo debilis retiformis: Schmidt, 1953: 63.

Bufo retiformis: Bogert, 1962: 8; Cochran & Goin, 1970: 101; Gorham, 1974: 83; Behler & King, 1979: 396; Frost, 1985: 58; Stebbins, 1985: 76; 2003: 217; Liner, 1994: 17; Elliott *et al.*, 2009: 166.

Anaxyrus retiformis: Frost *et al.*, 2006a: 363; Elliott *et al.*, 2009: 166; Dodd, 2013: 149.

Bufo (Anaxyrus) retiformis: **new combination**

Distribution: South-central Arizona, into adjacent Mexico. See map in Hulse (1978) or Dodd (2013). **U.S.**: AZ.

Common name: Sonoran Green Toad.

Commentary: Hulse (1978) provided a CAAR account.

Bufo (Anaxyrus) speciosus Girard, 1856, new combination

Bufo speciosus Girard, 1856a: 86.—**O**: **S** (sd., Cochran, 1961a: 37) USNM 2608 (Rio Grande City), 2610 (Brownsville), 131559 (Nuevo León); Kellogg (1932: 46) cited 3 specimens but did not include 131559.—**OT**: U.S., Texas, Cameron County, Brownsville; U.S., Texas, Starr County, Rio Grande City (as Ringgold Barracks); and Mexico, Nuevo León, Pesquiería Grande.—**Comment:** The OT was originally stated as the valley of the Río Bravo (= Rio Grande), and common in Nuevo León. Restricted to U.S, Texas, Cameron County, Brownsville (Smith & Taylor 1950a: 361); accepting the restriction would mean the one specimen of the S that is from that locality should be considered the L. Thus, we hereby designate USNM 2610 as lectotype/lectophorant of *B. speciosus* Girard, 1856.

Bufo speciosus: Baird, 1859e: 26; Garman, 1888b: 136; Cochran & Goin, 1970: 93; Gorham, 1974: 84; Conant, 1975: 313; Behler & King, 1979: 396; Frost, 1985: 60; Stebbins, 1985: 74; 2003: 215; Garrett & Barker, 1987:11; Liner, 1994: 17; Conant & Collins, 1998: 523; Dixon, 2000: 72; Lemos-Espinal & Smith, 2007a: 42, 240; Elliott *et al.*, 2009: 152.

Bufo spectabilis: Peters, 1874: 747.—**IS**.

Bufo lentiginosus speciosus: Garman, 1884a: 43, by implication.

Bufo compactilis speciosus: Cope, 1889b: 273; Smith, 1947a: 8; Schmidt, 1953: 63.

Anaxyrus speciosus: Frost *et al.*, 2006a: 363; Elliott *et al.*, 2009: 152; Dodd, 2013: 152.

Bufo (Anaxyrus) speciosus: **new combination**

Misapplication of names:

Bufo compactilis: Strecker, 1915: 51; Stejneger & Barbour, 1917: 28; 1943: 41.

Distribution: Central and western Texas, southeastern New Mexico, and western Oklahoma, south into Mexico. See map in Elliott *et al.* (2009)or Dodd (2013). **U.S.**: NM, OK, TX.

Common name: Texas Toad.

Commentary: For many years, this species was reported in literature as *Bufo compactilis* in North America. All records for North America under that name should be corrected to *Bufo speciosus*.

Bufo (Anaxyrus) terrestris (Bonnaterre, 1789), new combination

?"*Rana musica*" Linnaeus, 1766: 354.—**O**: none indicated.—**OT**: Surinam. We tentatively accept this locality as correct, thus this nomen cannot apply to *B. terrestris* of eastern North America. But some authors of the time used this name to apply to the species we currently treat as *B. terrestris*.

Bufo musicus: Sonnini & Latreille, 1801: 127. The species described under this name as being frequently seen by Bosc in Carolina, seems to be *B. terrestris*. We assume it is actually a misapplication of the Linnean name. Schmidt (1953: 65) assumed this synonymy and invalidly restricted the OT to Charleston, SC.

Bufo lentiginosus var. *musicus*: Günther, 1859b: 63.

Rana terrestris Bonnaterre, 1789: 8.—**O**: specimen figured as *Rana terrestris* by Catesby, 1743: 69, pl. 69.—**OT**: U.S., Carolina. Invalidly restricted to Charleston, SC, by Schmidt (1953: 65).

Bufo terrestris: Brocchi, 1882: 77; Stejneger & Barbour, 1917: 29; 1943: 43; Schmidt, 1953: 65; Cochran & Goin, 1970: 95; Gorham, 1974: 85; Conant, 1975: 307; Behler & King, 1979: 397; Frost, 1985: 62; Conant & Collins, 1998: 515; Elliott *et al.*, 2009: 138.

Bufo terrestris terrestris: Viosca, 1949: 10.

Anaxyrus terrestris: Frost *et al.*, 2006a: 363; Elliott *et al.*, 2009: 138; Dodd, 2013: 155.

Bufo (Anaxyrus) terrestris: **new combination**

Bufo rufus Schneider, 1799: 230.—**O**: frogs described by Bartram, 1791: 279, by indication.—**OT**: U.S., "septentrionalis Americanae" = within the Carolinas, Georgia, and Florida.—**SS**: Wright & Wright, 1949: 199.

Rana lentiginosa Shaw, 1802a: 173.—**O**: includes specimen figured as "land frog" by Catesby, 1743: 69, pl. 69.—**OT**: U.S., Carolina and Virginia.—**OS**: by sharing **O**.

Bufo lentiginosus: Holbrook, 1836: 79, 1839a: 7; Günther, 1859a: 63; Cope, 1886b: 516.

Telmatobius lentiginosus: Le Conte, 1855b: 426.

Chilophryne lentiginosa: Cope, 1862b: 358.

Incilius lentiginosus: Cope, 1863: 50.

Bufo lentiginosus lentiginosus: Cope, 1889b: 289.

Bufo erythronotus Holbrook, 1938b: 99.—**O**: includes specimen of pl. 21; not known to now exist.—**OT**: U.S., South Carolina, vicinity of Charleston.—**SS**: Girard, 1856a: 86, but Le Conte, 1855b: 430, disagreed.—**Comment**: Considered **ND** by Boulenger (1882a: 281) and Nieden (1923: 146).

Bufo lentiginosus pachycephalus Cope, 1889b: 288.—**O**: **S** (sd.,Cochran, 1961a: 35) USNM 14681 (2).—**OT**: U.S., Florida, Alachua County, Micanopy.—**SS**: Cochran (1961a: 35).

Distribution: Southeastern Louisiana east on the coastal plain to peninsular Florida, and northward into southeastern Virginia. See map in Blem (1979), Elliott *et al.* (2009) or Dodd (2013). **U.S.**: AL, FL, GA, LA, MS, NC, SC, VA.

Common name: Southern Toad.

Commentary: Blem (1979) provided a CAAR account.

Bufo (Anaxyrus) woodhousii Girard, 1856, **new combination**

Synonymy: See synonymy in the two subspecies below.

Distribution: Western U.S., except the Pacific coastal and Great Basin areas, south into Mexico. See map in Elliott *et al.* (2009). **U.S.**: AR, AZ, CO, ID, LA, MT, ND, NM, OR, SD, TX, UT, WY.

Common name: Woodhouse's Toad.

Content: Two subspecies are recognized. In a phylogenetic analysis, Masta *et al.* (2002) found two distinctive clades largely concordant with *B. w. woodhousii* and *B. w. australis;* Sullivan *et al.* (1996b) cited evidence to reject *velatus* as a valid subspecies; we here treat that nomen as a synonym of *B. fowleri* (*viz.*). Also see the Commentary for *B. fowleri* regarding the study by Sanders (1986).

Bufo (Anaxyrus) woodhousii woodhousii Girard, 1856, **new combination**

"Bufo dorsalis" Hallowell, 1852a: 181.—**O**: **H** (sd., Stejneger, 1890a: 117) USNM 2531.—**OT**: U.S., New Mexico [Territory], Zuni, Great, and Little Colorado Rivers; given by Stejneger, 1890a: 116, as San Francisco Mountain, New Mexico Territory = Arizona.—**SS**: ; Baird, 1859c: 44, with *B. woodhousii;* Yarrow, 1875: 521, with *B. lentiginosus woodhousei*; Boulenger, 1882a, with *Bufo americanus*. Homonymy with *Bufo dorsalis* Spix, 1824 (now *Bufo (Rhinella) ornatus* Spix, 1824); thus, the nomen is not available for this species.

Bufo woodhousii Girard, 1856a: 86.—**O**: **H** (sd., Kellogg, 1932: 73, and Cochran, 1961a: 37) USNM 2531.—**OT**: U.S., Arizona, Coconino County, San Francisco Mountain *fide* Stejneger, 1890a: 116.

Bufo woodhousii: Baird, 1859b: 20; 1859c: 44; 1859e: 27; Coues, 1875: 629; Stejneger & Barbour, 1917: 29; Frost, 1985: 65; Stebbins, 2003: 211; Brennan & Holycross, 2006: 36 (in reference to *B. w. australis*); Elliott *et al.*, 2009: 134.

Incilius woodhousei: Cope, 1863: 50.—**UE**.

Bufo lentiginosus var. *woodhousei*: Yarrow, 1875: 521.—**UE.**

Bufo lentiginosus woodhousei: Cope, 1889b: 278.—**UE**.

Bufo lentiginosus woodhousii: Stejneger, 1890a: 116; Strecker, 1915: 52.

Bufo lentiginosus woodhousi: Nieden, 1923: 126.—**UE**.

Bufo woodhousii woodhousii: Smith, 1934a: 456; Stejneger & Barbour, 1943: 44; Conant & Collins, 1998: 518; Dixon, 2000: 74.

Bufo compactilis woodhousii: Linsdale, 1940: 206.

Bufo woodhousei woodhousei: Schmidt, 1953: 66; Cochran & Goin, 1970: 97;Gorham, 1974: 86; Conant, 1975: 309; Behler & King, 1979: 398; Cook, 1983: 1; Stebbins, 1985: 72.—**UE**.

Bufo woodhousei: Sanders, 1986: 7.

Bufo woodhouseii woodhouseii: Garrett & Barker, 1987: 15.—**UE.**

Anaxyrus woodhousii: Frost *et al.*, 2006a: 363; Elliott *et al.*, 2009: 134; Dodd, 2013: 166.

Anaxyrus woodhousii woodhousii: Frost *et al.*, 2008: 3.

Bufo (Anaxyrus) woodhousii woodhousii: **new combination**

Bufo frontosus Cope, 1867a: 301.—**O**: not stated, if deposited in ANSP or USNM, now lost (Shannon & Lowe, 1955: 188).—**OT**: U.S., Arizona, near latitude 35° along the Colorado valley from Fort Mojave to Fort Yuma. There have been several unjustified restrictions.—**SS:** Synonymy with *B. woodhousii* by Yarrow, 1875: 520, and others. Some authors have suggested the name is a synonym of *Bufo cognatus* Say, 1822 (*e.g.*, Ellis & Henderson, 1913; Krupa, 1990). We tentatively retain it as a synonym of *B. woodhousii.*

Bufo lentiginosus frontosus: Cope, 1875b: 29; Coues, 1875: 627.

Bufo lentiginosus var. *frontosus*: Yarrow, 1875: 520.

Bufo aduncus Cope, 1889b: 457.—**O**: **H** (od.), USNM 14100.—**OT**: probably U.S., Texas, Cook County, Gainsboro.—**SS**: Cochran, 1961a: 31.

Bufo antecessor Sanders, 1987: 62.—**O**: **H** (od.), USNM 25322.—**OT**: U.S., Idaho, Nez Perce County, 9 miles northwest of Louiston (= Lewiston).—**SS**: *Bufo woodhousii* by Collins, 1989: 19.

Bufo planiorum Sanders, 1987: 87.—**O**: **H** (od.), USNM 2535.—**OT**: U.S., Montana, Yellowstone River.—**SS**: *Bufo woodhousii* by Collins, 1989: 19.

Distribution: Central Texas and eastern New Mexico, north to southern Montana and North Dakota, and west of the Great Basin, from central Arizona north through most of Utah and western Colorado, with populations in southwestern Idaho and northeastern Washington. See map in Dodd

(2013). **U.S.**: AR, CA, CO, IA, ID, KS, LA, MO, MT, ND, NE, NV, OR, SD, TX, UT, WA, WY.

Common name: Rocky Mountain Toad (SECSN) or Woodhouse's Toad (C&T).

Bufo (Anaxyrus) woodhousii australis Shannon & Lowe, 1955, **new combination**

Bufo woodhousei australis Shannon & Lowe, 1955: 185.—**O**: **H** (od.), UIMNH 67056 (ex FAS 6817).—**OT**: U.S., Arizona, Maricopa County, Tempe, irrigation ditch.

Bufo woodhousei australis: Gorham, 1974: 86; Conant, 1975: 310; Stebbins, 1985: 72.

Bufo woodhouseii australis: Garrett & Barker, 1987: 14.

Bufo woodhousii australis: Conant & Collins, 1998: 519; Dixon, 2000: 74; Lemos-Espinal & Smith, 2007b: 43, 241.

Anaxyrus woodhousii australis: Frost, McDiarmid, & Mendelson, 2008: 3.

Bufo (Anaxyrus) woodhousii australis: **new combination.**

Distribution: Southeastern Arizona, southwestern New Mexico, Trans-Pecos Texas, south into north central Mexico. **U.S.**: AZ, NM, TX.

Common name: Southwestern Woodhouse's Toad.

Subgenus **Incilius** Cope, 1863

Incilius Cope, 1863: 50.—**NS**: *Bufo coniferus* Cope, 1862, sd. Kellogg, 1932: 29.

Incilius: Frost *et al.*, 2009b: 218.

"Cranopsis" Cope, 1875c: 96.—**NS**: *Cranopsis fastidiosus* Cope, 1875. Unavailable, a junior homonym of *Cranopsis* Adams, 1860 (Mollusca); designation of *Bufo cognatus* as NS by Frost *et al.* (2006a: 222), in error.

Cranopsis: Cope, 1878c: 96; Frost *et al.*, 2006a: 363.

Ollotis Cope, 1875c: 98.—**NS** (mo.): *Ollotis coerulescens* Cope, 1875 (= *Cranopsis fastidiosus*), subjective junior synonym of *Incilius.*

Olliotis: Cope, 1878c: 98; Frost *et al.*, 2008: 9.

Cranophryne Cope, 1889b: 20.—**NN** for *Cranopsis*. (junior synonym of *Incilius*).

Distribution: Southern U.S., south to Ecuador.

Common name: Central American Toads (SECSN) or Middle American Toads (C&T).

Content: 33 species, of which two occur in North America.

Commentary: This provides a good example of why use of subgenera can help keep names more stable. Species in this taxon had been treated in the genus *Bufo* for well over 100 years. Frost *et al.* (2006a) showed this clade to be monophyletic and warranted being named as a distinctive taxon. They revived the name *Cranopsis,* proposing it as a genus separate from *Bufo.* As herpetologists began to try to become familiar with the new name, it was discovered that *Cranopsis* is a junior homonym and not available for an amphibian (Frost *et al.,* 2006b), and the synonym *Ollotis* was selected to replace *Cranopsis.* Then an older name *Incilius* was discovered as the proper senior synonym for this taxon (Frost *et al.,* 2009b), and herpetologists were again faced with learning to use the third new name for the taxon in three years. If the authors had elected to treat the clade as a subgenus of *Bufo,* the clade would still have been identifiable by the subgeneric nomen in parenthesis, but the genus would have retained the well-known name, and the name changes at the subgenus level would have had less effect on general stability. Thus we greatly favor utilization of subgenera, where appropriate, to avoid such problems. Note: Dodd (2013) uses *Ollotis* for the U.S. forms, suggesting that the nomen *Incilius* applies to a different generic lineage further south in Central America.

Bufo (Incilius) alvarius Girard, 1859, new combination

Bufo alvarius Girard, 1859: 26.—**S**: USNM 2571 (2, lost) and 2572 *fide* Kellogg, 1932:38.—**L**: USNM 2572 (Cope, 1889b: 267), by implication, reference to the specimen as "holotype;" sd. Fouquette (1968a: 71). See discussion in that work for details.—**OT**: valley of the Gila and Colorado Rivers. There were several corrections and restrictions, finalized by Fouquette (1968a: 70-72) as U.S, California, Imperial County, old Fort Yuma (on the north bank of the Colorado River, opposite its junction with the Gila River).

Bufo alvarius: Baird, 1859e: 26; Coues, 1875: 628; Cope, 1889b: 265; Stejneger & Barbour, 1917: 26; 1943: 39; Schmidt, 1953: 61; Cochran & Goin, 1970: 90; Gorham, 1974: 76; Behler & King, 1979: 386; Frost, 1985: 34; Stebbins, 1985: 69; 2003: 207; Liner, 1994:

16; Brennan & Holycross, 2006: 36; Elliott *et al.* 2009: 168.

Phrynoidis alvarius: Cope, 1862c: 358.

Cranopsis alvaria: Frost *et al.*, 2006a: 364.

Ollotis alvaria: Frost *et al.*, 2006b: 558, by implication; Elliott *et al.*, 2009: 168; Dodd, 2013: 177.

Incilius alvarius: Frost, 2009b: 418, senior synonym.

Bufo (Incilius) alvarius: **new combination.**

Distribution: Southern Arizona and adjacent corners of southwestern New Mexico and southeastern California, south into northwestern Mexico. See map in Fouquette (1970) and Fouquette *et al.* (2005), or Dodd (2013). **U.S.**: AZ, CA, NM.

Common name: Sonoran Desert Toad (earlier called Colorado River Toad).

Commentary: Fouquette (1970) provided a CAAR account.

Bufo (Incilius) nebulifer Girard, 1856, **new combination**

Bufo valliceps Wiegmann, 1833. Now a misapplication of the name.

Bufo valliceps: Peters, 1863a: 81; Strecker, 1915: 52; Stejneger & Barbour, 1917: 29; 1943: 43; Viosca, 1949: 10; Schmidt, 1953: 66; Cochran & Goin, 1970: 99; Conant, 1975: 311; Behler & King, 1979: 397; Sanders, 1986: 7; Dixon, 2000: 73; Elliott *et al.* 2009: 144.

Bufo valliceps valliceps: Gorham, 1974: 85; Garrett & Barker, 1987: 12.

"Bufo granulosus" Baird & Girard, 1852d: 173 (*nec* Spix, 1824).—**O**: **H** (od.), USNM; USNM 2595 *fide* Kellogg (1932: 160) and Cochran (1961a: 33).—**OT**: U.S., Texas, between Indianola (Calhoun County) and San Antonio (Bexar County. Smith & Taylor (1950a: 360) restricted the OT to San Antonio, and Schmidt (1953: 66) restricted it to Indianola, but as neither presented evidence that the H came from there, neither is valid. The nomen is a junior homonym, preoccupied by *Bufo granulosus* Spix, 1824, a species of Panama and South America, thus unavailable for this species.

Bufo nebulifer Girard, 1856a: 87.—**NN** for *B. granulous* Baird & Girard, 1852.—**Note:** This work is often cited as 1853 or 1854, but the cover page of the volume clearly indicates it was printed in 1856. Authorship is often cited as Baird & Girard, 1852, which is for the *B. granulosus* name; *nebulifer* is a substitute name by Girard 1856a.

Bufo nebulifera: Baird, 1859c: 44; 1859e: 25

Chilophryne nebulifera: Cope, 1862c: 358.

Incilius nebulifer: Cope, 1863: 50; Frost *et al.*, 2009b: 418.
Bufo nebulifer: Mulcahy & Mendelson, 2000: 173.—Removed from synonymy of *B. valliceps*. Lemos-Espinal & Smith, 2007a: 40, 239.
Cranopsis nebulifer: Frost *et al.*, 2006a: 364.
Ollotis nebulifer: Frost *et al.*, 2006b: 558; Elliott *et al.*, 2009: 144; Dodd, 2013: 180.
Bufo (Incilius) nebulifer: **new combination.**

Distribution: The Gulf coast, from Mississippi to Texas, inland in eastern and central Texas, south into Mexico, with disjunct populations in southern Arkansas and adjacent northeastern Louisiana. See map in Elliott *et al.* (2009) or Dodd (2013). **U.S.**: AR, LA, MS, TX.

Common name: Gulf Coast Toad (SECSN) or Coastal Plain Toad (C&T); the former has a long history of use and we feel it is preferable, as "Coastal Plain" is not specific as to which coast.

Commentary: Porter (1970) provided a CAAR account; however, it is treated as *B. valliceps,* and includes both that species and *B. nebulifer*. There is a large literature on this species under the name *Bufo valliceps,* particularly by W.F. Blair and his students (*e.g.,* Blair, 1972). Mulcahy & Mendelson (2000) found the more southerly Mexican population treated under that name to be a separate sister species of the more northern populations; the southern one retained the name *B. valliceps,* while the name *B. nebulifer* was recognized as the valid name for the North American population.

Subgenus **Rhinella** Fitzinger, 1826

"Oxyrhynchus" Spix, 1824: 49.—**NS**: not designated. Junior homonym of *Oxyrhynchus* Leach 1818 (Pisces).—**SS**: Cuvier, 1829: 112.

Rhinella Fitzinger, 1826: 39.—**NS** (mo.): *Bufo proboscideus* Spix, 1824,—**SS**: Cuvier, 1829: 112 (treated as a subgenus of *Bufo*); and Boulenger, 1882a: 281, as a synonym of *Bufo*.

Chaunus Wagler, 1828: 744.—**NS** (mo.): *Chaunus marmoratus* Wagler, 1828 (now *Bufo granulosus* Spix, 1824).—**SS**: Cuvier, 1829; Boulenger 1882a; with *Bufo*;. Chaparro *et al.*, 2007a: 211, with *Rhinella*.

Otilophis Cuvier, 1829: 112.—**NS**: *Rana margaritirera* Laurenti, 1768.—**SS**: Duméril & Bibron, 1841:662, with *Bufo*: Frost *et al.*, 2006a, with *Rhinella,* by implication.

Osilophus: Tschudi, 1838: 52.—**IS.**

Phryniscus Wiegmann, 1834b: 514.—**NS** (mo.): *Phryniscus nigricans* Wiegmann, 1834.—**SS:** Günther, 1859a: 43, with *Atelopus*; Boulenger, 1894b: 374, with *Bufo*; Frost *et al.,* 2006b with *Chaunus,* thus *Rhinella,* by implication.

Otolophus Fitzinger, 1843: 32.—**NS** (od.): *Bufo margaritifer* Daudin, 1802.—**SS**: Kellogg, 1932: 28 with *Bufo*; Frost *et al.,* 2006b, with *Rhinella,* by implication.

Trachycara Tschudi, 1845: 169.—**NS** (mo.): *Trachycara fusca* Tschudi, 1845.—**SS**: Peters, 1873c: 624, with *Bufo*; Frost *et al.,* 2006b, with *Rhinella,* by implication.

Stenodactylus Philippi, 1902: 40.—**NS** (mo.): *Stenodactylus ventralis* Philippi, 1902.—**ND,** Nieden, 1923: 146.—**SS**: Frost *et al.,* 2006b, with *Chaunus,* thus *Rhinella*, by implication.

"Aruncus" Philippi, 1902: 4.—**NS** (mo.): *Aruncus valdivianus* Philippi, 1902.—**ND,** Nieden, 1923: 146.—**SS**: Frost *et al.,* 2006b, with *Chaunus,* thus *Rhinella*, by implication.

Rhamphophryne Trueb, 1971: 6.—**NS** (od.): *Rhamphophryne acrolopha* Trueb, 1971.—**SS**: Chaparro *et al.,* 2007: 211, with *Rhinella.*

Atelophryniscus McCranie, Wilson & Williams, 1989: 2.—**NS** (od.): *Atelophryniscus chrysophorus* McCranie *et al.,* 1989.—**SS**: Pramuk and Lehr, 2005: 610 with *Bufo veraguensis*-group, thus with *Chaunus* or *Rhinella.*

Distribution: Southern Texas to southern South America, with one species widely introduced elsewhere. **U.S.:** TX.

Common name: South American Toads (SECSN) or Neotropical Toads (C&T).

Content: 77 species, mostly in South America, but one reaches North America.

Commentary: Pramuk *et al.* (2008) found this taxon to be the sister of *Bufo* subgenera *Anaxyrus* + *Incilius,* as treated here, and we treat *Rhinella* as another subgenus of *Bufo.* Again, after over 100 years of treatment under the genus *Bufo,* Frost *et al* (2006a) found this to be a monophyletic clade and elevated it to genus level, recognizing *Chaunus* as the appropriate name to revive for these species. But then, Chaparro *et al.* (2007) noted the older generic name *Rhinella* is a senior synonym so herpetologists had to learn a second new name for this taxon within two years. Again, this points out one of the advantages to stability of treating such clades as subgenera.

Bufo (Rhinella) marinus (Linnaeus, 1758), new combination

Rana marina Linnaeus, 1758: 211.—**O**: includes specimen illustrated by Seba, 1734: pl. 76, fig. 1, by indication.—**OT**: America (*sensu lato*); corrected to Surinam (Müller & Hellmich, 1936: 14).

Rana marina: Gmelin, 1789: 1049; Barnes, 1826: 271.

Bufo marinus: Schneider, 1799: 219; Barnes, 1826: 271; Gravenhorst, 1829: 54; Wagler, 1830: 207; Reinhardt & Lütken, 1863: 156; Jiménez de la Espada, 1875b: 195; Stejneger & Barbour, 1933: 29; 1943: 42; Schmidt, 1953: 64; Cochran & Goin, 1970: 98; Conant, 1975: 315; Behler & King, 1979: 393; Frost, 1985: 53; Garrett & Barker, 1987: 9; Liner, 1994: 17; Conant & Collins, 1998: 526; Dixon, 2000: 71; Lemos-Espinal & Smith, 2007a: 40, 238; Elliott *et al.*, 2009: 150.

Bufo marinis: Barbour & Noble, 1920: 425.—**UE**.

Bufo marinus marinus: Schmidt, 1932: 159; Gorham, 1974: 81.

Chaunus marinus: Frost *et al.*, 2006a: 364; Maneyro & Kwet, 2008: 99, by implication.

Rhinella marina: Chaparro *et al*, 2007: 211, by implication; Elliott *et al.* 2009: 150; Dodd, 2013: 186.

Rhinella marinus: Pramuk *et al.*, 2008: 76.—**UE.**

Bufo (Rhinella) marinus: **new combination**

Bufo brasiliensis Laurenti, 1768: 26.—**O**: including specimen illustrated by Seba, 1734: pl. 73, fig. 1, 2.—**OT**: Brazil.—**SS**: Nieden, 1923: 138.

Bufo brasiliensis: Shaw, 1802a: 160.

Rana gigas Walbaum, 1784: 239.—**O**: lost (formerly Endler Collection, *fide* Smith *et al.*, 1977: 423).—**OT**: Virginia, in error. Kellogg (1932: 54) considered this a **NN** for *Rana marina* Linnaeus.—**SS**: Smith *et al.*, 1977: 423.

"Rana humeris-armata" de la Cepède, 1788b: 458. Substitute name for *Rana marina* Linnaeus. Rejected as published in a nonbinomial work (ICZN, 2005: 55).

Rana humeris-armata: Bonnaterre, 1789: 6. Substitute name for *Rana marina* Linnaeus.

Bufo agua Sonnini & Latreille, 1801: 130.—**O**: MNHN, including specimen illustrated in Daudin, 1802b:100, pl. 37.—**OT**: Principally Brazil and Cuba.—**SS**: Daudin, 1802b: 99.

Bufo agua: Spix, 1824: 44 [Spix, 1824: 20, *fide* Bokermann (1966: 17)].

Bufo agua: Sonnini & Latreille, 1830a: 130.
Docidophryne agua: Fitzinger, 1843: 32.
Phrynoidis agua: Cope, 1862c: 358.
Bufo marinus agua: Gadow, 1901: 178.

Bufo horridus Daudin, 1802b: 97.—**O**: **H**: MNHN, specimen of pl. 36.—**OT**: not known.—**SS**: Günther, 1859a: 61, with *B. agua*; Boulenger, 1882a: 315.

Bombinator horridus: Merrem, 1820: 179.
Bufo humeralis Daudin, 1803: 205.—**O**: **S:** 2 in collection of madame Bonaparte, now lost.—**OT**: Guyane (= French Guiana), Cayenne.—**SS**: Raddi, 1823: 69; Gravenhorst, 1829: 54; Duméril & Bibron, 1841: 705; Boulenger, 1882a: 315.

Bombinator maculatus Merrem, 1820: 178.—**O**: based in part on specimens in Seba, 1734: pl 73, fig. 1,2.—**OS**: *Bufo brasiliensis* Laurenti, 1768, and *Bufo agua* Sonnini & Latreille, 1801.—**OT**: Brazil.

"Rana maxima" Merrem, 1820: 182.—**O**: not stated, nor known to exist.—**OT**: not stated nor known. Offered as a synonym of *Bufo horridus* Daudin, 1802. Preoccupied by *Rana maxima* Laurenti, 1768.

Bufo maculiventris Spix, 1824: 43.—**O**: **S**, 4 in ZSM, presumed lost (Hoogmoed & Gruber, 1983: 371), confirmed by Glaw & Franzen, 2006: 161. One **S** is specimen figured in pl. 14, fig. 1.—**OT**: Brazil, Solimoens; restricted to Brazil, Amazonas, between Tabatinga and the border with Peru and Colombia, along the Rio Negro (Bokermann, 1966: 21).—**SS**: Tschudi, 1838: 88; Schinz, 1833: 235 with *B. agua*. Duméril & Bibron, 1841: 704; Peters, 1872: 226; Boulenger, 1882a: 315.

Bufo lazarus Spix, 1824: 45.—**O**: **S** (Hoogmoed & Gruber, 1983: 371, sd.) 2, including specimen in pl. 17, fig 1; ZSM 2513/0 (2).—**OT**: Amazon River forests; referred by Bokermann (1966: 21) to be on the stretch of the Amazon between the mouth of the Rio Negro and the mouth of the Amazon at the Atlantic.—**SS**: Tschudi (1838: 88) with *Bufo agua*. Duméril & Bibron (1841: 704); Günther (1859a: 61); Boulenger (1882a: 315).

Docidophryne Lazarus: Fitzinger, 1861: 415.

Bufo albicans Spix, 1824: 47.—**O**: **S:** (Hoogmoed & Gruber, 1983: 371, sd.) ZSM 1140/0 and RMNH 2191 (2).—**L**: ZSM 1140/0 (Hoogmoed & Gruber, 1983: 371).—**OT**: Brazil, Amazonas, Rio Negro.—**SS**: Tschudi (1838: 88); Peters (1872: 226).

Bufo horribilis Wiegmann, 1833: 654.—**O**: **S**, ZNB 3479 (Misantla), 3480 (Veracruz), 3493 (Mexico) *fide* Peters (1863a: 81) and Kellogg (1932: 54).—**OT**: Mexico, Misantla; Mexico, Veracruz; and Mexico.

Restricted to Mexico, Veracruz, Veracruz (Smith & Taylor, 1950: 351).—**Comment:** We will make that action valid by here designating the lectotype/lectophorant of *Bufo horribilis* Wiegmann, 1833, to be ZNB 3480, the S from that locality.—**SS**: Nieden (1923: 138); Schmidt (1953: 64); Duellman (1961: 23).

Bufo marinus var. *horribilis*: Peters, 1873b: 618.

Bufo marinus horribilis: Shannon & Werler, 1955: 368; Lynch & Fugler, 1965: 8.

Bufo marinus var. *fluminensis* Jiménez de la Espada, 1875b: 199.—**O**: **S**: MNCN.—**OT**: Ecuador, la Cuenca del Guayas or Rio de Guayaquil; Brazil, Fazenda imperial de Santa Cruz, 14 leagues from the capital; Bahia; Rio de Janeiro and la Tijuca; Tabatinga, the Amazon and border with Peru; Ecuador, Chonana; and, Babahoyo. [as noted by Frost (2009-13), some of these localities are outside the known range of the species].

Bufo marinus var. *napensis* Jiménez de la Espada, 1875b: 201.—**O**: **S**: MNCN.—**OT**: Ecuador: Cotapino, near Santa Rosa de Napo; Archidona de Quíjos; and San Jose de Moti.

Bufo pithecodactylus Werner, 1899b: 481.—**O**: **H**, ZFMK 27999 (ex ZIUG), by implication (Böhme & Bischoff, 1984: 178).—**OT**: Colombia, La Union.—**SS**: Nieden, 1923: 138.

Bufo pythecodactylus: Rivero, 1961: 27.—**IS**.

Bufo angustipes Taylor & Smith, 1945: 553.—**O**: **H** (od.), USNM 116513.—**OT**: Mexico, Chiapas, La Esperanza.—**SS**: Duellman (1961: 23).

Distribution: Southern Texas southward to central South America; introduced in Florida and Hawaii (Lever, 2003). Also introduced widely through much of the world (Australia, Japan, New Guinea, Taiwan, Philippines, many Pacific islands), largely as a biological control of sugar cane beetles, often with adverse results. Deliberate introductions into Florida, for pest control, began before 1936, but did not result in colonization; but accidental introduction perhaps about 1954 did result in successful colonization, and with other introductions following, the species is now reproducing in many areas of the state, but particularly abundant in southernmost Florida and some of the keys (Meshaka *et al.*, 2004). Introduction into Hawaii is well documented. They were introduced in Oahu from Puerto Rico in March and April, 1932, and immediately became established. In 1933, the species was introduced to four other islands (Hawaii, Kauai, Maui, and Molokai) (Pemberton, 1933, 1934; Oliver & Shaw,

1953; and Lever, 2003). See maps in Easteal (1986) and Dodd (2013). **U.S.**: *FL, *HI, TX.

Common name: Cane Toad (also previously widely known as the Marine Toad or Giant Toad).

Content: Subspecies currently not usually recognized.

Commentary: Easteal (1986) provided a CAAR account. Some taxonomic accounts have referred other names to the synonymy of *B. marinus* (e.g., *paracnemis, platensis, schneideri,* from Uruguay and southern Brazil) but these are now considered separate species, or synonyms of such species (Maneyro & Kwet, 2008).

*Family **Dendrobatidae** Cope, 1865

Phyllobatae Fitzinger, 1843: 32.—**NG:** *Phyllobates* Duméril & Bibron, 1841.

Phyllobatidae: Parker, 1933: 12.

Phyllobatinae: Ardila-Robayo, 1979: 385.

Eubaphidae Bonaparte, 1850: [2].—**NG:** *Eubaphus* Bonaparte, 1831.

Hylaplesiina Günther, 1858a: 69.—**NG:** *Hylaplesia* Boie, 1827.

Hylaplesiida: Knauer, 1878: 112.

Dendrobatidae Cope, 1865a: 103.—**NG:** *Dendrobates* Wagler, 1830.—**Comment:** Placed on the official list of Family-group names in zoology, with precedence over Phylobatae when used as synonyms (ICZN 2009: 103).

Dendrobatidae: Cope, 1875b: 8.

Dendrobatinae: Gadow, 1901: 139.

Dendrobatoidae: Dubois, 1992: 309. As epifamily.

Colostethidae Cope, 1867b: 191.—**NG:** *Colostethus* Cope, 1866.

Calostethina: Mivart, 1869: 293.—**IS.** As subfamily.

Colostethidae: Cope, 1875b: 7.

This synonymy, as well as the generic and specific one, is based on Frost (2009-13), with little evaluation, because the taxon is non-native.

Distribution: Nicaragua to the Guianas, Amazonian Bolivia, and southeastern Brazil. *One species introduced in North America (Hawaii).

Common name: Poison Dart Frogs.

Content: Three subfamilies, containing 12 genera, with 178 species.

Commentary: In the past, relationships of this family have been suggested to be with ranoids or "leptodactyloids". The latter seems now to be favored. In the molecular analysis of Frost *et al.* (2006a: 58), dendrobatids may be interpreted as a sister taxon to centrolenids + bufonids + most "leptodactyloid" groups, although imbedded within other "leptodactyloids."

*Genus **Dendrobates** Wagler, 1830

"Hysaplesia" Boie *in* Schlegel, 1826: 239.—**NS:** *Hysaplesia borbonica* Tschudi, 1838. **Comment:** Frost (2009-13) lists *Hyla punctata* as NS; Boie lists several species, including *punctata*; *H. borbonica* is first listed, now in Leptophryne (Bufonidae), so that would seem the logical choice. But if Frost was the first reviser, then his choice stands. Suppressed for priority (ICZN 2009: 103).

"Hylaplesia": Boie *in* Schlegel, 1827: 294.—**IS**, thus also unavailable.

Dendrobates Wagler, 1830: 202.—**NS:** *Hyla tinctoria* Daudin (=*Rana tinctoria* Cuvier, 1797; sd. Duméril & Bibron, 1841: 651). Placed on the List of Generic Names in Zoology (ISZN 2009: 103).

Eubaphus Bonaparte, 1831b: 75.—**NS:** *Rana tinctoria* Shaw, mo.—**SS:** Dubois, 1982: 272.

Dendromedusa Gistel, 1848: xi.—**NN** for *Hysaplesia* Boie, 1826.

Distribution: Nicaragua to Brazil. *Introduced in the U.S. (Hawaii).

Common name: Poison Dart Frogs (SECSN) or Poison Frogs (C&T).

Content: Five species, of which one has been introduced to the U.S. in Hawaii.

Commentary: Frost (2009-13) lists this genus and six others (with 59 species) in a subfamily Dendrobatinae.

Dendrobates auratus (Girard, 1856)

Phyllobates auratus Girard, 1856b: 226.—**O: H:** not stated, but USNM 10307 *fide* Dunn, 1941: 88, and Cochran, 1961a: 69.—**OT:** Panama, Taboga Island.

Hylaplesia aurata: Cope, 1863: 49.

Dendrobates tinctorius var. *auratus*: Steindachner, 1864: 261 (implied).

Dendrobates trivittatus var. *aurata*: Peters, 1873b: 618.

Dendrobates auratus: Dunn, 1931: 393; Cochran, 1961b: 107; Gorham, 1974: 113; Frost, 1985: 97; Dodd, 2013: 809.

Dendrobates tinctorius auratus: Laurent, 1942: 12.

Dendrobates latimaculatus Günther, 1859a: 125.—**O: H:** BM 52.12.11.8 (Silverstone, 1975: 40).—**OT:** Panama, Darien.—**SS:** Taylor, 1952b: 635; Silverstone, 1975: 40.

Hylaplesia tinctoria latimaculata: Dunn, 1941: 88.

Distribution: Nicaragua to Colombia. *Introduced and established in Hawaii (Oahu) (McKeown, 1996) See map in Dodd (2013). **U.S.:** *HI.

Common name: Green-and-black Poison Dart Frog (SECSN) or Green and Black Poison Frog (C&T).

Commentary: The "black" coloration in this species is often a bronze to brownish, hence the common name may be misleading; it has sometimes been called the Golden Poison Dart Frog (*e.g.,* Cochran, 1961b), and in fact, the specific nomen means "golden."

Family **Hylidae** Rafinesque, 1815

Synonymy: see below under Hylini.

Distribution: North America, Middle America, South Ame-rica, the West Indies, Eurasia, northern Africa, Australia.

Common name: Treefrogs.

Content: Three subfamilies, containing 901 species in 46 genera. One subfamily has five genera that are found in North America. Another subfamily has one genus introduced in North America.

Commentary: North America contains only 3.8% of the species diversity of this family.

Subfamily **Hylinae** Rafinesque, 1815

Synonymy: see below under Hylini.

Distribution: As for the family.

Common name: Treefrogs.

Content: 646 species, in 40 genera, of which 33 species in five genera are found in North America. Four tribes are recognized, of which two have members in North America. North America contains only 5.1% of the species diversity of this subfamily.

Tribe **Hylini** Rafinesque, 1815

Hylarinia Rafinesque, 1815: 78.—**NG**: *Hylaria* Rafinesque, 1814.

Hylina: Gray, 1825: 213.—**NG**: *Hyla* Laurenti, 1768: 32.—**OS**: Dubois, 1984b: 36.

Hyladae: Boie, 1828: 363.

Hylenae: Gray, 1829: 203.

Hyladina: Bonaparte, 1838a: 195.

Hylae: Tschudi, 1838: 25.

Hyladini: Bonaparte, 1839a: 225.

Hylaina: Bonaparte, 1845a: 378.

Hylidae: Bonaparte, 1850: [2]; Boulenger, 1882a: 330; Bolkay, 1919: 349, 356; Blackburn & Wake, 2011: 42.

Hylina: Bonaparte, 1850: [2]; Knauer, 1878: 109.

Hyloidea: Stannius, 1856: 5; Dubois, 1983: 272.

Hyloides: Bruch, 1862: 221.

Hylida: Knauer, 1878: 109; Bayer, 1885: 18.

Hylidi: Acloque, 1900: 489.

Hylinae: Gadow, 1901: 139; Faivovich *et al*., 2005: 1; Blackburn & Wake, 2011: 42.

Hylini: Faivovich *et al*., 2005: 3.

Dryophytae Fitzinger, 1843: 31.—**NG**: *Dryophytes* Fitzinger, 1843: 31.—**SS**: Dubois, 1984b: 36.

Calamitina Gravenhorst, 1843: 393.—**NG**: *Calamita* Schneider, 1799: 14.

Calamitae: Leunis, 1844: 145.—**NG**: *Calamites* Wagler, 1830: 200.

Acridina Mivart, 1869: 292.—**NG**: *Acris* Duméril & Bibron, 1841: 506.—**SS**: Dubois, 1984b: 36.

Acridinae: Kuhn, 1965: 96.

Triprioninae Miranda-Ribeiro, 1926: 64.—**NG**: *Triprion* Cope, 1866.—**SS**: Dubois, 1984b: 36.

Distribution: North and Middle America to Colombia, Europe and the Middle East, Asia.

Common name: Treefrogs.

Content: 170 species in 16 genera, of which 32 species in four genera are found in North America.

Commentary: The other genera of the tribe which are outside of North America are primarily Middle American. The only species found in the Old World are members of the genus *Hyla*. North America contains 18.8% of the species diversity of the tribe.

Genus **Acris** Duméril & Bibron, 1841

Acris Duméril & Bibron, 1841: 506.—**NS**: *Rana gryllus* Le Conte, 1825: 282 (sd. Fitzinger, 1843: 31).

Distribution: U.S. east of the Rocky Mountains, north into southern Ontario, south into northern Mexico. **Can:** ON. **U.S.:** AL, AR, CO, DE, GA, FL, IA, IL, IN, KS, KY, LA, MD, MI, MN, MO, MS, NC, NE, NJ, NM, NY, OH, OK, PA, SC, SD, TN, TX, VA, WI, WV.

Common name: Cricket Frogs.

Content: Three species, all occurring in North America.

Commentary: Faivovich *et al.* (2005) found this genus a sister taxon to *Pseudacris* (which is supported by the analysis of Pyron & Wiens, 2011), and that clade together to be the sister group of all other Hylini, thus occupying the ancestral position for the tribe.

Acris blanchardi Harper, 1947

Acris gryllus blanchardi Harper, 1947: 39.—**O**: **H**: CM 26607 (od.).—**OT**: U.S., Missouri, Christian County, Ozark, meadow near Smallen's Cave.

Hyla ocularis blanchardi: Mittleman, 1947b: 472.

Acris gryllus blanchardi: Schmidt, 1953: 68.

Acris crepitans blanchardi: Conant, 1958: 277; Cochran & Goin, 1970: 112; Gorham, 1974: 88; Garrett & Barker, 1987: 16; Liner, 1994: 16; Dixon, 2000: 65; Stebbins, 2003: 219.

Acris blanchardi: Gamble *et al.*, 2008: 121; Crother, 2012: 11; Dodd, 2013: 205.

Acris gryllus paludicola Burger, Smith, & Smith, 1949: 131.—**O**: **H**: UIMNH 872 (od.).—**OT**: U.S., Texas, Jefferson County, 5 miles west of Sabine Pass. Populations of *paludicola* representatives were not separable from other *blanchardi* studied by Gamble *et al.* (2008: 122), and they rejected the subspecies.

Acris gryllus paludicola: Schmidt, 1953: 68.

Acris crepitans paludicola: Duellman, 1977: 2; Garrett & Barker, 1987: 18; Conant & Collins, 1998: 530; Dixon, 2000: 65; Rose *et al.*, 2006: 430.

Distribution: Most of Texas, and part of eastern New Mexico, east to Louisiana and Arkansas, north to southeastern Wisconsin, Iowa, Missouri, southern Michigan, into most of Ohio, northern Kentucky, and western West Virginia. Formerly known from Pelee Island in Lake Erie (Ontario), apparently extinct there now. See map in Gamble *et al.* (2008) or Dodd (2013). **Can.**: ON?. **U.S.**: AR, CO, IA, IL, IN, KS, KY, LA, MI, MN, MO, MS, NE, NM, OH, OK, SD, TN, TX, WI, WV.

Common name: Blanchard's Cricket Frog. *French Canadian name:* Rainette grillon de Blanchard.

Commentary: In early literature, this species was commonly identified as *A. g. gryllus* (*e.g.,* Davis & Rice, 1883: 27). Rose *et al.* (2006) showed that the described subspecies *paludicola* is not related to *A. gryllus,* as originally described, but is largely indistinguishable from *A. blanchardi,* at that time considered a subspecies of *A. crepitans.* They recommended maintaining *paludicola* as a subspecies of *A. crepitans* pending further study. The molecular analysis by Gamble *et al.* (2008) found *blanchardi* to be a species distinct from *A. crepitans* or *A. gryllus,* but a sister species of the former (supported by the analysis of Pyron & Wiens, 2011). They also found the frogs designated as *paludicola* were not distinctive enough from *blanchardi* to justify recognition as a subspecies. Listed in Ontario as Extirpated, and not known there since 1987 (Fisher et al., 2007).

Acris crepitans Baird, 1856

?"Rana pumila" Le Conte, 1825: 282.—**O**: not stated, fate unknown, probably lost.—**OT**: North America.—**SS**: Frost, 2009-13 (with uncertainty) regarded this as a **ND**.—**Comment**: Designation of a N would allow settling the status of this nomen and to restrict its OT. It could possibly be used if a new subspecies needed to be recognized and if no other nomen were available for it.

Acris crepitans Baird, 1856: 59.—**O**: **S,** not stated, probably including a specimen in USNM or ANSP; current fate unknown, probably lost.—**OT**: U.S., "Northern States generally".—**Comment**: Several restrictions of the OT were made, some unreasonable, or with no evidence provided that the O actually came from there: Smith & Taylor (1950a: 359) to New York, Albany County, Albany, disputed by Duellman (1970: 647) as outside the known range; Schmidt (1953: 68) to West Virginia, Potomac River at Harper's Ferry. However, the restriction by Smith *et al.* (1995: 14) to U.S., New York, Nassau County, Long Island, the Locusts near Oyster Bay, carries an explanation of why that is probably where the S were collected, and is here accepted as a valid restriction of OT.

Acris crepitans: Baird, 1859c: 44, 1859e: 28, but includes *A. gryllus* as a synonym; Stejneger & Barbour, 1943: 44; Viosca, 1949: 10; Behler & King, 1979: 400; Frost, 1985: 120; Stebbins, 1985: 77; 2003: 218; Lemos-Espinal & Smith, 2007a: 45, 242; Gamble *et al.*, 2008: 122; Elliott *et al.*, 2009: 114; Crother, 2012: 11; Dodd, 2013: 219.

Acris gryllus crepitans: Cope, 1875b: 30; 1889b: 326; Davis & Rice, 1883: 27; Strecker, 1915: 49; Schmidt, 1953: 68.

Acris crepitans crepitans: Conant, 1958: 276; Cochran & Goin, 1970: 112; Gorham, 1974: 88; Garrett & Barker, 1987: 17; Conant & Collins, 1998: 528; Dixon, 2000: 65; Stebbins, 2003: 218-219 (by implication).

Distribution: Southeastern New York southward through eastern Virginia, the central Carolinas, northern Georgia, southern Tennessee and Arkansas, southward to the Gulf coast, into eastern Texas. See map in Gamble *et al.* (2008), Elliott *et al.* (2009) or Dodd (2013). **U.S.**: AL, AR, CO, DE, GA, FL, KY, LA, MD, MS, NC, NJ, NY, OH, PA, SC, TN, TX, VA, WV.

Common name: Eastern Cricket Frog (SECSN) or Northern Cricket Frog (C&T).

Content: Formerly, *A. blanchardi* was treated as a subspecies of this form, but Gamble *et al.* (2008) found it to be a distinct species. The only other population treated as a subspecies was *A. c. paludicola,* but that population is now a part of *A. blanchardi*. So currently there are no subspecies recognized.

Commentary: Before the 1940's, this form was commonly treated as a synonym of *A. gryllus* (*e.g.,* Stejneger & Barbour, 1917-1933). Gamble *et al.* (2008) found this to be the sister species of *A. blanchardi,* and also found a molecularly distinctive eastern and western clade within the species, and recommended further molecular and call studies to determine if they are distinctive enough to be named.

Acris gryllus (Le Conte, 1825)

Synonymy: See the subspecies *Acris gryllus gryllus,* below.

Distribution: Southeastern quarter of the U.S., east of the Mississippi River. See map in Gamble *et al.* (2008), Elliott *et al.* (2009) or Dodd (2013). **U.S.:** AL, FL, GA, LA, MS, NC, SC, TN, VA.

Common name: Southern Cricket Frog.

Content: Two subspecies are commonly recognized. We tentatively continue to recognize these, but see the commentary following.

Commentary: Gamble *et al.* (2008) found this species to be distinct from *A. crepitans* and *A. blanchardi,* and to be the sister taxon of that pair. They also found differences in eastern and western clades of *A. gryllus,* but these do not match the defined subspecies. They recommend further molecular and call studies to better establish the taxonomy of these frogs.

Acris gryllus gryllus (Le Conte, 1825)

?"*Hyla ocularis*" Latreille *in* Sonnini & Latreille, 1801: 187.—**O**: MNHN *fide* Daudin, 1802a: 33, probably lost; presumed to include specimen figured as *Hyla oculata,* pl. 4, fig.2 (Daudin, 1802a).—**OT**: U.S., Carolina.—**Comments**: 1) Authorship often cited as Bosc, or Bosc & Daudin. Harper (1940) provided an interesting discussion of the history of this nomen, but that discussion has relevance only in the domain of taxonomic history, not of zoological nomenclature. In the latter domain, the author of a nomen is not the person who coined it, even if written in an unpublished manuscript, and even if that person later claimed for authorship of that nomen, but the person who made it available through publication and respecting the rules of nomen availability according to the *Code*. Furthermore, a nomenclatural author is not a person, but a signature (for more details, see Dubois, 2008a). The nomen *Hyla ocularis*, although coined in a manuscript by Bosc, was made nomenclaturally available through its publication by Latreille (*in* Sonnini & Latreille, 1801: 187), who is therefore its nomenclatural author. 2) Schmidt (1953: 71) published a restricted **OT** for this nominotypical subspecies, but as this restriction is not accompanied by evidence that the **O** actually came from there, it is invalid. 3) Mittleman (1946) argued that the description applies to the species currently recognized as *Acris gryllus,* rather than to that currently treated as *Pseudacris ocularis*. Neill (1950b: 156) disputed that view, and considered the name to represent an unrecognizable composite. We agree with Neill and consider the nomen *Hyla ocularis* Latreille, 1801, to be a ND, and as such not applicable to either this species nor the *Pseudacris* species.

Hyla oculata: Daudin, 1802a: 20, pl. 4, fig. 2.—**IS.**

Calamita ocularis: Merrem, 1820: 172.

Auletris ocularis: Wagler, 1830: 201.

Rana gryllus Le Conte, 1825: 282.—**O**: **S**, status controversial (see *Comments*).—**OT**: USA, Georgia.—**Comments**: [1] Dunn (1938: 153) listed 13 specimens: USNM 3564 (5), USNM 5909 (7), ANSP 1989.90, from Le Conte's collection and from Georgia as "the nearest approach to types that we are likely to get"! Cochran (1961a: 74) gave a slightly different list of 14 specimens (USNM 3564 (7), USNM 5909 (7)) as "probable cotypes". All North American authors have ignored the fact that Duméril & Bibron (1841: 509) also mentioned specimens in the Paris Museum from Georgia obtained from Le Conte, which are also entitled to be considered syntypes. Two of these are still present in that collection, under numbers MNHN 4564 and 1989.3521. [2] Dunn (1938: 153) proposed to restrict the **OT** to "Georgia, Liberty County, Riceboro", the location of a plantation of Le Conte, but as this restriction was not accompanied by evidence that any **S** actually came from there, it is invalid.

Hylodes gryllus: Holbrook, 1838b: 75.

Acris gryllus: Duméril & Bibron, 1841: 507; Knauer, 1878: 111; Stejneger & Barbour, 1917: 30; 1943: 45; Bishop, 1928a: 167; Viosca, 1949: 10; Frost, 1985: 120; Elliott *et al.*, 2009: 118; Dodd, 2013: 226.

Acris gryllus subspecies *gryllus*: Cope, 1875b: 30.

Acris gryllus gryllus: Cope, 1889b: 325, 329; Schmidt, 1953: 67; Conant, 1958: 276; Cochran & Goin, 1970: 111; Gorham, 1974: 89; Behler & King, 1979: 401; Conant & Collins, 1998: 528; Crother, 2012: 11.

Acris gryllus var. *bufonia* Boulenger, 1882a: 337.—**O**: **H** (mo.), BM 45.11.9.35 (Duellman, 1977: 2), adult ♀ (Boulenger, 1882a: 337).—**OT**: USA, Louisiana, New Orleans.—**SS**: Duellman (1977: 2).

Distribution: As for the species, but excluding most of peninsular Florida. **U.S.**: AL, FL, GA, KY, LA, MS, NC, SC, TN, VA.

Common name: Coastal Plain Cricket Frog.

Acris gryllus dorsalis (Harlan, 1827)

Rana dorsalis Harlan, 1827: 340.—**O**: **S** 7 specimens, mentioned as originally in ANSP, now lost (Duellman, 1977: 3).—**OT**: Florida.—**Comment**: Netting & Goin (1945: 305) restricted the OT to "Lower (northern) 100 miles of the Saint John's River", citing evidence that the S series was undoubtedly collected along that stretch of the river by one of two expeditions; we find this a valid restriction.

Acris gryllus dorsalis: Netting & Goin, 1945: 305; Schmidt, 1953: 68; Conant, 1958: 276; Cochran & Goin, 1970: 111; Gorham, 1974: 89; Behler & King, 1979: 401; Conant & Collins, 1998: 528.

Acris acheta Baird, 1856: 59.—**O**: number unknown; fate unknown, probably lost.—**OT**: U.S., Florida, Monroe County, Key West.—**SS**: Boulenger, 1882a: 336.

Acris gryllus achetae: Garman, 1884a: 43.—**UE**.

Distribution: Most of peninsular Florida. **U.S.**: FL.

Common name: Florida Cricket Frog.

Commentary: A poorly defined subspecies. See Commentary for the species.

Genus **Hyla** Laurenti, 1768

"Ranella" Garsault, 1764: 18.—**IS** (by first reviser, Dubois & Bour, 2010b: 28).

"Ranetta" Garsault, 1764: pl. 672.—**NS**: *Hyla arborea* (Dubois & Bour, 2010b: 28, sd.). Also selected as correct alternate spelling.—**OS:** (senior) of *Hyla* Laurenti, 1768, but invalid by Article 23.9.1 of the Code (see Dubois & Bour, 2010b:28).

Hyla Laurenti, 1768: 32.—**NS**: *Hyla viridis* Laurenti, 1768: 33 (Stejneger, 1907: 75, sd.).—**SS** of *Rana arborea* Linnaeus, 1758: 213; (first synonymy by de la Cepède, 1788b: 310-311).

Calamita Schneider, 1799: 151.—**NS**: *Rana arborea* Linnaeus, 1758: 213 (Stejneger, 1907: 75, sd.).—**SS**: Merrem, 1820: 169.

Hylaria Rafinesque, 1814b: 103.—**NN** for *Hyla* Laurenti, 1768: 32.—**OS**: Dubois, 1984b: 15.

Calamites Wagler, 1830: 200 [*nec* Guettard, 1770: 404].—**NN** for *Calamita* Schneider, 1799: 151.—**SS**: Dubois, 1984b: 14.—**Comment**: for the nomenclatural status of this nomen, see Dubois (1987a: 42-43).

Hyas Wagler, 1830: 201 [*nec* Leach, 1814: 431; *nec* Gloger, 1827: 279].—**NS** (mo.): *Rana arborea* Linnaeus, 1758: 213.—**SS**: Bonaparte, 1832: 25; Wiegmann, 1832: 201.

Dendrohyas Wagler, 1830: 342.—**NN** for *Hyas* Wagler, 1830: 201.—**SS**: Bonaparte, 1832: 25.

Discodactylus Wagler *in* Michahelles, 1833: 888.—**NN** for *Calamita* Schneider, 1799: 151.—**SS**: Dubois, 1987c: 42.—**Comment**: for the nomenclatural status of this nomen, see Dubois (1987c: 42).

Distribution: North and Middle America, Europe, the Middle East, and Asia.

Common name: Holarctic Treefrogs (SECSN) or Treefrogs (C&T).

Content: 35 species, of which there are ten (28.6%) in North America.

Commentary: Faivovich *et al.* (2005) found members of this genus to be the most derived in the tribe. The clades of *Hyla* form a sister taxon to the clade including *Smilisca*. These and more primitive genera form a clade that is a sister-group to the one containing *Acris* and *Pseudacris*. Faivovich *et al.* (2005) found the Eurasian clade, the *Hyla arborea*-group, is a sister (ancestral) to all North American *Hyla;* they also recognized several species groups for North American *Hyla,* which we recognize as subgenera. The nominotypical subgenus *Hyla* is appropriate for the extralimital *Hyla arborea* group as recognized by Faivovich et al. (2005) and supported by Pyron & Wiens (2011).

Subgenus ***Dryophytes*** Fitzinger, 1843

Dryophytes Fitzinger, 1843: 31.—**NS** (od.): *Hyla versicolor* Le Conte, 1825: 281.

Distribution: Much of the southern U.S., south to Guatemala, and in eastern Asia.

Common name: None in use.

Content: 14 species, in two species-groups. Seven occur in North America.

Commentary: Further study may clarify the relationships of two species and perhaps indicate the two contained species-groups be assigned separate subgeneric status. *Hyla femoralis* shares relationships with both species-groups and Faivovich *et al.* (2005) left it unassigned, whereas Hua *et al.* (2009) found it to be a sister to other members of the *versicolor*-group, supported by Pyron & Wiens (2011). *H. andersonii* shares morphological features with the *H. cinerea* group (subgenus *Epedaphus* as treated here), but genetically it was found to be closer to the *H. eximia* group (Faivovich *et al.*, 2005). But Hua *et al.* (2009) found *H. andersonii* to be a member of the *versicolor*-group, along with *H. femoralis,* also supported by Pyron & Wiens (2011). But the latter also found the *cinerea* group to form a sister clade to the one including the *versicolor* and *eximia* groups. Thus, we include the species of the *eximia* and *versicolor* groups recognized by Faivovich *et al.*, plus *H. femoralis* in this subgenus, which are the same species supported by the analysis of Hua *et al.* (2009). The following are treated as belonging to the *eximia* group: *H. arenicolor, H. wrightorum,* along with their Middle American relatives. We currently treat the following as members of the *versicolor* group: *H. andersonii, H. avivoca, H. chrysocelis, H. femoralis, H. versicolor.*

North America contains 50% of the species diversity of the subgenus.

Hyla (Dryophytes) andersonii Baird, 1856 **new combination**

Hyla andersonii Baird, 1856: 60.—**O**: **H:** USNM 3600 (mo.) (Cochran, 1961a: 51).—**OT**: U.S., South Carolina, Anderson County, Anderson.—**Comments**: 1) Schmidt (1953: 69) restricted the OT to Aiken County, SC, but this action is invalid, as no evidence was provided that the H actually came from there. 2) Some authors (*e.g.*, Neill, 1957b; Gosner & Black, 1967) have even disputed occurrence of the species in that part of South Carolina.

Hyla andersonii: Cope, 1889b: 365; Stejneger & Barbour, 1917: 31; 1943: 48; Conant & Collins, 1998: 531; Frost, 1985: 125; Faivovich *et al.*, 2005: 102; Elliott *et al.*, 2009: 52; Dodd: 2013: 235.

Hyla andersoni: Schmidt, 1953: 69; Conant, 1958: 280; Cochran & Goin, 1970: 105; Gorham, 1974: 94; Behler & King, 1979: 402.—**UE.**

Hyla (Dryophytes) andersonii: **new combination.**

Distribution: Disjunct areas in Atlantic Coastal Plain pine barrens, in New Jersey, North and South Carolina, and the Florida panhandle and adjacent Alabama, with questionable records in Pennsylvania and Georgia. See map in Gosner & Black (1967), and updated maps in Means (2005), Elliott *et al.* (2009) or Dodd (2013). **U.S.**: AL, FL, GA?, NC, NJ, PA?, SC.

Common name: Pine Barrens Treefrog.

Commentary: Gosner & Black (1967) provided a CAAR account, in which they indicated that the OT in South

Carolina must be an error. However, more recent records from that state make it likely the original data are correct. See discussion of distribution by Means (2005). Faivovich *et al.* (2005) showed this species in the ancestral position for the *eximia*-group, as a sister to all other members of the group. However, in contrast, Hua *et al.* (2009) found this species in the ancestral position for the *versicolor*-group, and included it as a member of that group. Further study is warranted.

Hyla (Dryophytes) arenicolor Cope, 1866 **new combination**

"Hyla affinis" Baird, 1856: 61 [*nec Hyla affinis* Spix, 1824: 33].—**O: L**: USNM 11410a (Gorman, 1960: 218, sd.).—**OT**: Mexico, Northern Sonora.—**Comment**: Smith & Taylor (1950a: 354) and Gorman (1960: 218) proposed two different **OT** restrictions for this species, but as these restrictions were not accompanied by evidence that the **O** actually came from there, neither is valid.

Hyla affinis: Baird, 1859e: 29.

"Hylarana fusca" Baird, 1859e: 35 [*nec Hyla fusca* Laurenti, 1768: 34].—**O: H**: specimen shown in fig. 10-13 of plate 37 in Baird (1859e, mo.); fate unknown, most probably lost.—**OT**: Presumably United States-Mexico boundary region.—**SS**: Frost & McDiarmid in Frost (2009-13).—**Comment**: Nomen tentatively placed in this synonymy on the basis of the illustration and the presumption that the specimen actually came from the United States-Mexico boundary region.

Hyla arenicolor Cope 1866a: 84.—**NN** for *Hyla affinis* Baird, 1856.

Hyla arenicolor: Yarrow, 1875: 524; Cope, 1889b: 369; Strecker, 1915: 50; Stejneger & Barbour, 1917: 32; 1943: 48; Schmidt, 1953: 69; Cochran & Goin, 1970: 104; Gorham, 1974: 94; Conant, 1975: 325; Behler & King, 1979: 402; Frost, 1985: 126; Stebbins, 1985: 79; 2003: 221; Garrett & Barker, 1987: 19; Liner, 1994: 22; Conant & Collins, 1998: 538; Dixon, 2000: 66; Brennan & Holycross, 2006: 40; Lemos-Espinal & Smith, 2007b: 45, 243; Dodd, 2013: 239.

Hyla (Dryophytes) arenicolor: **new combination.**

Hyla copii Boulenger, 1887b: 53.—**O: S:** BM 1947.2.23.26-27 (Condit, 1964: 89, sd.).—**OT**: USA, Texas, El Paso.—**SS**: Cope, 1888: 80.

Hyla coper: Cope, 1888: 80.—**UE**.

Hyliola digueti Mocquard, 1899: 154.—**O: S**: MNHN 1898.257-258 (Guibé, 1950: 19, od.).—**OT**: Mexico, Nayarit ("territoire de

Tepic").—**SS**: Kellogg (1932: 152).—**Comment**: Kellogg (1932: 158) stated that three additional specimens in the Paris Museum, MNHN 1901.343-345, were also **S**, but that is incorrect. The original description stated that only two specimens were available (Mocquard, 1899: 166). According to the Paris Museum handwritten catalogue, the three additional specimens, also collected by Léon Diguet, were sent to that Museum on 5 June 1901, i.e., long after publication of the description.

Distribution: Trans-Pecos Texas west to most of Arizona, south into Mexico, north to southern Colorado and Utah. Elliott *et al.* (2009) and Dodd (2013) have maps. **U.S.**: AZ, CO, NM, TX, UT.

Common name: Canyon Treefrog.

Commentary: Barber (1999) found evidence that populations in Arizona and New Mexico may include two more full species. One clade was found to be more closely related to *H. eximia* than to the other populations of *H. arenicolor,* but the locality for the *H. eximia* sample was not given; perhaps it was from an Arizona population of *H. wrightorum,* which have been identified by some authors as *H. eximia,* or perhaps from more southern populations of *H. eximia. Those data are critical to the analysis.*

Faivovich *et al.* (2005) showed this species to be in a terminal clade of the Hylini along with the Mexican species *H. eximia* and *H. euphorbiaceae* (so presumably *H. wrightorum* as well), along with the Asian species *H. japonica* and the Middle American *H. walkeri.* Bryson *et al.* (2010) identified a Mexican population that represents an undescribed cryptic species, apparently ancestral to the clade including *H. arenicolor* and its sister clade of *H. wrightorum* + *H. eximia.*

Hyla (Dryophytes) avivoca Viosca, 1928 **new comb.**

Hyla avivoca Viosca, 1928: 89.—**O**: **H** (od.), USNM 75017.—**OT**: U.S., Louisiana, St. Tammany Parish, outskirts of Mandeville.

Hyla avivoca: Stejneger & Barbour, 1933: 33; 1943: 48; Viosca, 1949: 10; Cochran & Goin, 1970: 106; Conant, 1975: 324; Behler & King, 1979: 403; Frost, 1985: 127; Conant & Collins, 1998: 538; Elliott *et al.,* 2009: 68; Dodd 2013: 245.

Hyla avivoca avivoca: P.W. Smith, 1953: 152; Conant, 1958: 282; Gorham, 1974: 95.

Hyla (Dryophytes) avivoca: **new combination.**

Hyla phaeocrypta ogechiensis Neill, 1948b: 175.—**O**: **H**: WTN 18007, probably now FSM.—**OT**: U.S., Georgia, Burke County, Ogeechee River at Midville.—**SS**: P.W. Smith, 1953: 152.

Hyla phaeocrypta ogechiensis: Schmidt, 1953: 72.

Hyla avivoca ogechiensis: P.W. Smith, 1953: 172; Conant, 1958: 282; Cochran & Goin, 1970: 106; Gorham, 1974: 95.

Misapplication of names:

Hyla phaeocrypta: Viosca, 1923: 96.

Hyla phaeocrypta phaeocrypta: Neill, 1948b: 175; Schmidt, 1953: 72.

Distribution: Mississippi River valley, southeastern Louisiana northward through all of Mississippi into southern Illinois and Indiana and western Tennessee and Kentucky, and eastward through eastern and southern Alabama, the Florida panhandle, central Georgia and southeastern South Carolina; scattered localities in central Louisiana, central Arkansas, and southeastern Oklahoma. See map in Redmer (2005), Elliott *et al.* (2009)or Dodd (2013). **U.S.**: AL, AR, FL, GA, IL, IN, KY, LA, MS, OK, SC, TN.

Common name: Bird-voiced Treefrog.

Content: Neill (1948b) described the two subspecies listed in the synonymy, as races of *H. phaeocrypta,* and these were assigned to *H. avivoca* by P.W. Smith (1953). Smith (1966b) continued to recognize these subspecies and his map showed the two to be allopatric. The differences seem slight and newer records show a more continuous distribution, and most workers (*e.g.,* CNAH) no longer recognize geographic races in this species; however they are listed by Frost *et al.* 2012).

Commentary: Smith (1966b) provided a CAAR account. Based on the commentary for the two related species above, we treat this as a sister-taxon to the *versicolor* + *chrysoscelis* pair. For a while this species was referred to by the nomen *phaeocrypta* (*e.g.,* Schmidt, 1953: 72), but P.W. Smith (1953) showed the nomen *Hyla phaeocrypta* to be a synonym of *H. versicolor*.

Hyla (Dryophytes) chrysoscelis Cope, 1880 **new combination**

Hyla femoralis chrysoscelis Cope, 1880: 29.—**O**: **H**, ANSP 13672 (Malnate, 1971: 123).—**N**: TNHC 37293 (Smith, Fitzgerald, & Guillette, 1992: 152).—**OT**: U.S., Texas, Bastrop County, 2 miles west of Colorado River on Highway 969.—**NP.**—**Comment**: Nomen placed on the *Official List of Specific Names in Zoology* (ICZN, 1993a)

Hyla versicolor chrysoscelis: Strecker, 1910: 117; 1915: 50; Stejneger & Barbour, 1917: 34; 1943: 51; Strecker & Williams, 1928: 11; Viosca, 1949: 10; Schmidt, 1953: 73; Conant, 1958: 282.

Hyla chrysoscelis: Johnson, 1966: 361; Cochran & Goin, 1970: 105; Gorham, 1974: 96; Conant, 1975: 323; Behler & King, 1979: 404; Frost, 1985: 131; Garrett & Barker, 1987: 20; Conant & Collins, 1998: 536; Dixon, 2000: 66; Elliott *et al.*, 2009: 66; Dodd, 2013: 250.

Hyla (Dryophytes) chrysoscelis: **new combination**

Hyla versicolor sandersi Smith & Brown, 1947: 48.—**O**: **H** (od.), USNM 123978.—**OT**: U.S., Texas, Atascosa County, 8 miles southwest of Somerset.—**SS**: Johnson, 1966: 361.

Hyla versicolor sandersi: Conant, 1958: 282.

Distribution: Much of eastern U.S., into southern Manitoba; difficult to assess because of confusion of identity with *H. versicolor*. See map in Cline (2005a; U.S. only), Holloway *et al.* (2006), Elliott *et al.* (2009) or Dodd (2013). **Can.**: MB. **U.S.**: AL, AR, DE, FL, GA, IA, IL, IN, KS, KY, LA, MD, MI, MN, MO, MS, NC, ND, NE, OH, OK, SC, SD, TN, TX, VA, WI, WV.

Common name: Cope's Gray Treefrog. *French Canadian name:* Rainette criarde.

Commentary: Fitzgerald *et al.* (1981) determined that the holotype/holophorant of this species is actually a specimen of the tetraploid *H. versicolor,* and petitioned the Commission to allow the diploid species to retain the name *chrysoscelis,* accompanied by the description of a neotype/neophoront by Smith *et al.* (1992), and that was so declared (ICZN, 1993a). This species was not sampled by Faivovich *et al.* (2005), but it is a diploid cryptic sister species of *H. versicolor*.

Hyla (Dryophytes) femoralis Daudin, 1800 **new combination**

Hyla femoralis Daudin, 1800: [81].—**O**: **S**, several specimens (SVL from 8 to 14 lines) presumably originally in MNHN, including the specimen shown in Daudin (1800: pl. [5] fig. 1; 1802b: 2: pl. 3 fig. 1; 1803: 8, pl. 93, fig. 1); fate unknown, probably lost.[**Note:** the pages and plates are not numbered. We count i = title page, to viii = last page before first species account; 1 = first page of first species account, 21 = last page of text; plates and the blank pages facing them are not counted in pagination. Plates are counted sequentially as they appear].—**OT**: Carolina.—**Comments**: [1] Stejneger & Barbour (1917: 33), followed by all subsequent editions of their checklist, credited this nomen to Latreille *in* Sonnini & Latreille (1801: 181), but, as established by Harper (1940), this nomen had previously been made available by Daudin (1800). [2] Despite other arguments of Harper (1940), authorship of the nomen must be credited to Daudin, not to Bosc, as there is no evidence in Daudin (1800-1803) and Sonnini & Latreille (1801) that the description of the species in Daudin (1800) was not written by Daudin himself after study of specimens rather than copied from the manuscript description of Bosc; quite to the contrary, the description of the species published later by Bosc (1803: 185) is much shorter and less detailed than that of Daudin (1800). [3] Schmidt (1953: 71) restricted the **OT** to "Charleston, South Carolina" but as that restriction is not accompanied by evidence that the S actually came from there, it is invalid.

Hyla femoralis: Sonnini & Latreille, 1801: 181; 1830a: 181; Cope, 1889b: 371; Stejneger & Barbour, 1917: 33; 1943: 50; Viosca, 1949: 10; Schmidt, 1953: 71: Conant, 1958: 280; Cochran & Goin, 1970: 105; Gorham, 1974: 98; Behler & King, 1979: 407; Frost, 1985: 134; Conant & Collins, 1998: 534; Elliott *et al.*, 2009: 60; Dodd, 2013: 274.

[*Calamita*] *femoralis*: Merrem, 1820: 171.

Auletris femoralis: Le Conte, 1855b: 428.

Hyla (Dryophytes) femoralis: **new combination**

Distribution: From southeastern Louisiana, mostly below the fall line, east along the Gulf coast to most of Florida and northward along the Atlantic coast to Virginia, possibly Maryland, with apparently disjunct populations in central Alabama and Mississippi. See map in Hoffman (1988), Elliott *et al.*(2009) or Dodd (2013). **U.S.**: AL, FL, GA, LA, MS, NC, SC, VA.

Common name: Pine Woods Treefrog.

Commentary: Faivovich *et al.* (2005) found this to share relationships with the *H. eximia* and *H. versicolor* groups, and left it unassigned. Hua *et al.* (2009) found this species the sister taxon of a clade including *H. versicolor, H. chrysocelis, and H. avivoca,* and included it in the *versicolor* species-group. Hoffman (1988) provided a CNAH account.

Hyla (Dryophytes) versicolor LeConte, 1825 new combination

"Hyla verrucosa" Daudin *in* Sonnini & Latreille, 1801: 186.—**O: H** MNHN ; fate unknown, probably lost.—**OT**: unknown.—**NO.—Comment**: Boulenger, 1882a, considered this a possible senior synonym of *H. versicolor*.

Calamita verrucosus: Merrem, 1820: 172.

Hyla versicolor Le Conte, 1825: 281.—**O**: not stated, presumed lost.—**N:** AMNH 84483 (Smith, Fitzgerald, & Guillette, 1992: 152).—**OT**: original, U.S, northern states; **OT** of N: U.S., New Jersey, Bergen County, Alpine.—**NP.—Comments**: 1) As mentioned by Duméril & Bibron (1841: 549), a specimen of this species from "New York" was donated by Le Conte to the Paris Museum, probably before May 1832, *i.e.*, shortly after the original description. It–is still present under the number MNHN 4805. It was most likely part of the original type-series, but this cannot be demonstrated. Nevertheless, as a specimen seen and given by the original author, and with a rather precise locality, it would have been a better choice for a N than a specimen collected later and elsewhere. 2) This nomen was placed on the *Official List of Specific Names in Zoology* in Opinion 1716 (ICZN, 1993a).

Hyla versicolor: Tschudi, 1838: 33; Cope, 1889b: 373; Werner, 1909: 51; Strecker, 1915: 49; Cochran & Goin, 1970: 105; Gorham, 1970: 6; 1974: 107; Conant, 1975: 323; Behler & King, 1979: 404; Frost, 1985: 156; Garrett & Barker, 1987: 23; Conant & Collins, 1998: 536; Dixon, 2000: 67; Elliott *et al.*, 2009: 62; Dodd, 2013: 294.

Dendrohyas versicolor: Tschudi, 1838: 33

Hyla versicolor versicolor: Stejneger & Barbour, 1917: 34; 1943: 51; Viosca, 1949: 10; Schmidt, 1953: 73; Conant, 1958: 282.

Hyla (Dryophytes) versicolor: **new combination**

Hyla richardii Baird, 1856: 60.—**O**: **H** (mo.): MCZ 2128 (Barbour & Loveridge, 1929: 280).—**OT**: U.S., Massachusetts, Essex County, Cambridge, Mount Auburn.—**SS**: Cope, 1889b: 373; Barbour & Loveridge, 1929: 280.

Hyla richardi: Cope, 1889b: 373.—**UE**.

Hyla versicolor phaeocrypta Cope, 1889b: 375.—**O**: **H:** USNM 12074.—**OT**: U.S., Illinois, Wabash County, Mount Carmel.—**SS**: Viosca, 1928: 89; P.W. Smith, 1953: 169.—**Comment**: Nomen considered a senior synonym of *Hyla avivoca* Viosca, 1928, by Mittleman (1945a: 31).

Hyla phaeocrypta: Mittleman, 1945a: 31; Schmidt, 1953: 71.

Hyla phaeocrypta phaeocrypta: Neill, 1948b: 175.

Distribution: Central and northeastern U.S. and adjacent Canada. This species is morphologically virtually identical to the partially sympatric *H. chrysoscelis,* so this distribution is subject to modification as populations become more certainly identified. See map and discussion in Holloway *et al.* (2006). Elliott *et al.* (2009) and Dodd (2013) have maps, and Cline (2005b) has a map of the U.S. distribution. **Can.**: MB, NB, ON, QC. **U.S.**: AL, AR, CT, DE, IA, IL, IN, KS, LA, MA, MD, ME, MI, MN, MO, ND, NH, NJ, NY, OH, OK, PA, RI, TN?, TX, VA, VT, WI, WV.

Common name: Gray Treefrog (SECSN) or Eastern Gray Treefrog (C&T). *French Canadian name:* Rainette versicolore.

Commentary: This is a tetraploid species, morphologically indistinguishable from the diploid *H. chrysoscelis,* but populations can be easily identified by the distinctly different male advertisement call or by chromosomal examination (see review by Holloway *et al.,* 2006). Faivovich *et al.* (2005) found this is a sister-species of *H. avivoca.* Holloway *et al.* (2006) showed there were multiple origins of the tetraploid population from three now-extinct species (not involving *H. chrysoscelis* or *H. avivoca*) and now all maintain gene flow but are reproductively isolated from their sympatric, closely-related congeners.

Hyla (Dryophytes) wrightorum Taylor, 1939 **new combination**

Hyla wrightorum Taylor, 1939c: 436.—**O: H** (od.) UMMZ 79141.—**OT**: U.S., Arizona, Apache County, 11 miles south of Springerville.

Hyla wrightorum: Stejneger & Barbour, 1943: 51; Sullivan, 1986: 378; Duellman, 2001b: 793, 980; Gergus *et al.*, 2004: 767; Brennan & Holycross, 2006: 40; Lemos-Espinal *et al.*, 2004: 6, 79; Elliott *et al.*, 2009: 70; Dodd, 2013: 309.

Hyla eximia wrightorum: Schmidt, 1953: 71: Liner, 1994: 23.

Hyla regilla wrightorum: Jameson *et al.*, 1966: 594; Cochran & Goin, 1970: 109; Gorham, 1974: 104.

Hyla (Dryophytes) wrightorum: **new combination**

Misapplied names:

Hyla eximia: Yarrow, 1875: 524; Stejneger & Barbour, 1917: 33; Duellman, 1970: 499, 2001a: 499; Stebbins, 1985: 81; 2003: 224.

Distribution: Disjunct: central Arizona to southwestern New Mexico, and southeastern Arizona south into Mexico; the single record in southeastern Arizona is believed to be an introduction. See maps in Elliott *et al.* (2009) and Dodd (2013). **U.S.**: AZ, NM.

Common name: Arizona Treefrog.

Commentary: Regarded as a close relative of the Mexican species, *H. eximia,* but not sampled by Faivovich *et al.* (2005); this would apparently be included among the most derived cluster of species in the tribe.

Subgenus **Epedaphus** Cope, 1885

Epedaphus Cope, 1885: 383.—**NS**: *Hyla gratiosa* LeConte, 1857: 146 (mo.).

Distribution: Southeastern U.S.

Common name: No name in use. Ubiquitous Southeastern Treefrogs seems appropriate.

Content: Contains three species that are prominent members of the anuran fauna of the southeastern U.S., where they are endemic.

Commentary: This subgenus corresponds to the *Hyla cinerea*-group of Faivovich *et al., (*2005), who found it to be a sister group to the subgenus *Dryophytes* as treated here.

Hyla (Epedaphus) cinerea (Schneider, 1799) **new combination**

Calamita cinereus Schneider, 1799: 174.—**O**: **S**, at least two specimens: 1) the specimen from "Carolina" used by Pennant (1792) for his description of the "cinereous frog", current status unknown, probably lost; 2) a specimen from "India orientalis" in Bloch's collection, possibly now in ZMB.—**OT**: 1) U.S., Carolina; 2) "India orientalis"—**Comment**: Schmidt (1953: 69) restricted the **OT** to "Charleston, South Carolina" but as this restriction was not accompanied by evidence that a S actually came from there, the action is invalid. The India locality must be an error, thus a less restrictive correction to Carolina is in order.

Hyla cinerea cinerea: Garman, 1890: 189; Strecker & Williams, 1928: 10; Stej-neger & Barbour, 1943: 49; Viosca, 1949: 10; Conant, 1958: 279.

Hyla cinerea: Strecker, 1915: 49; Stejneger & Barbour, 1917: 32; Schmidt, 1953: 69; Cochran & Goin, 1970: 102; Gorham, 1974: 96; Conant, 1975: 320; Behler & King, 1979: 405; Frost, 1985: 131; Garrett & Barker, 1987: 21; Conant & Collins, 1998: 532; Dixon, 2000: 66; Elliott *et al.*, 2009: 44; Dodd, 2013: 262.

Hyla (Epedaphus) cinerea: **new combination**

Calamita carolinensis Schneider, 1799: 174.—**O**: **H** (mo.), specimen used by Pennant (1792) in the description of his "cinereous frog", current status unknown, probably lost.—**OT**: Carolina.—**Comment**: Frost (2009-13) rightly remarked that this nomen had not been published by Pennant (1792), but it was erroneously credited to that author by Schneider (1799: 174) and many that followed. Schneider considered this nomen as an invalid senior synonym of his *Calamita cinereus*. This nomen was used as valid by Günther (1859a: 105), as *Hyla carolinensis*. As it was associated there to a description, it then became available, but it must be credited to Schneider (1799) by virtue of Art. 11.6.1.

Hyla carolinensis: Günther, 1859a: 105; Cope, 1889b: 366.

Hyla lateralis Daudin, 1800: [181], pl. [11, fig. 1].—**O**: **S**, several specimens, including one illustrated in pl. 11, and several mentioned by other authors under different nomina; current status unknown, probably lost.—**OT**: U.S., South Carolina, Charleston; and Virginia.—**SS**: Daudin, 1800: 18.—**Comment**: Schmidt (1953: 71) restricted the **OT** to "Charleston, South Carolina" but as this restriction is not accompanied by designation of a L or N from that locality, it is invalid.

Hyla lateralis: Sonnini & Latreille, 1801: 180; 1830b: 180.
Calamita lateralis: Merrem, 1820: 171.

Rana bilineata Shaw, 1802a: 136.—**O**: **H** (mo.), figure by Catesby (1743: pl. 71), as *Rana viridis, linea utrinque longitudinali flava*.—**OT**: Warm and temperate parts of North America.—**SS**: Merrem, 1820: 171.—**Comment**: Schmidt (1953: 70) restricted the **OT** to "Charleston, South Carolina" but as this restriction is not accompanied by evidence that the H actually came from there, it is invalid.

Hyla bilineata: Sonnini & Latreille, 1801: 179.

Hyla blochii Daudin, 1802: 43.—**O**: **S:** in Bloch Museum, Berlin, now ZMB (see *Comment*).—**OT**: India orientalis.—**SS**: Merrem, 1820: 171.—**Comment**: The S was initially used by Schneider (1799: 174) for his description of a specimen of *Calamita cinereus* from "India orientalis" (*not* his *Bufo cinereus* as stated by Frost, 2009-13).

Hyla semifasciata Hallowell, 1857f: 307.—**O**: **S**, ANSP 2024-2025 (Malnate, 1971: 352).—**OT**: U.S., Texas.—**Comment**: Schmidt (1953: 70) restricted the **OT** to "vicinity of Houston, Texas" but as this restriction is not accompanied by evidence that the S actually came from there, it is invalid.

Hyla semifasciata: Baird, 1859e: 28.
Hyla carolinensis semifasciata: Cope, 1875b: 31
Hyla cinerea semifasciata: Garman, 1890: 189

Hyla evittata Miller, 1901: 76.—**O**: **H** (od.) USNM 26291.—**OT**: U.S., Virginia, Four Mile Run, near Alexandria.—**SS**: Dunn, 1918a: 21.—**Comment:** Reed (1956: 328) notes that Alexandria and Four Mile Run are located outside of any county boundaries.

Hyla evittata: Stejneger & Barbour, 1917: 33.
Hyla cinerea evittata: Dunn, 1918a: 21; Stejneger & Barbour, 1923: 30; 1943: 49; Schmidt, 1953: 70; Conant, 1958: 279; Cochran & Goin, 1970: 103. Subspecies rejected by Reed (1956: 332).

Distribution: Delaware southward along Atlantic coast to Peninsular Florida, east along Gulf coast to eastern and southern Texas; northward to southeastern Missouri (introduced), southwestern Indiana, southern Illinois, western Kentucky, and western Tennessee. See map in Redmer & Brandon (2003), Elliott *et al.* (2009) or Dodd (2013). **U.S.**: AL, AR. DE, FL, GA, IL, KY, LA, MD, *MO, MS, NC, OK, SC, TN, TX, VA.

Common name: Green Treefrog.

Content: No subspecies are currently recognized.

Commentary: Redmer & Brandon (2003) provided a CAAR account. This is a sister species to *Hyla gratiosa* (Faivovich *et al.*, 2005).

Hyla (Epedaphus) gratiosa Le Conte, 1856 new combination

Hyla gratiosa Le Conte, 1856: 146.—**O**: **L:** ANSP 2089 (see *Comments*).—**OT**: USA, Georgia, "lower country of Georgia".—**Comments**: [1] Cochran's (1961a: 54) mention of six "cotypes" is no doubt in error, as Le Conte's (1856) original description mentioned only three specimens. Malnate (1971: 351) stated that the specimen ANSP 2089 (sex and size not provided) was the "holotype" of this species. Under Art. 74b of the 1985 edition of the *Code*, this statement amounted to a valid designation of lectotype "by inference of holotype". However, under Art. 74.6 of the 1999 edition, this is no longer true, as Le Conte (1856: 146) had clearly stated that his description was based on "three specimens". Under the *Code* now in force, the latter therefore remain collectively the onomatophore ("type"/ type series) of this nominotypical taxon. To ensure stability applies, we hereby designate the specimen mentioned by Malnate (1971), *i.e.*, ANSP 2089, as the lectotype (lectophoront) of this taxon. [2] Schmidt (1953: 71) restricted the **OT** to "Liberty County, Georgia" but as this restriction is not accompanied by evidence that the **S** actually came from there, it is invalid.

Epedaphus gratiosus: Cope, 1885: 383.

Hyla gratiosa: Cope, 1889b: 377; Stejneger & Barbour, 1917: 33; 1943: 50; Viosca, 1949: 10; Schmidt, 1953: 71; Conant, 1958: 283; Cochran & Goin, 1970: 103; Gorham, 1974: 99; Behler & King, 1979: 408; Frost, 1985: 136; Conant & Collins, 1998: 533; Elliott *et al.*, 2009: 48; Dodd, 2013: 280.

Hyla (Epedaphus) gratiosa: **new combination.**

Distribution: Southeastern U.S., north into the Delmarva peninsula and New Jersey. See map in Caldwell (1982) or Dodd (2013). **U.S.**: AL, DE, FL, GA, KY, LA, MD, MS, NC, NJ, SC, TN, VA.

Common name: Barking Treefrog.

Commentary: Caldwell (1982) provided a CAAR account. This is the sister species of *H. cinerea* (Faivovich *et al.*, 2005). Smith *et al.* (2005) found this species part of a clade with *H.*

versicolor and *H. avivoca,* but Hua *et al.* (2009) indicated that the sample of *H. gratiosa* used in that study was not that species, and their reanalysis agreed with the results of Faivovich *et al.* (2005).

Hyla (Epedaphus) squirella Daudin, 1800 **new combination**

Hyla squirella Daudin, 1800: [7], pl. [5, fig. 2].—**O**: **S:** includes the specimen shown in plate 5; fate unknown, probably lost.—**OT**: U.S., la Caroline.—**Comments**: 1) See *Comment 1,* above, under *H. femoralis*; this also applies to authorship of this species. 2) Harper (1940: 709) restricted the OT to Charleston, South Carolina, but it was not accompanied by evidence that the S actually came from there, hence the action is invalid.

Hyla squirella: Sonnini & Latreille, 1801: 181; 1830a: 181; Cope, 1889b: 363; Strecker, 1915: 49; Stejneger & Barbour, 1917: 34; 1943: 51; Viosca, 1949: 10; Schmidt, 1953: 72; Conant, 1958: 280; Cochran & Goin, 1970: 104; Gorham, 1974: 96; Behler & King, 1979: 409; Frost, 1985: 154; Conant & Collins, 1998: 535; Elliott *et al.*, 2009: 56; Dodd, 2013: 288.

Calamita squirella: Merrem, 1820: 171.

Hyla (Epedaphus) squirella: **new combination**

Hyla delitescens Le Conte, 1825: 281.—**O**: **S** (sd.), ANSP 1948-1958 (Malnate, 1971: 350).—**OT**: U.S., Georgia.—**SS**: Duellman, 1977: 103.

Hyla flavigula Glass, 1946: 101.—**O**: **H** (od.) TCWC 1192.—**OT**: U.S., Texas, Aransas County, Aransas National Wildlife Refuge.—**SS**: Neill, 1949a: 78.

Distribution: Gulf coastal area, Texas to Florida, then northward in Atlantic coastal area to southeastern Virginia. See map in Martof (1975d), Elliott *et al.* (2009) or Dodd (2013). **U.S.**: AL, FL, GA, LA, MS, NC, SC, TX, VA.

Common name: Squirrel Treefrog.

Commentary: Martof (1975d) provided a CAAR account. Faivovich *et al.* (2005) found this to be a sister taxon of the pair *H. cinerea* + *H. gratiosa,* and the reanalysis of Smith *et al.* (2005) by Hua *et al.* (2009) supports that relationship.

Genus **Pseudacris** Fitzinger, 1843

Synonymy: See below under subgenus *Pseudacris.*

Distribution: Western Canada and much of eastern Canada from Northwest Territory to much of Ontario and southern Quebec, and most of the US, south into Baja California.

Common name: Chorus Frogs. In the past they were also called Swamp Treefrogs.

Content: 18 species, all in North America.

Commentary: Although Cope recognized *Pseudacris* as a senior synonym of *Chorophilus* in 1889, he and others continued to use the Baird name until Stejneger & Barbour (1917) again pointed out the priority of the name *Pseudacris.*

Subgenus **Pseudacris** Fitzinger, 1843

Pseudacris Fitzinger, 1843: 31.—**NS** (od.): *Rana nigrita* Le Conte, 1825.

"Chorophilus" Baird, 1856: 60.—**NS** (od.): *Rana nigrita* Le Conte, 1825.—**OS**: Cope, 1889b: 331.

Helocaetes Baird, 1856: 60.—**NS**: (sd. Schmidt, 1953: 73) *Hyla triseriata* Wied-Neuwied, 1838.—**SS**: Cope, 1862a: 157.

Heloecetes: Baird, 1859e: 28.—**IS**.

Distribution: Most of the eastern two-thirds of the U.S., plus adjacent Canada.

Common name: Trilling Chorus Frogs (Moriarty & Cannatella, 2004).

Content: Nine species, including seven species of the *nigrita*-group, plus their sister clade composed of *P. brachyphona* + *P. brimleyi* (Moriarty & Cannatella, 2004; Moriarty Lemmon *et al.*, 2008). The **nigrita group** includes *P. clarki, P. feriarum, P. fouquettei, P. kalmi, P. maculata, P. nigrita, P. triseriata;* while the **brachyphona group** includes *P. brimleyi* and *P. brachyphona.*

Pseudacris (Pseudacris) brachyphona (Cope, 1889) **new combination**

C[horophilus] feriarum brachyphonus Cope, 1889b: 341.—**O**: **H** (mo.), fate unknown, probably lost.—**OT**: U.S., western Pennsylvania, Westmoreland or Armstrong County, near the Kiskiminitas River [separates the counties].

Pseudacris brachyphona: Walker, 1932: 379; Stejneger & Barbour, 1933: 31; 1943: 45; Schmidt, 1953: 74;

Conant, 1958: 291; Cochran & Goin, 1970: 113; Gorham, 1974: 110; Behler & King, 1979: 411; Frost, 1985: 171; Conant & Collins, 1998: 547; Elliott *et al.*, 2009: 102; Dodd, 2013: 313.

Hyla (Pseudacris) brachyphona: Dubois, 1984a: 85.

Pseudacris (Pseudacris) brachyphona: **new combination.**

Distribution: Northern Alabama, northern Georgia, east-central Tennessee, eastern Kentucky, western Virginia, most of West Virginia, eastern Ohio, and southwestern Pennsylvania, plus apparently disjunct populations in central Kentucky. See map in Hoffman (1980), Elliott *et al.* (2009) or Dodd (2013). **U.S.:** AL, GA, KY, MS, OH, PA, TN, VA, WV.

Common name: Mountain Chorus Frog.

Commentary: Hoffman (1980) provided a CAAR account. Moriarty & Cannatella (2004) showed that this is the sister species to *P. brimleyi,* and the two together form a sister clade to the *P. nigrita*-group. The analysis of Pyron & Wiens (2011) indicates the *brimleyi+brachyphona* pair as a sister to all other species of this subgenus.

Pseudacris (Pseudacris) brimleyi Brandt & Walker, 1933 **new combination**

Pseudacris brimleyi Brandt & Walker, 1933: 2.—**O**: **H** (od.), UMMZ 74361,—**OT**: U.S., North Carolina, Beaufort County, near Washington.

Pseudacris brimleyi: Stejneger & Barbour, 1943: 45; Schmidt, 1953: 74; Conant, 1958: 291; Cochran & Goin, 1970: 113; Gorham, 1974: 110; Behler & King, 1979: 412; Frost, 1985: 171; Conant & Collins, 1998: 546; Elliott *et al.*, 2009: 100; Dodd, 2013: 319.

Hyla (Pseudacris) brimleyi: Dubois, 1984a: 85.

Pseudacris (Pseudacris) brimleyi: **new combination.**

Distribution: The Atlantic coast from eastern Georgia, through the Carolinas, into Virginia. Hoffman (1983) has a good map, as does Dodd (2013). **U.S.**: GA, NC, SC, VA.

Common name: Brimley's Chorus Frog.

Commentary: Hoffman (1983) provided a CAAR account. This is a sister species of *P. brachyphona* (Moriarty & Cannatella, 2004).

Pseudacris (Pseudacris) clarkii (Baird, 1856) **new combination**

Helocaetes clarkii Baird, 1856: 60.—**O**: **S**, original number(s) unknown, but USNM 3313, 3315, 3317 reported by Duellman (1977: 108); Cochran (1961a: 50) reported that USNM 3313 is the only specimen still existing; sd., as co-type.—**OT**: U.S., Texas, Galveston and Indianola;—**Comment:** Schmidt (1953: 74) restricted the OT to Texas, Galveston County, Galveston; although the only remaining S is from Galveston (the others from Indianola and between Indianola and San Antonio), this does not automatically make it the L. To stabilize the nomenclatural status of this nomen, we hereby designate USNM 3313 as the lectotype (lectophorant), and the OT becomes its locality of collection: U.S., Texas, Galveston County, Galveston.

Heloecetes clarkii: Baird, 1859e: 28.

Chorophilus triseriatus clarkii: Cope, 1875b: 30.

Pseudacris triseriata clarkii: Burt, 1932: 80.

Pseudacris nigrita clarkii: Stejneger & Barbour, 1933: 31; 1943: 46; Burt, 1936a: 774.

Pseudacris clarki: Smith, 1934a: 462; Schmidt, 1953: 74; Conant, 1958: 290; 1975: 331; Gorham, 1974: 110; Dixon, 2000: 68.—**UE**

Hyla (Pseudacris) clarkii: Dubois, 1984a: 85.

Pseudacris clarkii: Frost, 1985: 171; Garrett & Barker, 1987: 24; Liner, 1994: 25; Conant & Collins, 1998: 546; Dodd, 2013: 328.

Pseudacris (Pseudacris) clarkii: **new combination**

Distribution: Central portions of Texas, Oklahoma, and Kansas. See map in Pierce & Whitehurst (1990). **U.S.**: KS, OK, TX.

Common name: Spotted Chorus Frog.

Commentary: Pierce & Whitehurst (1990) provided a CAAR account. This is a essentially a sister species of *P. maculata,* and these two together are a sister taxon to the other congeners of the *nigrita* group (Moriarty Lemmon *et al.,* 2007b; supported by Pyron & Wiens, 2011).

Pseudacris (Pseudacris) feriarum (Baird, 1856) **new combination**

Helocaetes feriarum Baird, 1856: 60.—**O**: **S**: including UMMZ 3857, and USNM 3592 (11) (Cochran, 1961a: 50, sd.).—**OT**: U.S., Pennsylvania, Cumberland County, Carlisle.

Pseudacris feriarum: Cope, 1862a: 157; Stejneger & Barbour, 1917: 31; 1943: 46; Dodd, 2013: 348.

Chorophilus feriarum feriarum: Cope, 1889b: 341.

Chorophilus nigritus feriarum: Hay, 1892: 460.

Hyla feriarum: Noble, 1923a: 5.

Hyla (Pseudacris) feriarum: Myers, 1926: 283.

Pseudacris nigrita feriarum: Stejneger & Barbour, 1933: 31; Neill, 1949b: 227; Smith & Smith, 1952: 167; Schmidt, 1953: 75.

Pseudacris triseriata feriarum: Schwartz, 1957: 11; Conant, 1958: 289; Gorham, 1974: 110; Behler & King, 1979: 415; Garrett & Barker, 1987: 27; Conant & Collins, 1998: 544; Dixon, 2000: 69.

Hyla (Pseudacris) triseriata feriarum: Dubois, 1984a: 86.

Pseudacris feriarum feriarum: Hedges, 1986: 11.

Pseudacris (Pseudacris) feriarum: **new combination.**

Distribution: Eastern Mississippi, most of Tennessee and Kentucky, southeastern Missouri, mostly above the fall line in Alabama and Georgia, but south along the Chattahoochee-Apalachicola Rivers into Florida, northeastward, upland from the coast through the central Carolinas and much of Virginia, into central Pennsylvania. See map in Moriarty Lemmon *et al.* (2008), Elliott *et al.* (2009) or Dodd (2013). **U.S.**: AL, AR, FL, GA, IL, KY, MD, MO, MS, NC, PA, SC, TN, VA.

Common name: Upland Chorus Frog.

Content: There are no subspecies currently recognized, but in the past this form has been treated as a subspecies of *P. nigrita* or *P. triseriata,* and forms now considered full species were once treated as subspecies of this species. Hedges (1986) treated *P. kalmi* as a subspecies of *P. feriarum.*

Commentary: This is the sister species of *P. triseriata;* the two together have a sister taxon relationship to *P. kalmi* (Moriarty & Cannatella, 2004). The analysis of Pyron & Wiens (2011) supports a sister relationship with *P. triseriata,* and indicates the pair in turn form a sister clade to one containing *P. kalmi* and *P. nigrita.*

Pseudacris (Pseudacris) fouquettei Moriarty Lemmon, Lemmon, Collins, & Cannatella, 2008 **new combination**

Pseudacris fouquettei Moriarty Lemmon, Lemmon, Collins, & Cannatella, 2008: 4.—**O**: **H**: TNHC 62265.—**OT**: U.S., Louisiana, East Baton Rouge Parish, NW of Baywood on Lee Price Road, 1.4 mi W of jct. with SR 37; N 30.7147, W 90.8919.

Pseudacris fouquettei: Dodd, 2013: 357.

Pseudacris (Pseudacris) fouquettei: **new combination**

Distribution: The Gulf coast, from western Mississippi, Louisiana and eastern Texas north to eastern Oklahoma, Arkansas and extreme southeastern Missouri. See map in Moriarty Lemmon *et al.* (2008), Elliott *et al.* (2009) or Dodd (2013). **U.S.:** AR, LA, MO, MS, OK, TX.

Common name: Cajun Chorus Frog.

Commentary: Moriarty & Cannatella (2004) found this a sister species to *P. nigrita;* see *Commentary* for that species. But the analysis of Pyron & Wiens (2011) indicates this species is sister to all other species of the subgenus except *P. brimleyi* and *P. brachyphona.* Often reported from Texas as *P. triseriata* (*e.g.*, Strecker, 1915: 48; Dixon, 2000: 69) or *P. nigrita triseriata* (*e.g.,* Viosca, 1949: 10) or *P. t. feriarum* (*e.g.,* Garrett & Barker, 1987: 27).

Pseudacris (Pseudacris) kalmi Harper, 1955 **new combination**

"Chorophilus triseriatus corporalis" Cope, 1875b: 30.—**NU.**—**O**: none yet identified.—**OT**: U.S., New Jersey.—**SS**: Frost, 2009-13.—**Comment**: Schmidt (1953: 75) considered this a *nomen nudum*; he and Duellman (1977: 172) considered it a synonym of *Pseudacris feriarum.*

Pseudacris nigrita kalmi Harper, 1955a: 1.—**O**: **H** (od.), CM 33917.—**OT**: U.S., New Jersey, Burlington County, 5 miles east of Moorestown, Centreton (on south side of Rancocas Creek, designated as Bougher on Mount Holly quadrangle of U.S. Geological Survey, 1898).

Pseudacris triseriata kalmi: Schwartz, 1957: 11; Conant: 1958: 288; Gorham, 1974: 110; Conant & Collins, 1998: 543.

Hyla (Pseudacris) triseriata kalmi: Dubois, 1984a: 86.

Pseudacris feriarum kalmi: Hedges, 1986: 11.

Pseudacris kalmi: Collins, 1997: 12; Moriarty Lemmon *et al.*, 2007b: 1074; Dodd, 2013: 367.

Pseudacris (Pseudacris) kalmi: **new combination.**

Distribution: New Jersey and the Delmarva peninsula. See map in Moriarty Lemmon *et al.* (2007b), Elliott *et al.* (2009) or Dodd (2013). **U.S.**: DE, MD, NJ.

Common name: New Jersey Chorus Frog.

Commentary: As seen in the synonymy, this form was considered a subspecies of other chorus frog species, but Moriarty Lemmon *et al.* (2007b) demonstrated its apparent distinctness as a full species. It is a sister taxon to the pair, *P. triseriata* + *P. feriarum* (Moriarty & Cannatella, 2004), or a sister to *P. nigrita* (Pyron & Wiens, 2011).

Pseudacris (Pseudacris) maculata (Agassiz, 1850) **new combination**

Hylodes maculatus Agassiz, 1850: 378.—**O**: **S**, MCZ 38 (2 specimens, including adult of Fig.1 and 2 in pl. 6) (Barbour & Loveridge, 1929: 281).—**OT**: U.S., Michigan, and Canada, Ontario, region of Lake Superior; corrected to Can., Ontario, vicinity of Fort William.—**Comment:** Schmidt (1953: 75) restricted the OT to precise localities, without any evidence that the S came from there, hence that action is invalid.

Pseudacris nigrita maculata: P.W. Smith, 1956: 171.

Pseudacris triseriata maculata: Schwartz, 1957: 11; Conant, 1958: 288; Cook, 1964: 186; Gorham, 1974: 110; Conant & Collins, 1998: 544.

Hyla (Pseudacris) triseriata maculata: Dubois, 1984a: 87.

Pseudacris maculata: Platz, 1989: 709; Moriarty Lemmon *et al.*, 2007b: 1068; Dodd, 2013: 371.

Pseudacris (Pseudacris) maculata: **new combination**

"Chorophilus septentrionalis" Boulenger, 1882a: 335.—**O**: **H**: BM 50.4.22.8 (Duellman, 1977: 172, mo.).—**OT**: Canada, Northwest Territory, Great Bear Lake.—**SS**: P.W. Smith, 1956: 171.—Junior secondary homynym, preoccupied by *Hyla septentrionalis* Duméril & Bibron, 1841.

Pseudacris septentrionalis: Stejneger & Barbour, 1917: 17.

Hyla septentrionalis: Noble, 1923: 5.

Pseudacris nigrita septentrionalis: Wright & Wright, 1933: 250; Stejneger & Barbour, 1933: 32; 1943: 46.

Hyla canadensis Noble, 1923: 5.—**NN** for *C. septentrionalis* Boulenger, 1882a.

Misapplication of names:

Chorophilus triseriatus triseriatus: Yarrow, 1875: 523.

Distribution: Much of southern Canada from Quebec to British Columbian and Yukon, south through the Midwest states to include Illinois and Missouri, Kansas, Colorado, plus the northwestern half of New Mexico and into east-central Arizona, parts of Utah and Idaho, and a disjunct area in northern New York and Vermont and adjacent Ontario and Quebec. See map in Moriarty Lemmon *et al.* (2007b), Elliott *et al.* (2009) or Dodd (2013). **Can.**: AB, BC, MB, NT, ON, QC, SK, YT. **U.S.**: AZ, CO, IA, ID, IL, IN, KS, MN, MO, MT, ND, NE, NM, OK, SD, UT, WI, WY.

Common name: Boreal Chorus Frog. *French Canadian name:* Rainette faux-grillon boréale.

Commentary: Schmidt (1953) restricted the type locality to the vicinity of Sault Sainte Marie in Michigan and Ontario (without explanation). Cook (1964) corrected the OT to Canada, Ontario, vicinity of Fort William, based on information given in the original description and follow-up examinations of type material by other authors. Duellman (1977) also later restricted the OT to the north shore of Lake Superior, but the Cook correction should stand. This species is not genetically clearly separable from *P. clarki*, although they are morphologically quite distinct, hence they are essentially sister species (Moriarty Lemmon *et al.*, 2007a). The analysis of Pyron & Wiens (2011) also found these two were sister species.

Pseudacris (Pseudacris) nigrita (Le Conte, 1825) **new combination**

Rana nigrita Le Conte, 1825: 282.—**O**: **S**, original number unknown, but including ANSP 2211-2214 (Malnate, *in* Gates, 1988, based on Harper's, 1935, identification of those specimens as S), and possibly USNM 5935, 234482-234486, collected by Le Conte (Gates, 1988).—**OT**: not stated, but the probable S were collected from Georgia.—**Comment**: The OT is uncertain; Harper (1935: 289) restricted it to U.S., Georgia, Liberty County, Riceboro—the location of one of LeConte's plantations; he also noted that Cope (1889b: 339) received 2 specimens from Le Conte; that would seem to constitute adequate evidence that the S series likely came from there and the restriction is accepted as valid. 2) Le Conte donated a specimen to MNHN, which may be a S, from "Georgia;" (Duméril & Bibron 1841: 509).

Acris nigrita: Duméril & Bibron, 1841: 509.

Cystignathus nigritus: Holbrook, 1842: 107.
Chorophilus nigritus: Baird, 1856: 60; Cope, 1889b: 337.
Pseudacris nigrita: Günther, 1859a: 97; Stejneger & Barbour, 1917: 31; Frost, 1985: 171; Moriarty & Cannatella, 2004: 417; Dodd, 2013: 385.
Hyla nigrita: Noble, 1923: 5.
Pseudacris nigrita nigrita: Stejneger & Barbour, 1933: 31; 1943: 46; Schmidt, 1953: 74; Conant, 1958: 290; Gorham, 1974: 110; Conant & Collins, 1998: 545.
Pseudacris (Pseudacris) nigrita: **new combination**

Chorophilus verrucosus Cope, 1878c: 87 (*nec Hyla verrucosa* Daudin, 1801).—**O**: **H**: ANSP 10773 (Malnate, 1971: 353, mo.).—**OT**: U.S., Florida, Volusia County, Volusia [east side of St. John's River, about 5 miles southeast of Lake George].—**SS**: Cope, 1889b: 338. Also a junior secondary homonym of *Hyla verrucossa* Daudin, 1801.
Hyla verrucosa: Brocchi, 1881: 43 (*nec Hyla verrucosa* Daudin, 1801).
Pseudacris verrucosus: Noble, 1923: 1.
Pseudacris nigrita verrucosa: Brady & Harper, 1935: 108; Stejneger & Barbour, 1943: 47; Schmidt, 1953: 75; Conant, 1958: 290; Gorham, 1974: 110; Conant & Collins, 1998: 545.

Hyla (Pseudacris) nigrita floridensis: Dubois, 1984a: 88.—**NN** for *Chorophilus verrucosus* Cope, 1878.

Distribution: The Gulf coastal area from southeastern Mississippi to Florida, northward up the Apalachicola-Chattahoochee River system, throughout peninsular Florida, and up the Atlantic coast into North Carolina. See map in Moriarty Lemmon *et al.* (2007b) or Elliott *et al.* (2009). The map in Girard) is good except the westernmost records are not this species; Dodd (2013) has an updated map. **U.S.**: AL, FL, GA, MS, NC, SC, VA.

Common name: Southern Chorus Frog.

Content: No subspecies are currently recognized. Members of the species from peninsular Florida were formerly treated as *P. n. verrucosa,* but Moriarty & Cannatella (2004) rejected recognition of the subspecies as unnecessary and uninformative.

Commentary: Gates (1988) provided a CAAR account. The studies of Moriarty Lemmon *et al.* (2007b, 2008) indicated that this is a sister taxon of *P. fouquettei*; the pair form a sister taxon to the clade of *P. kalmi* + *P. triseriata* + *P. feriarum*. However, the analysis of Pyron & Wiens (2011) indicates a sister species relationship with *P. kalmi,* with that pair having a sister relationship to the pair *P. feriarum* + *P. triseriata,* with *P. fouquettei* much more basal.

Pseudacris (Pseudacris) triseriata (Wied-Neuwied, 1838) **new combination**

Hyla triseriata Wied-Neuwied, 1838: 249.—**O**: originally in Wied collection, present location unknown.—**OT**: U.S., Indiana, Mount Vernon at Ohio River; corrected to Indiana, Posey County, between Rush Creek and Big Creek along route from New Harmony to Mt. Vernon, by Harper (1955b: 155); based on evidence from Wied's writings on the subject (Wied-Neuwied, 1865: 118).

Helocaetes triseriatus: Baird, 1856: 60.

Chorophilus triseriatus: Cope, 1875b: 30; 1889a: 342.

Chorophilus nigritus triseriatus: Hay, 1892: 470.

Pseudacris triseriata: Stejneger & Barbour, 1917: 31; Frost, 1985: 172; Stebbins, 1985: 78; 2003: 219; Moriarty & Cannatella, 2004: 409-420; Elliott *et al.*, 2009: 86; Dodd, 2013: 421.

Hyla triseriata: Noble, 1923: 5.

Hyla (Pseudacris) triseriata: Myers, 1926: 283.

Pseudacris nigrita triseriata: Stejneger & Barbour, 1933: 32; 1943: 47; Burt, 1936a: 775; Smith & Smith, 1952: 173; Schmidt, 1953: 75.

Pseudacris triseriata triseriata: Schwartz, 1957: 11; Conant, 1958: 288; Cochran & Goin, 1970: 115; Gorham, 1974: 110; Behler & King, 1979: 415; Conant & Collins, 1998: 542.

Pseudacris (Pseudacris) triseriata: **new combination**.

Distribution: Generally the midwest and Great Lakes region, into western Pennsylvania and western New York, northeastward to southern Ontario. See map in Moriarty Lemmon *et al.* (2007b) or Elliott *et al.* (2009). **Can.**: NL, ON, QC. **U.S.**: KY, IL, IN, MI, NY, OH, PA.

Common name: Western Chorus Frog (SECSN) or Midland Chorus Frog (C&T). *French Canadian name:* Rainette faux-grillon de l'Ouest.

Content: There are no subspecies recognized. At various times in its nomenclatural history, this form has been treated as a subspecies of *P. nigrita,* and *P. nigrita* was once treated as a subspecies of this species, in error. Other species of the genus have sometimes been treated as subspecies of this species.

Commentary: This is the sister species of *P. feriarum* (Moriarty & Cannatella, 2004; Pyron & Wiens, 2011). Listed as at risk by Quebec.

Subgenus **Hyliola** Mocquard, 1899

Hyliola Mocquard, 1899: 337.—**NS** (sd., Stejneger, 1907: 76): *Hyla regilla* Baird & Girard, 1852.

Distribution: Western North America from southern Alaska through British Columbia, Washington, Oregon, Idaho, western Montana, California, Nevada, western Arizona, and south through Baja California (Mexico).

Common name: None in general use; Western Chorus Frogs is appropriate.

Content: We resurrect the nomen to contain this clade. Four species are currently recognized.

Commentary: Members of this subgenus form a clade that is a sister taxon to all other *Pseudacris* (Moriarty Lemmon *et al.,* 2007b; supported by Pyron & Wiens, 2011).

Pseudacris (Hyliola) cadaverina (Cope, 1866) **new combination**

"Hyla nebulosa" Hallowell, 1855a: 96 (*nec Hyla nebulosa* Spix, 1824, of Brazil).—**O**: **S**: ANSP 1987, 1988.—**OT**: U.S., California, Los Angeles County, Tejon Pass.

Hyla cadaverina: Cope, 1866a: 84. **NN** for *H. nebulosa* Hallowell, 1855, preoccupied by *H. nebulosa* Spix, 1824.

Hyla cadaverina: Gorham, 1974: 95; Frost, 1985: 129; Stebbins, 1985: 80; 2003: 221.

Pseudacris cadaverina: Hedges, 1986: 11; Liner, 1994: 26; Dodd, 2013: 322.

Pseudacris (Hyliola) cadaverina: **new combination**

"Hyla californiae" Bogert, 1958: 11.—**NU**.

Hyla californiae Gorman, 1960: 214.—**O**: **H** (od.) MVZ 31773.—**OT**: Mexico, Baja California, Partido del Norte, "Alaska" [La Rumorosa].

Hyla californiae: Cochran & Goin, 1970: 105.

Misapplication of name:

Hyla regilla: Schmidt, 1953: 72.

Distribution: Coastal southern California, south into northern Baja California. See map in Gaudin (1979) or Dodd (2013). **U.S.**: CA.

Common name: California Treefrog (SECSN) or California Chorus Frog (C&T).

Commentary: Schmidt (1953: 72) treated this as a junior synonym of *Hyla regilla* as did Behler & King (1979: 408). Gaudin (1979) provided a CAAR account as a species of *Hyla*. Moriarty & Cannatella (2004) found this to be a sister species of *P. regilla,* while Recuero *et al.* (2006a) used it as an out-group so it appeared to be sister to the remainder of the subgenus.

Pseudacris (Hyliola) hypochondriaca (Hallowell, 1854) **new combination**

Synonymy: See below under *Pseudacris (Hyliola) hypochondriaca hypochondriaca.*

Distribution: Southern California and Nevada, plus adjacent western Arizona, south through Baja California (Mexico). **U.S.**: AZ, CA, NV.

Common name: Baja California Treefrog (SECSN) or Baja California Chorus Frog (C&T).

Content: Recuero *et al.* (2006a) recognized two genetically well-defined subspecies, and we tentatively follow them. The CNAH website does not recognize either subspecies, and Dodd (2013) is critical of the analysis and perhaps unwarranted recognition of this population and perhaps *P. sierra*. Further study of these populations is needed.

Commentary: This species is the sister of *P. sierra* (Recuero *et al.,* 2006a, as corrected 2006b).

Pseudacris (Hyliola) hypochondriaca hypochondriaca (Hallowell, 1854) **new combination**

Hyla scapularis var. *hypochondriaca* Hallowell, 1854a: 97.—**O**: **S** (sd. Test, 1899: 479) USNM 3235 (9).—**OT**: U.S., California, Los Angeles County, Tejon Pass.—**Comment**: *H. scapularis* is a junior synonym of *H. regilla*; see that account below.

Hyla regilla hypochondriaca: Gorham, 1974: 104.

Pseudacris regilla hypochondriaca: Liner, 1994: 26.

Pseudacris hypochondriaca hypochondriaca Recuero *et al.*, 2006a: 302.

Pseudacris (Hyliola) hypochondriaca hypochondriaca: **new combination.**

Hyla regilla deserticola Jameson *et al.*, 1966: 582—**O**: **H**: SDNHM 54176.—**OT**: Mexico, Baja California, San Borjas.

Distribution: Southern California, Nevada, western Arizona, and the northern half of the Baja California peninsula. **U.S.**: AZ, CA, NV.

Common name: As for the species.

Commentary: Schmidt (1953: 72) treated this species in the synonymy of *Hyla regilla,* by citing both *H. scapularis* and *H. cadaverina* as junior synonyms of *H. regilla.* Likewise, Cochran & Goin (1970: 109) treated this as a synonym of *H. regilla regilla.* Behler & King (1979: 408) also included it in the synonymy of *H. regilla.* Stebbins (1985: 80; 2003: 222-223) also seems to have treated this in the synonymy of *H. regilla.* Recuero *et al.* (2006a) show the two subspecies are well defined sister clades, and might well be recognized as full species. Note: In their book, Elliott *et al.* (2009) apparently treated this species as a part of *P. cadaverina.*

Pseudacris (Hyliola) regilla (Baird & Girard, 1852) **new comb.**

Hyla regilla Baird & Girard, 1852a: 174.—**O**: **S** (sd.): USNM 9182 (Puget Sound), USNM 15409 (Sacramento River) (Cochran 1961a: 58), and MCZ 2149 (Barbour & Loveridge, 1929: 280; Puget Sound, exchanged from USNM).—**OT**: U.S.: Washington, Puget Sound, and Oregon, Sacramento River.—**L**: USNM 9182 (Jameson *et al.*, 1966: 553): U.S., Washington, Puget Sound.

Hyla regilla: Baird, 1859d: 12; Cooper, 1860: 304; Boulenger, 1882a: 374; Stejneger & Barbour, 1943: 50; Schmidt, 1953: 72; Frost, 1985: 149; Stebbins, 1985: 80; 2003: 222.

Hyliola regilla: Mocquard, 1899: 337, 339.

Pseudacris regilla: Hedges, 1986: 11; Recuero *et al.*, 2006b: 511; Dodd, 2013: 400.

Pseudacris (Hyliola) regilla: **new combination**

Hyla scapularis Hallowell, 1854e: 183.—**O**: **H**: ANSP 1978 (Malnate, 1971, sd.).—**OT**: U.S., Oregon Territory.—**SS**: Baird & Girard, 1853a: 301.—**Comment**: Schmidt (1953: 72) restricted the OT to U.S., Washington,

Vancouver, without presenting evidence the specimen came from there, thus that action is invalid.

Hyla regilla pacifica Jameson, Mackey, & Richmond, 1966: 591—**O: H**: CAS 101007.—**OT**: U.S., Oregon, Lincoln County, Big Creek State Park, 4 miles south of Waldport.

Pseudacris pacifica Recuero *et al.*, 2006a: 302.

Hyla regilla cascadae Jameson, Mackey, & Richmond, 1966: 379.—**O**: **H**: CAS 101038.—**OT**: U.S., Oregon, Deschutes County, one-half mile south of Bend; R12E, T18S, S9NW, elevation 3750 feet.

Distribution: Southern Alaska (introduced) southward through British Columbia (introduced), Washington, western Oregon and into northern California. **Can.**: *BC. **U.S.**: *AK, CA, OR, WA.

Common name: Pacific Treefrog (SECSN) or Pacific Chorus Frog (C&T). *French Canadian name:* Rainette du Pacifique.

Content: Jameson *et al.* (1966) recognized a number of geographic races as named subspecies. Some of these have now been given full species status, and the others seem not separable genetically, so the species as currently treated has no subspecies.

Commentary: There were several restrictions of the type locality (see Frost, 2009-13), perhaps none justified, but as a lectotype (lectophoront) has been designated it is now academic. However, Jameson *et al.* (1966), in designating a new "holotype" (lectotype/lectophorant) erred in stating that the type locality of the specimen selected is Fort Vancouver, Washington. The locality for that specimen is "Puget Sound." Schmidt's (1953) restriction was to "Sacramento River, California," and that restriction is invalid as he cited no evidence that the Puget Sound collecting locality was erroneous. Thus the original onymotope (type locality), Puget Sound and Sacramento River should stand, but for the lectotype designation, which now sets the OT as the origin of the L, which is also Puget Sound by our best understanding. This species is the sister taxon of *P. cadaverina* (Moriarty & Cannatella, 2004) or sister to the clade formed by *P. hypochondriaca* + *P. sierra* (Recuero *et al.*, 2006a, as corrected 2006b). Recuero *et al.*, 2006a, erred in applying the name *P. pacifica,* and later corrected it to *P. regilla* (Recuero *et al.*, 2006b). The natural range of the species is in northwestern U.S.; introduced into Canada and Alaska. The relationship of this species to its subgeneric relatives needs further study.

—

Pseudacris (Hyliola) sierra (Jameson, Mackey & Richmond, 1966) **new combination**

Hyla regilla sierra Jameson *et al.*, 1966: 605.—**O**: **H**: CAS 100991.—**OT:** U.S., California, 1.25 miles SSE of Tioga Pass Ranger Station (east of entrance to Yosemite National Park); R25E, T1N, S31, elevation 9600 feet.

Hyla regilla sierra: Gorham, 1974: 104.

Pseudacris sierra Recuero *et al.*, 2006b: 511 (correction in name for Central Clade).

Pseudacris (Hyliola) sierra: **new combination**

Misapplication of names:

Pseudacris regilla Recuero *et al.*, 2006a: 302 (name in error for the Central Clade).

Distribution: Central California, eastern Oregon, Idaho, western Montana, probably northwestern Utah, but more study is needed to determine the exact boundaries of the species. **U.S.**: CA, ID, MT, OR, UT?.

Common name: Sierran Treefrog (SECSN) or Sierra Chorus Frog (C&T).

Commentary: This is the sister species of *P. hypochondriaca* (Recuero *et al.*, 2006a, as corrected 2006b; those authors erred in applying the name *P. regilla* to this species, then corrected it to *P. sierra* in the second paper). As noted above, the recognition of this and other populations of this western group may be premature, and further study is needed.

Subgenus **Limnaoedus** Mittleman & List, 1953

Limnaoedus Mittleman & List, 1953: 81.—**NS** (od.): *Hylodes ocularis* Holbrook, 1838b,—**SS**: Boulenger, 1882a: 333, with *Hyla ocularis* Latreille, 1801: 187.

Parapseudacris Hardy & Burrows, 1986: 80.—**NS** (od.): *Hyla crucifer* Wied-Neuwied, 1838.

Distribution: Most of eastern North America, from eastern Texas to Manitoba, to the east coast.

Common name: Spring Peepers and Grass Frogs.

Content: Two species, endemic to North America.

Commentary: Franz & Chantell (1978) provided a generic CAAR account. Originally proposed as a genus for *P. ocularis,* when it was being treated as a species of *Hyla*. We resurrect it as a subgenus for that species, plus its sister species, *P. crucifer*. The analysis of Pyron & Wiens (2011) indicates this subgenus is a sister to the subgenera *Pseudacris* and *Pycnacris,* as treated here.

Pseudacris (Limnaoedus) crucifer (Wied-Neuwied, 1838) **new combination**

Hyla crucifer Wied-Neuwied, 1838: 275.—**O**: **H**: originally Wied collection; fate not known, probably lost.—**OT**: U.S., Kansas, Leavenworth County, Fort Leavenworth (apparently an error as the species as known does not occur anywhere near there).

Hyla crucifera: Myers, 1927: 338.—**UE.**

Hyla crucifera crucifera: Harper, 1939a: 1.—**UE.**

Hyla crucifer: Stejneger & Barbour, 1917: 32; Viosca, 1949: 10; Gorham, 1970: 6; Conant, 1975: 319; Behler & King, 1979: 406; Frost, 1985: 132.

Hyla crucifer crucifer: Stejneger & Barbour, 1943: 49; Schmidt, 1953: 70; Conant, 1958: 278; Cochran & Goin, 1970: 108; Gorham, 1974: 97.

Parapseudacris crucifer: Hardy & Burrows, 1986: 80.

Pseudacris crucifer: Hedges, 1986: 11; Cocroft, 1994: 433 (by implication); Conant & Collins, 1998: 540; Dixon, 2000: 68; Elliott *et al.,* 2009: 76; Dodd, 2013: 331.

Pseudacris crucifer crucifer: Garrett & Barker, 1987: 25; Stevenson & Crowe, 1992: 86.

Pseudacris (Limnaoedus) crucifer: **new combination**

Hyla crucifera bartramiana Harper, 1939a: 1.—**O**: **H**: ANSP 21526.—**OT**: U.S., Georgia, Charlton County, 4 miles west of Folkston. Subspecies rejected by Moriarty & Cannatella (2004: 417).

Pseudacris crucifer bartramiana: Stejneger & Barbour, 1943: 50; Schmidt, 1953: 70; Conant, 1958: 279; Cochran & Goin, 1970: 108; Gorham, 1974: 97; Stevenson & Crowe, 1992: 86.

Hylodes pickeringii Holbrook, 1839a: 131, pl. 37.—**O**: **S**: includes specimen in pl. 27; apparently no longer exists.—**OT**: U.S., Massachusetts (especially Salem), New York, Philadelphia.—**Comment:** Schmidt (1953: 70) gave the OT as Massachusetts, Essex County, Danvers; if this is interpreted as a restriction, it is invalid, as no evidence was cited, or if it is interpreted as quoting the original OT, it is in error.

Hylodes pickeringii: Storer, 1839: 240.

Hyla pickeringii: Holbrook, 1840: 135.
Acris pickeringii: Jan, 1857: 53; Günther, 1859a: 71.
Hyla pickeringi: Davis & Rice, 1883: 28.—**UE.**
Hyliola pickeringii: Mocquard, 1899: 339.

Distribution: Widespread in eastern North America, from eastern Texas north to southeastern Manitoba, eastward to the Atlantic coast; but absent from peninsular Florida. See map in Elliott *et al.* (2009) or Dodd (2013). **Can.**: MB, NB, NS, ON, PE, QC. **U.S.**: AL, AR, CT, DC, DE, FL, GA, IA, IL, IN, KS, KY, LA, MA, MD. ME, MI, MN, MO, MS, NC, NH, NJ, NY, OH, OK, PA, RI, SC, TN, TX, VA, VT, WI, WV.

Common name: Spring Peeper. *French Canadian name:* Rainette crucifère.

Content: No subspecies are recognized here. However, as noted in the synonymy, Harper (1939a) proposed a southern subspecies as a geographic race, but it has been rejected by Austin *et al.* (2002) and Moriarty & Cannatella (2004).

Commentary: Frost (2009-13) has additional comments on the nomenclature of this species. This is the sister species of *P. ocularis,* and the subgenus (as treated here) is the sister taxon of the subgenus *Pseudacris* as treated here (Moriarty & Cannatella, 2004).

Pseudacris (Limnaoedus) ocularis (Holbrook, 1838) **new combination**

?*Hyla occularis* Sonnini & Latreille, 1830a [v2]: 186. They indicate a *Hyla ocularis,* not giving an author, but not clear if intended as a new species, and refer to Bosc finding it in Carolina, on trees (*P. ocularis* is not found on trees, nor is *Acris,* but in grassy habitat). They apparently refer to *H. ocularis* Latreille 1801 (see below), but it is still unclear what species that might be.

Hylodes ocularis Holbrook, 1838b: 79.—**O**: not stated or known.—**OT**: U.S., South Carolina and Georgia. See the discussion of the nomen *Hyla ocularis* Latreille, 1802, under the taxon *Acris g. gryllus,* above. As we now regard that name as a *nomen dubium,* it is unavailable for this taxon. Other references to that nomen under different generic designation, given in the *Acris* account, seem clearly to refer to the Latreille name. The first description of the species of this account seems to be that of Holbrook (1838b), so we indicate that here, a change from previous authorship generally seen as Bosc, or Bosc & Daudin, or Latreille.

Cystignathus ocularis: Cope, 1875b: 27.
Chorophilus ocularis: Cope, 1889b: 348.
Pseudacris ocularis: Stejneger & Barbour, 1917: 31; Noble, 1923: 1; Burt, 1936: 771; Carr, 1940: 57; Hedges, 1986: 11; Cocroft, 1994: 433; Silva, 1997: 611 (by implication); Conant & Collins, 1998: 550; Elliott *et al.*, 2009: 80.
Hyla ocularis: Noble, 1923: 5 (by implication); Harper, 1939b: 139; Schmidt, 1953: 71; Conant, 1958: 283; Cochran & Goin, 1970: 108.
Limnaoedus ocularis: Mittleman & List, 1953: 83; Gorham, 1974: 97; Conant, 1975: 326; Behler & King, 1979: 410; Frost, 1985: 158.
Pseudacris (Limnaoedus) ocularis: **new combination**

Chorophilus angulatus Cope, 1875b: 30.—**O**: not stated.—**OT**: South Carolina.—**SS**: Boulenger, 1882a: 333.—**Comment**: Schmidt (1953: 71) restricted the OT to Charleston, South Carolina, but as that action was not accompanied by evidence the original specimen(s) came from there, it is invalid.

Distribution: Florida except western panhandle, southern Georgia, north on Atlantic coastal plain into southeastern Virginia. See map in Franz & Chantell (1978) or Elliott *et al.* (2009). **U.S.**: FL, GA, NC, SC, VA.

Common name: Little Grass Frog.

Commentary: Mittleman (1946) maintained that the species described in Sonnini & Latreille (1801) was actually *Acris gryllus* (see account for *A. g. gryllus*). Others (e.g., Neill, 1950b) disagreed. We follow Neill's finding that the nomen *Hyla ocularis,* Latreille, 1801, is based on an uncertain composite, and consider it a ND, thus unavailable. Franz & Chantell (1978) provided a CAAR account. Moriarty & Cannatella (2004) showed this species is a sister to *P. crucifer,* and these two together are a sister taxon to the subgenus *Pseudacris* as treated here.

Subgenus **Pycnacris new subgenus**

Type-species (nucleospecies), by present designation: *Rana ornata* Holbrook, 1836: 97.—Etymology: from Greek πυκνός (*pycnos*), "strong, mighty" and άκρίς (*acris*), "locust", in reference to robust ("fat") (false) cricket frogs.

Diagnosis: Body shape stout, robust ("fat") as compared to any other species of the genus, which are more slender. Intercalary cartilage reduced, wafer-shaped, with shallow curving articulating surfaces, as compared to other species of the genus in which the cartilage is either cuboidal or has the distal articulating surface more concave than the proximal. Adults are larger (males 30-44 mm snout-vent length) than other congeners. Advertisement call of males is a series of short "whistles," rising in frequency, not a series of distinct pulses (*i.e.*, not "trilled") as it is in all other congeners except *P. (Limnaeodus) crucifer*. *P. crucifer* has a longer "whistle" at a higher frequency.

Distribution: Eastern Texas northward into Illinois and Missouri, and eastward across the Gulf coastal plain and up the Atlantic coastal plain to North Carolina, but absent from peninsular Florida.

Content: Two species form this subgenus.

Common name: None in general use, but the taxon has been called the Fat Chorus Frogs (*e.g.*, Moriarty & Cannatella, 2004) Stout Chorus Frogs might be a better term.

Commentary: This taxon is a sister to the subgenus *Pseudacris* as treated here (Moriarty & Cannatella, 2004).

Pseudacris (Pycnacris) ornata (Holbrook, 1836) **new comb.**

Rana ornata Holbrook, 1836: 97, pl. 16.—**O**: includes specimen figured in pl. 16; not known to exist.—**OT**: U.S., South Carolina, Charleston County, about 4 miles from Charleston, between Cooper and Ashley Rivers.

Cystignathus ornatus: Holbrook, 1842: 103; Günther, 1859a: 29.

Chorophilus ornatus: Le Conte, 1855b: 428; Cope, 1889b: 333.

Pseudacris ornata: Stejneger & Barbour, 1917: 31; 1943: 47; Viosca, 1949: 10; Schmidt, 1953: 76; Cochran & Goin, 1970: 117; Gorham, 1974: 110; Conant, 1975: 332; Behler.& King, 1979: 413; Frost, 1985: 171; Conant & Collins, 1998: 548; Elliott *et al.*, 2009: 82; Dodd, 2013: 395.

Hyla ornata: Noble, 1923: 5.

Hyla (Pseudacris) ornata: Dubois, 1984a: 85.

Pseudacris (Pycnacris) ornata: **new combination**

Litoria occidentalis Baird & Girard, 1853a: 301.—**O**: unknown;—**OT**: stated as U.S., California, San Francisco, but in error (Stejneger & Barbour, 1933).—**SS**: Schmidt (1953: 76).—**Comment**: Schmidt, 1953: 76, restricted the OT to Liberty County, Georgia, but presented no evidence that the specimen(s) came from there, therefore the action is invalid.

Chorophilus occidentalis: Cope, 1889b: 335.

Pseudacris occidentalis: Stejneger & Barbour, 1917: 30.

Hyla occidentalis: Noble, 1923: 5.

Pseudacris nigrita occidentalis; Stejneger & Barbour, 1933: 32. Probably a misapplication of the name, as they gave the range as Arkansas and Texas.

Chorophilus copii Boulenger, 1882a: 334—**O**: **H:** BM 58.2.23.55 (Duellman, 1977: 170, sd.).—**OT**: U.S., Georgia.

Hyla copii: Günther, 1901a: 266; Noble, 1923: 5.

Pseudacris copii: Noble, 1923: 1.

Hyla weberi Noble, 1923: 5.—**NN** for *Chorophilus copii* Boulenger, 1882a.

Distribution: Gulf and Atlantic coastal plains from southeastern Louisiana to eastern North Carolina; absent from peninsular Florida. See map in Elliott *et al.* (2009) or Dodd (2013). **U.S.**: AL, FL, GA, LA, MS, NC, SC.

Common name: Ornate Chorus Frog.

Commentary: This species is the sister of *P. streckeri* and together they compose the subgenus *Pycnacris*. Glorioso (2010) provided a CAAR account.

Pseudacris (Pycnacris) streckeri Wright & Wright, 1933 **new combination**

Synonymy: See below under *Pseudacris (Pycnacris) streckeri streckeri*.

Distribution: East-central Texas, central Oklahoma, and northwestern Arkansas, with relictual populations in Illinois and Missouri (see subspecies). Smith (1966a) has a map, as do Elliott *et al.* (2009). **U.S.**: AR, IL, MO, OK, TX.

Common name: Strecker's Chorus Frog.

Content: Two subspecies are often identified, and we recognize them here.

Commentary: P.W. Smith (1966a) provided a CAAR account. The two species of this subgenus are a sister taxon to the subgenus *Pseudacris*.

Pseudacris (Pycnacris) streckeri streckeri Wright & Wright, 1933 **new combination**

Pseudacris streckeri Wright & Wright, 1933: 26, 102, pl. 35.—**O**: **S** (sd., Smith, 1966a), presumably CU 2485 (5); missing, probably lost.—**OT**: Referred to a locality of Helotes Creek in Texas; presumed S is from U.S., Texas, Bexar County, Somerset, as revised (Smith, 1966a), based on CU data.—**Comment:** Schmidt (1953: 76) erred in designating Texas, McLennan County, Waco, as OT.

Pseudacris streckeri: Burt, 1936a: 771; Stejneger & Barbour, 1943: 47; Frost, 1985: 172; Elliott *et al.*, 2009: 84; Dodd, 2013: 416.

Pseudacris streckeri streckeri: P.W. Smith, 1951: 190; Schmidt, 1953: 76; Cochran & Goin, 1970: 118; Gorham, 1974: 110; Conant, 1975: 333; Behler & King, 1979: 414; Garrett & Barker, 1987: 26; Conant & Collins, 1998: 549; Dixon, 2000: 68-69.

Hyla (Pseudacris) streckeri: Dubois, 1984a: 85.

Pseudacris (Pycnacris) streckeri streckeri: **new combination**

Misapplied names:

Chorophilus ornatus: Strecker, 1915: 48.

Distribution: East-central Texas, much of Oklahoma, and northwestern Arkansas, into south-central Kansas. See map in Dodd (2013). **U.S.**: AR, LA, OK, TX.

Common name: Strecker's Chorus Frog.

Pseudacris (Pycnacris) streckeri illinoensis Smith, 1951 **new combination**

Pseudacris streckeri illinoensis P.W. Smith, 1951: 190.—**O**: **H** (od.), INHS 5982.—**OT**: U.S., Illinois, Morgan County, 3 miles north of Meredosia.

Pseudacris streckeri illinoisensis: Schmidt, 1953: 76.—**IS.**

Pseudacris streckeri illinoensis: Cochran & Goin, 1970: 118; Gorham, 1974: 110; Conant, 1975: 333; Duellman, 1977: 171; Behler & King, 1979: 414; Conant & Collins, 1998: 549.

Hyla (Pseudacris) streckeri illinoensis: Dubois, 1984a: 86.

Pseudacris illinoensis: Collins, 1991: 42; Dodd, 2013: 363.

Pseudacris (Pycnacris) streckeri illinoensis: **new combination**

Distribution: Two relictual populations, widely disjunct from main population (nominotypical subspecies), one in west-central

Illinois and another in southeastern Missouri, extreme southern Illinois, northeastern Arkansas. See map in P.W. Smith (1966a), Elliott *et al.* (2009) or Dodd (2013). **U.S.**: AR, IL, MO.

Common name: Illinois Chorus Frog.

Commentary: Smith (1966a) treated this subspecies in a CAAR account. Collins (1991) treated this form as a full species, but provided no justification. Moriarty & Cannatella (2004) indicated these populations are not yet ready for full species status without more critical study. Treatment by other authors varies: the CNAH website does not recognize this form as a species nor does it recognize subspecies of *P. streckeri*; Frost (2009-13) recognizes it as a full species, as do Pyron & Wiens (2011) and Dodd (2013). We continue to treat it as a subspecies, pending further study.

Genus **Smilisca** Cope, 1865

Smilisca Cope, 1865b: 194.—**NS** (mo.): *Smilisca daulinia* Cope, 1865 (now *Smilisca baudinii* (Duméril & Bibron, 1841)).

Pternohyla Boulenger, 1882c: 326.—**NS** (mo.): *Pternohyla fodiens* Boulenger, 1882c: 326.—**SS**: Faivovich *et al.*, 2005: 106.

Distribution: Southern Texas and southern Arizona, south through Middle America to northwestern South America.

Common name: Mexican Treefrogs.

Content: Eight species, two of which reach North America.

Commentary: Duellman (1968a) provided a generic CAAR account for *Smilisca*, and Trueb (1969) for *Pternohyla*, before they were deemed synonyms. This genus is the most derived of a clade of four genera (or five *fide* Pyron & Wiens, 2011); that clade is the sister taxon of *Hyla* (Faivovich *et al.*, 2005), supported by Pyron & Wiens (2011). *Pternohyla* is paraphyletic with respect to *Smilisca*, according to the results of the study by Faivovich *et al.* (2005).

Smilisca baudinii (Duméril & Bibron, 1841)

Hyla baudinii Duméril & Bibron, 1841: 564.—**O**: **H**: MNHN 4798, adult ♂, SVL 54.8 mm (Kellogg, 1932: 160; Guibé, 1950: 18; Duellman, 1977: 176).—**OT**: Mexico.—**Comment**: Smith & Taylor (1950a: 347) and Schmidt (1953: 69) published two distinct **OT** restrictions for this nominal taxon, but as neither is accompanied by evidence that the **H** actually came from there, both are invalid.

Hyla baudinii: Reinhardt & Lütken, 1863: 207; Stejneger & Barbour, 1917: 32.

Smilisca baudinii: Cope, 1866b: 127; 1887a: 13; 1889b: 379; Strecker, 1915: 50; Gorham, 1974: 111; Frost, 1985: 173; Garrett & Barker, 1987: 28; Liner, 1994: 28; Conant & Collins, 1998: 539; Lemos-Espinal & Smith, 2007a: 46, 244; Elliott *et al.*, 2009: 124; Dodd, 2013: 432.

Smilisca daudinii: Cope, 1871d: 205.—**IS**.

Hyla baudini: Brocchi, 1877a: 124; Schmidt, 1953: 69.—**UE**.

Hyla baudinii baudinii: Stejneger & Barbour, 1923: 29; 1943: 49; Viosca, 1949: 10.

Smilisca baudini baudini: Smith, 1947b: 408.—**UE**.

Smilisca baudini: Duellman & Trueb, 1966: 289; Cochran & Goin, 1970: 109; Conant, 1975: 326; Behler & King, 1979: 416; Dixon, 2000: 69.—**UE**.

Hyla vanvlietii Baird, 1856: 61.—**O**: **H**: Status uncertain (see *Comment*).—**OT**: USA, Texas, Cameron County, Brownsville, elev. 15 m.—**SS**: Cope, 1866b: 127.—**Comment**: Uncertainty exists regarding the original **H** of this species. Kellogg (1932: 160) stated that this specimen was "presumably" USNM 3239 and was now lost. In contrast, Cochran (1961a: 60) mentioned with a query the juvenile USNM 3256 as a possible "cotype" of this species, and Duellman (1968b: 1), without further discussion, considered that specimen as the **H**.

Hyla vanvlietii: Baird, 1859e: 29.

Hyla vanvloeti: Boulenger, 1882a: 371, 494.—**IS/UE**.

Hyla vanvlieti: Schmidt, 1953: 69.—**UE**.

Hyla vociferans Baird 1859e: pl. 38, fig. 11-13.—**O**: **H**: specimen shown in pl. 38 fig. 11-13; fate unknown, probably lost.—**OT**: unknown.—**SS**: Cope, 1887a: 13.

Hyla muricolor Cope, 1862c: 359.—**O**: **H** (mo.), USNM 25097 (Kellogg, 1932: 160; Cochran, 1961a: 56, with a query).—**OT**: Mexico, Veracruz, 27 km ENE by road of Huatusco, Hacienda Mirador, elev. 1020 m.—**SS**: Boulenger, 1882a: 371.

Smilisca daulinia Cope, 1865b: 194.—**O**: **H** (mo.), "a skeleton in the private anatomical museum of Hyrtl, Professor of Anatomy in the University of Vienna"; fate unknown, probably lost.—**OT**: unknown, but likely to be in "tropical America", as it was described in a paper dealing with herpetology of that region.—**SS**: Cope, 1866a: 85.—**Comments**: [1] The statement by Cope (1866a: 85) that this nomen "must probably be considered an erroneous orthography" of *Hyla baudinii* Duméril & Bibron, 1841 is not supported, because the latter nomen was not mentioned in the original publication, where the new nomen is based on an identified specimen which was not part of the specimens used by Duméril & Bibron (1841) for their description. [2] Smith & Taylor (1950a: 347) proposed a "restricted type locality" in Mexico for this species, but as it was not accompanied by evidence that the **H** actually came from there, it is invalid.

Hyla pansosana Brocchi, 1877a: 125.—**O**: **H**: MNHN 6313, SVL 48 mm (Guibé, 1950: 23).—**OT**: Guatemala, Alta Verapaz, Panzós, Río Polochie, elev. 36 m.—**SS**: Boulenger, 1882a: 371.

Hyla beltrani Taylor, 1942a: 306.—**O**: **H:** UIMNH 25046 (ex EHT-HMS 29563).—**OT**: Mexico, Chiapas, Tapachula, elev. 140 m.—**SS**: Duellman & Trueb, 1966: 291.

Hyla manisorum Taylor, 1954: 630.—**O**: **H**: UKMNH 34927, adult ♀, SVL 76 mm.—**OT**: Costa Rica, Provincia Limón, Bataan, elev. 15 m.—**SS**: Duellman & Trueb, 1966: 291.

Distribution: Costa Rica north to Mexico, along both coasts, and into southern Texas on the Gulf coast. See map in Duellman (1968b), Elliott *et al.* (2009) or Dodd (2013). **U.S.**: TX.

Common name: Mexican Treefrog.

Commentary: Duellman (1968b) provided a CAAR account. Faivovich *et al.* (2005) found this to be a sister taxon to all other *Smilisca*. Listed as Threatened by Texas.

Smilisca fodiens (Boulenger, 1882)

Pternohyla fodiens Boulenger, 1882c: 326.—**O**: **H**: BM 1947.2.24.26 (ex BM 82.11.27.8 *fide* Condit, 1964: 98; Duellman, 1977: 173).—**OT**: Mexico, Sinaloa, Presidio.

Pternohyla fodiens: Cope, 1887a: 12; Cochran & Goin, 1970: 109; Gorham, 1974: 111; Behler & King, 1979: 416; Frost, 1985: 172; Stebbins, 1985: 78; 2003: 220.

Smilisca fodiens: Faivovich *et al.*, 2005: 106; Brennan & Holycross, 2006: 38; Elliott *et al.*, 2009: 122; Cox *et al.*, 2012: 226; Dodd, 2013: 428.

Hyla rudis Mocquard, 1899: 163.—**O**: **H**: MNHN 1897.0217, juvenile (Kellogg, 1932: 136).—**OT**: Mexico, Jalisco, Guadalajara.—**SS**: Kellogg, 1932: 136.—**Comment**: The collection number of the **H** in the Paris Museum collection has been controversial: Kellogg (1932: 136) gave it as "No. 373a, parchment label No. 97-217"; Guibé (1950: 25) gave it as 97-217; and Duellman (1977: 173) and Trueb (1969: 77.3) gave it as 373a. The actual complete number is MNHN 1897.0217. The number 373a was the number of this specimen in the oldest catalogue, before the reorganization and renumbering of the herpetological collection of the Paris Museum which took place in 1864.

Distribution: South-central Arizona, south into western coastal Mexico. See map in Trueb (1969), Elliott *et al.* (2009) or Dodd (2013). **U.S.**: AZ.

Common name: Lowland Burrowing Treefrog (SECSN) or Northern Casquehead Frog (C&T).

Commentary: Trueb (1969) provided a CAAR account, under the genus *Pternohyla*. Faivovich *et al.* (2005) found this to be a sister taxon to all other *Smilisca* except *S. baudinii*. Cox *et al.* (2012), in a molecular study, found rather distinctive clades on either side of a geographic barrier in Mexico, but did not recommend any taxonomic changes, pending further study.

*Tribe **Lophyohylini** Miranda-Ribeiro, 1926

Lophiohylinae Miranda-Ribeiro, 1926: 64.—**NG**: *Lophiohyla* Miranda-Ribeiro, 1926: 64, **UE** of *Lophyohyla* Miranda-Ribeiro, 1923: 5.—**Comment**: Faivovich *et al.* (2005: 107) resurrected this nomen for a tribe of the subfamily *Hylinae*. For this nomen they used the spelling *Lophiohylini*, derived from the original subfamilial nomen. However, the latter nomen was based on the generic nomen *Lophiohyla* Miranda-Ribeiro, 1926, which is an unjustified emendation of the original nomen *Lophyohyla* Miranda-Ribeiro, 1923 (see Dubois, 1984c: 21-22). Under Art. 35.4 of the current *Code*, a family-series nomen based upon an unjustified emendation of a generic nomen is to be considered an unjustified emendation itself and must be corrected to the spelling formed from the stem of the original nomen: so here the tribe nomen must be corrected to *Lophyohylini*.

Lophiohylini: Faivovich *et al.*, 2005: 107.—**IS.**

Trachycephalinae Lutz, 1969: 275.—**NG**: *Trachycephalus* Tschudi, 1838: 74.—**SS**: Faivovich *et al.*, 2005: 107.

Distribution: Southern Florida (perhaps introduced) and many Caribbean islands and throughout most of the Neotropics, from Mexico to Argentina.

Common name: None in general use. Neotropical Casque-headed Treefrogs may be appropriate.

Content: According to the revision by Faivovich *et al.* (2005), there are eight genera containing 65 species. A single genus and species is found within North America, and is usually treated as introduced.

*Genus **Osteopilus** Fitzinger, 1843

Osteopilus Fitzinger, 1843: 30.—**NS**: *Trachycephalus marmoratus* Duméril & Bibron, 1841: 538 (od.).

Calyptahyla Trueb & Tyler, 1974: 19.—**NS**: *Trachycephalus lichenatus* Gosse, 1851: 362 (od.).—**SS:** (Trueb & Tyler, 1974: 41) with *Hyla crucialis* Harlan, 1826a: 64.—**SS**: Henderson & Powell, 2003: 13, with *T. lichenatus*.

Distribution: Cuba and the Greater Antilles, Jamaica, and the Bahamas, with one species reaching southern Florida.

Common name: West Indian Treefrogs.

Content: Eight species, with one, possibly introduced, reaching North America.

***Osteopilus septentrionalis** (Duméril & Bibron, 1841)

"Trachycephalus marmoratus" Duméril & Bibron, 1841: 538.—**O**: **H**: MNHN 4612.—**OT**: Cuba. Preoccupied by *Hyla marmorata* Laurenti, 1768.

Hyla septentrionalis Duméril & Bibron, 1841: 538.—**NN** for *Trachycephalus marmoratus.*

Hyla septentrionalis: Stejneger & Barbour, 1923: 35 (noted as recently introduced); Myers, 1950: 206; Schmidt, 1953: 236 (regarded as introduced); Cochran & Goin, 1970: 102 (treated as introduced); Gorham, 1974: 105; Conant, 1975: 325 (treated as native).

Hyla septentrionalis septentrionalis: Barbour, 1937: 94.

Hyla dominicensis septentrionalis: Mertens, 1938: 333.

Osteopilus septentrionalis: Behler & King, 1979: 410 (treated as introduced); Frost, 1985: 168; Schwartz & Henderson, 1991: 148; Conant & Collins, 1998: 538 (noted as immigrant); Faivovich *et al.*, 2005; Elliott *et al.,* 2009: 120 (treated as invasive); Dodd, 2013: 820 (treated as non-native).

Trachycephalus insulsus Cope, 1863: 43.—**O: S**: USNM 6265 (2; destroyed), 6266 (2; now renumbered 12166 and 167237).—**OT**: Cuba.

Hyla insulsa: Jaume, 1966: 15.

Trachycephalus wrighti Cope, 1863: 45.—**O**: **H**: USNM 5174.—**OT**: Cuba, Oriente Province, District of Guantanamo.

Hyla schebestana Werner, 1917: 36.—**O**: **H**: ZMH (unnumbered; apparently destroyed).—**OT**: Cuba.

Misapplied names:

Hyla dominicensis: Peterson *et al.*, 1952: 63.

Distribution: Widely distributed in Cuba, and on islands to the north and south of Cuba, and in southern Florida and the Keys. *Tentatively we treat this species as introduced, but it may have dispersed to Florida naturally (see "Distribution" in Duellman & Crombie, 1970). The latter authors have a map of the species range, as does Dodd (2013). Note: Dodd (2013) indicates reports of this form into northern Florida and to the west in Texas and Colorado, and scattered reports as far north as Minnesota; none of these beyond Florida are yet confirmed as established populations. McKeown (1996) listed it as established in Hawaii; it has been taken there but breeding populations are not known. **U.S.:** *?FL.

Common name: Cuban Treefrog.

Commentary: Myers (1950) presented a detailed history of the nomenclature for this species up to that time. Duellman & Crombie (1970) provided a CAAR account, and also included a detailed nomenclatural history of this species, treated in the genus *Hyla*.

*Subfamily **Pelodryadinae** Günther, 1858

Pelobii Fitzinger, 1843: 31.—**NG**: *Pelobius* Fitzinger, 1843.

Pelodryadidae Günther, 1858a: 346.—**NG**: *Pelodryas* Günther, 1858.

Pelodryadidae: Savage, 1973: 354.

Pelodryadinae: Dowling & Duellman, 1974-1978: 37.1; Blackburn & Wake, 2011: 42.

Chiroleptina Mivart, 1869: 294.—**NG**: *Chiroleptes* Günther, 1858.

Cycloraninae Parker, 1940: 12.—**NG**: *Cyclorana* Steindachner, 1867.

Cycloranini: Lynch, 1971: 76.

Nyctimystinae Laurent, 1975: 183.—**NG**: *Nyctimystes* Stejneger, 1916.

Distribution: Native to Australia and Papua-New Guinea. Some species have been introduced to other regions of the world, including one genus and species now found in North America.

Common name: Australian Treefrogs (widely used; not listed by SECSN or C&T).

Content: One genus (*Litoria*) with 188 species, one of which may be introduced into North America.

***Litoria** Tschudi, 1838

Litoria Tschudi, 1838: 36.—**NS** (mo.): *Litoria freycineti* Tschudi, 1838.

Ranoidea Tschudi, 1838: 35.—**NS** (mo.): *Ranoidea jacksoniensis* Tschudi, 1838.—**SS**: Steindachner, 1867b: 61, with *Hyla*.—**Comment**: Wells & Wellington (1985: 5) resurrected the nomen with no explanation, but that action has been generally ignored.

Lepthyla Duméril & Bibron, 1841: 504. Substitute name for *Litoria* Tschudi.

Dryopsophus Fitzinger, 1843: 30.—**NS** (od.): *Hyla citropa* Duméril & Bibron, 1841.—**SS**: Fitzinger (1843: 30), with *Hyla*.—**Comments:** 1) Fitzinger designated *Hyla citropa* Péron as NS, but this is interpreted as the species and author given here. 2) Wells & Wellington (1985: 5) resurrected the nomen with no explanation, but it has been generally ignored.

Euscelis Fitzinger, 1843: 30.—**NS** (od.): *Litoria lesueurii* Duméril & Bibron, 1841.—**SS**: Gorham (1974: 93), with *Hyla*.

Pelobius Fitzinger, 1843: 31.—**NS** (od.): *Litoria freycineti* Tschudi, 1838.—**SS**: Tyler, 1971: 353, with *Litoria*.—**Comment:** Fitzinger designated the NS as *Litoria freycineti* Duméril & Bibron, 1841, but this is interpreted as the species and author as given here.

Polyphone Gistel, 1848: xi. Substitute name for *Ranoidea* Tschudi, 1838.

"Chiroleptes" Günther, 1859a: 34.—**NS** (mo.): *Alytes australia* Gray, 1842. Junior homonym of *Chiroleptes* Kirby, 1837 (Insecta).—**SS**: Boulenger (1882a: 267) with *Litoria*.

Pelodryas Günther, 1859a: 119.—**NS** (mo.): *Rana caerulea* White, 1790.—**SS**: Boulenger (1882a: 338) with *Hyla*.

Cyclorana Steindachner, 1867b: 29.—**NS** (mo.): *Cyclorana novaehollandiae* Steindachner, 1867.—**SS**: Frost *et al.* (2006: 205) as a subgenus of *Litoria*.

Phractops Peters, 1867: 30.—**NS** (mo.): *Phractops alutaceus* Peters, 1867.—**SS**: Günther, 1868: 145, with *Cyclorana*.

Chirodryas Keferstein, 1867: 358.—**NS** (mo.): *Chirodryas raniformis* Keferstein, 1867.—**SS**: Keferstein (1868a: 279), with *Hyla*.

Hylomantis Peters, 1880: 224.—**NS** (mo.): *Hylomantis fallax* Peters, 1880.—**SS**: Cogger & Lindner (1974: 67), with *Litoria*.—**Comment:** Boulenger (1882a) regarded *Hylomantis* a synonym of *Hyperolius*.

Dryomantis Peters, 1882a: 8. Replacement name for *Hylomantis* Peters, 1880.—**Comment**: Wells & Wellington (1985: 5) resurrected the nomen with no explanation, but it has been generally ignored.

Mitrolysis Cope, 1889b: 312.—**NS** (mo.): *Chiroleptes alboguttatus* Günther, 1858.—**SS**: Tyler (1974: 27), with *Litoria*; Parker (1940: 20), with *Cyclorana*.

Fanchonia Werner, 1893: 81.—**NS** (mo.): *Fanchonia elegans* Werner, 1893.—**SS**: Werner (1897: 266), with *Hyla*.

Nyctimystes Stejneger, 1916: 85.—**NS** (mo.): *Nyctimantis papua* Boulenger, 1897.—**SS**: Frost *et al.* (2006: 205), with *Litoria*.

Coggerdonia Wells & Wellington, 1985: 4.—**NS** (od.): *Hyla adelaidensis* Gray, 1841.

Colleeneremia Wells & Wellington, 1985: 4.—**NS** (od.): *Hyla rubella* Gray, 1842.

Brendanura Wells & Wellington, 1985: 4.—**NS** (od.): *Chiroleptes alboguttatus* Günther, 1867.

Neophractops Wells & Wellington, 1985: 5.—**NS** (od.): *Chiroleptes platycephalus* Günther, 1873.

Llewellynura Wells & Wellington, 1985: 5.—**NS** (od.): *Hyla dorsalis microbelows* Cogger, 1966.

Mahonabatrachus Wells & Wellington, 1985: 5.—**NS** (od.): *Hyla meiriana* Tyler, 1969.

Mosleyia Wells & Wellington, 1985: 5.—**NS** (od.): *Hyla nannotis* Andersson, 1916.

Pengilleyia Wells & Wellington, 1985: 5.—**NS** (od.): *Litoria tyleri* Martin, Watson, Gartside, Littlejohn & Loftus-Hills, 1979.

Rawlinsonia Wells & Wellington, 1985: 5.—**NS** (od.): *Hyla ewingi* Duméril & Bibron, 1841.

Saganura Wells & Wellington, 1985: 6.—**NS** (od.): *Hyla burrowsi* Scott, 1942.

Sandyrana Wells & Wellington, 1985: 6.—**NS** (od.): *Hyla infrafrenata* Günther, 1867.—**Comment:** Frost (2009-13) notes that all the generic nomina proposed by Wells & Wellington, 1985, are junior synonyms of *Litoria*, "by acclamation."

Distribution: As for the subfamily. *One species possibly introduced into North America.

Common name: Australian Treefrogs.

Content: 188 species, of which one is introduced and possibly established in North America.

Commentary: As this genus and its family, and the introduced species are non-native, we adopt the synonymies of Frost (1909-13) with essentially no critical evaluation.

*Litoria caerulea (White, 1790)

Rana Caerulea White, 1790: 248 (*fide* Frost, 2009-13; not seen).—**O**: formerly Leverian Museum, now probably lost.—**OT**: New South Wales, Australia.—**Comment:** Sherborn (1891) argued that other authors actually provided manuscripts to White, which he used in describing and naming the animals and plants taken in his voyage, and thus probably Shaw *in* White, 1790 would be the appropriate author for this name. However, that is a misinterpretation of the *Code.* As noted above in the account for *Acris g. gryllus,* under the synonym *"Hyla ocularis,"* the author of a nomen is not the person who coined it, even if written in an unpublished manuscript, and even if that person later claimed for authorship of that nomen, which Sherborn notes is true for some of White's names, rather the person who made it available through publication. Had White attributed a name to another author in his book, that author could properly be credited for the name. But White did not specifically indicate authorship for this name (nor others), thus the correct author is White, as has been properly indicated by authors citing this nomen.

Rana coerulea: Daudin, 1802: 70.—**IS**.
Hyla coerulea: Oken, 1815-1816: 223.—**IS**.
Calamites coerulea: Wagler, 1830: 145.—**IS**.
Pelodryas caeruleus: Günther, 1859a: 119.
Hyla caerulea: Boulenger, 1882a: 383; Gadow, 1901: 198; Gorham, 1974: 95.
Hyla caerulea caerulea: Copland, 1957: 30.
Litoria caerulea: Tyler, 1971: 352; Frost, 1985: 181.
Pelodryas caerulea: Savage, 1986: 42.

Rana austrasiae Schneider, 1799: 150. Substitute name for *Rana caerulea* White, 1790.

Hyla cyanea Daudin, 1803: 43.—**O**: none designated.—**OT**: New Holland (Australia).—**SS**: Daudin (1803: 43).

Calamita cyanea: Fitzinger, 1826: 64; Fitzinger, 1861: 413; Cope, 1867b: 201.
Hyla (Calamites) cyaneus: Fitzinger, 1843: 30.

Hyla irrorata De Vis, 1884: 128.—**O: S**: QM J12870-12880.—**OT**: Gympie, Queensland, Australia.—**SS**: Covacevich (1974: 53).—**Comment**: Wells & Wellington (1985: 6) resurrected the nomen with no explanation.

Litoria irrorata: Tyler, 1971: 353.
Pelodryas irrorata Wells & Wellington, 1985: 6.

Distribution: Eastern and northern Australia and nearby islands, New Guinea, introduced to New Zealand. Introduced to Florida, with several small populations established along the Gulf coast (*fide* Lever, 2003, and Kraus, 2009). But Krysko *et al.* (2011) presented evidence that there are no established breeding populations in Florida. Thus, this may not yet be a successful invasive species. **U.S.**: *?FL.

Common name: Great Green Treefrog (has also been called White's Treefrog and Blue Treefrog; not listed by SECSN or C&T).

Family **Leptodactylidae** Werner, 1838 (1896)

"Cystignathi" Tschudi, 1838: 26, 78.—**NG**: *Cystignathus* Wagler, 1830, an obligate synonym of *Leptodactylus,* thus rendering the family name unavailable.

Cystignathidae: Günther, 1858a: 346; 1859a: 26; Bolkay, 1919: 350, 356.

Leiuperina Bonaparte, 1850: [2].—**NG**: *Leiuperus* Duméril & Bibron, 1841

Plectromantidae Mivart, 1869: 291.—**NG**: *Plectromantis* Peters, 1862

Adenomeridae Hoffman, 1878: 613.—**NG**: *Adenomera* Steindachner, 1867.

Leptodactylidae Werner, 1896b: 357.—**NG**: *Leptodactylus* Fitzinger, 1826.—**NN** for Cystignathi. Now a conserved name. For purposes of priority, it takes the date 1838.

Leptodactylidae: Frost et al., 2006: 207; Blackburn & Wake, 2011: 42.

Pseudopaludicolinae Gallardo, 1965: 84.—NG: *Pseudopaludicola* Miranda-Ribeiro, 1926.

Distribution: The neotropics, from southernmost North America, though Middle and South America.

Common name: Neotropical Thin-toed Frogs.

Content: Four genera with 100 species. One genus and species is in North America.

Commentary: Formerly the largest family of the order, but with species of the now separate family Eleutherodactylidae removed, as well as several other extralimital taxa now treated as separate families, this now constitutes a rather small family. Frost *et al.* (2006) placed this family in an unranked taxon, Diphybatrachia, with the family Centrolenidae (extralimital) as a sister family; the family corresponds closely to the former Leptodactylinae, excluding a few forms.

North America contains only about 1% of the species diversity of this family.

Genus **Leptodactylus** Fitzinger, 1826

Synonymy: See under subgenus *Leptodactylus,* below.

Distribution: Southern Texas through Middle and South America, and many Caribbean islands.

Common name: Neotropical Grass Frogs (SECSN) or Tropical Frogs (C&T).

Content: There are 89 species, with one in North America.

Commentary: Frost et al. (2006) proposed placing some species of the genus in a subgenus *Lithodytes* Fitzinger, 1843; the lone North American species is in the nominotypical subgenus.

Subgenus **Leptodactylus** Fitzinger, 1826

Leptodactylus Fitzinger, 1826: 38.—**NS**: *Rana typhonia* Latreille, 1801 (= *Rana fusca* Schneider, 1799), sd.

"Cystignathus" Wagler, 1830: 202.—**NS**: *Rana mystacea* Spix, 1824 (Lynch, 1971: 187, sd.).—**SS**: Tschudi, 1838: 78 [not treated here as a valid genus].

Gnathophysa Fitzinger, 1843: 31.—**NS**: *Rana labyrinthica* Spix, 1824 (od.).—**SS**: Jiménez de la Espada, 1875b: 36.

Plectromantis Peters, 1862a: 232.—**NS**: *Plectromantis wagneri* Peters, 1862 (mo.).—**SS**: Nieden, 1923: 479.

Adenomera Steindachner, 1867b: 37.—**NS**: *Adenomera marmorata* Steindachner, 1867 (mo.).—**SS**: Frost *et al.*, 2006: 207.

Entomoglossus Peters, 1870a: 647.—**NS**: *Entomoglossus pustulatus* Peters, 1870 (mo.).—**SS**: Boulenger, 1882a: 237.

"Pachypus" Lutz, 1930: 22.—**NS**: none designated. Preoccupied by *Pachypus* D'Alton, 1840 (Mammalia) and *Pachypus* Cambridge, 1873 (Arachnida).

"*Cavicola*" Lutz, 1930: 22.—**NS**: none designated. Preoccupied by *Cavicola* Ancey, 1887 (Mollusca).

Parvulus Lutz, 1930: 22.—**NS**: *Leptodactylus nanus* Müller, 1922 (Parker, 1932: 342, sd.). Proposed as a subgenus of *Leptodactylus*.

Vanzolinius Heyer, 1974: 88.—**NS** (od.): *Leptodactylus discondactylus* Boulenger, 1883.—**SS**: Frost *et al.*, 2006: 207.

Distribution: As for the genus.

Common name: No common name in use for the subgenus; names for the genus may be used here as well.

Content: 81 species, including one in North America (1.2%).

Commentary: Frost *et al.* (2006) include eight species in the subgenus *Lithodytes*.

Leptodactylus (Leptodactylus) fragilis (Brocchi, 1877)

Cystignathus fragilis Brocchi, 1877b: 182.—**O**: **H** (sd., Kellog, 1932: 85, and Guibé, 1950:30) MNHN 6316.—**OT**: Tehuantepec, Mexico.

Leptodactylus fragilis: Brocchi, 1881: 19; Heyer, 1978: 31, recognized this as the correct name for the species; Frost, 1985: 243; Garrett & Barker, 1987: 30; Elliott *et al.*, 2009: 298; Dodd, 2013: 436.

Leptodactylus cf *labialis-fragilis*: Flores-Villela, 1993: 260.

Leptodactylus (Leptodactylus) fragilis: Frost et al., 2006: 207-208, by implication.

Misapplication of names:

Cystignathus labialis: Cope, 1879b: 265. (*nec* Cope, 1878).

Leptodactylus labialis: Ives, 1892: 461; Wright & Wright, 1938; Stejneger & Barbour, 1943: 52; Schmidt, 1953: 59; Conant, 1958: 302; Cochran & Goin, 1970: 120; Heyer, 1971: 104; Gorham, 1974: 63; Behler & King, 1979: 420; Dubois & Heyer, 1992: 584; Liner, 1994: 25; Conant & Collins, 1998: 508; Gorzula & Señares, 1999: 58; Dixon, 2000: 63.

Leptodactylus mystaceus labialis: Shreve, 1957: 246.

Leptodactylus labialus: Wilczynski *et al.*, 2002: 149.—**UE**.

Leptodactylus albilabris: Boulenger, 1881: 30; Taylor, 1932a: 243; Stejneger & Barbour, 1933 (*nec* Günther, 1859).

Leptodactylus albilebris: Pratt, 1935: 164.—**IS**.

Leptodactylus melanotus: Dunn & Emlen, 1932: 22 (part).

Distribution: Southern Texas, southward through Caribbean coastal and southern Pacific coastal Mexico and southern Mexico through Central America, northern Colombia and reaching Venezuela. See map in Heyer *et al.* (2006), or Dodd (2013) for North America. May have been extirpated from its North American range in Texas; the state lists the species as endangered. **U.S.**: TX.

Common name: Mexican White-lipped Frog (SECSN) or White-lipped Frog (C&T).

Commentary: Heyer (1971) provided a CAAR account as *L. labialis,* then (Heyer *et al.*, 2006) a revised CAAR account as *L. fragilis.* Most literature on this species uses the name *Leptodactylus labialis,* but Heyer (1978) found that name to be a junior synonym of *L. mystacinus,* and the correct name for this species to be *L. fragilis.* He reversed himself (Dubois & Heyer, 1992), but then later re-established the name again (Heyer, 2002).

Epifamily **Ranoidia** Rafinesque, 1814

Synonymy: See family Ranidae, below.

Distribution: Cosmopolitan except for most of Australia.

Content: 16 families, with 168 genera and 2240 species. In North America there are two families, represented by three genera with 33 species.

Common name: True frogs and narrowmouth toads.

Commentary: This is a major group in the Old World. There are only two families which have managed to disperse to the New World, with relatively little diversification occurring in North America. This continent contains only 1.5% of the species diversity of the superfamily.

Family **Microhylidae** Günther, 1858

Hylaedactyli Fitzinger, 1843: 33.—**NG**: *Hylaedactylus* Duméril & Bibron, 1841.

Hylaedactylidae: Bonaparte, 1850: 1; Günther, 1858a: 346.

Hylaedactylina: Bonaparte, 1850: 1.

Hylaedactylida: Knauer, 1878: 112.

Gastrophrynae Fitzinger, 1843: 33.—**NG**: *Gastrophryne* Fitzinger, 1843.—**SS**: Parker, 1934: 71.

Gastrophrynidae: Metcalf, 1923: 25.

Gastrophryninae: Metcalf, 1923: 274

Gastrophrynini: Dubois, 2005b: 15.

Micrhylidae Günther, 1858a: 346.—**NG**: *Microhyla* Duméril & Bibron, 1841.—**IS.**

Michrylidae: Fatio, 1872: 230.—**IS.**

Microhylinae: Noble, 1931: 451.

Microhylidae: Parker, 1934: i; Blackburn & Wake, 2011: 43.

Microhyloidea: Laurent, 1967: 208; Duellman, 1975: 5; Dubois, 1992: 309.

Microhylini: Dubois, 2005b: 15.

Engystomidae Bonaparte, 1850: [2].

Engystomina: Bonaparte, 1850: [2].

Engystomatidae Günther, 1858a: 346; 1859a: 51; Boulenger, 1882a: 146; Bolkay, 1919: 349; 356; Bourret, 1927: 262.

Engystomidae: Cope, 1865a: 100; 1887a: 18.

Engystomatidarum: Boulenger, 1887b: 50.—**NG**: *Engystoma* Fitzinger, 1826.—**SS**: Parker, 1934: 71.

Kalophryninae Noble, 1931: 451.

Kaloulinae: Noble, 1931: 71.

Kaloulidae: Parker, 1934: 16.

Cacopinae Noble, 1931: 532.—**NG**: *Cacopus* Günther, 1858.—**SS**: Parker, 1934: 71.

Otophryninae: Wassersug & Pyburn, 1987: 166.—**NG**: *Otophryne* Boulenger, 1900.—**SS**: Wild, 1995: 845.

Distribution: North and South America, subsaharan Africa, India, and Korea south into northern Australia.

Common name: Microhylid Frogs and Toads

Content: 487 species (in 55 genera) in about 11 subfamilies (there is some controversy in identifying subfamilies), with 3 species in one subfamily in North America.

Commentary: The analysis by Scott (2005) places microhylids, along with extralimital hemisotids, as basal in the large, complex Ranoid clade. North America contains only 0.6% of the species diversity of this family.

Subfamily **Gastrophryninae** Fitzinger, 1843

Gastrophrynae Fitzinger, 1843: 33.—**NG**: *Gastrophryne* Fitzinger, 1843.

Gastrophrynidae: Metcalf, 1923: 25.

Gastrophryninae: Metcalf, 1923: 274; Blackburn & Wake, 2011: 43.

Gastrophrynini: Dubois, 2005b: 15.

Engystomatidarum Boulenger, 1887b: 50.—**NG**: *Engystoma* Fitzinger, 1826.—**SS**: Parker, 1934: 71.

Distribution: North America to South America.

Common name: Narrowmouth Toads.

Content: 52 species in nine genera; 3 species in 2 genera are in North America.

Commentary: Parker (1934) recognized the family Microhylidae, and included the North American and Asian species in the genus *Microhyla*. Carvalho (1954) removed the American species to the genus *Gastrophryne,* but did not recognize subfamilial units. Metcalf (1923) recognized the family under the name Gastrophrynidae, and included the subfamily Gastrophryninae much as currently constituted. Dubois (2005b) recognized a broader Microhylinae and included a tribe Gastrophrynini, much the same as the current subfamily. Frost *et al.* (2006) treated most American genera as members of a subfamily Gastrophryninae, and we adopt that treatment here. They also suggested that this subfamily was most closely related to the Madagascar subfamily Cophylinae. Van der Meijden *et al.* (2007) found this subfamily to be a sister to the African arboreal subfamily Phrynomerinae, and their reanalysis of the microhylid data used by Frost *et al.* indicates the gastrophrynines form a sister group to almost all the other microhylids sampled, except for one microhyline and one phrynomerine.

North America contains 5.8% of the species diversity of the subfamily.

Genus **Gastrophryne** Fitzinger, 1843

"Microps" Wagler, 1828: 744.—**NS** (mo.): *Microps unicolor* (SS: *Rana ovalis* Schneider), *nec* Megerle, 1823 (Coleoptera). Junior homonym, so unavailable.

"Stenocephalus" Tschudi, 1838: 86.—**NS**: same as above, *nec* Latreille, 1825 (Hemiptera). Junior homonym, so unavailable.

"Engystoma" Duméril & Bibron, 1841: 738.—**NS**: same as above, *nec* Fitzinger, 1826 (od.). Junior homonym, unavailable.—**Comment**: *Engystoma* Fitzinger, 1826, was based on *Rana gibbosa* Linnaeus, which is now *Breviceps* (Ranidae) so cannot be used for a genus in this family.

Gastrophryne Fitzinger, 1843: 33.—**NS** (od.): *Engystoma rugosum* Duméril & Bibron, 1841 (now *Gastrophryne carolinensis* Holbrook).

Distribution: Southern U.S. south to Costa Rica. See map in Nelson (1973c).

Common name: North American Narrow-mouthed Toads (SECSN) or Narrowmouth Toads (C&T). North and Middle American Narrowmouth Frogs would be more accurate, but quite cumbersome.

Content: Five species, two of which (40%) are in North America.

Commentary: American species of this genus were generally treated in the genus *Microhyla* from the early

1930's, based on Noble (1931) and Parker (1934), until 1954 (Carvalho). The two North American species probably were formed from an ancestral population split by advancing ice during the Pleistocene isolating one population in peninsular Florida and the other in southern Texas and Mexico; when glaciers retreated the disjunct populations dispersed northward and made secondary contact. There was controversy as to whether the two should be treated as subspecies or full species, but Blair (1955b) provided evidence they were reproductively isolated and thus distinct species. Nelson (1972c) supported the view that the two North American members of this genus are separate species and each other's nearest relative. A reexamination of Blair's work was done by Loftus-Hills & Littlejohn (1992) who also supported separate species. Nelson (1973c) provided a CAAR account of the genus. The genus seems most closely related to *Hypopachus,* the only other genus in North America; van der Meijden *et al.* (2007), showed the two are sister taxa in their molecular phylogenetic analysis, which Carvalho (1954), Nelson (1972b, c, 1973a, b), Bogart & Nelson (1976) and others had maintained based on studies of adult and larval morphology, serum proteins, karyology, and advertisement calls.

Gastrophryne carolinensis (Holbrook, 1935)

Engystoma carolinense Holbrook, 1835: pl. 10 (pages not numbered).—**O**: **S:** ANSP 14455-14457 (Malnate, 1971: 349 and Nelson, 1972a: 1, sd.). Also Holbrook, 1936: 83: pl. XI mislabelled X, sd.—**OT**: (see Adler (1976: xxxvi) never found north of Charleston, South Carolina, extending west to the lower Mississippi. Data with syntypes is "South Carolina;" restricted to South Carolina, Charleston (Stejneger & Barbour, 1933: 43).

Stenocephals carolinensis: Tschudi, 1838: 86.

Engystoma carolinense: Boulenger, 1891: 453; Cope, 1889b: 385; Strecker, 1909: 116.

Gastrophryne carolinensis: Stejneger, 1910: 166; Stejneger & Barbour, 1917: 40; 1943: 61; Strecker & Williams, 1928: 13; Cochran & Goin, 1970: 88; Gorham, 1974: 119; Conant, 1975: 334; Behler & King, 1979: 383; Frost, 1985: 380; Garrett & Barker, 1987:

34; Conant & Collins, 1998: 551; Dixon, 2000: 78; Elliott *et al.*, 2009: 276; Dodd, 2013: 439.

Gastrophryne carolinense: Strecker, 1915: 46.

Engystoma carolinensis: Nieden, 1926: 64.

Microhyla carolinensis: Parker, 1934: 126; Viosca, 1949: 10; Schmidt, 1953: 76.

Gastrophryne carolinensis carolinensis: Carvalho, 1954: 13.

Engystoma rugosum Duméril & Bibron, 1841: 744.—**O**: **H:** MNHN 5032 (Guibé, 1950: 62, sd.).—**OT**: meridional parts of North America. Restricted (Guibé, 1950: 62) to Louisiana, New Orleans, based on collection data. Schmidt (1953: 77) restricted the OT to South Carolina, Charleston, but with no justification so that action was invalid.—**SS**: Boulenger, 1891: 453.

Gastrophryne rugosum: Fitzinger, 1843: 33.

Distribution: Eastern Texas and eastern Oklahoma, east to the Atlantic coast, including all of Florida and north as far as most of eastern Virginia and the Delmarva peninsula. See map in Nelson (1972a) or Dodd (2013). **U.S.**: AL, AR, DE, FL, GA, IL, KY, LA, MD, MO, MS, NC, OK, SC, TN, TX, VA.

Common name: Eastern Narrow-mouthed Toad (SECSN) or Eastern Narrowmouth Toad (C&T).

Commentary: Nelson (1972a) provided a CAAR account. As noted in the commentary for the genus, this seems now definitively established as the sister species of *G. olivacea* (below), and the two are separate species, rather than subspecies of one species. Makowsky *et al.* (2009) studied molecular genetics of this species and found a surprising lack of variation throughout its geographic range.

Gastrophryne olivacea (Hallowell, 1857)

Engystoma olivaceum Hallowell, 1857d: 252.—**O**: **H:** probably ANSP 2745 (Nelson, 1972b, sd.).—**OT**: not stated, but title of the description indicated either Kansas or Nebraska. Restricted to Kansas, Geary County, Fort Riley (Smith & Taylor, 1950a: 358). and to Kansas, Douglas County, vicinity of Lawrence, by Schmidt (1953: 77) but neither gave any evidence the H came from there so those actions are invalid.

Microhyla olivacea: Parker, 1934: 201.

Gastrophryne olivacea: Smith, 1933b: 217; Stejneger & Barbour, 1943: 61; Nelson, 1972b: 1; Conant, 1975: 335; Behler & King, 1979: 384; Frost, 1985: 380;

Garrett & Barker, 1987: 35; Sullivan *et al.*, 1996a: 43; Conant & Collins, 1998: 553; Dixon, 2000: 78; Stebbins, 2003: 243; Brennan & Holycross, 2006: 38; Elliott *et al.*, 2009: 280; Dodd, 2013: 448.

Microhyla carolinensis olivacea: Hecht & Matalas, 1946: 5; Schmidt, 1953: 77.

Gastrophryne carolinensis olivacea: Carvalho, 1954: 13.

Gastrophryne olivacea olivacea: Conant, 1958: 295; Chrapliwy *et al.*, 1961: 89; Cochran & Goin, 1970: 88; Gorham, 1974: 119; Stebbins, 1985: 95; Liner, 1994: 22; Lemos-Espinal & Smith, 2007a: 53, 250.

Engystoma texense Girard, 1860: 169.—**O**: **S:** USNM 2644 (2) (Cochran, 1961a: 48, sd.).—**OT**: "procured in Texas." Listed as U.S., Texas, Medina County, Rio Seco (Cochran, 1961a: 48).—**SS**: Boulenger, 1882a: 162, with *Engystoma carolinenses (sensu lato)*; Smith, 1933b: 217, with *G. olivacea.*

Gastrophryne texana: Stejneger, 1910: 166.

Gastrophryne texense: Strecker, 1909, 117; 1915: 47.

Gastrophryne texensis: Stejneger & Barbour, 1917: 40.

Engystoma texensis: Nieden, 1926: 65.

Engystoma areolata Strecker, 1909: 118.—**O**: **H:** USNM 38999 (Cochran, 1961a: 48, sd.).—**OT**: U.S., Texas, Victoria County, Guadalupe River bottom.—**SS**: Burt, 1938: 23.—**Comment**: Nelson (1972b) indicated that the H may be a hybrid, *G. olivacea* X *G. carolinensis.*

Gastrophryne areolata: Stejneger, 1910: 166; Strecker, 1915: 47; Stejneger & Barbour, 1917: 40; 1943: 61.

Engystoma areolata: Nieden, 1926: 65.

Microhyla areolata: Parker, 1934: 201.

Microhyla mazatlanensis Taylor, 1943a: 355.—**O**: **H**: FMNH 100040 (*ex* EHT-HMS 1236).—**OT**: Mexico, Sinaloa, 2 miles east of Mazatlán.—**SS**: Hecht & Matalas, 1946: 5.

Microhyla carolinensis mazatlanensis: Hecht & Matalas, 1946: 5.

Gastrophryne carolinensis mazatlanensis: Carvalho, 1954: 13.

Microhyla olivacea mazatlanensis: Langebartel & Smith, 1954: 126; Gorham, 1974: 119.

Gastrophryne mazatlanensis: Wake, 1961: 88.

Gastrophryne olivacea mazatlanensis: Chrapliwy *et al.*, 1961: 81; Cochran & Goin, 1970: 89; Stebbins, 1985: 95.

Misapplication of names:

Engystoma carolinense: Boulenger, 1882a: 162 (part).

Gastrophryne carolinensis: Lowe, 1964: 167 (part).

Distribution: Most of Texas, Oklahoma, and Kansas, southern Nebraska, eastern Missouri, eastern Arkansas, into northern Mexico and then northward to southern Arizona and New Mexico, and a disjunct population in Colorado. See map in Nelson (1972b) or Dodd (2013). **U.S.**: AR, AZ, CO, KS, MO, NE, NM, OK, TX.

Common name: Western Narrow-mouthed Toad (SECSN) or Western Narrowmouth Toad (C&T).

Content: In the past this was sometimes treated as a subspecies of *G. carolinensis*. Taylor (1943a) described *G. mazatlanensis* (as *Microhyla*), and Carvalho (1954) treated it as a subspecies of this species; Wake (1961) treated it as a distinct species, but Nelson (1972b) rejected it as not definable as a subspecies. Currently no subspecies are recognized. However, there is enough morphological and call variation to suggest there may be another cryptic species to be recognized among the populations now treated under this name.

Commentary: Nelson (1972b) provided a CAAR account.

Genus **Hypopachus** Keferstein, 1867

Hypopachus Keferstein, 1867: 351.—**NS**: *Hypopachus Seebachii* Keferstein, 1867 (mo.) (=*Engystoma variolosum* Cope, 1866b).

Distribution: Disjunct distribution from southern Texas to Costa Rica.

Common name: Sheep Frog.

Content: Two species, of which one occurs in North America.

Commentary: See commentary on *Gastrophryne,* for discussion of the two genera being sister taxa. However, using mostly morphological characters, Wild (1995) found *Hypopachus* the sister taxon to a clade containing most of the other New World microhylid genera, with *Gastrophryne* deeply nested in that clade.

Hypopachus variolosus (Cope, 1866)

Engystoma variolosum Cope, 1866b: 131.—**O**: **H**: USNM 6486.—**OT**: Arriba, Costa Rica.—**Comment:** Savage (1974: 77) commented that the collector who sent the specimen to Cope apparently loosely applied the term "arriba" to the Mesita Central of Costa Rica.

Systoma variolosum: Cope, 1867b:194.

Hypopachus variolosus: Cope, 1875b: 101; Cochran & Goin, 1970: 88; Gorham, 1974: 121; Conant, 1975: 336; Behler & King, 1979: 385; Frost, 1985: 382; Garrett & Barker, 1987: 36; Liner, 1994: 25; Conant & Collins, 1998: 554; Dixon, 2000: 79; Elliott *et al.*, 2009: 282; Dodd, 2013: 455.

Hypopachus seebachi Keferstein, 1867: 352.—**O**: **S**: ZFMK 28389-28392 (Böhme & Bischoff, 1984:182, sd.) and ZMB 5926 (Bauer, Günther & Robeck, 1996: 264).—**OT**: Costa Rica.—**SS**: Cope, 1871a: 167; Boulenger, 1882a: 159.

Hypopachus inguinalis Cope, 1871a: 166.—**O**: **H**: USNM; possibly USNM 6792 (3) (Stuart, 1963).—**OT**: Guatemala, Vera Paz, near ruins of Coban, but labelled "Guatemala" by the collector; emended (Nelson, 1973a: 12) to Guatemala, Baja Verapaz, near San Gerónimo (the home of the collector).—**SS**: Cope, 1887a: 18.

Engystoma inguinalis: Brocchi, 1877b: 189.

Hypopachus variolosus inguinalis: Cope, 1887a: 18

Hypopachus inguinalis: Stuart, 1963; Nelson, 1973a: 11-12.

Hypopachus oxyrrhinus Boulenger, 1883c: 344.—**O**: **S** BM 1882.12.5.8 and 1883.4.5.3 (Kellogg, 1932: 184, sd.).—**OT**: Mexico, Sinaloa, Presidio; Emended to Presidio de Mazatlan, Sinaloa (Parker, 1934: 114).—**SS**: Günther, 1900: 211; Nelson, 1974: 261.

Hypopachus oxyrhynchus: Boulenger, 1890b: 21.—**IS**.

Hypopachus oxyrhinus: Parker, 1934:—**IS**.

Hypopachus oxyrrhinus oxyrrhinus: Shannon & Humphrey, 1958: 94.

Hypopachus cuneus Cope, 1889b: 395.—**O**: **S**: includes USNM 15676 (Cochran, 1961a: 62, sd.).—**OT**: U.S., Texas, Duval County, Nueces, neighborhood of San Diego. Schmidt (1953: 77) gave it as Texas, Duval County, San Diego.—**SS**: Boulenger, 1890b: 21, with *H. oxyrrhinus* (as *H. oxyrhynchus*); Günther, 1900: 211; Nelson, 1974: 261.

Hypopachus cuneus: Strecker, 1915: 47; Stejneger & Barbour, 1917: 40; 1943: 62.

Hypopachus cuneus cuneus: Taylor, 1940b: 516; Schmidt, 1953: 78; Conant, 1958: 296; Gorham, 1974: 120.

Hypopachus globulosus Schmidt, 1939: 2.—**O**: **H**: FMNH 4641.—**OT**: Honduras, Lake Ticamaya, east of San Pedro Sula.—**SS**: Nelson, 1974: 261.

Hypopachus caprimimus Taylor, 1940b: 526.—**O**: **H**: FMNH 100086 (*ex.* EHT-HMS 18149) (Marx, 1976: 56).—**OT**: Mexico, Guerrero, Agua del Obispo, in pine forest.—**SS**: Nelson, 1974: 262.

Hypopachus cuneus nigroreticulatus Taylor, 1940b: 518.—**O**: **H:** FMNH 100064 (*ex.* EHT-HMS) (Marx, 1976: 56).—**OT**: Mexico, Campeche, Encarnación.—**SS**: Nelson, 1974: 262.

Hypopachus ovis Taylor, 1940b: 520.—**O**: **H**: FMNH 100078 (*ex* EHT-HMS 1050) (Marx, 1976: 56).—**OT**: Mexico, Nayarit, Tepic.—**SS**: Shannon & Humphrey, 1958, with *H oxyrrhinus*; Nelson, 1974: 262.

Hypopachus oxyrrhinus ovis: Shannon & Humphrey, 1958: 94.

Hypopachus alboventer Taylor, 1940b: 522.—**O**: **H**: FMNH 100085 (*ex* EHT-HMS 18615) (Marx, 1976: 56.—**OT**: Mexico, Morelos, 8 miles east of Cuernavaca.—**SS**: Nelson, 1974: 262.

Hypopachus alboventer alboventer: Davis, 1955. 71; Shannon & Humphrey, 1958: 94.

Hypopachus maculatus Taylor, 1940b: 524.—**O**: **H**: FMNH 100087 (*ex* EHT-HMS 1023) (Marx, 1976: 56).—**OT**: Mexico, Chiapas, near San Ricardo.—**SS**: Nelson, 1974: 262.

Hypopachus championi Stuart, 1940: 19.—**O**: **H**: UMMZ 85533.—**OT**: Guatemala, Baja Verapaz, San Gerónomo, about 1 km south, temporary pool on desert flats.—**SS**: Nelson, 1973a: 12, with *H. inguinalis*: Nelson, 1974: 262.

Hypopachus alboventer reticulatus Davis, 1955: 71.—**O**: **H:** TCWC 10855.—**OT**: Mexico, Guerrero, 2 miles south of Almolonga, elev. 5600 feet.—**SS**: Nelson, 1974: 262,

Hypopachus alboventer reticulatus: Shannon & Humphrey, 1958: 94.

Hypopachus reticulatus: Lynch, 1965b: 369.

Hypopachus oxyrrhinus taylori Shannon & Humphrey, 1958: 89.—**O**: **H**: FAS 11309 (now UIMNH).—**OT**: Mexico, Nayarit, 3 miles east San Blas.—**SS**: Nelson, 1974: 262.

Distribution: Southern Texas to Costa Rica, mostly below 1500 m elev. See map in Dodd (2013). **U.S.**: TX.

Common name: Sheep Frog.

Content: As evidenced in the synonymy, many populations have been described, some as distinct species, some as subspecies. But Nelson (1974) described the variation and found it to be clinal and not warranting any subspecies designations; none are currently recognized. However, Frost *et al.* (2008) suggested that more than one species still may be hidden under this name.

Family **Ranidae** Rafinesque, 1814

Ranaridia Rafinesque, 1814b: 102. **NG**: *Ranaridia* Rafinesque, 1814, **UE** of *Rana* Linnaeus, 1758.

Ranarinia: Rafinesque, 1815: 78.

Ranae: Goldfuss, 1820: 131; Leunis, 1844: 145; Tschudi, 1838: 25.

Ranadae: Gray, 1825: 213.

Ranina: Gray, 1825: 214; Bonaparte, 1832: 10, 1840b: 394; 1850: [2]; Gravenhorst, 1845: 43; Günther, 1859a: 4.

Ranoidea: Fitzinger, 1826: 37 (explicit family); Wied-Neuwied, 1865: 105; Laurent, 1967: 208; Lynch, 1973: 162; Duellman, 1975: 5.

Ranidae: Boie, 1828: 363; Bonaparte 1832: 10, 1838b: 392, 1840a: 100, 1840b: 394, 1850: [2]; Günther, 1859a: 4; Boulenger, 1882a: 3; 1888b: 204; Bolkay, 1919: 350, 356; Blackburn & Wake, 2011: 44.

Ranaria: Hemprich, 1820: xix.

Raniadae: M.A. Smith, 1831: 18.

Raniformes: Duméril & Bibron, 1841: 491. (explicit family); Cope, 1864a: 181.

Ranini: Bronn, 1849: 684; Dubois, 1987c: 35.

Raninae: Bronn, 1849: 684.

Ranoides: Bruch, 1862: 221.

Ranida: Haeckel, 1866: 132; Knauer, 1878: 104; Bayer, 1885: 18.

Ranidi: Acloque, 1900: 490.

Ranoidae: Dubois, 1992: 309 (as epifamily)

Ranoidia: Dubois, 2005b: 17 (as epifamily)

Limnodytae Fitzinger, 1843: 31.—**NG**: *Limnodytes* Duméril & Bibron, 1841.

Limnodytini: Dubois, 1981: 231.

Amolopsinae Yang, 1991: 172 (*fide* Frost, 2009-13, and others; not seen).—**NG**: *Amolops* Cope, 1865.

Amolopinae: Zhang & Wen, 2000: 69 (*fide* Frost, 2009-13, and others; not seen).

Stauroini Dubois, 2005b: 17.—**NG**: *Staurois* Cope, 1865.

Distribution: Cosmopolitan, except most of Australia, and southern South America.

Common name: True frogs.

Content: Fourteen genera and 347 species, of which one genus (with three subgenera) and 30 species are found in North America, following our treatment.

Commentary: Frost *et al.* (2006) raised about ten subfamilies which were formerly included in a broader family Ranidae to full family status, leaving only a few genera that had been considered members of a subfamily Raninae. We follow that treatment here and recognize a family Ranidae without subdivision into subfamilies or tribes; however, the North American members would be recognized in this family regardless of whether we followed Frost *et al.*, or most earlier taxonomies.

North America contains 8.6% of the species diversity of this family.

*Genus **Glandirana** Fei, Ye & Huang, 1990

Glandirana Fei, Ye & Huang, 1990: 146, *fide* Frost, 2009-13, not seen.—**NS** (od.): *Rana minima* Ting & Ts'ai, 1979.

Rugosa Fei, Ye & Huang, 1990: 145.—**NS** (od.): *Rana rugosa* Temminck & Schlegel, 1838.—**SS**: Frost *et al.* (2006).

Distribution: Native to Asia (China, Korea, Japan, Russia), but one species introduced to Hawaii.

Common name: Wrinkled Frogs (SECSN) or Rugose True Frogs (C&T).

Content: Four species, of which one is introduced to North America in Hawaii.

Commentary: The native range of this taxon is extralimital. Frost *et al.* (2006) treat it as a separate genus. We follow that treatment here, as it is imbedded in another clade two levels above that of the genus *Rana*. We are not particularly concerned with the details of its relationship, as it is a non-native taxon. Our synonymies of the genus and species are adopted from Frost (2009-13) with little evaluation.

***Glandirana rugosa** (Schlegel, 1838)

Rana rugosa Schlegel *in* Temminck & Schlegel, 1838: 110.—**O**: **S:** specimens of pl. 3, fig. 3 & 4; + RMNH 2064 (9) (Frost, 1985: 512, sd.); BM specimens (Boulenger, 1882a:35).—**OT**: Japan. Stejneger (1907: 123) indicated probably Nagasaki.

Rana (Rana) rugosa: Nakamura & Ueno, 1963: 49 (*fide* Frost, 2009-13; not seen); Dubois, 1987a: 42.

Rana rugosa: Cochran & Goin, 1970: 87; Gorham, 1974: 151; Frost, 1985: 512.

Rugosa rugosa: Fei, Ye & Huang, 1990: 145 (*fide* Frost, 2009-13; not seen).

Rana (Rugosa) rugosa: Dubois, 1992: 322.

Glandirana rugosa: Frost *et al.*, 2006: 368; Dodd, 2013: 819.

Distribution: Native to Japan. Introduced to Hawaii and well established on Hawaii, Kauai, Maui, and Oahu. **U.S.**: *HI.

Common name: Japanese Wrinkled Frog (SECSN) or Wrinkled Frog (C&T).

Commentary: Apparently introduced to Hawaii about 1895 (Svihla, 1936; Oliver & Shaw, 1953; McKeown, 1996; Lever, 2003; Kraus, 2009).

Genus **Rana** Linnaeus, 1758

Synonymy: See under the subgenus *Rana,* below.

Distribution: North, Middle, and South America except temperate portions, and much of Eurasia.

Common name: Brown Frogs (SECSN) or Pacific True Frogs (C&T).

Content: At least two subgenera, three as treated here, with a total of 101 species, of which three subgenera and 30 species are found in North America.

Commentary: Until recently, all North American ranid frogs were considered members of the genus *Rana*. Hillis & Wilcox (2005) analyzed relationships among the American species, and elected to retain the species in a genus *Rana,* but named a number of "subgenera" corresponding to clades elucidated by their study. Some of those taxa were indicated to be nested at different ranks within others, so that caused names to be unavailable under the *Code*. The analysis by Frost *et al.* (2006) supported that study, finding North American species fall into two rather distinctive clades, which they treated as separate genera, *Rana* and *Lithobates*. Discussion of extralimital members of the family is beyond the scope of this work. However, as with our treatment of bufonid genera, where a taxonomic choice can be made for treatment of two clades as either separate genera or subgenera of one genus, without upsetting monophyly, we elect the latter, so here we recognize a genus *Rana,* with two North American subgenera, *Rana* and *Lithobates*. Frost *et al.* (2006: 249) showed in cladogram format that their genera *Rana* and *Lithobates* form one monophyletic unit consisting of two sister clades. Frost *et al.* elected to treat the members of those two clades as separate genera; we elect to treat them as subgenera, in the senior genus *Rana*. Pyron & Wiens (2011) retained *Rana* as the generic name for all North American ranids, without adopting any subdivision of the genus.

An additional advantage of retaining the generic name *Rana* for all North American species was pointed out by Hillis (2007). He searched electronically for articles on *Rana catesbeiana* and found almost 2000, but a search for *Lithobates catesbeiana* returned none. With more than five additional years since Frost *et al.* (2006) split the genus, we searched again on the two nomina, but restricted the time frame to the years 2007 to mid-2011. *Rana catesbeiana* returned 672 articles, while *Lithobates catesbeiana* returned 23. At the very least, this reflects either resistance to or delay in awareness of the name change, and again suggests that stability of names is better served by our treatment of the two names as subgenera rather than genera. Use of the subgenus name is optional when listing the species in an article, so those of us interested in specifying the subgenus will do so, while those that want to continue to use the old familiar name, with or without the parenthetical subgenus, may do so as well, perhaps mainly in non-taxonomic works.

Under this treatment, about 30% of the species diversity of the genus is in North America.

Subgenus **Rana** Linnaeus, 1758

Rana Linnaeus, 1758: 210.—**NS** (sd., Fleming, 1822: 304): *Rana temporaria* Linnaeus, 1758.

Ranaria Rafinesque, 1814b: 102.—**NN** for *Rana* Linnaeus, 1758.

Crotaphitis Schultze, 1891: 176.—**NS**: *Rana arvalis* Nilsson, 1907 (Dubois & Ohler, 1995: 183, sd.).—**Comment**: Earlier designation of *R. temporaria* Linnaeus, 1758, by Stejneger, 1907, was in error.—**SS**: Nikolskii, 1918: 7.

Amerana Dubois, 1992: 322.—**NS** (od.): *Rana boylii* Baird, 1856. As subgenus.

Aurorana Dubois, 1992: 322.—**NS** (od.): *Rana aurora* Baird & Girard, 1852. As subgenus.

"Pseudoamolops" Jiang, Fei, Ye, Zeng, Ahen, Xie, & Chen, 1997: 67. As subgenus of *Amolops,* but no NS given, thus name is unavailable (see Dubois, *et al.,* 2005).

Pseudoamolops Fei, Ye, & Jiang, 2000: 25.—**NS**: *Rana sauteri* Boulenger, 1909.—**SS**: Frost *et al.* (2006), with *Rana.*

"Laurasiarana" Hillis & Wilcox, 2005: 305, 311.—**NS**: *Rana aurora* Baird & Girard, 1852.—**OS**: Dubois (2007a: 3), with *Aurorana* Dubois, 1992; also **NU**.

Distribution: Western North America and much of Eurasia.

Common name: Brown Frogs (SECSN) or Pacific True Frogs (C&T).

Content: 48 species, of which 8 are in North America.

Commentary: The cladistic analyses of Hillis & Wilcox (2005) and Frost *et al.* (2006: 249) showed this subgenus is a sister taxon to *Lithobates,* treated here as another subgenus of *Rana,* but by Frost *et al.* as a separate genus.

All North American species of the subgenus *Rana* are members of the *Rana boylii* species group. It seems most closely related to the Eurasian *Rana temporaria* group. This group was assigned to a subgenus *Amerana* (Dubois, 1992), and Hillis & Wilcox (2005) used that nomen for a sub-subgeneric taxon or clade, which violates the *Code.* We have elected to treat the subgenus *Rana* as more inclusive, containing the Eurasian and western North American species, and the present rules of the *Code* do not allow any categories in the genus series other than genus and subgenus. We hope that the Commission will consider allowing expansion of the genus series to an infinite hierarchy as in the family-series, but at present our taxonomic freedom at the genus level is restrained by the rules of the *Code*. Our listing reflects a "best guess" based on the observation by Hillis & Wilcox (2005) and Vredenburg *et al.* (2007) that the arrangement we follow here does not conflict with their studies nor any other recent studies, none of which included samples of every species recognized here.

Under our treatment, North America contains about 16.7% of the species diversity of the subgenus.

Rana (Rana) aurora Baird & Girard, 1852

Rana aurora Baird & Girard, 1852a: 174.—**O**: **S** USNM 11711 (4) (Cochran, 1961a: 72, sd.).—**OT**: U.S., Washington, Puget Sound.

Rana temporaria aurora: Cope, 1875b: 32.

Rana agilis aurora: Cope, 1886b: 521; 1889b: 439.

Rana aurora: Stejneger & Barbour, 1917: 36; Boulenger, 1920: 448; Frost, 1985: 482; Liner, 1994: 27; Shaffer, Fellers, Voss, Oliver, & Pauly, 2004a: 2667; Elliott *et al.*, 2009: 238.

Rana aurora aurora: Camp, 1917: 116; Stejneger & Barbour, 1923: 34; 1943: 54; Schmidt, 1953: 84; Cochran & Goin, 1970: 81; Gorham, 1974: 141; Behler & King, 1979: 369; Stebbins, 1985: 83; 2003: 226.

Rana (Rana) aurora: Dubois, 1987a: 61, by implication.

Rana (Aurorana) aurora: Dubois, 1992: 322.

Rana (Laurasiarana) aurora: Hillis & Wilcox, 2005: 311.

Rana (Amerana) aurora: Dubois, 2006c: 830.

Distribution: Vancouver Island and southwestern British Columbia, south along the Pacific coast into northwestern California, and an introduced population in Alaska. See map in Altig & Dumas (1972), only the range of *R. a. aurora,* as treated there. There are also maps in Elliott *et al.* (2009) and Dodd (2013). There apparently has been some reduction in abundance of the species in some parts of its range (Pearl, 2005). **Can.**: BC. **U.S.**: *AK, CA, OR, WA.

Common Name: Northern Red-legged Frog. *French Canadian name:* Grenouille à pattes rouges du Nord.

Content: R. draytoni was formerly considered a subspecies of this species, but has been elevated to a distinct species, so there are currently no subspecies recognized.

Commentary: Altig & Dumas (1972) provided a CAAR account, but included *R. draytoni* as a subspecies. Green (1986a) found *aurora* karyologically closest to *R. draytoni,* but using electrophoresis (1986b) found *aurora* closest to *R. cascadae. Rana draytoni* does not seem to be the closest relative of *R. aurora*; rather, *R. cascadae* is well supported by most recent studies as the sister of this species. Listed in British Columbia as Special Concern.

Rana (Rana) boylii Baird, 1856

Rana boylii Baird, 1856: 62.—**O**: **S:** USNM 3370 (2) (Cochran, 1961a: 72, sd.).—**OT**: od., California (interior); corrected to U.S., El Dorado (Cope, 1889b: 447), S from El Dorado County (Cochran, 1961a: 72); restricted to U.S., California, El Dorado County, vicinity of Coloma along south fork of American River (Jennings, 1988a: 59; based on collection data).

Rana boylii: Baird, 1859d: 12; Boulenger, 1920: 469; Frost, 1985: 484; Stebbins, 1985: 86; 2003: 231; Dodd, 2013: 697.

Rana boylii boylii: Camp, 1917: 117; Stejneger & Barbour, 1917: 36; 1943: 55.

Rana boyli boyli: Schmidt, 1953: 84; Gorham, 1974: 141.—**UE.**

Rana boylei: Zweifel, 1955: 207; Cochran & Goin, 1970: 77; Behler & King, 1979: 371; Liner, 1994: 27.—**UE.**

Rana (Rana) boylii: Dubois, 1987a: 41.

Rana (Amerana) boylii: Dubois, 1992: 322.

Rana (Laurasiarana) boylii: Hillis & Wilcox, 2005: 317.

Rana pachyderma Cope, 1884: 25.—**O: S** ANSP 14569-14570 (Ashland), 14571-14575 (McCloud River), but 14569-14570 lost (Malnate, 1971: 351, sd.).—**OT**: U.S., California, Shasta County, Baird, at McCloud River, and U.S., Oregon, Jackson County, Ashland, at northern base of Siskiyou Mountains.—**SS**: Cope, 1889b: 444.

Distribution: Originally coastal Oregon southward to coastal southern California, inland to central California montane regions, with a disjunct population in Baja California (see map in Zweifel, 1968c, Fellers, 2005a, and Dodd, 2013), but now nearly extinct in the western half of their Oregon range and in most of southern California (Fellers, 2005a). **U.S.**: CA, OR.

Common name: Foothill Yellow-legged Frog.

Commentary: Zweifel (1968c) provided a CAAR account. This is a sister species to the pair, *R. pretiosa-R luteiventris,* and the clade formed by those three together is a sister taxon to the other species of the *boylii*-group (Hillis & Wilcox, 2005).

Rana (Rana) cascadae Slater, 1939

Rana cascadae Slater, 1939: 145.—**O**: **H:** USNM 10868 (ex CPS 2883).—**OT**: U.S., Washington, Rainier National Park, Elysian Fields, elev. 5700 ft.

Rana aurora cascadae: Stejneger & Barbour, 1943: 54; Schmidt, 1953: 85; Behler & King, 1979: 369.

Rana cascadii: Vincent, 1947: 20.—**UE**.

Rana cascadae: Cochran & Goin, 1970: 81; Gorham, 1974: 141; Frost, 1985: 485; Stebbins, 1985: 85; 2003: 230; Elliott *et al.*, 2009: 236; Dodd, 2013: 707.

Rana (Rana) cascadae: Dubois, 1987a: 41, by implication.

Rana (Aurorana) cascadae: Dubois, 1992: 322.

Rana (Laurasiarana) cascadae: Hillis & Wilcox, 2005: 317.

Rana (Amerana) cascadae: Dubois, 2006c: 830.

Distribution: Olympic and Cascade Mountains of Washington, Oregon, and California, in lake and meadow habitats generally above 1000 feet. Most California populations are now extinct (Pearl & Adams, 2005). See maps in Altig & Dumas (1971), Pearl & Adams (2005), Elliott *et al.* (2009), and Dodd (2013). **U.S.**: CA, OR, WA.

Common name: Cascades Frog.

Commentary: Altig & Dumas (1971) provided a CAAR account. This species is sister to *R. aurora*. Pearl & Adams (2005) reviewed conservation status and noted 50% to 99% decline in California, and notable declines in Oregon, where the species remains relatively abundant. It seems surprising that this species is not federally listed.

***Rana (Rana) draytonii** Baird & Girard, 1852

Rana Draytonii Baird & Girard, 1952a: 174.—**O**: **S:** USNM 11497 (6) (Cochran, 1961a: 73, sd.).—**OT**: U.S., California, San Francisco, and Washington/Oregon, Columbia River.—**Comment**: Schmidt (1953: 85) restricted the OT to the vicinity of San Francisco, but provided no evidence that the S all came from there, thus that action is invalid. If the specimen(s) from San Francisco can be identified, one of them could be designated a lectotype (lectophorant), thereby also validaating Schmidt's restriction.

Rana draytoni draytoni: Cope, 1889b: 441.—**UE.**

Rana aurora draytoni: Camp, 1917: 124; Stejneger & Barbour, 1943: 54; Schmidt, 1953: 85; Cochran & Goin, 1970: 81; Gorham, 1974: 141; Behler & King, 1979: 361; Liner, 1994: 27.—**UE.**

Rana draytonii: Stejneger & Barbour, 1917: 37; Boulenger, 1920: 446; Dodd, 2013: 715.

Rana aurora draytonii: Stejneger & Barbour, 1923: 34; Stebbins, 1985: 83; 2003: 226.

Rana (Rana) draytonii: Dubois, 1987a: 41, by implication.

Rana (Aurorana) draytonii: Dubois, 1992: 322; Shaffer *et al.*, 2004a: 2667.

Rana (Laurasiarana) draytonii: Hillis & Wilcox, 2005: 311.

Rana (Amerana) draytonii: Dubois, 2006c: 830.

Rana Lecontii Baird & Girard, 1853a: 301.—**O**: **S** (sd., Cochran, 1961a: 74) USNM 3362 (2).—**OT**: U.S., California, San Francisco County, San Francisco.—**SS**: Cope, 1886b: 521, 1889b: 441.

Rana lecontei: Brocchi, 1881: 14.—**UE.**

"*Rana nigricans*" Hallowell, 1855a: 96.—**O**: **S:** USNM 3366, 3376 (Cochran, 1961a: 76, sd.); *nec Rana nigricans* Agassiz, 1850.—**OT**: U.S., California, Kern County, El Paso Creek.—**SS**: Cope, 1889b: 441. Unavailable junior homonym.

Rana longipes Hallowell, 1859b: 20.—**NN** for *Rana nigricans* Hallowell, 1854.—**SS**: Cope (1883: 28) with *R. temporaria aurora.*

Epirhexis longipes: Yarrow, 1883a: 179.

Distribution: Pacific coast of most of California, into northern Baja California, inland into the San Joaquin valley and Sierra Nevada, to elevations of about 2440 meters (see map in Dodd, 2013). It was apparently introduced into California from Baja California in 1915 (see Kraus, 2009: 192, and his citations); also introduced into Nevada in the 1940s. However, it is now extinct in most of southern California and the Sierra Nevada in California (Fellers, 2005b). **U.S.**: *CA, *NV.

Common Name: California Red-legged Frog.

Commentary: Altig & Dumas (1972) provided a brief CAAR account as a subspecies of *R. aurora*. This was formerly treated as a subspecies of *R. aurora,* but was elevated to a separate species (Shaffer *et al.,* 2004a). It does not seem to be the sister species of *R. aurora,* rather a sister to the clade including four species, ((*aurora, cascadae*) (*muscosa, sierrae*)). This species is federally listed as threatened throughout its range, but probably warrants listing as endangered (see discussion, Fellers, 2005b).

Rana (Rana) luteiventris Thompson, 1914

Rana pretiosa luteiventris Thompson, 1914: 53.—**O**: **H**: UMMZ 43037.—**OT**: U.S., Nevada, Elko County, Anne Creek. Corrected (Kluge, 1983a: 60) to Lower Annie Creek near Carlin, where the creek is crossed by the Southern Pacific Railroad.

Rana pretiosa luteiventris: Camp, 1917: 124; Stejneger & Barbour, 1917: 39; 1943: 59; Schmidt, 1953: 85; Cochran & Goin, 1970: 79; Gorham, 1974: 150.

Rana (Rana) luteiventris: Dubois, 1987a: 41, by implication.

Rana luteiventris: Cuellar, 1996: 145; supported by Green *et al.,* 1997: 1-8; Stebbins, 2003: 229; Dodd, 2013: 723.

Rana (Laurasiarana) luteiventris: Hillis & Wilcox, 2005: 311.

Rana (Amerana) luteiventris: Dubois, 2006c: 830.

Distribution: Washington and Oregon, east of the Cascades, east into Montana and western Wyoming, north to Yukon and southeastern Alaska, and southward to northern Nevada and central Utah. Reaser & Pilliod (2005) have a range map for the U.S. distribution, and Elliot *et al.* (2007) have a map of the Canadian distribution. Dodd (2013) has a map of the entire distribution. **Can.**: AB, BC, YT. **U.S.**: AK, ID, MT, NV, OR, UT, WA, WY.

Common Name: Columbia Spotted Frog. *French Canadian name:* Grenouille maculée de Columbia.

Commentary: Turner & Dumas (1972) had a brief CAAR account, as a subspecies of *R. pretiosa.* Cuellar (1996) suggested that *R. pretiosa* was composed of two crytic species, then Green *et al.* (1997) formally recognized this form as a distinct sister species, separate from *R. pretiosa.* The pair are related to *R. boylii* as a sister taxon (Hillis & Wilcox, 2005). Reaser & Pilliod (2005) reviewed the conservation status and it is apparent that *R. luteiventris* should be federally listed, as it is in significant decline in many parts of its range. Funk *et al.* (2008) found at least three highly divergent clades of *luteiventris* but declined to name them pending further studies. Weller & Green (1997) report the species in decline in Alberta, and extirpated from the southwestern part of the province, but no decline seen elsewhere in Canada.

Rana (Rana) muscosa Camp, 1917

Rana boylii muscosa Camp, 1917: 118.—**O**: **H:** MVZ 771.—**OT**: U.S., California, Los Angeles County, near Pasadena, Arroyo Seco Canyon, elev. about 1300 feet.

Rana boylii muscosa: Stejneger & Barbour, 1917: 36; 1943: 55.

Rana boyli muscosa: Schmidt, 1953: 84.

Rana muscosa: Zweifel, 1955: 229; Cochran & Goin, 1970: 78; Gorham, 1974: 148; Behler & King, 1979: 376; Frost, 1985: 506; Stebbins, 1985: 86; 2003: 233; Elliott *et al.*, 2009: 250.

Rana (Rana) muscosa: Dubois, 1987a: 41, by implication.

Rana (Amerana) muscosa: Dubois, 1992: 322.

Rana (Laurasiarana) muscosa: Hillis & Wilcox, 2005: 311.

Distribution: Formerly a disjunct distribution, in southern California, in the San Bernardino, San Jacinto, and San Gabriel Mountains, and in the Sierra Nevada of California and Nevada. It is now extinct in most of that range, surviving only in some isolated populations in the Sierra Nevada (Vredenburg *et al.,* 2005, 2007). Elliott *et al.* (2009) have a map of the former range of this and *R. sierrae,* and Dodd (2013) has a map showing current and former range. **U.S.**: CA, NV.

Common Name: Mountain Yellow-legged Frog (SECSN) or Sierra Madre Yellow-legged Frog (C&T).

Commentary: Zweifel (1968a) provided a CAAR account, which includes *R. sierrae,* before its separation as a distinct species. Macey *et al.* (2001) suggested from their molecular study that there were substantial differences between the two disjunctive populations, possibly indicating species distinctness, and it was then confirmed that *R. sierrae* is a cryptic sister species of *R. muscosa* (Vredenburg *et al.*, 2005), and the pair has a sister taxon relationship to the pair of species above (*R. aurora, R. cascadae*). This species is federally listed as endangered, and Vredenberg *et al.* (2007) found that over 95% of populations of *R. muscosa* are now extinct.

Rana (Rana) pretiosa Baird & Girard, 1853

Rana pretiosa Baird & Girard, 1853b: 378.—**O**: **S:** USNM 11409 (5) (Cochran, 1961a: 76, sd.).—**L**: Green, Kaiser, Sharbel, Kearsley, & McAllister, 1997: 5, selected one of the S-series, then the others were renumbered 498959-498962.—**OT**: U.S., Washington (stated as Oregon), Puget Sound.

Rana pretiosa: Cooper, 1860: 304; Boulenger, 1919: 414; 1920: 452; Behler & King, 1979: 378; Frost, 1985: 511; Stebbins, 1985: 84; 2003: 228; Dodd, 2013: 739.

Rana temporaria pretiosa: Cope, 1889b: 432.

Rana pretiosa pretiosa: Thompson, 1914: 53; Camp, 1917: 124; Stejneger & Barbour, 1917: 38; 1943: 58; Schmidt, 1953: 85; Cochran & Goin, 1970: 79; Gorham, 1974: 150.

Rana (Rana) pretiosa: Dubois, 1987a: 41, by implication.

Rana (Aurorana) pretiosa: Dubois, 1992: 322.

Rana (Laurasiarana) pretiosa: Hillis & Wilcox, 2005: 311.

Rana (Amerana) pretiosa: Dubois, 2006c: 830.

Distribution: Cascade Mountains of Oregon and adjacent California, northward to northwestern Washington and adjacent British Columbia, but now perhaps extinct in Canada and most of Oregon and Washington (Pearl & Hayes, 2005; also see their map). Dodd (2013) has a map indicating current and former range. **Can.**: BC. **U.S.**: CA, OR, WA.

Common Name: Oregon Spotted Frog. *French Canadian name:* Grenouille maculée de l'Orégon.

Commentary: Turner & Dumas (1972) provided a CAAR account, but it includes *R. luteiventris,* now recognized as a separate species. Cuellar (1996) raised the subspecies *luteiventris* to separate species status; Green *et al.* (1996) indicated there were two cryptic species, then Green *et al.* (1997) split *R. luteiventris* as a separate sister species, in agreement with Cuellar, but not having seen that previous article. Pearl & Hayes (2005) discussed the extinctions of populations in much of its former range. *R. pretiosa* is not yet federally listed in the U.S., but the declines suggest that it should be; it is listed as Endangered in British Columbia.

Rana (Rana) sierrae Camp, 1917

Rana boylii sierrae Camp, 1917: 120.—**O**: **H:** MVZ 3734 (ex Camp 1901).—**OT**: U.S., California, Inyo County, Sierra Nevada, 2 miles southeast of Kearsarge Pass at Matlack Lake, elev. 10,500 feet.

Rana boylii sierrae: Stejneger & Barbour, 1917: 36; 1943: 55.

Rana boyli sierrae: Schmidt, 1953: 84; Gorham, 1974: 141.

Rana sierrae: Vredenburg *et al.*, 2007: 370; Elliott *et al.*, 2009: 250; Dodd, 2013: 747.

Distribution: Sierra Nevada of California and western Nevada, west to the type locality. Zweifel (1968a) had a map as a subspecies of *R. muscosa*. Also a map in Elliott *et al.* (2009), showing former distribution of this and *R. muscosa*. Dodd (2013) has a current map. **U.S.**: CA, NV.

Common Name: Sierra Nevada Yellow-legged Frog.

Commentary: This taxon was not recognized in any of the field guides that treated other western relatives (*e.g.*, Cochran & Goin, 1970; Behler & King, 1979; Stebbins, 1985, 2003); we assume they considered it synonymous with *R. muscosa,* and not distinctive enough to have subspecific recognition. Camp (1917) noted that other names misapplied to this taxon included *Rana aurora* (Stejneger, 1893b: 225), and *Rana pretiosa* (Stejneger, 1893b: 226; Yarrow & Henshaw, 1878, Lake Tahoe). It was separated from *R. muscosa,* as a distinct sister species, by Vrendenburg *et al.* (2007). Zweifel (1968a) provided a partial CAAR account, included as part of *R. muscosa.* Vredenburg *et al.* (2007) noted that over 93% of populations of this species are now extinct; it should now be federally listed as endangered, as a distinct species.

Subgenus **Lithobates** Fitzinger, 1843

Lithobates Fitzinger, 1843: 31.—**NS**: *Rana palmipes* Spix, 1824. As subgenus by Dubois (1992: 330).

Rana (Lithobates): Dubois, 1992: 330; Pauly, Hillis, & Cannatella, 2004: 116. As a subgenus.

Lithobates: Frost *et al.*, 2006: 369; Che *et al.*, 2007. As a genus.

"Ranula" Peters, 1859: 402.—**NS**: *Ranula gollmeri* Peters, 1859 (mo.). Preoccupied by *Ranula* Schumacher, 1817, so unavailable as a junior homynym.—**SS**: Peters, 1873c: 622, with *Rana,* and with *Lithobates* by implication.

Pohlia Steindachner, 1867b: 15.—**NS**: *Rana palmipes* Spix, 1824 (mo.).—**SS**: Boulenger, 1882a: 7.

Trypheropsis Cope, 1868b: 117.—**NS**: *Ranula chrysoprasina* Cope, 1866 (od.).—**SS**: O'Shaughnessy, 1879: 17, with *Hylarana*; Boulenger, 1882a: 7, with *Rana.*

Levirana Cope, 1894c: 197.—**NS**: *Levirana vibicaria* Cope, 1894 (mo.).—**SS**: Boulenger, 1920: 462, by implication.

Chilixalus Werner, 1899a: 117.—**NS**: *Ixalus warszewitschii* Schmidt, 1858 (mo.).—**SS**: Dubois, 1999: 85, with *Trypheropsis.*

Sierrana Dubois, 1992: 330.—**NS**: *Rana sierramadrensis* Taylor, 1939. As subgenus of *Rana.*

Zweifelia Dubois, 1992: 330.—**NS**: *Rana tarahumarae* Boulenger, 1917. As subgenus of *Rana.*

Aquarana Dubois, 1992: 331.—**NS**: *Rana catesbyeiana* Shaw, 1802. As subgenus of *Rana.*

Pantherana Dubois, 1992: 331.—**NS**: *Rana pipiens* Schreber, 1782. As subgenus of *Rana.*

"Novirana" Hillis & Wilcox, 2005: 311.—**NS**: *Rana pipiens* Schreber, 1782.—**OS**: Dubois (2007a: 392) with *Pantherana* Dubois, 1992; also a *nomen nudum.*

"Torrentirana" Hillis & Wilcox, 2005: 312.—**NS**: *Rana tarahumarae* Boulenger, 1917.—**OS**: Dubois (2007a: 392) with *Zweifelia* Dubois, 1992.

"Stertirana" Hillis & Wilcox, 2005: 311.—**NS**: *Rana montezumae* Baird, 1856. A *nomen nudum* (Dubois, 2007a: 392), conditionally unavailable.

Lacusirana Hillis & Wilcox, 2005: 311.—**NS**: *Rana megapoda* Taylor, 1942.

Nenirana Hillis & Wilcox, 2005: 311.—**NS**: *Rana areolata* Baird & Girard, 1852.

Scurrilirana Hillis & Wilcox, 2005: 311.—**NS**: *Rana berlandieri* Baird, 1859.

Distribution: Eastern North America, southward through Middle America and South America, except the southern temperate regions.

—

Common name: American Water Frogs (SECSN) or North American True Frogs (C&T).

Content: 49 species, of which 21 occur in North America; presumably this will soon be 50 species, 22 of which occur in North America, as a new species of the *R. pipiens* group has been reported in urban situations in New York and adjacent states, but has not yet been described (Newman *et al.,* 2012).

Commentary: We adopt an arrangement of species-groups similar to that of Hillis & Wilcox (2005), which was similar to that of earlier authors. Within this subgenus, they recognized four major clades: the *catesbeiana*-group, the *pipiens*-group, the *tarahumarae*-group, and the *palmipes*-group. There are no North American members of the *palmipes* group. The **catesbeiana group** includes *R. catesbeiana, R. clamitans, R. grylio, R. heckscheri, R. okalossae, R. septentrionalis,* and *R. virgatipes,* plus possibly *R. sylvatica* (all North American). The **tarahumarae group** includes several Middle American species, including one, *R. taramumarae,* which formerly reached North America; see Commentary for that species. The **pipiens-group** was further subdivided into three subgroups: the *areolata*-subgroup, the *montezumae*-subgroup, and the *berlandieri*-subgroup. In their study, Hillis & Wilcox (2005) found the *pipiens*-group to be monophyletic. Their *montezumae* subgroup contained only one North American species as treated here, and *R. pipiens* was not specified a subgroup, but is a sister to the *montezumae* subgroup, together forming a monophyletic clade that is a sister to the rest of the *pipiens*-group. The arrangement we adopt here assigns their *montezumae* subgroup + *R. pipiens* to constitute the subgroup we call the **pipiens subgroup**; that subgroup includes about six species, of which two are North American. Another member of the subgroup has been suggested to now be extinct, *R. fisheri,* but we list it as well (see Commentary for that species), so the subgroup includes *R. chiricahuensis, R. fisheri,* and *R. pipiens,* as North American members. The *areolata* and *berlandieri* subgroups are sisters, and the *pipiens* subgroup is a sister to that pair. The **areolata subgroup** (treated as an unranked clade "*Nenirana*" by Hillis & Wilcox, 2005; inferred to be a subgenus by some, *e.g.,* Engbrecht *et al.,* 2011) includes four species, all North American: *R areolata, R. capito, R. palustris,* and *R. sevosa.*

The **berlandieri subgroup** includes about 20 species, five of which are North American: *R. berlandieri, R. blairi, R. onca, R. sphenocephala, and R. yavapaiensis*. The *tarahumarae* and *palmipes* groups are sisters, and that clade is a sister taxon to the *pipiens*-group. The *catesbeiana*-group is in an ancestral position as sister to all the other species groups of *Lithobates*. Finally, the *boylii*-group (subgenus *Rana*), along with *R. sylvatica* and the extralimital *R. temporaria* occupy an ancestral sister taxon position to *Lithobates,* according to some authors (see Commentaries for the appropriate species).

North America contains about 44% of the species diversity of the subgenus.

Rana (Lithobates) areolata Baird & Girard, 1852, **new combination**
Synonymy: See under the subspecies below.
Distribution: Most of the range is of many disjunct populations, from Eastern Texas and adjacent Louisiana, eastward through most of Mississippi, and northward to include much of eastern Oklahoma and Kansas, northern Missouri, southern Iowa to western parts of Indiana, Tennessee and Kentucky. See map in Altig & Lohoefener (1983), but realize it includes three other taxa as subspecies; see only the two currently recognized subspecies there. Dodd (2013) has a more updated map. **U.S.:** AR, IA, IL, IN, KS, KY, LA, MO, MS, OK, TN, TX.

Common Name: Crawfish Frog.

Content: Two subspecies are currently recognized. Three others were formerly recognized, but one has been elevated to a full species (*Rana capito*) and includes another as a synonym (*R. a. aesopus*), and another has also been elevated to full species status (*Rana sevosa*).

Commentary: Altig and Lohoefener (1983) provided a CAAR account, but they included two species that are now recognized as distinct, *R. capito* and *R. sevosa*. Hillis & Wilcox (2005) found this to be a sister species of *R. capito* + *R. sevosa*. Parris & Redmer (2005) reviewed the conservation status and noted that several states have listed the species as endangered or of special concern, though no populations are yet federally listed.

Rana (Lithobates) areolata areolata Baird & Girard, 1852, **new combination**

Rana areolata Baird & Girard, 1852d: 173.—**O**: **S:** USNM (2).—**L**: USNM 3304 (implied, Cochran, 1961a: 72, but Harper, 1935, designated it first).—**OT**: U.S., Texas, Calhoun County, Indianola, and Arizona, Rio San Pedro of the Gila River (certainly an error); restriction by L designation, to the Texas locality.

Rana areolata areolata: Cope, 1875b: 32; 1889b: 410; Stejneger & Barbour, 1943: 53; Viosca, 1949: 10; Schmidt, 1953: 78; Neill, 1957a: 47; Cochran & Goin, 1970: 83; Gorham, 1974: 140; Conant, 1975: 349; Behler & King, 1979: 368; Garrett & Barker, 1987: 41 Conant & Collins, 1998: 571; Dixon, 2000: 74-75; Elliott *et al.*, 2009: 208.

Rana virescens areolata: Barbour & Cole, 1906: 153.

Rana areolata: Baird, 1859e: 28; Strecker, 1915: 46; Stejneger & Barbour, 1917: 35; Boulenger, 1920: 465; Frost, 1985: 181.

Rana (Rana) areolata: Dubois, 1987a: 42.

Rana (Pantherana) areolata: Dubois, 1992: 332.

Rana (Novirana) areolata: Hillis & Wilcox, 2005: 305.

Lithobates areolatus: Frost *et al.*, 2006: 369; Elliott *et al.*, 2009: 208; Dodd, 2013: 461.

Lithobates (Lithobates) areolatus: Dubois, 2006c: 829; 2006d: 325.

Rana (Nenirana) areolata: Hillis, 2007: 335, by implication.

Lithobates areolatus areolatus: Frost *et al.*, 2008: 7; Dodd, 2013: 461.

Rana (Lithobates) areolata areolata: **new combination**

Rana octoplicata Werner, 1893: 83.—**O**: **H:** in NMW.—**OT**: North America.—**SS**: Boulenger, 1893b: 37.

Distribution: Eastern Texas, southeastern Oklahoma, southwestern Arkansas and northern and central Louisiana. See map for the species in Dodd (2013), and for this subspecies in Altig and Lohoefener (1983). **U.S.**: AR, LA, OK, TX.

Common Name: Southern Crawfish Frog.

Rana (Lithobates) areolata circulosa Rice & Davis, 1878, **new combination**

Rana circulosa Rice & Davis, 1878: 355.—**O**: **H:** CHAS 160 (Goin & Netting, 1940: 143, sd.).—**OT**: U.S., Indiana, Benton County.—**Comment:** Stejneger & Barbour (1943: 54) erred in correcting the OT to northern Illinois; this was corrected by Mittleman (1947b: 474), to the original. Schmidt (1953) further restricted the OT but with no evidence the H came from there, that action is invalid.—**SS**: Boulenger, 1919: 415, with *R. areolata*.

Rana areolata circulosa: Davis & Rice, 1883: 28; Cope, 1889b: 413; Stejneger & Barbour, 1943: 54; Schmidt, 1953: 78; Neill, 1957a: 47; Cochran & Goin 1970: 85; Gorham, 1974: 140; Conant, 1975: 349; Behler & King, 1979: 368; Conant & Collins, 1998: 571; Elliott *et al.*, 2009: 208.

Lithobates areolatus circulosus: Frost *et al.*, 2008: 7; Dodd, 2013: 461.

Rana (Lithobates) areolata circulosa: **new combination.**

Distribution: Northern Mississippi, most of Arkansas, northeastern Oklahoma, eastern Kansas, much of Missouri and adjacent Iowa eastward to western Indiana and western Tennessee and Kentucky. **U.S.:** AR, IA, IL, IN, KS, KY, MO, MS, OK, TN.

Common Name: Northern Crawfish Frog.

Rana (Lithobates) berlandieri Baird, 1859, **new combination**

Rana berlandieri Baird, 1859e: 27.—**O**: **S** USNM 3293 (9), 131513 (Cochran, 1961a: 72, sd.); also MCZ 155 (2), exchanged from USNM (Barbour & Loveridge, 1929: 326).—**L**: USNM 131513 (Pace, 1974: 25).—**OT**: U.S., southern Texas; corrected to Texas, Brownsville (Kellogg, 1932: 205); OT of L, U.S., Texas, Brownsville.

Rana berlandieri: Baird, 1859c: 45; Conant, 1975: 346; Behler & King, 1979: 370; Frost, 1985: 483; Garrett & Barker, 1987: 42; Liner, 1994: 27; Conant & Collins, 1998: 569; Dixon, 2000: 75; Brennan & Holycross, 2006: 42; Lemos-Espinal & Smith, 2007a: 54, 251; Elliott *et al.*, 2009: 226.

Rana halecina berlandieri: Cope, 1875b: 32.

Rana virescens berlandieri: Cope, 1889b: 398 (part).

Rana pipiens berlandieri: Schmidt, 1941: 487; 1953: 83; Mittleman & Gier, 1942: 13; Stejneger & Barbour, 1943: 58; Conant, 1958: 302; Cochran & Goin, 1970: 82; Gorham, 1974: 150.

Rana berlandieri berlandieri: Sanders, 1973: 87.

Rana (Rana) berlandieri: Dubois, 1987a: 41, by implication.

Rana (Pantherana) berlandieri: Dubois, 1992: 331.

Rana (Novirana) berlandieri: Hillis & Wilcox, 2005: 305.

Lithobates berlandieri: Frost *et al.*, 2006: 369; Elliott *et al.*, 2009: 226; Dodd, 2013: 466.

Rana (Scurrilirana) berlandieri: Hillis, 2007: 335, by implication.

Rana (Lithobates) berlandieri: **new combination**

Rana halecina austricola Cope, 1886b: 517.—**O**: **S:** Mexican specimens of *Rana halecina* referred to in several publications, including Brocchi (1881: 10), and Günther (1859a) (by indication).—**L**: Kellogg, 1932: 207, specimen of *Rana lecontei* figured by Brocchi (1881: Pl. 4, fig. 1).—**OT**: Mexico, Veracruz, by L-designation.—**Comment**: Smith & Taylor (1950a: 351) and Schmidt (1953: 83) also restricted the OT, without justification, and apparently unaware of Kellogg's earlier acton, so those later actions are invalid.

Rana virescens austricola: Cope, 1889b: 398.

Rana virescens var. *austricola*: Ives, 1892: 461.

Rana austricola: Ruthven, 1912a: 305.

Rana halecina var. *austricola*: Boulenger, 1919: 413.

Rana pipiens austricola: Smith, 1947b: 409.

Rana virescens virescens Cope, 1889b: 401.

Rana virescens brachycephala Cope, 1889b: 403.

Rana pipiens brachycephala: Stejneger & Barbour, 1943: 58.

Rana berlandieri brownorum Sanders, 1973: 87.

Rana brownorum: Hillis *et al.*, 1983: 134.

Distribution: South-central to western Texas and adjacent southeastern New Mexico, south into Mexico. Also introduced to southwestern Arizona and adjacent California, where they are well-established, and some more western parts of Mexico. See map in Platz (1991). **U.S.**: *AZ, *CA, NM, TX.

Common Name: Rio Grande Leopard Frog.

Commentary: Platz (1991) provided a CAAR account. This species is in a clade with some Mexican species, which has a sister clade containing *R. blairi* (Hillis & Wilcox, 2005), its nearest North American relative.

Rana (Lithobates) blairi Mecham, Littlejohn, Oldham, Brown, & Brown, 1973, **new combination**

Rana blairi Mecham *et al.*, 1973: 3.—**O**: **H:** UMMZ 131690.—**OT**: U.S., Texas, Lubbock County, 1.6 km W New Deal.

Rana blairi: Behler & King, 1979: 370; Frost, 1985: 183; Garrett & Barker, 1987: 43; Conant & Collins, 1998: 568; Dixon, 2000: 75; Elliott *et al.*, 2009: 224.

Rana (Rana) blairi: Dubois, 1981: 233; 1987a: 41, by implication.

Rana (Pantherana) blairi: Dubois, 1992: 332.

Rana (Novirana) blairi: Hillis & Wilcox, 2005: 305.

Lithobates blairi: Frost *et al.*, 2006: 369; Elliott *et al.*, 2009: 224; Dodd, 2013: 472.

Lithobates (Lithobates) blairi: Dubois, 2006c: 829; 2006d: 325.

Rana (Scurrilirana) blairi: Hillis, 2007: 335, by implication.

Rana (Lithobates) blairi: **new combination.**

Distribution: U.S. central and southern plains, including northern Texas into eastern New Mexico and southeastern Arizona, western Oklahoma, Kansas into eastern Colorado, eastern Nebraska, southeastern South Dakota, western and southern Iowa into central Illinois and Indiana. See map in Brown (1992) or Dodd (2013). **U.S.**: AR, AZ, CO, IA, IL, IN, KS, MO, NE, NM, SD, TX.

Common Name: Plains Leopard Frog.

Commentary: Brown (1992) provided a CAAR account. As noted above, this species' closest North American relative is *R. berlandieri* (Hillis & Wilcox, 2005). Earlier, Hillis & Davis (1986) indicated that *R. blairi* was a member of a *pipiens* subgroup, more closely related to *R. pipiens* and *R. sphenocephala*. Results of a study by Tennessen & Blouin (2010), tend to agree more closely with Hillis & Davis (1986). Crawford *et al.* (2005) reviewed the conservation status and noted reductions in abundance and disappearance in areas now occupied by introduced *Rana catesbeiana*.

Rana (Lithobates) capito Le Conte, 1855, **new combination**

Rana capito Le Conte, 1855b: 425.—**O**: **H:** USNM 5903 (Harper, 1935: 79, sd.).—**OT**: U.S., Georgia, rice field ditches; validly restricted (Harper, 1935: 79) to U.S., Georgia, Liberty County, Riceborough.

Rana areolata capito: Cope, 1875b: 32; 1889a: 415; Neill, 1957a: 47; Conant, 1958: 305; Gorham, 1974: 140; Behler & King, 1979: 368.

Rana capito capito: Wright & Wright, 1942: 173; Schmidt, 1953: 79; Cochran & Goin, 1970: 85; Conant & Collins, 1998: 572.

Rana capito: Stegneger & Barbour, 1943: 55; Case, 1978: 308; Collins, 1991: 43; Young & Crother, 2001: 382; Elliott *et al.*, 2009: 210.

Rana (Rana) capito: Dubois, 1987a: 41, by implication.

Rana (Novirana) capito: Hillis & Wilcox, 2005: 305.

Lithobates capito: Frost *et al.*, 2006: 369; Dodd, 2013: 479.

Lithobates (Lithobates) capito: Dubois, 2006c: 829; 2006d: 325; Elliott *et al.*, 2009: 210.

Rana (Nenirana) capito: Hillis, 2007: 335, by implication.

Rana (Lithobates) capito: **new combination.**

Rana areolata aesopus Cope, 1886b: 517.—**O**: **H** (od.) USNM 4743.—**OT**: U.S., Florida, Alachua County, Micanopy.—**SS**: Boulenger, 1919: 415.

Rana aesopus: Stejneger & Barbour, 1917: 36.

Rana areolata aesopus: Neill, 1957a: 52; Conant, 1958: 305.

Rana capito stertens Schwartz & Harrison, 1956: 135.—**O**: **H:** CHM 55.146.12.—**OT**: U.S., South Carolina, Berkeley County, 6 mi. N Cainboy.—**Comment**: Neill (1957a) rejected the subspecies.

Distribution: Southeastern U.S., from eastern North Carolina to southern Florida and southern Alabama, plus disjunct populations in southern Mississippi and central Tennessee. See map in Altig & Lohoefener (1983), as *Rana areolata capito* plus *R. a. aesopus*. **U.S.**: AL, FL, GA, MS, NC, SC, TN.

Common Name: Gopher Frog (SECSN) or Carolina Gopher Frog (C&T). The former seems preferable because of the greater range of the species.

Content: Formerly, the race *aesopus* was recognized as a subspecies of *R. areolata,* then when *R. capito* was recognized as a distinct species, some workers recognized *aesopus* as a subspecies of *R. capito.* But the study by Young & Crother

(2001) did not support recognition of any subspecies of this form, so *aesopus* is now a junior synonym of *R. capito*.

Commentary: Altig & Lohoefener (1983) provided the only CAAR account of this form, as *R. areolata capito* plus *R. a. aesopus*; it needs to be updated. Case (1978) recognized this as a distinct species, rather than a subspecies of *R. areolata,* as it had previously been treated. This was supported by the work of Young & Crother (2001). The study of Hillis & Wilcox (2005) showed a sister relationship to *R. sevosa*. Jensen & Richter (2005) reviewed conservation status and indicated that the species is in decline in most parts of its range; it is listed as endangered, threatened, or of special concern in all states in its range.

Rana (Lithobates) catesbeiana Shaw, 1802, new combination

Rana catesbeiana Shaw, 1802a: 106.—**O**: specimen of pl. 33, probably based on Catesby's drawing; not known to exist.—**OT**: many parts of North America, specifically in Carolina and Virgina.—**Comment**: Kellogg (1932: 197) restricted the OT to South Carolina, based on his certainty that Catesby captured the frog used for the drawing somewhere in South Carolina, but Shaw had not seen any specimens himself. Smith & Taylor (1950a: 360) and Schmidt (1953: 79) further restricted it to Charleston, SC, but provided no evidence that the O came from there, thus those actions are invalid.

Rana catesbeiana: Storer, 1825: 276; Baird, 1859c: 45; 1859e: 27; Boulenger, 1920: 418; Stejneger & Barbour, 1917: 37; 1943: 56; Viosca, 1949: 10; Schmidt, 1953: 79; Cochran & Goin, 1970: 75; Gorham, 1970: 8; 1974: 141; Conant, 1975: 338; Behler & King, 1979: 372; Frost, 1985: 485; Stebbins, 1985: 92; 2003: 240; Garrett & Barker, 1987: 76; Liner, 1994: 27; Conant & Collins, 1998: 555; Dixon, 2000: 76; Lemos-Espinal *et al.,* 2004b: 6, 79; Brennan & Holycross, 2006: 76; Elliott *et al.,* 2009: 186.

Rana catesbiana: Cope, 1889b: 424; Dickerson, 1907: 89.—**IS.**

Rana catesbyana: Werner, 1909: 86; Smith, 1978: 66.—**IS.**

Rana catesbyiana: Smith, 1978: 7.—**IS.**

Rana (Rana) catesbeiana: Dubois, 1987a: 41, by implication.

Rana (Aquarana) catesbeiana: Dubois, 1992: 331.

Rana (Novirana) catesbeiana: Hillis & Wilcox, 2005: 305.

Lithobates catesbeianus: Frost *et al.*, 2006: 369; Elliott *et al.*, 2009: 186; Dodd, 2013: 486.

Lithobates (Aquarana) catesbeianus: Dubois, 2006d: 325.

Rana (Aquarana) catesbeiana: Hillis, 2007: 335, by implication.

Rana (Lithobates) catesbeiana: **new combination**

"Rana pipiens" Daudin, 1802b: 58. Misapplication of nomen, *nec Rana pipiens* Schreber, 1782.—**O**: **S**: includes specimen of pl. 18 and 5 others stated as deposited in MNHN (now lost?).—**OT**: L'Amerique Septentrionale, et sur-tout dans la Caroline ... la Virginie.—**Comment**: Primary junior homonym of *Rana pipiens* Schreber, 1782 (familiar to Daudin); said by Daudin to be the "Grenouille mugisante" of de la Cepède (*Rana catesbeiana*).

Rana mugiens Merrem, 1820: 175.—**O**: based partly on pl. 72 in Catesby (1743), *Rana catesbeiana* Shaw (1802), and *Rana pipiens* of Latreille (1825), as well as others mentioned by Kalm (1761) as "Manteskühe" and "Ochsenfrösche", thus a mixture of several different species (*fide* Frost, 2009-13).—**OT**: America septentrionali. Restricted to vicinity of New York city by Schmidt (1953: 79) but with no evidence the O were from there, hence invalid.—**SS**: Schinz, 1822: 164.

Rana mugiens: Schlegel, 1858: 56.

Rana scapularis Harlan, 1826a: 59.—**O**: none stated, probably originally in ANSP, now not known to exist.—**OT**: U.S., Pennsylvania. Restricted to vicinity of Philadelphia, PA (Schmidt, 1953: 79) but with no evidence the O came from there, hence invalid.—**SS**: Duméril & Bibron (1841: 370), with *R. catesbeiana* and *R. mugiens*.

Rana conspersa Le Conte, 1855b: 425.—**O**: **S:** includes ANSP 2918 (Malnate, 1971: 350, sd.).—**OT**: U.S., Pennsylvania. Corrected to U.S., Georgia, Liberty County, Riceborough (Schmidt, 1953: 79), based on Le Conte living in that area, but with no evidence the O came from there, hence invalid; and the OT as given is in range.—**SS**: Boulenger, 1882a: 36.

Rana nantaiwuensis Hsü, 1930: 19.—**O**: **H**: in museum of the University of Amoy, now apparently lost.—**OT**: China, Fujian Province, Xiamen Shi, Nantaiwu, Amoy.—**Comment**: Apparently the author was unaware of the introduction of bullfrogs to China, as well as many other parts of the world, much earlier.—**SS**: Zhao & Adler (1993: 140).

"Rana mugicus" Angel, 1947: 253—**O:** none given. Frost (2009-13) listed this as a synonym of *pipiens*, not a misspelling of *mugiens*, and noted it was based on specimens of *pipiens*. On the page 253 that he cited, there is merely a reference to the names *pipiens*, *mugicus*, and *catesbeiana*, but no description or anything to imply he was introducing the name there; perhaps there was a prior description elsewhere in the book, or in another publication, but we could not find such. May be **IS** of *R. mugiens*.

Distribution: Native to eastern North America, including Nova Scotia, New Brunswick, Quebec, and Ontario, south to Florida (except the southern portion), and west to the central plains. Introduced to other parts of North America (e.g., Arizona, California, Nevada, Oregon, Washington, British Columbia, all of Hawaii), and into Mexico. Widely introduced in Europe, Asia, Mexico, South America and the Caribbean (see Lever, 2003). Wherever this large predator has been introduced, native frog populations have declined. Elliott *et al.* (2009) have a map of the North American distribution, except introduced British Columbia populations; Fisher *et al.* (2007) have a map for Canada. Dodd (2013) maps the entire North American distribution. Including introductions, this species is found in every U.S. state except North Dakota. **Can.**: *BC, NB, NS, ON, QC. **U.S.:** AL, AR, *AZ, *CA, *CO, CT, DE, FL, GA, *HI, *IA, *ID, IL, IN, *KS, KY, LA, MA, MD, ME, MI, *MN, MO, MS, *MT, NC, *NE, NH, NJ, *NM, *NV, NY, OH, *OK, *OR, PA, RI, SC, *SD, TN, TX, *UT, VA, VT, *WA, WI, WV, *WY.

Common name: Bullfrog. *French Canadian name:* Ouaouaron.

Commentary: The study by Hillis & Wilcox (2005) found this species included in a clade with *R. clamitans, R. grylio, R. heckscheri, R. okaloosae, R. septentrionalis,* and *R. virgatipes*. They found *R. heckscheri* to be the sister of this species, and those two together as a sister to *R. clamitans* + *R. ocaloosae. R. grylio* was found to be a sister to the clade including those four.

Rana (Lithobates) chiricahuensis Platz & Mecham, 1979, **new combination**

Rana chiricahuensis Platz & Mecham, 1979.—**O**: **H:** AMNH 100372.—**OT**: U.S., Arizona, Cochise County, Coronado National Forest, 6 km W of Portal, at Herb Martyr Lake, elev. 1768 m.

Rana chiricahuensis: Frost, 1985: 486; Stebbins, 1985: 89; Liner, 1994: 27; Lemos-Espinal *et al.*, 2004b: 6, 79; Brennan & Holycross, 2006: 44; Elliott *et al.*, 2009: 230.

Rana (Rana) chiricahuensis: Dubois, 1987a: 41, by implication.

Rana (Pantherana) chiricahuensis: Dubois, 1992: 332.

Rana (Novirana) chiricahuensis: Hillis & Wilcox, 2005: 305.

Lithobates chiricahuensis: Frost *et al.*, 2006: 369; Elliott *et al.*, 2009: 230; Dodd, 2013: 515.

Lithobates (Lithobates) chiricahuensis: Dubois, 2006d: 325.

Rana (Lacusirana) chiricahuensis: Hillis, 2007: 335.

Rana (Lithobates) chiricahuensis: **new combination**

Rana subaquavocalis Platz, 1993: 155.—**O**: **H:** AMNH 136096.—**OT**: U.S., Arizona, Cochise County, 7 km southwest of Sierra Vista, at Ramsey Canyon, elev. 1622 m., 31° 26' 59" N, 110° 18' 13" W.—**SS**: Goldberg, Field, & Sredl (2004a: 313).

Rana (Novirana) subaquavocalis: Hillis & Wilcox, 2005: 305.

Lithobates (Lithobates) subaquavocalis: Dubois, 2006d: 325.

Distribution: Discontinuous, with three main populations, one in eastern New Mexico and western Arizona (although some specimens there have been identified as *Rana fisheri*; Hekkala *et al.*, 2011), another in southeastern Arizona into southwestern New Mexico, and two others in northern Mexico and southward, at relatively high elevations, generally above 1000 m. The species may have been extirpated in New Mexico. See map in Platz & Mecham (1984), but a fourth population indicated there in central Arizona has now been identified as *R. fisheri* (see that account). Dodd (2013) has a recent map. **U.S.**: AZ, NM.

Common Name: Chiricahua Leopard Frog (called Wrights' Leopard Frog by the authors of the species description, honoring the discoverers of one of the isolated populations).

Content: No subspecies have been described, but Platz (1993) noted that there seem to be other cryptic species included under this name, pending further study. Goldberg *et al.* (2004a) confirmed this, identifying at least two distinct genomes for the frogs currently treated under this name. They also synonymized another species with *R. chiricahuensis* on the basis of that study (see below).

Commentary: Platz & Mecham (1984) provided a CAAR account. *Rana subaquavocalis* was described as a similar species that had become reproductively isolated by its calling behavior, exclusively underwater. The main population at Ramsey Canyon seems indeed to call exclusively underwater, on the bottom of deep pools, where the call can only be heard by other underwater frogs in the same habitat. Typical *R. chiricahuensis,* as most other ranids, characteristically broadcasts and receives audio information through the air, so females of either species would hear the calls of only conspecific males. Goldberg *et al.* (2004a) found that typical *R. chiricahuensis* and frogs identified as *R. subaquavocalis* did not differ significantly in their molecular analysis, and the spectral features of the calls broadly overlapped. In most situations, this would certainly justify synonymy of the two. But the separation could be so recent that the DNA makeup has not yet diverged, yet the populations may still be reproductively isolated by the underwater calling behavior of *R. subaquavocalis,* justifying its retention as a distinct species. However, the convincing point made by Goldberg *et al.,* was that, apparently at localities other than Ramsey Canyon, they or others cited had observed *R. subaquavocalis* calling above water, and cited another example of *R. chiricahuensis* calling underwater. As the two are virtually indistinguishable by morphology or call, we would be interested in knowing more details of how the identities of the observed frogs were determined, but for the present we will accept those observations as denying two separate species.

Most of the species of the subgroup are restricted to Mexico. *R. chiricahuensis* has a sister species in Mexico, and has two other Mexican species in a sister group (Hillis & Wilcox, 2005).

Also see the account of *R. fisheri*; it is possible that *R. chiricahuensis* may ultimately be found to be synonymous with *R. fisheri,* or perhaps *R. chiricahuensis* may be reduced to a subspecies of *R. fisheri,* with further study.

This species is federally listed as threatened.

Rana (Lithobates) clamitans Latreille *in* Sonnini & Latreille, 1801

Synonymy: See the subspecies below.

Distribution: Eastern Texas, Oklahoma, and Kansas, east to the Atlantic coast, and northward to southeastern Canada, to include Manitoba, Quebec, and southern Ontario, plus Nova Scotia and Labrador, with isolated populations in Newfoundland. Absent from most of peninsular Florida and the Illinois prairie. Introduced in parts of Utah, Washington, Hawaii, and British Columbia. See map in Stewart (1984), which does not include introduced populations. Fisher *et al.* (2007) have a map of the Canadian distribution, including introductions there. **Can.**: *BC, MB, NB, *NL, NS, ON, QC, PE. **U.S.**: AL, AR, DE, FL, GA, *HI, *IA, IL, IN, KY, LA, MA, MD, ME, MI, MN, MO, MS, NC, NH, NJ, NY, OK, PA, RI, SC, TN, TX, *UT, VA, VT, *WA, WI, WV.

Common name: Green Frog (SECSN) or Bronze Frog (C&T). *French Canadian name:* Grenouille verte.

Content: Two subspecies are currently recognized, though they are poorly differentiated.

Commentary: Stewart (1968) provided a CAAR account. Hillis & Wilcox (2005) found *R. okaloosae* is the sister species of *R. clamitans,* and the two together have a sister relationship to *R. catesbieana* + *R. heckscheri.*

Rana (Lithobates) clamitans clamitans Latreille, 1801, **new combination**

Rana clamitans Latreille *in* Sonnini & Latreille, 1801: 157.—**O**: **H:** MNHN 1397 (Guibé, 1950: 35, sd.).—**OT**: U.S., South Carolina, near Charleston. **Comment:** Authorship is sometimes given as Daudin (and occasionally as Bosc), but Harper (1940) and Stewart (1984) have clarified the correct dates for the descriptions and that Latreille is the proper author.

Rana clamata: Daudin, 1802a: 54; Cope, 1889b: 419; Blatchley, 1897b: 183.—**UE**.

Rana clamitans: Bosc, 1803: 436; Sonnini & Latreille, 1830a: 157; Baird, 1859c: 45; Davis & Rice, 1883: 28; Boulenger, 1920: 425; Stejneger & Barbour, 1917: 37; 1943: 56; Viosca, 1949: 10; Schmidt, 1953: 79; Frost, 1985: 486; Stebbins, 1985: 93; 2003: 242.

Rana clamator: Le Conte, 1855b: 424.—**IS**.

Rana (Rana) clamitans: Guibé, 1950: 35; Dubois, 1987c: 41, by implication.

Rana clamitans clamitans: Conant, 1958: 299; Cochran & Goin, 1970: 75; Gorham, 1974: 142; Behler & King, 1979: 373; Garrett & Barker, 1987: 45; Dixon, 2000: 76; Elliott *et al.*, 2009: 190.

Rana (Aquarana) clamitans: Dubois, 1992: 331; Hillis, 2007: 335, by implication.

Rana (Novirana) clamitans: Hillis & Wilcox, 2005: 305.

Lithobates clamitans: Frost *et al.*, 2006: 369; Elliott *et al.*, 2009: 190; Dodd, 2013: 522.

Lithobates (Aquarana) clamitans: Dubois, 2006c: 829.

Lithobates clamitans clamitans: Frost, McDiarmid & Mendelson, 2008: 8; Dodd, 2013: 522.

Rana (Lithobates) clamitans clamitans: **new combination**

Rana fontinalis Le Conte, 1825: 282.—**O**: not stated nor known to exist (possibly ANSP or USNM).—**OT**: not stated; restricted to U.S., New York, New York City (Schmidt, 1953: 79), but as no evidence was presented that the O came from there, the action is invalid.—**SS**: Duméril & Bibron, 1841: 374.

Rana flaviviridis Harlan, 1826a: 58.—**O**: not stated nor known to exist.—**OT**: U.S., the middle states, abounding near Philadelphia; restricted to vicinity of Philadelphia (Schmidt, 1953: 80).—**SS**: Duméril & Bibron, 1841: 374, with *Rana fontinalis*.

Rana horiconensis Holbrook, 1838: 91.—**O**: specimen figured pl. 16, not now known to exist.—**OT**: U.S., New York, outlet of Lake George.—**SS**: Le Conte, 1855b: 424.

Rana nigricans Agassiz, 1850: 379.—**O**: **S**: MCZ 7 (3) (Barbour & Loveridge, 1929: 330, sd.).—**OT**: Can., Ontario, Lake Superior, localities along northern shores.—**SS**: Baird, 1859c: 45; Barbour & Loveridge, 1929: 330.

Distribution: Eastern Texas to the Atlantic coast, below the Fall line; absent from most of peninsular Florida (see map in Stewart, 1968). Dodd (2013) has a map for the species, without subspecific boundaries. **U.S.**: AL, FL, GA, LA, MS, NC, OK, SC, TX, VA.

Common name: Bronze Frog.

Rana (Lithobates) clamitans melanota (Rafinesque, 1820)

Ranaria (Rana) melanota Rafinesque, 1820: 5.—**O**: not stated nor known to exist.—**OT**: U.S., New York and Vermont, Lake Champlain; and New York, Lake George.—**SS**: Cope, 1886b: 519.

Rana clamitans melanota: Rhoads, 1895: 394; Conant, 1958: 300; Cochran & Goin, 1970: 77; Gorham, 1974: 142; Behler & King, 1979: 373; Elliott *et al.*, 2009: 190.

Lithobates clamitans melanota: Frost, McDiarmid & Mendelson, 2008: 7.

Rana horiconensis Holbrook, 1838b: 91.—**O:** not known to exist.—**OT:** U.S., New York, outlet of Lake George.

Subspecies not recognized:

Rana clamitans: Schmidt, 1953: 79.

Distribution: As for the species, but that portion essentially above the Fall line (see map in Stewart, 1968). **Can.**: *BC, MB, NB, *NL, NS, ON, QC, PE. **U.S.**: AL, AR, DE, GA, *IA, IL, IN, KY, LA, MA, MD, ME, MI, MN, MO, MS, NC, NH, NJ, NY, OK, PA, RI, SC, TN, *UT, VA, VT, *WA, WI, WV.

Common name: Northern Green Frog (SECSN) or Green Frog (C&T). *French Canadian name:* Grenouille verte.

Rana (Lithobates) fisheri Stejneger, 1893, **new combination**

Rana fisheri Stejneger, 1893b: 227.—**O**: **H:** USNM 18957 (Cochran, 1961a: 73, sd.).—**OT**: U.S., Nevada, Clark County, Vegas Valley.

Rana fisheri: Stejneger & Barbour, 1943: 56; Behler & King, 1979: 374; Frost, 1985: 491.

Rana pipiens fisheri: Stebbins, 1951: 365; Schmidt, 1953: 83; Cochran & Goin, 1970: 82; Gorham, 1974: 150.

Rana onca fisheri: Stebbins, 1985: 91; 2003: 238.

Rana (Rana) fisheri: Dubois, 1987a: 41, by implication.

Rana (Pantherana) fisheri: Dubois, 1992: 331.

Rana (Novirana) fisheri: Hillis & Wilcox, 2005, 312.

Lithobates fisheri: Frost, *et al.*, 2006: 369; Dodd, 2013: 547.

Lithobates (Lithobates) fisheri: Dubois, 2006c: 829.

Rana (Lacusirana) fisheri: Hillis, 2007: 335, by implication.

Rana (Lithobates) fisheri: **new combination.**

Misapplied names:

Rana chiricahuensis: Platz & Mecham, 1979, as applied to the central Arizona population.

Distribution: Extinct at the OT. As indicated below, the species is currently restricted to an area in central Arizona near the Mogollon Rim, entering western New Mexico (see map, Hekkala *et al.*, 2011; also Dodd, 2013). Earlier, Stuart *et al.* (2008) suggested the species is totally extinct and gave the historic range including a map. **U.S.**: AZ, NM, NV (formerly, now extinct there).

Common name: Vegas Valley Leopard Frog.

Commentary: Identity of this taxon is controversial, but Hekkala *et al.* (2011) demonstrated that populations previously identified as *Rana chiricahuensis* in central Arizona, are actually molecularly identical to museum specimens of *R. fisheri* collected in the Vegas Valley between 1910 and 1939. Goldberg *et al.* (2004a) and Hillis & Wilcox (2005) previously found the central Arizona population to be separable from the rest of *R. chiricahuensis* molecularly, and Hekkala *et al.* (2011) supported that finding. More study is needed to determine whether *R. fisheri* is a distinct species from *R. chiricahuensis,* or whether those populations are all referable to the same species. Earlier, Jennings, (1988b) suggested that *R. fisheri* may be synonymous with *Rana onca,* but the study of Hekkala *et al.* (2011) refutes that. The species is currently federally listed as extinct, but that needs to be revised now to threatened or endangered.

Rana (Lithobates) grylio Stejneger, 1902, **new combination**

Rana grylio Stejneger, 1902a: 212.—**O**: **H:** USNM 27443.—**OT**: U.S., Mississippi, Hancock County, Bay St. Louis.

Rana grylio: Stejneger & Barbour, 1917: 38; 1943: 57; Boulenger, 1920: 421; Viosca, 1949: 10; Schmidt, 1953: 80; Conant, 1958: 298; Cochran & Goin, 1970: 74; Gorham, 1974: 145; Behler & King, 1979: 374; Frost, 1985: 494; Garrett & Barker, 1987: 46; Conant & Collins, 1998: 557; Dixon, 2000: 77; Elliott *et al.*, 2009: 194.

Rana (Rana) grylio: Dubois, 1987a: 41, by implication.

Rana (Aquarana) grylio: Dubois, 1992: 311.

Rana (Novirana) grylio: Hillis & Wilcox, 2005: 305.

Lithobates grylio: Frost *et al.*, 2006: 369; Elliott *et al.*, 2009: 194; Dodd, 2013: 551.

Rana (Aquarana) grylio: Hillis, 2007: 335, by implication.

Rana (Lithobates) grylio: **new combination.**

Distribution: Eastern Texas to all of Florida along the Gulf coast, then northward along the Atlantic coast to South Carolina. See map in Altig & Lohoefener (1982) or Dodd (2013). Also introduced to Caribbean islands and China. **U.S.**: AL, FL, GA, LA, MS, SC, TX.

Common name: Pig Frog.

Commentary: Altig & Lohoefener (1982) provided a CAAR account. Hillis & Wilcox (2005) found this to be a sister species to the clade including *R. catesbeiana, R. heckscheri, R. clamitans,* and *R. okaloosae.*

Rana (Lithobates) heckscheri Wright, 1924, **new combination**

Rana heckscheri Wright, 1924: 143.—**O**: **H**: CU 1025 (lost); Sanders (1984) identified fig. 5 & 6 of pl. 38 (Wright & Wright, 1932) as of the H.—**OT**: U.S., Florida, Nassau County, Callahan, Alligator Swamp.

Rana heckscheri: Stejneger & Barbour, 1933: 40; 1943: 57; Schmidt, 1953: 80; Cochran & Goin, 1970: 75; Gorham, 1974: 145; Conant, 1975: 339; Behler & King, 1979: 375; Frost, 1985: 494; Conant & Collins, 1998: 556; Elliott *et al.*, 2009: 206.

Rana (Rana) heckscheri: Dubois, 1987a: 41, by implication.

Rana (Aquarana) heckscheri: Dubois, 1992: 331.

Rana (Novirana) heckscheri: Hillis & Wilcox, 2005: 305.

Lithobates heckscheri: Frost *et al.*, 2006: 369; Elliott *et al.*, 2009: 206; Dodd, 2013: 556.

Lithobates (Aquarana) heckscheri: Dubois, 2006c: 829.

Rana (Aquarana) heckscheri: Hillis, 2007: 335, by implication.

Rana (Lithobates) heckscheri: **new combination.**

Distribution: Southern Mississippi to northern Florida along the Gulf coast, northward along the Atlantic coast to southern North Carolina. See map in Sanders (1984) or Dodd (2013). Also introduced into China. **U.S.**: AL, FL, GA, MS, NC, SC.

Common name: River Frog.

Commentary: Sanders (1984) provided a CAAR account. Hillis & Wilcox (2005) found this to be a sister species to *R. catesbeiana*.

Rana (Lithobates) okaloosae Moler, 1985, **new combination**

Rana okaloosae Moler, 1985: 379.—**O**: **H**: FSM 53964.—**OT**: U.S., Florida, Okaloosa County, Eglin Air Force Base, along Malone Creek, Sec 24-T2N-R25W, elev. 13 m.

Rana (Rana) okaloosae: Dubois, 1987a: 42, by implication.

Rana (Aquarena) okaloosae: Dubois, 1992: 331.

Rana okaloosae: Conant & Collins, 1998: 559.

Rana (Novirana) okaloosae: Hillis & Wilcox, 2005: 305.

Lithobates okaloosae: Frost *et al.*, 2006: 369.

Lithobates (Aquarana) okaloosae: Dubois, 2006c: 829.

Rana (Aquarana) okaloosae: Hillis, 2007: 335.

Rana (Lithobates) okaloosae: **new combination.**

Distribution: In the western panhandle of Florida, in Okaloosa, Santa Rosa, and Walton Counties. See map in Moler (1993) or Dodd (2013). **U.S.:** FL.

Common name: Florida Bog Frog.

Commentary: Moler (1993) provided a CAAR account. Listed by Florida as a species of special concern. Austin *et al.* (2003) found that samples of this species are phylogenetically imbedded with populations of *R. clamitans,* suggesting that this may be a synonym of that species or a subspecies. However, the strikingly different advertisement call indicates it is reproductively isolated from *R. clamitans.* Hillis & Wilcox (2005) found this species is a sister of *R. clamitans.* Further evidence of distinctness was provided by Conlon *et al.* (2007), based on comparative analysis of skin secretions.

Rana (Lithobates) onca Cope *in* Yarrow, 1875, **new combination**

Rana onca Cope *in* Yarrow, 1875: 528.—**O**: **H:** USNM 25331 (Cochran, 1961a: 76, sd.).—**OT**: U.S., Utah. Restricted (Tanner, 1929: 49) to Utah, Washington County, along the Virgin River; Pace (1974: 29) determined it was probably collected in the vicinity of St. George, Washington County, Utah.

Rana draytoni onca: Cope, 1886b: 521; 1889a: 443.

Rana onca: Stejneger & Barbour, 1917: 38; Boulenger, 1920: 428; Behler & King, 1979: 376; Platz, 1984: 1; Frost, 1985: 508; Brennan & Holycross, 2006: 44; Elliott *et al.,* 2009: 228.

Rana pipiens onca: Wright & Wright, 1949: 506.

Rana onca onca: Stebbins, 1985: 91, by implication; 2003: 238, by implication.

Rana (Rana) onca: Dubois, 1987a: 41, by implication.

Rana (Pantherana) onca: Dubois, 1992: 331.

Rana (Novirana) onca: Hillis & Wilcox, 2005: 305.

Lithobates onca: Frost *et al.,* 2006: 369; Elliott *et al.,* 2009: 228; Dodd, 2013: 565.

Lithobates (Lithobates) onca: Dubois, 2006d: 325.

Rana (Scurrilirana) onca: Hillis, 2007: 335, by implication.

Rana (Lithobates) onca: **new combination.**

Distribution: The Virgin River Valley region of Utah, and adjacent Arizona and Nevada. See map in Jennings (1988b). Bradford *et al.* (2005) have a map and described the current distribution of remaining populations. There are also maps in Elliott *et al.* (2009) and Dodd (2013). **U.S.**: AZ, NV, UT.

Common name: Relict Leopard Frog.

Commentary: Schmidt (1953: 83) treated this name as a synonym of *R. pipiens brachycephala* (which is now treated as a synonym of *R. pipiens*). Jennings (1988b) provided a CAAR account, and suggested the species is extinct. Jaeger *et al.* (2001) discussed its existence, distribution, and relationships, and Bradford *et al.* (2005) discussed its conservation status. Both those authors and Pfeiler & Markow (2008) provided evidence of a close relationship, or possible synonymy with *R. yavapaiensis*. Hillis & Wilcox (2005) considered it a sister species to *R. yavapaiensis*. Conservation status has been proposed but as yet the species has not been listed.

Rana (Lithobates) palustris Le Conte, 1825, new combination

Rana palustris Le Conte, 1825: 282.—**O**: not stated nor known to exist.—**OT**: not stated; Schmidt (1953:83) designated it as U.S., Pennsylvania, vicinity of Philadelphia, but with no evidence presented that the O came from there, and no basis given for the action, that action is invalid.

Rana (Rana) palustris: Guérin-Méneville, 1838: 16; Dubois, 1987a: 41, by implication.

Rana palustris: Davis & Rice, 1883: 28; Stejneger & Barbour, 1917: 38; 1943: 57; Boulenger, 1920: 444; Viosca, 1949: 10; Schmidt, 1953: 83; Cochran & Goin, 1970: 83; Gorham, 1970: 12; Conant, 1975: 347; Behler & King, 1979: 377; Frost, 1985: 509; Garrett & Barker, 1987: 47; Conant & Collins, 1998: 569; Dixon, 2000: 77; Elliott *et al.*, 2009: 214.

Rana palustris palustris: Hardy, 1964: 91; Gorham, 1974: 149.

Rana (Pantherana) palustris: Dubois, 1992: 332.

Rana (Novirana) palustris: Hillis & Wilcox, 2005: 305.

Lithobates palustris: Frost *et al.*, 2006: 369; Elliott *et al.*, 2009: 214; Dodd, 2013: 568.

Lithobates (Lithobates) palustris: Dubois, 2006c: 829.

Rana (Nenirana) palustris: Hillis, 2007: 335.

Rana (Lithobates) palustris: **new combination**

Rana pardalis Harlan, 1826a: 59.—**O**: not stated nor known to exist.—**OT**: U.S., Pennsylvania, vicinity of Philadelphia.—**Comment**: Harlan proposed the name as a junior synonym of *Rana palustris* Le Conte.—**SS**: Harlan (1835: 104).

Rana palustris mansuetii Hardy, 1964: 91.—**O**: **H**: USNM 150535.—**OT**: U.S., North Carolina, Robeson County, approximately one mile southwest of Maxton, at Maxton Pond. Subspecies rejected by Schaaf & Smith (1970: 240).

Rana palustris mansuetii: Gorham, 1974: 149.

Distribution: Eastern North America, but absent from most of Alabama, Georgia and Florida; to southern Ontario and southern Quebec to Nova Scotia in the north. See map in Schaaf & Smith (1971) plus Fisher *et al.* (2007) for Canada, and Dodd (2013) for the complete distribution. **Can.**: NB, NS, ON, QC. **U.S.**: AL, AR, CT, DE, FL, GA, IA, IL, IN, KS, KY, LA, MA, MD, ME, MI, MN, MO, NC, NH, NJ, NY, OH, OK, PA, RI, SC, TN, TX, VA, VT, WI, WV.

Common name: Pickerel Frog. *French Canadian name:* Grenouille des marais.

Content: Hardy (1964) described subspecies, but Schaaf & Smith (1970) found that geographic variation did not match appropriately and rejected them. Pace (1974) found that in parts of its range where this species overlapped with *R. pipiens,* it was often difficult to identify a specimen as one species or the other. Although she did not directly state that hybridization might be occurring, it seems to be inferred. Currently no subspecies are recognized.

Commentary: Schaaf & Smith (1971) provided a CAAR account. Hillis & Wilcox (2005) found this to be a sister species to the clade including *R. areolata, R. capito,* and *R. sevosa*. Listed in Quebec as at risk.

Rana (Lithobates) pipiens Schreber, 1782, **new combination**

Rana pipiens Schreber, 1782: 182.—**O**: not stated nor known to exist.—**N:** UMMZ 71365 (Pace, 1974: 16).—**OT**: U.S., Carolina, New Jersey, New York. Kellogg (1932: 204) restricted the OT to U.S., New Jersey, Gloucester County, Racoon. Kauffeld (1936: 11) pointed out that Kellogg apparently misunderstood the locality information in Schreber's description, and clarified that the actual OT is the New York locality. Schmidt (1953: 82) restricted the OT to U.S., New York, White Plains, with brief explanation, apparently unaware of Kauffeld's correction.—**OT of N:** U.S., New York, Thompkins County, Etna, Fall Creek.

Rana pipiens: Sonnini & Latreille, 1801: 153; 1830a: 153; Fitzinger, 1826: 64; Stejneger & Barbour, 1917: 38; Gorham, 1970: 11; Conant, 1975: 344; Behler & King, 1979: 377; Dunlap & Platz, 1981: 878; Frost, 1985: 510; Stebbins, 1985: 88; 2003: 234; Conant & Collins, 1998: 566; Dixon, 2000: 27; Brennan & Holycross, 2006: 42; Elliott *et al.,* 2009: 216.

Rana pipiens pipiens: Test, 1893: 57; Wright & Wright, 1933: 34; Stejneger & Barbour, 1943: 57; Schmidt, 1953: 82; Conant, 1958: 300; Cochran & Goin, 1970: 82; Gorham, 1974: 150.

Rana (Rana) pipiens: Dubois, 1987a: 41, by implication.

Rana (Pantherana) pipiens: Dubois, 1992: 332; Hillis, 2007: 335.

Rana (Novirana) pipiens: Hillis & Wilcox, 2005: 305.

Lithobates pipiens: Frost *et al.,* 2006: 369; Elliott *et al.,* 2009: 216; Dodd, 2013: 578.

Lithobates (Lithobates) pipiens: Dubois, 2006c: 829.

Rana (Lithobates) pipiens: **new combination**

Rana virescens brachycephala Cope, 1889b: 403.—**O**: **S**: USNM 3363 (15); see Cochran (1961a: 76) and Kellogg (1932: 208).—**OT**: U.S., Montana, Yellowstone River

Rana pipiens brachycephala: Test, 1893: 57; Schmidt, 1953: 83; Cochran & Goin, 1970: 82.—**SS:** Dunlap & Platz (1981: 875) with *R. pipiens.*

Rana brachycephala: Kauffeld, 1937: 86; Stejneger & Barbour, 1939.—**SS:** Dunlap & Platz, 1981: 878, by implication.

Rana burnsi Weed, 1922: 108.—**O**: **H:** FMNH 3065.—**OT**: U.S., Minnesota, Kandiyohi County, New London.—**SS**: Kellogg, 1932: 209.

Rana pipiens burnsi Myers, 1927: 338.—**SS:** Kellogg, 1932: 209, and Dunlap & Platz, 1981: 878, with *R. pipiens* Schreber.

Rana burnsorum: Michels & Bauer, 2004: 86. Unjustified emendation, but academic, as the name is a junior synonym.

Rana kandiyohi Weed, 1922: 109.—**O**: **H**: FMNH 3066.—**OT**: U.S., Minnesota, Kandiyohi County, New London.—**SS**: Kellog (1932: 209) with *Rana brachycephala*; Kauffeld, 1937: 84.

Rana pipiens kandiyohi: Wells, 1964: 52.

Rana noblei Schmidt, 1925: 1.—**O**: **H**: AMNH 5285.—**OT**: China, Yunnan, Yunnanfu, in error; based on a mislabeled specimen.—**SS**: Schmidt, 1953: 83.

Misapplication of names:

Rana ocellata: Sonnini & Latreille, 1801: 156.

"Rana halecina:" Daudin, 1802a: 63. Interpretted *Rana halecina*: Kalm, 1761, as synonym of *R. pipiens* Schreber; however, the nomen is a junior primary homonym of *Rana halecina* Linnaeus, 1766 (=*Rana virescens* Kalm; =*Rana ocellata* Linnaeus; currently *Leptodactylus ocellatus* (Linnaeus)); Baird, 1859c: 45; Cooper, 1860: 304; Davis & Rice, 1883: 28; Boulenger, 1920: 433.

Rana halecina halecina: Cope, 1875b: 32.

Distribution: Much of the U.S., except absent from the westernmost states and the southern states from Texas to southern Nebraska, east to the Atlantic coast; and much of Canada, except the northernmost parts of the provinces. They have disappeared from much of their range over the last several years (Rorabaugh, 2005); that author has a map of the U.S. distribution. Elliott *et al.* (2009) and Dodd (2013) have maps of the total North American distribution. **Can.**: AB, BC, MB, NB, *NL, NS, NT, ON, QC, PE, SK, YT. **U.S.:** all **except**: AK, AL, AR, DE, FL, GA, HI, KS, LA, MO, MS, NC, NV, OK, SC, WA.

Common name: Northern Leopard Frog. *French Canadian name:* Grenouille léopard du Nord.

Content: Currently no subspecies are recognized.

Commentary: For many years all "leopard frogs" were covered under this name, throughout North America and Mexico. Now only a relatively small portion of that original range includes the current species *Rana pipiens*. This species is sister to the rest of the *pipiens*-subgroup, as treated here (Hillis & Wilcox, 2005), including *R. chiricahuensis* (and *R. subaquavocalis*), and at least three Mexican species. However, a study by Tennessen & Blouin (2010) found *R. pipiens* seemed more closely related to *R. blairi*. Rorabaugh (2005) reviewed conservation status and indicated that declines are alarming, especially in western and central parts of the range. In Canada, listed as endangered in British Columbia and of Special Concern in Northwest Territories. No U.S. populations are yet federally listed, but some states have listed the species as of special concern.

Rana (Lithobates) septentrionalis Baird, 1856, **new combination**

Rana septentrionalis Baird, 1856: 61.—**O**: not stated nor known to exist.—**OT**: U.S., northern Minnesota. Schmidt (1953: 80) further restricted this but provided no evidence the O came from there, so the action is invalid.

Rana septentrionalis: Cope, 1889b: 416; Stejneger & Barbour, 1917: 39; 1943: 59; Boulenger, 1920: 423; Schmidt, 1953: 80; Cochran & Goin, 1970: 77; Gorham, 1970: 10; 1974: 151; Conant, 1975: 342; Behler & King, 1979: 379; Frost, 1985: 514; Conant & Collins, 1998: 562; Elliott *et al.*, 2009: 196.

Rana (Rana) septentrionalis: Dubois, 1987a: 41.

Rana (Aquarana) septentrionalis: Dubois, 1992: 331.

Rana (Novirana) septentrionalis: Hillis & Wilcox, 2005: 317.

Lithobates septentrionalis: Frost *et al.*, 2006: 369; Elliott *et al.*, 2009: 196; Dodd, 2013: 608.

Lithobates (Aquarana) septentrionalis: Dubois, 2006c: 829.

Rana (Aquarana) septentrionalis: Hillis, 2007: 335, by implication.

Rana (Lithobates) septentrionalis: **new combination.**

Distribution: Northeastern North America; northern Minnesota and Wisconsin east to northern New York on to northern Maine, northward to Manitoba east to southern Newfoundland. See map in Hedeen (1977) or Dodd (2013). **Can.**: MB, NB, *NL, NS, ON, QC. **U.S.**: ME, MI, MN, NH, NY, VT, WI.

Common name: Mink Frog. *French Canadian name:* Grenouille du Nord.

Commentary: Hedeen (1977) provided a CAAR account. The study by Hillis & Wilcox (2005) found this species to be a sister group, basal to the clade including *R. grylio, R. catesbeiana, R. heckscheri, R. clamitans,* and *R. okaloosae.*

Rana (Lithobates) sevosa Goin & Netting, 1940, **new combination**

Rana sevosa Goin & Netting, 1940: 137.—**O**: **H:** CM 16809.—**OT**: U.S., Louisiana, St. Tammany Parish, Slidell.

Rana capito sevosa: Wright & Wright, 1942: 186; Schmidt, 1953: 79; Cochran & Goin, 1970: 349; Conant & Collins, 2003: 572.

Rana sevosa: Stejneger & Barbour, 1943: 59; Young & Crother, 2001: 382; Elliott *et al.,* 2009: 210.

Rana areolata sevosa: Viosca, 1949: 10; Conant, 1975: 349; Gorham, 1974: 140; Behler & King, 1979: 368.

Rana (Novirana) sevosa: Hillis & Wilcox, 2005: 305.

Lithobates sevosus: Frost *et al.,* 2006: 369; Elliott *et al.,* 2009: 210; Dodd, 2013: 617.

Lithobates (Lithobates) sevosus: Dubois, 2006c: 829.

Rana (Nenirana) sevosa: Hillis, 2007: 335, by implication.

Rana (Lithobates) sevosa: **new combination.**

Distribution: Originally the Gulf coastal area from southeastern Louisiana to southwestern Alabama, with an isolated population reported in central Tennessee. Now extinct in most of its range (Richter & Jensen, 2005; also see their map). See map in Altig & Lohoefener (1983), as *R. areolata sevosa.* Dodd (2013) has an updated map. **U.S.**: AL, LA, MS, *?TN.

Common name: Dusky Gopher Frog.

Commentary: Altig & Lohoefener (1983) provided a brief CAAR account as a subspecies of *R. areolata.* Young & Crother (2001) recognized this as a distinct species; it was formerly treated as a subspecies of *R. areolata.* Hillis & Wilcox (2005) found this to be a sister species to *R. capito,* and both were most closely related to *R. areolata.* This species is federally listed as endangered in Alabama, Louisiana, and Mississippi, and has apparently been extirpated except in a restricted area of southern Mississippi.

Rana (Lithobates) sphenocephala Cope, 1886, **new combination**

"Rana virescens plantis tetradactylis" Kalm, 1761: 46.—**O**: not stated nor known to exist.—**OT**: U.S., New Jersey (Pace, 1974: 11, sd.).—**Comment**: unavailable nonbinomial name (Stejneger, 1893b: 228).—**SS**: Stejneger & Barbour (1917: 38).

Rana utricularius Harlan, 1826a: 60.—**O**: not stated nor known to exist.—**N**: ANSP 2803 (Pace, 1974: 18).—**OT**: U.S., Pennsylvania and New Jersey; OT of N, U.S., Pennsylvania, Philadelphia.—**Comment**: 1) Brown *et al.* (1977) treated the name as a ND, probably a synonym of *Rana pipiens,* but Zug (1982: 80) disagreed. 2) Brown *et al.* (1982) made a case that Pace violated Art. 75b of the *Code,* thus the action is invalid. 3) the name *Rana sphenocephala* Cope, 1886, was given priority over this name (ICZN, 1992).

Rana utricularia: Conant, 1975: 345.

Rana utricularia utricularia: Pace, 1974: 23.

Rana sphenocephala utricularius: Collins, 1997: 13.

Rana (Novirana) sphenocephala utricularia: Hillis & Wilcox, 2005: 305.

Lithobates (Lithobates) sphenocephalus utricularius: Dubois, 2006a: 325.

Rana (Scurrilirana) sphenocephala utricularia: Hillis, 2007: 335.

Lithobates sphenocephalus utricularius: Frost, McDiarmid & Mendelson, 2008: 8.

"Rana oxyrhynchus" Hallowell, 1857b: 141.—**O**: originally probably in ANSP (lost).—**N**: USNM 56130 (Pace, 1974: 18).—**OT**: U.S., Florida, 300 miles from Key West, near St. John's River, in a sulphur spring. **OT** of **N**, U.S., Florida, Volusia County, Enterprise. Nomen unavailable, preoccupied by *Rana oxyrhynchus* Smith [or Sundeval? as given by Schmidt 1953: 82], 1849 (currently *Ptychadena oxyrhynchus*).—**SS**: Cope (1886b: 517) with *R. halecina*; Boulenger (1919: 413) with *R. utricularius,* thus with *R. sphenocephala.*

Rana halecina sphenocephala Cope, 1886b: 517.—**NN** for *Rana oxyrhynchus* Hallowell, 1856. Priority given over *R. utricularius* (see that nomen above).

Rana virescens sphenocephala: Cope, 1889b: 99.

Rana sphenocephala: Strecker, 1915: 46; Stejneger & Barbour, 1917: 39; Strecker & Williams, 1928: 12; Cochran & Goin, 1970: 83; Behler & King, 1979: 379;

Frost, 1985: 515; Garrett & Barker, 1987: 48; Dixon, 2000: 78; Elliott *et al.*, 2009: 220.

Rana pipiens sphenocephala: Stejneger & Barbour, 1943: 58; Viosca, 1949: 10; Schmidt, 1953: 82; Conant, 1958: 301; Gorham, 1974: 150.

Rana utricularia sphenocephala: Pace, 1974: 22; Conant & Collins, 1998: 568.

Rana (Rana) sphenocephala: Dubois, 1987a: 42, by implication.

Rana (Pantherana) sphenocephala: Dubois, 1992: 332.

Rana sphenocephala sphenocephala: Collins, 1997: 13.

Rana (Novirana) sphenocephala sphenocephala: Hillis & Wilcox, 2005: 305.

Lithobates sphenocephalus: Frost *et al.*, 2006: 369; Elliott *et al.*, 2009: 220; Dodd, 2013: 621.

Lithobates (Lithobates) sphenocephalus sphenocephalus: Dubois, 2006c: 325.

Rana (Scurrilirana) sphenocephala sphenocephala: Hillis, 2007: 335, by implication.

Lithobates sphenocephalus sphenocephalus: Frost, McDiarmid & Mendelson, 2008: 8.

Rana (Lithobates) sphenocephala: **new combination**

Rana virescens virescens Cope, 1889b: 399.—**O**: not stated, but specimen figured seems similar to *R. sphenocephala, fide* Brown *et al.* (1977: 200).—**OT**: not stated. This was inadvertent validation of previously unavailable name (*fide* Frost, 2009-13); see Brown *et al.* (1977)—the name could be applicable to *R. blairi, R. pipiens, R. sphenoscephala*, or other species.

Distribution: Eastern U.S., from eastern Texas north to eastern Kansas, eastward to most of Florida and to southern New York, but not in the Appalachian area. Introduced in California. See map in Brown *et al.* (1982), Elliott *et al.* (2009) or Dodd (2013). **U.S.**: AL, AR, *CA, DE, FL, GA, IA, IL, IN, KS, KY, LA, MD, MO, MS, NC, NJ, OK, SC, TN, TX, VA.

Common name: Southern Leopard Frog (SECSN) or Florida Leopard Frog (C&T); the former seems preferable, as the latter implies a more restricted range, but fits the nominate subspecies better.

Content: Two subspecies, as indicated in the synonymy, are sometimes recognized, although their distinctiveness is controversial—see Brown et al (1977), Zug (1982), Wassersug

(1982), Uzzell (1982), Brown *et al.* (1982). We opt to await further clarification of the relationships of the Florida population (which is currently treated as *Rana (Lithobates) sphenocephala sphenocephala* by some workers) and the rest of the frogs treated as this species (currently treated as *Rana (Lithobates) sphenocephala utricularia* by those workers). Perhaps there will be insufficient differentiation to warrant recognition as subspecies, or perhaps the data will determine that two or more subspecies be recognized, or maybe the molecular data will demonstrate there are two separate species. At present we reserve judgement, and do not recognize two distinct nominate populations.

Commentary: This species is in a sister clade to that including the two species immediately above, and along with several Mexican species that clade is sister to a larger clade of about nine species, including *R. yavapaiensis* and *R. onca* (Hillis & Wilcox, 2005).

The ICZN acted in a compromising response (International Commission on Zoological Nomenclature, 1992: 171-173) to a request from Brown *et al.* (1977) to suppress the nomina *Rana utricularius* Harlan and *Rana virescens* Cope as *nomina dubia* or *nomina oblita,* and to conserve the nomen *Rana sphenocephala* Cope. They had argued that Pace (1974) had revived the name *Rana utricularia* Harlan, 1826, which had not been used since its description, and had no type material. Pace had also designated a neotype (neophoront) in a way that may not have followed the *Code*. After six years, several herpetologists responded negatively, opposing the request; these included Zug (1982), Wassersug (1982), and Uzzell (1982), who argued that the requested action could add confusion to the nomenclature, rather than stabilize it, suggesting that there was on-going research activity on the relationships of the named subspecies, seeking to clarify their relationships, and the commission should not rush to a decision. Brown *et al.* (1982) responded with further evidence of usage of the junior name and lack of use of the older one. After another 10 years there were still no further publications clarifying the biological questions about the taxonomic relations of the populations, and the commission made its ruling. They did not suppress the *utricularius* nomen as requested, rather they gave precedence to the junior name (*sphenocephala*) over the older one (*utricularia*), so that if they are treated as synonyms, *sphenocephala* is the valid one. They did not act on the request to suppress the *virescens* nomen, so it remains available. This means that the correct name for the species is *Rana sphenocephala,* but that if subspecies are recognized, or the populations are found to be separate species, the nomen *utricularia* is available for one of them. Both names were placed on the official list of species names.

Rana (Lithobates) sylvatica Le Conte, 1825, **new combination**

? *Rana canagica* Pallas, 1831: 12.—**O**: none stated, nor any known to exist, but probably deposited in ZISP.—**OT**: Aleutian islands between Camtschatcam and America, and in continental Russia, corrected to U.S., Alaska, Kanaga Island (Kuzmin, 1996: 51).—**SS**: Lindholm, 1924: 46, with *R. sylvatica.* Myers (1930a: 62) disagreed and suggested it is a senior synonym of *Bufo boreas* (see that account). Said to be **ND** (Stejneger & Barbour, 1933: vi), but Kuzmin (1996: 51) treated it as a **NO**.

Rana sylvatica Le Conte, 1825: 282.—**O**: not stated nor known to exist.—**OT**: not stated; Schmidt (1953: 81) designated it as U.S., New York, vicinity of New York city, but presented no evidence the O was from there, hence that action is invalid.

Rana temporaria var. *sylvatica*: Günther, 1859a: 17; Jordan, 1878: 188.

Rana temporaria silvatica: Cope, 1875b: 32; Davis & Rice, 1883: 28.—**IS**.

Rana silvatica: Cope, 1889b: 115; Boulenger, 1920: 458.—**IS**.

Rana sylvatica: Stejneger & Barbour, 1917: 39; Conant, 1958: 303; Cochran & Goin, 1970: 78; Gorham, 1970: 10; 1974: 152; Behler & King, 1979: 380; Frost, 1985: 516; Stebbins, 1985: 83; 2003: 227; Conant & Collins, 1998: 563; Elliott *et al.* 2009: 198.

Rana sylvatica sylvatica: Schmidt, 1938: 378; 1953: 81; Stejneger & Barbour, 1943: 59.

Rana (Rana) sylvatica: Dubois, 1987a: 41.

Rana (Novirana) sylvatica: Hillis & Wilcox, 2005: 305.

Lithobates sylvaticus: Frost *et al.*, 2006: 369; Elliott *et al.*, 2009: 198; Dodd, 2013: 637.

Lithobates (Aquarana) sylvaticus: Dubois, 2006c: 829.
Rana (Lithobates) sylvatica: **new combination**

Rana pennsylvanica Harlan, 1826a: 58.—**O**: not stated, probably originally in ANSP; described as a junior synonym of *R. sylvatica* Le Conte.—**OT**: Pennsylvania and New Jersey.—**SS**: Duméril & Bibron, 1841: 363.

Rana cantabrigensis Baird, 1856: 62.—**O**: MCZ, Agassiz collection, specimens uncertain (Barbour & Loveridge, 1929: 327).—**OT**: U.S., Massachusetts, Cambridge; said to be in error (Howe, 1899: 369, presumed from western or northwestern North America, but occurs there so could be correct); Schmidt (1953: 81) designated an OT of Can., Saskatchewan, Moose Jaw, but without any evidence it came from there, that action is invalid.—**SS**: Günther, 1859a: 17.
Rana temporaria cantabrigensis: Cope, 1875b: 32.
Rana cantabridgensis cantabridgensis: Cope, 1886b: 520.—**IS**.
Rana cantabrigensis cantabrigensis: Cope, 1889a: 436; Stejneger & Barbour, 1917: 36.
Rana cantabrigensis: Boulenger, 1920: 455; Stejneger & Barbour, 1933: 40.
Rana sylvatica cantabrigensis: Schmidt, 1938: 378; 1953: 81; Stejneger & Barbour, 1943: 60. Subspecies rejected by Martof & Humphries, 1959: 350.

Rana cantabridgensis latiremis Cope, 1886b: 520.—**O**: **S:** USNM (4), including 13723 (od.); USNM 13723-13726 (Cochran, 1961a: 73, sd.).—**OT**: U.S., Alaska, Lake Alloknagits; corrected to Alaska, north of Nushagak, Lake Aleknagik (Cochran, 1961a: 73). Subspecies rejected by Boulenger, 1891: 453.
Rana cantabrigensis latiremis: Cope, 1889a: 435; Stejneger & Barbour, 1917: 37.
Rana sylvatica latiremis: Schmidt, 1938: 378; Stejneger & Barbour, 1943: 60. Subspecies rejected by Martof & Humphries, 1959: 350.

Rana cantabrigensis evittata Cope, 1889a: 435.—**O**: **H**: USNM 5169.—**Comment**: Cope designated the one specimen, but Cochran (1961a: 72) listed 23 additional specimens as "syntypes."—**OT**: U.S., Washington, Puget Sound.—**SS**: Boulenger, 1919: 414.

Rana sylvatica cherokiana Witschi, 1954: 764.—**O**: **H**: USNM 134417.—**OT**: U.S., North Carolina, Cherokee County, Murphy, elevation 462 m. Subspecies rejected by Martof & Humphries, 1959: 350.

Rana maslini Porter, 1969: 213.—**O**: **H:** USNM 166435.—**OT**: U.S., Wyoming, Albany County, 0.25 miles north of Fox Park, in sub-alpine park, elev. 9000 ft.—**SS**: Bagdonas & Pettus,1976: 105; synonymy rejected by Collins (1997: 12), with no explanation.

Distribution: Northern Georgia to eastern Missouri, and north through the Appalachians to include all of the northeastern and Great Lakes states, then most of Canada and most of Alaska, with isolated populations in northeastern Alabama, northwestern Arkansas and adjacent Missouri, Kansas, northern Colorado and adjacent Wyoming, and northern Idaho. This is the only herp species found in every Canadian province and territory (but introduced in Newfoundland), and the only North American frog whose range extends north of the Arctic Circle. See map in Martof (1970) or Dodd (2013). **Can.**: AB, BC, MB, NB, *NL, NS, NT, NU, ON, PE, QC, SK, YT. **U.S.**: AK, AL, AR, CO, DE, GA, ID, *IL, IN, KY, MA, MD, ME, MI, MN, MO, NC, ND, NH, NJ, NY, OH, OK, PA, RI, TN, VA, VT, WI, WV, WY.

Common name: Wood Frog. *French Canadian name:* Grenouille des bois.

Commentary: Martof (1970) provided a CAAR account. There is a long history of controversy regarding the relationship of this species to other ranids. Hillis & Davis (1986) found it to be sister to the extralimital *R. temporaria,* and that pair in turn was an ancestral sister to the *R. boylii* species-group. That would place it in the subgenus *Rana*. Pytel (1986) also found it to be most closely related to the Eurasian *R. temporaria* group. Hillis & Wilcox (2005) found it to be a sister taxon to the *R. catesbeiana*-group, which places it in the subgenus *Lithobates,* but their support for that relationship was weak. Frost *et al.* (2006) included it in *Lithobates,* essentially adopting Hillis & Wilcox' (2005) arrangement. Che *et al.* (2007) found it to be a sister to the pair *R. catesbeiana* + *R. heckscheri*. The phylogenetic position of *R. sylvatica* still remains somewhat uncertain at this time, but we tentatively consider it a member of the subgenus *Lithobates*. Zeyl (1993) examined variation among samples from much of the U.S. range and some of the Canadian distribution, using gel electrophoresis. He found several genetic distances were similar to those between species of other congeneric anurans, but despite the high divergence found, none of the samples seemed to represent distinct species.

Rana (Lithobates) tarahumarae Boulenger, 1917, **new combination**

Rana tarahumarae Boulenger, 1917: 416.—**O**: **S:** BM 1947.2.1.63-1947.2.1.64 (ex 1914.1.28.148-149) and 1947.2.28.76-1947.2.28.79 (ex 1911.12.12.36-39) (Kellogg, 1932: 214, sd.).—**OT**: Mexico, Chihuahua, Sierra Tarahumaré, Ioquiro and Barranca del Cobre. Restricted (Smith & Taylor, 1950: 327) to Chihuahua, Yoquivo. Modified (Schmidt, 1953: 81) to Mexico, Sonora, Sierra Tarahumare, but with no evidence, so invalid.

Rana tarahumarae: Stejneger & Barbour, 1933: 42; 1943: 60; Boulenger, 1920: 468; Schmidt, 1953: 81; Cochran & Goin, 1970: 77; Gorham, 1974: 152; Behler & King, 1979: 381; Frost, 1985: 517; Stebbins, 1985: 87; 2003: 234; Liner, 1994: 28; Lemos-Espinal *et al.*, 2004b: 6, 79; Brennan & Holycross, 2006: 46; Elliott *et al.*, 2009: 234.

Rana (Rana) tarahumarae: Dubois, 1987a: 41, by implication.

Rana (Zweifelia) tarahumarae: Dubois, 1992: 330.

Rana (Novirana) tarahumarae: Hillis & Wilcox, 2005: 305.

Lithobates tarahumarae: Frost *et al.*, 2006: 369; Elliott *et al.*, 2009: 234; Dodd, 2013: 669.

Lithobates (Lithobates) tarahumarae: Dubois, 2006c: 830.

Rana (Zweifelia) tarahumarae: Hillis, 2007: 335.

Rana (Lithobates) tarahumarae: **new combination.**

Distribution: South-central Arizona, south into Mexico. Most of the range is extralimital. See map in Zweifel (1968b), Elliott *et al.* (2009) or Dodd (2013). The species is apparently extinct in Arizona (hence, in North America; Rorabaugh & Hale, 2005). **U.S.**: AZ.

Common name: Tarahumara Frog.

Commentary: Zweifel (1968b) provided a CAAR account. Hillis & Wilcox (2005) found this to be in a clade of five Mexican species, which in turn is a sister to the large *pipiens-group.* The species formerly reached North America but is federally listed as extirpated in the U.S. Hale *et al.* (2005) indicated that the Arizona populations died off by 1983, probably due to chytridiomycosis; Sonoran populations in warmer, more southern localities are infected with the fungus but survive, apparently because of lack of stress of colder waters to the north. There have been reports of reintroductions, but it is unclear how successful those have been. We list it in hopes that North American populations may be re-established.

—

Rana (Lithobates) virgatipes Cope, 1891, new combination

Rana virgatipes Cope, 1891: 1017.—**O**: **S:** ANSP 10759-10764 (6) (Malnate, 1971: 353, sd.).—**OT**: U.S., New Jersey, Atlantic County, cut-off of a tributary of the Great Egg Harbor River; revised (Fowler, 1906b: 662) to New Jersey, Atlantic County, above May's Landing, Mare Run, tributary of Great Egg Harbor.

Rana virgatipes: Fowler, 1906b: 662; Stejneger & Barbour, 1917: 39; 1943: 60; Boulenger, 1920: 429; Schmidt, 1953: 80; Cochran & Goin, 1970: 74; Gorham, 1974: 153; Conant, 1975: 340; Behler & King, 1979: 381; Frost, 1985: 520; Conant & Collins, 1998: 558; Elliott *et al.*, 2009: 202.

Rana (Rana) virgatipes: Dubois, 1987a: 41, by implication.

Rana (Aquarana) virgatipes: Dubois, 1992: 331.

Rana (Novirana) virgatipes: Hillis & Wilcox, 2005: 305.

Lithobates virgatipes: Frost *et al.*, 2006: 369; Elliott *et al.*, 2009: 202; Dodd, 2013: 674.

Lithobates (Lithobates) virgatipes: Dubois, 2006c: 829.

Rana (Aquarana) virgatipes: Hillis, 2007: 335.

Rana (Lithobates) virgatipes: **new combination.**

Distribution: Atlantic coastal plain, from southeastern Georgia and adjacent Florida, to New Jersey (see map in Gosner & Black, 1968, or Dodd, 2013). **U.S.**: FL, DE, GA, MD, NC, NJ, SC, VA.

Common name: Carpenter Frog (earlier often called Sphagnum Frog).

Commentary: Gosner & Black (1968) provided a CAAR account. Hillis & Wilcox (2005) found this a basal sister to the clade including *R. septentrionalis* and the *R. catesbeiana* group.

Rana (Lithobates) yavapaiensis Platz & Frost, 1984, **new combination**

Rana yavapaiensis Platz & Frost, 1984: 940.—**O**: **H:** AMNH 117632.—**OT**: U.S., Arizona, Yavapai County, Tule Creek, 34° 00' N, 112° 16' W, elev. 670 m.

Rana yavapaiensis: Frost, 1985: 521; Stebbins, 1985: 91; 2003: 239; Liner, 1994: 28; Brennan & Holycross, 2006: 44; Elliott *et al.*, 2009: 232.

Rana (Rana) yavapaiensis: Dubois, 1987a: 41.

Rana (Pantherana) yavapaiensis: Dubois, 1992: 331.

Rana (Novirana) yavapaiensis: Hillis & Wilcox, 2005: 305.

Lithobates yavapaiensis: Frost *et al.*, 2006: 369; Elliott *et al.*, 2009: 232; Dodd, 2013: 681.

Lithobates (Lithobates) yavapaiensis: Dubois, 2006c: 830.

Rana (Scurrilirana) yavapaiensis: Hillis, 2007: 335, by implication.

Rana (Lithobates) yavapaiensis: **new combination.**

Distribution: Southeastern to central Arizona, adjacent western New Mexico, and a disjunct population in northwestern Arizona and adjacent Utah and Nevada. There are old records of the species along the southern Colorado River and into southeastern California. See map in Platz (1988). **U.S.**: AZ, NM, NV, UT.

Common name: Lowland Leopard Frog.

Commentary: Platz (1988) provided a CAAR account. This species is indicated (Hillis & Wilcox, 2005) to be a sister to *R. onca*. Both are imbedded in a clade of mostly Middle American species, which forms a sister clade to the one including the other North American members of the *berlandieri* subgroup.

LITERATURE CITED

ABEL, O. 1919. Die Stämme der Wirbeltiere. Berlin, Vereinigung wissenschaftlicher Verleger: i-xviii, 1-914.

ACLOQUE, A. 1900. Faune de France contenant la description de espèces indigènes, disposees en tableaux analytiques, et illustree de figures representant les types caractéristiques des genres. Tome 1 Mammifères, oiseaux, poissons, reptiles, batraciens, protochordes. Paris, Baillière: i-vii, 1-548.

ADAMS, L. A., AND H. T. MARTIN. 1929. A new urodele from the Pliocene of Kansas. *American Journal of Science* (5) 17: 504-520.

ADAMS, M. J. 2005. Ascaphidae: *Ascaphus montanus*, Montana (Rocky Mountain) Tailed Frog. *In*: LANNOO, M. J. (ed.), Amphibian Declines: The Conservation Status of United States Species. Part 2: Species Accounts. Berkeley, University of California Press: 382.

ADAMS, M. J., AND C. A. PEARL. 2005. *Ascaphus truei*, (Coastal) Tailed Frog. *In*: LANNOO, M. J. (ed.), Amphibian Declines: The Conservation Status of United States Species. Part 2: Species Accounts. Berkeley, University of California Press: 382-384.

ADLER, K. 1965. *Plethodon longicrus. Catalogue of American Amphibians and Reptiles.* American Society of Ichthyologists and Herpetologists (18): 1.

ADLER, K. 1976. New genera and species described in Holbrook's "North American Herpetology". *In*: ADLER, K. (ed.), Holbrook's North American Herpetology, 2nd edition (1842), reprint. Ithaca, Society for the Study of Amphibians and Reptiles: xxix-xliii.

ADLER, K., AND D. M. DENNIS. 1962. *Plethodon longicrus*, a new salamander (Amphibia: Plethodontidae) from North Carolina. *Special Publication of the Ohio Herpetological Society* (4): 1-14, pl. 1-2.

AGASSIZ, L. 1842-1846. Nomenclator zoologicus, continens nomina systematica generum animalium tam viventium quam fossilium, secundum ordinem alphabeticum disposita, adjectis auctoribus, libris, in quibus reperiuntur, anno editionis, etymologia et auctore. Fasciculus VI. Nomina systematica generum reptilium. (Addenenda et corrigenda). Soloduri, Jent & Gassman: 1-8.

AGASSIZ, L. 1850. Lake Superior: its physical character, vegetation, and animals, compared with those of other and similar regions. Boston, Gould, Kendall and Lincoln: [i-vii + 1 pl.], i-xii, 1-428, pl. 1-8.

ALEXANDRINO, J., S. J. E. BAIRD, L. LAWSON, J. R. MACEY, C. MORITZ, AND D. B. WAKE. 2005. Strong selection against hybrids at a hybrid zone in the *Ensatina* ring species complex and its evolutionary implications. *Evolution* 59 (6): 1124-1347.

ALLEN, E. R., AND W. T. NEILL. 1949. A new subspecies of salamander (genus *Plethodon*) from Florida and Georgia. *Herpetologica* 5(2): 112-114.

ALTIG, R., AND P. C. DUMAS. 1971. *Rana cascadae. Catalogue of American Amphibians and Reptiles.* Society for the Study of Armphibians and Reptiles (105): 1-2.

ALTIG, R., AND P. C. DUMAS. 1972. *Rana aurora. Catalogue of American Amphibians and Reptiles.* Society for the Study of Amphibians and Reptiles (160): 1-4.

ALTIG, R., AND R. LOHOEFENER. 1982. *Rana grylio. Catalogue of American Amphibians and Reptiles.* Society for the Study of Amphibians and Reptiles (286): 1-2.

ALTIG, R., AND R. LOHOEFENER. 1983. *Rana areolata. Catalogue of American Amphibians and Reptiles.* Society for the Study of Amphibians and Reptiles (324): 1-4.

ALTIG, R., R. W. MCDIARMID, K. A. NICHOLS, AND P. C. USTACH. 1998. A key to the anuran tadpoles of the United States and Canada. Contemporary Herpetology Information Series 2: unpaginated.

ANDERSON, J. A., AND S. G. TILLEY. 2003. Systematics of the *Desmognathus ochrophaeus* complex in the Cumberland Plateau of Tennessee. *Herpetological Monographs* 17: 75-110.

ANDERSON, J. D. 1965. *Ambystoma annulatum. Catalogue of American Amphibians and Reptiles.* American Society of Ichthyologists and Herpetologists (19): 1-2.

ANDERSON, J. D. 1967a. *Ambystoma texanum. Catalogue of American Amphibians and Reptiles*. American Society of Ichthyologists and Herpetologists (37): 1-2.

ANDERSON, J. D. 1967b. *Ambystoma opacum. Catalogue of American Amphibians and Reptiles*. American Society of Ichthyologists and Herpetologists (46): 1-2.

ANDERSON, J. D. 1967c. *Ambystoma maculatum. Catalogue of American Amphibians and Reptiles*. American Society of Ichthyologists and Herpetologists (51): 1-4.

ANDERSON, J. D. 1968. *Rhyacotriton* and *R. olympicus. Catalogue of American Amphibians and Reptiles*. American Society of Ichthyologists and Herpetologists (68): 1-2.

ANDERSON, J. D. 1969. *Dicamptodon* and *D. ensatus. Catalogue of American Amphibians and Reptiles*. American Society of Ichthyologists and Herpetologists (76): 1-2.

ANGEL, F. 1947. Vie et moeur des amphibiens. Paris, Payot: 1-317.

Anonymous. 1869. Robert Kennicott. *Transactions of the Chicago Academy of Sciences* 1 (2): 131-226.

ANTHONY, C. D. 2005. *Plethodon ouachitae*, Rich Mountain Salamander. *In*: LANNOO, M. J. (ed.), Amphibian Declines: The Conservation Status of United States Species. Berkeley, University of California Press: 831-833.

APPLEGARTH, J. S. 1994. Wildlife surveying and monitoring methods: amphibians and reptiles of the Eugene district. Eugene, Oregon, U.S. Department of the Interior, Bureau of Land Management: 1-59.

ARDILA-ROBAYO, M. C. 1979. Status sistematico del genero *Geobatrachus* Ruthven, 1915 (Amphibia: Anura). *Caldasia* (Bogotá) 11 (59): 383-495. [not seen; *fide* Frost (2009-13)].

ASHTON, R. E., JR. 1990. *Necturus lewisi. Catalogue of American Amphibians and Reptiles*: Society for the Study of Amphibians and Reptiles (456): 1-2.

ASHTON, R. E., JR., AND R. FRANZ. 1979. *Bufo quercicus. Catalogue of American Amphibians and Reptiles*: Society for the Study of Amphibians and Reptiles (222): 1-2.

AUSTIN, J. D., S. C. LOUGHEED, P. E. MOLER, AND P. T. BOAG. 2003. Phylogenetics, zoogeography, and the role of dispersal and vicariance in the evolution of the *Rana catesbeiana* (Anura: Ranidae) species group. *Biological Journal of the Linnean Society* 80: 601-624.

AUSTIN, J. D., S. C. LOUGHEED, L. NEIDRAUER, A. A. CHEK, AND P. T. BOAG. 2002. Cryptic lineages in a small frog: the post-glacial history of the spring peeper, *Pseudacris crucifer* (Anura: Hylidae). *Molecular Phylogenetics and Evolution* 25: 316-329.

BAEZ, A. M., AND L. A. PUGENER. 2003. Ontogeny of a new Palaeogene pipid frog from southern South America and xenopodinomorph evolution. *Zoological Journal of the Linnean Society* 139 (3): 439-476.

BAGDONAS, K. R., AND D. PETTUS. 1976. Genetic compatibility in wood frogs (Amphibia, Anura, Ranidae). *Journal of Herpetology* 10 (2): 105-112.

BAILEY, J. R. 1937. Notes on plethodont salamanders of the southeastern United States. *Occasional Papers of the Museum of Zoology, University of Michigan* (364): 1-10.

BAILEY, M. A., AND P. E. MOLER. 2003. *Necturus alabamensis*. *Catalogue of American Amphibians and Reptiles*. Society for the Study of Amphibians and Reptiles (761): 1-2.

BAIRD, A. B., J. K. KREJCA, J. R. REDDELL, C. E. PEDEN, M. J. MAHONEY, D. M. HILLIS, AND M. J. LANNOO. 2006. Phylogeographic structure and color pattern variation among populations of *Plethodon albagula* on the Edwards Plateau of central Texas. *Copeia* 2006(4): 760-768.

BAIRD, S. F. 1850a ("1849"). Revision of the North American tailed-batrachia, with descriptions of new genera and species. *Journal of the Academy of Natural Sciences of Philadelphia* (2) 1 (4): 281-292.

BAIRD, S. F. 1850b ("1849"). Descriptions of four new species of North American salamanders, and one new species of scink [sic]. *Journal of the Academy of Natural Sciences of Philadelphia* (2) 1 (4): 292-294.

BAIRD, S. F. 1856 ("April 1854"). Descriptions of new genera and species of North American frogs. *Proceedings of the Academy of Natural Sciences of Philadelphia* 7 (2): 59-62.

BAIRD, S. F. 1859a. Reptiles (explanation of plates). *In*: Reports of explorations and surveys to ascertain the most practicable and economical route for a railroad from the Mississippi River to the Pacific Ocean, made under the direction of the Secretary of War, in 1853-6, by Lieutenant

Q.W. Whipple and assisted by J.C. Ives. Volume 10: Part 3. Washington, B. Tucker, Printer: 7-16, pl. 24-36.

BAIRD, S. F. 1859b ("1857"). Report on reptiles collected on the survey (No. 4). In: (ed.), Report of Lieut. E.G. Beckwith, Third Artillery, upon explorations for a railroad route, near the 38th and 39th parallels of north latitude. by Captain J.W. Gunnison, Corps of Topographical Engineers, and near the forty first parallel of north latitude, by Lieut. E.G. Beckwith, Third Artillery. Zoological Report. In: Reports of Explorations and Surveys, to ascertain the most practicable and economical route for a railroad from the Mississippi River to the Pacific Ocean, made under the directon of the Secretary of War, in 1853-5, according to acts of Congress of March 3, 1853, May 31, 1854, and August 5, 1854. Volume 10: Part 4, Zoological Report. Washington, B. Tucker, Printer: 17-20, pl. 17-24.

BAIRD, S. F. 1859c. Report upon the reptiles of the route (No. 4). In: Reports of explorations for a railroad route (near the thirty-fifth parallel of north latitude) from the Mississippi River to the Pacific Ocean by Lieutenant Q.W. Whipple and assisted by J.C. Ives, Corps of Typographical Engineers. 1853-'54. In: Reports of Explorations and Surveys, to ascertain the most practicable and economical route for a railroad from the Mississippi River to the Pacific Ocean, made under the directon of the Secretary of War, in 1853-5, according to acts of Congress of March 3, 1853, May 31, 1854, and August 5, 1854. Volume 10: Part 6—Zoological Report. Washington, B. Tucker, Printer: 37-45, pl. 25-27.

BAIRD, S. F. 1859d ("1857"). Report upon reptiles collected on the survey (No. 4). In: (ed.), Report of Lieut. Henry L. Abbot, Corps of Topographical Engineers, upon exploratons for a railroad route, from the Sacramento Valley to the Colmbia River, made by Lieut. R.S. Williamson, Corps of Topgraphical Engineers, assisted by Lieut. Henry L. Abbot, Corps of Topographical Engineers. In: Reports of Explorations and Surveys, to ascertain the most practicable and economical route for a railroad from the Mississippi River to the Pacific Ocean, made under the directon of the Secretary of War, in 1853-5, according to acts of Congress of March 3, 1853, May 31, 1854, and August 5, 1854. Volume 10: Part 4, Zoological Report. Washington, B. Tucker, Printer: 9-13, pl. 11, 28, 30, 44.

BAIRD, S. F. 1859e. Reptiles of the boundary, with notes by the naturalists of the survey. *In*: United States and Mexican boundary survey, made under the order of Lieut. Col. W. H. Emory, major first cavalry, and United States Commissioner. Volume 2, Part 2 (3). *Edited by* EMORY, W. H. 1-35, pl. 1.

BAIRD, S. F., AND C. GIRARD. 1852a. Descriptions of new species of reptiles, collected by the U.S. exploring expedition under the command of Capt. Charles Wilkes, U.S.N. First part.-Including the species from the Western coast of America. *Proceedings of the Academy of Natural Sciences of Philadelphia* 6: 174-177.

BAIRD, S. F., AND C. GIRARD. 1852b. Characteristics of some new reptiles in the museum of the Smthsonian Institution. *Proceedings of the Academy of Natural Sciences of Philadelphia* 6: 68-70.

BAIRD, S. F., AND C. GIRARD. 1852d. Characteristics of some new reptiles in the museum of the Smithsonian Institution: Third part. Containing the batrachians in the collection made by J. H. Clark, esq., under Col. J. D. Graham, on the United States and Mexican Boundary. *Proceedings of the Academy of Natural Sciences of Philadelphia* 6: 173.

BAIRD, S. F. AND C. GIRARD. 1853a. List of reptiles collected in California by Dr. John L. Le Conte, with description of new species. *Proceedings of the Academy of Natural Sciences of Philadelphia* 6: 300-302.

BAIRD, S. F., AND C. GIRARD. 1853b. [untitled: two new tailless batrachians from Oregon]. *Proceedings of the Academy of Natural Sciences of Philadelphia* 6: 378-379.

BAKER, C. L. 1945. The natural history and morphology of Amphiumae. *Report of the Reelfoot Lake Biological Station* (69): 55-91.

BAKER, C. L. 1947. The species of amphiumae. *Journal of the Tennessee Academy of Science* 22: 9-21.

BAKER, J. K. 1957. *Eurycea troglodytes*: a new blind cave salamander from Texas. *Texas Journal of Science* 9(3): 328-336.

BARBER, P. H. 1999. Phylogeography of the Canyon Treefrog, *Hyla arenicolor* (Cope) based on mitochondrial DNA sequence data. *Molecular Ecology* 8: 547-562.

BARBOUR, R. W. 1950. A new subspecies of the salamander *Desmognathus fuscus*. *Copeia* 1950(4): 277-278.

BARBOUR, T. 1910. *Eleutherodactylus ricordii* in Florida. *Proceedings of the Biological Society of Washington* 23: 100.

BARBOUR, T. 1937. Third list of Antillean amphibians and reptiles. *Bulletin of the Museum of Comparative Zoology* 82 (2): 77-166.

BARBOUR, T., AND L. J. COLE. 1906. Reptilia, Amphibia, and Pisces. *IN*: Vertebrates from Yucatan. *Bulletin of the Museum of Comparative Zoology* 50 (5): 146-159, pl. 1-2.

BARBOUR, T., AND A. LOVERIDGE. 1929. Typical reptiles and amphibians in the Museum of Comparative Zoology. *Bulletin of the Museum of Comparative Zoology* 69 (10): 205-360.

BARBOUR, T., AND A. LOVERIDGE. 1946. First supplement to typical reptiles and amphibians. *Bulletin of the Museum of Comparative Zoology* 96 (2): 59-214.

BARBOUR, T., AND G. K. NOBLE. 1920. Some amphibians from northwestern Peru, with a revision of the genera *Phyllobates* and *Telmatobius*. *Bulletin of the Museum of Comparative Zoology* 63 (8): 396-427, pl. 1-3.

BARNES, D. H. 1826. An arrangement of the genera of batracian animals, with a description of the more remarkable species; including a monograph of the doubtful reptils [sic]. *American Journal of Sciences and Arts* 11: 268-297.

BARTON, B. S. 1808. Some account of the *Siren lacertina*. and other species of the same genus of amphibious animals. In a letter from professor Barton, of Philadelphia, to Mr. John Gottlob Schneider, of Saxony. Philadelphia, (privately published): [i], 1-34, 1 pl.

BARTON, [B. S.]. 1809. Some account of a new species of North American lizard. *Transactions of the American Philosophical Society* 6: 108-112, pl. 4, fig. 6.

BARTON, B. S. 1814 ("1812"). A memoir concerning an animal of the class Reptilia, or Amphibia, which is known, in the United-States, by the names of alligator and hell-bender.

Philadelphia, privately published; printed by Griggs & Dickinson: i-iv, 5-26, +1 pl.

BARTRAM, W. 1791. Travels in North & South Carolina, Georgia, east and west Florida, the Cherokee country, the extensive territories of the Muscogulges or Creek Confederacy and the country of the Chactaws: containing an account of the soil and natural productions of those regions, together with observations on the manners of the Indians. Philadelphia, James & Johnson: i-xxxiv, 1-522 + pl. 1-9.

BAUER, A. M., D. A. GOOD, AND R. GÜNTHER. 1993. An annotated type catalogue of the caecilians and salamanders (Amphibia: Gymnophiona and Caudata) in the Zoological Museum, Berlin. *Mitteilungen aus dem Zoologischen Museum in Berlin* 69 (2): 285-306.

BAUER, A. M., R. GÜNTHER, AND H. E. ROBECK. 1996. An annotated type catalogue of the hemisotid, microhylid, myobatrachid, pelobatid and pipid frogs in the Zoological Museum, Berlin. *Mitteilungen aus dem Zoologie Museum in Berlin* 72: 259-275.

BAYER, F. 1885 ("1884"). O kostře žab z čeledi pelobatid. Příspěvek srovnávací osteologicii obojživelníkův [On the frog skeleton from the pelobatid family. A comparative amphibian osteological contribution]. *Abhandlungen der königlich-böhmischen Geselschaft der Wissenschaften (Z Pojednání Král. České Společnosti Nauk Řady* 6. Díl 12, Čls. 13) (6) 12 (13): 2-24, pl. 1-2.

BEACHY, C. K. 2005. *Gyrinophilus porphyriticus*, Spring Salamander. *In*: LANNOO, M. J. (ed.), Amphibian Declines: The Conservation Status of United States Species. Part 2: Species Accounts. Berkeley, University of California Press: 776-778.

BEAMER, D. A., AND T. LAMB. 2008. Dusky salamanders (*Desmognathus*, Plethodontidae) from the coastal plain: multiple independent lineages and their bearing on the molecular phylogeny of the genus. *Molecular Phylogenetics and Evolution* 47: 143-153.

BEAMER, D. A., AND M. J. LANNOO. 2005a. *Plethodon chattahoochee*, South Mountain Gray-cheeked Salamander. *In*: LANNOO, M. J. (ed.), Amphibian Declines:

The Conservation Status of United States Species. Part 2: Species Accounts. Berkeley, University of California Press: 793-794.

BEAMER, D. A., AND M. J. LANNOO. 2005b. *Plethodon chlorbryonis*, Atlantic Coast Slimy Salamander. *In*: LANNOO, M. J. (ed.), Amphibian Declines: The Conservation Status of United States Species. Part 2: Species Accounts. Berkeley, University of California Press: 795-796.

BEAMER, D. A., AND M. J. LANNOO. 2005c. *Plethodon cylindraceus,* White-spotted Slimy Salamander. *In*: LANNOO, M. J. (ed.), Amphibian Declines: The Conservation Status of United States Species. Part 2: Species Accounts. Berkeley, University of California Press: 800-802.

BEAMER, D. A., AND M. J. LANNOO. 2005d. *Plethodon glutinosus*, Northern Slimy Salamander. *In*: LANNOO, M. J. (ed.), Amphibian Declines: The Conservation Status of United States Species. Part 2: Species Accounts. Berkeley, University of California Press: 810-812.

BEAMER, D. A., AND M. J. LANNOO. 2005e. *Plethodon grobmani*, Southeastern Slimy Salamander. *In*: LANNOO, M. J. (ed.), Amphibian Declines: The Conservation Status of United States Species. Part 2: Species Accounts. Berkeley, University of California Press: 811-812.

BEAMER, D. A., AND M. J. LANNOO. 2005f. *Plethodon meridianus*, South Mountain Gray-cheeked Salamander. *In*: LANNOO, M. J. (ed.), Amphibian Declines: The Conservation Status of United States Species. Part 2: Species Accounts. Berkeley, University of California Press: 822-823.

BEAMER, D. A., AND M. J. LANNOO. 2005g. *Plethodon metcalfi,* Southern Gray-cheeked Salamander. *In*: LANNOO, M. J. (ed.), Amphibian Declines: The Conservation Status of United States Species. Part 2: Species Accounts. Berkeley, University of California Press: 823-825.

BEAMER, D. A., AND M. J. LANNOO. 2005h. *Plethodon mississippi,* Mississippi Slimy Salamander. *In*: LANNOO, M. J. (ed.), Amphibian Declines: The Conservation Status of United States Species. Part 2: Species Accounts. Berkeley, University of California Press: 825-826.

BEAMER, D. A., AND M. J. LANNOO. 2005i. *Plethodon montanus*, Northern Gray-cheeked Salamander. *In*: LANNOO, M. J. (ed.), Amphibian Declines: The Conservation Status of United States Species. Part 2: Species Accounts. Berkeley, University of California Press: 826-828.

BEAMER, D. A., AND M. J. LANNOO. 2005j. *Plethodon ocmulgee*, Ocmulgee Slimy Salamander. *In*: LANNOO, M. J. (ed.), Amphibian Declines: The Conservation Status of United States Species. Part 2: Species Accounts. Berkeley, University of California Press: 831.

BEAMER, D. A., AND M. J. LANNOO. 2005k. *Plethodon savannah*, Savannah Slimy Salamander. *In*: LANNOO, M. J. (ed.), Amphibian Declines: The Conservation Status of United States Species. Part 2: Species Accounts. Berkeley, University of California Press: 837-838.

BEHLER, J. L., AND F. W. KING. 1979. The Audubon Society Field Guide to North American Reptiles and Amphibians. New York, A. A. Knopf: 1-743, incl. 657 pl.

BELL, T. 1839. A history of British reptiles. London, Van Voorst: i-xxiv, 3-142.

BERNARDO, J. 1994. Experimental analysis of allocation in two divergent natural salamander populations. *American Naturalist* 143: 14-38.

BESHARSE, J. C., AND J. R. HOLSINGER. 1977. *Gyrinophilus subterraneus*, a new troglobitic salamander from southern West Virginia. *Copeia* 1977 (4): 624-634.

BISHOP, S. C. 1924. Notes on salamanders. *Bulletin of the New York State Museum. Report of the Director for 1923* 253: 87-103.

BISHOP, S. C. 1928a. Notes on some amphibians and reptiles from the southeastern states, with a description of a new salamander from North Carolina. *Journal of the Elisha Mitchell Scientific Society* 43 (3/4): 153-170.

BISHOP, S. C. 1928. A new subspecies of the red salamander from Louisiana. *Occasional Papers of the Boston Society of Natural History* 5: 247-249, pl. 15.

BISHOP, S. C. 1934. Description of a new salamander from Oregon, with notes on related species. *Proceedings of the Biological Society of Washington* 47: 169-171.

BISHOP, S. C. 1937. A remarkable new salamander from Oregon. *Herpetologica* 1 (3): 92-95.

Bishop, S. C. 1941. Notes on salamanders with descriptions of several new forms. *Occasional Papers of the Museum of Zoology, University of Michigan* (451): 1-27.

Bishop, S. C. 1942. An older name for a recently described salaman-der. *Copeia* 1942(4): 256.

Bishop, S. C. 1943. Handbook of Salamanders. Ithaca, Comstock: i-xiv, 1-555.

Bishop, S. C. 1944. A new neotenic plethodont salamander, with notes on related species. *Copeia* 1944 (1): 1-5.

Bishop, S. C. 1945. The identity of *Siredon harlanii* Duméril, Bibron, and Duméril and *Axolotes maculata* Owen. *Herpetologica* 3 (1): 24.

Bishop, S. C., and B. D. Valentine. 1950. A new species of *Desmognathus* from Alabama. *Copeia* 1950: 39-43.

Bishop, S. C., and M. R. Wright. 1937. A new neotenic salamander from Texas. *Proceedings of the Biological Society of Washington* 50: 141-144.

Blackburn, D. C. and D. B. Wake. 2011. Class Amphibia Gray, 1925. *In*: Zhang, Z.-Q. (ed.), Animal biodiversity: an outline of higher level classification and survey of taxonomic richness. *Zootaxa* (3148) Monograph. Magnolia Press, Aukland NZ: 39-55.

Blair, A. P. 1957. Amphibians. *In*: Blair, W. F., A. P. Blair, P. Brodcorb, F. R. Cagle, and G. A. Moore (eds.), Vertebrates of the United States. New York, McGraw-Hill: 211-271.

Blair, A. P. 1961. Metamorphosis of *Pseudotriton palleucus* with iodine. *Copeia* 1961 (3): 499.

Blair, A. P. 1967. *Plethodon ouachitae. Catalogue of American Amphibians and Reptiles*: American Society of Ichthyologists and Herpetologists (40): 1-2.

Blair, A. P., and H. L. Lindsay. 1965. Color pattern variation and distribution of two large *Plethodon* salamanders endemic to the Ouachita Mountains of Oklahoma and Arkansas. *Copeia* 1965 (3): 331-335.

Blair, W. F. 1955a. Differentiation of mating call in spadefoots, genus *Scaphiopus*. *Texas Journal of Science* 7 (2): 183-188.

Blair, W. F. 1955b. Mating call and stage of speciation in the *Microhyla olivacea-M. carolinensis* complex. *Evolution* 9 (4): 469-480.

Blair, W. F. 1957. Mating call and relationships of *Bufo hemiophrys* Cope. *Texas Journal of Science* 9 (1): 99-108.

Blair, W. F. 1959. Genetic compatibility and species groups in U.S. toads (*Bufo*). *Texas Journal of Science* 11 (4): 427-453.

Blair, W. F. 1972. Evolution in the genus *Bufo*. Austin, University of Texas Press: i-viii, 1-459, pl. 1-6.

Blanford, W. T. 1870. Observations on the Geology and Zoology of Abyssinia: made during the progress of the British expedition to that country in 1867-68. London, Macmillan & Co.: i-xii, 1-487.

Blatchley, W. S. 1897a. Indiana caves and their fauna. *In*: *Annual Report of the State Geologist for 1896. Indianapolis, Department of Geological and Natural Resources, Indiana*: 121-175, pl. IV-XII.

Blatchley, W. S. 1897b. The fauna of Indiana caves. *In*: *Annual Report of the State Geologist for 1896. Indianapolis, Department of Geological and Natural Resources*: 175-185.

Blatchley, W. S. 1901. On a small collection of batrachians with descriptions of two new species. In: *Annual Report of the State Geologist for 1900. Indianapolis, Department of Geological and Natural Resources*: 759-763.

Blem, C. R. 1979. *Bufo terrestris. Catalogue of American Amphibians and Reptiles.* Society for the Study of Amphibians and Reptiles (223): 1-4.

Bogart, J. P., and M. W. Klemens. 2008. Additional distributional records of *Ambystoma laterale, A. jeffersonianum* (Amphibia: Caudata) and their unisexual kleptogens in northeastern North America. *American Museum Novitates*: 1-58.

Bogart, J. P., L. A. Lowcock, C. W. Zeyl, and B. K. Mable. 1987. Genome constitution and reproductive biology of hybrid salamanders, genus *Ambystoma*, on Kelley's Island in Lake Erie. *Canadian Journal of Zoology* 65 (9): 2188-2201.

Bogart, J. P., and C. E. Nelson. 1976. Evolutionary implications from karyotypic analysis of frogs of the families Microhylidae and Rhinophrynidae. *Herpetologica* 32 (2): 199-208.

Bogert, C. M. 1954. Amphibians and reptiles of the world. *In*: Drimmer, F. (ed.), The animal kingdom: the strange and wonderful ways of mammals, birds, reptiles, fishes and insects. A new authentic natural history of the wildlife of the world. Volume 2. New York, Greystone Press: Book III: 1189-1390.

Bogert, C. M. 1958. Sounds of North American frogs: The biological significance of voice in frogs. Folkways Records and Service Corp. 1-16.

Bogert, C. M. 1962. Isolation mechanisms in toads of the *Bufo debilis* group in Arizona and western Mexico. *American Museum Novitates* (2100): 1-37.

Böhme, W., and W. Bischoff. 1984. Amphibien und Reptilien in Die Wirbeltiersammlungen des Museums Alexander Koenig. *Bonner Zoologische Monographien* 19: 151-213.

Boie, H. 1828. Bemerkungen über die Abtheilungen im natürlichen Systeme und deren Characteristik. *Isis von Oken* 21 (4): 351-364.

Bokermann, W. C. A. 1966. Lista anotada das localidades tipo de anfíbios Brasileiros. *São Paulo, Serviço de Documentação-RUSP*: 1-183.

Bolkay, S. J. 1919. Osnove uporedne osteologije anurskih batrahija: sa dodatkom o porijeklu Anura i sa skicom naravnoga sistema istih [Elements of the comparative osteology of the tailless Batrachia: with an appendix on the probable origin of the Anurans and a sketch of their natural system. English resume pp. 353-356]. Glasnika Zemaljskog Muzeja u Bosni i Hercegovini (Sarajevo) 31: 277-357.

Bonaparte, C. L. 1831b. Saggio d'una distribuzione metodica degli animali vertebrati. Roma, Presso Antonoio Boulzaler: 1-144.

Bonaparte, C. L. 1832 ("1831"). Saggio di una distribuzione metodica degli animali vertebrati. Roma, Presso Antonio Boulzaler: 1-78.

Bonaparte, C. L. 1838a. *Hyla viridis. Raganella arborea. In*: Iconographia della fauna italica per le quattro classi degli animali vertebrati. Roma, Salviucci: [193-198], pl. [46].

Bonaparte, C. L. 1838b. Amphibiorum tabula analytica. *Nuovi Annali delle Scienze Naturali* (Bologna) 1: 391-397.

Bonaparte, C. L. 1838c. *Rana temporaria*. Ranocchia rossa. *In*: Iconographia della fauna italica per le quattro classi degli animali vertebrati. Roma, Salviucci: [203-204], pl. [46].

Bonaparte, C. L. 1839a. *Bufo vulgaris*. Rospo comune. *In*: Iconografia della fauna italica per le quattro classi degli animali vertebrati. Roma, Salviucci: [223-228], pl. [49].

Bonaparte, C. L. 1839b. *Euproctus platycephalus*. Euprotto del Rusconi. *In*: Iconografia della fauna italica per le quattro classi degli animali vertebrati. Roma, Salviucci: [259-266], pl. [54].

Bonaparte, C. L. 1839c. Synopsis Vertebratorum Systematis. *Nuovi Annali delle Scienze naturali, Bologna* 2: 105-133.

Bonaparte, C. L. 1839e ("1832-1841"). Iconografia della fauna italica per le quattro classi degli animali vertebrati. Tomo 2. Amphibi. Roma, Salviucci: 270 pp, unnumbered.

Bonaparte, C. L. 1840a. Prodromus systematis herpetologiae. *Nuovi Annali delle Scienze naturali, Bologna* 4: 90-101.

Bonaparte, C. L. 1845a. Specchio generale dei sistemi erpetologico ed anfibiologico. Atti della sesta Riunione degli Scienzati italiani, tenuta in Milano nel Settembre del MDCCCXLIV. Atti della sesta Riunione degli Scienzati italiani (tenuta in Milano nel Settembre del MDCCCXLIV) 6: 376-378.

Bonaparte, C. L. 1845b. Specchio generale dei sistemi erpetologico, anfibiologico ed ittiologico. Milano, Luigi di Giacomo Pirola: 1-11.

Bonaparte, C.-L. 1850. Herpetologiae et Amphibiologiae, editio altera reformata. *In:* Bonaparte, C.-L. Conspectus systematum: Mastozoölogiae, editio altera reformata. Ornithologiae, editio reformata additis synonymi Grayania et Selysanis. Herpetologiae et Amphibiologiae, editio altera reformata. Ichthyologiae, editio reformata. Lugduni Batavorum, Apud E. J. Brill Academiae Typographum: 2 pp (unnumbered).

Bonett, R. M., and P. T. Chippindale. 2004. Speciation, phylogeography and evolution of life history and morphology in plethodontid salamanders of the *Eurycea multiplicata* complex. *Molecular Ecology* 13: 1189-1203.

Bonett, R. M., and P. T. Chippindale. 2005. *Eurycea quadridigitata*, Dwarf Salamander. *In*: Lannoo, M. J. (ed.), Amphibian Declines: The Conservation Status of

United States Species. Berkeley, University of California Press: 759-760.

BONETT, R. M., P. T. CHIPPINDALE, P. E. MOLER, R. W. VAN DEVENTDER, AND D. B. WAKE. 2009. Evolution of gigantism in amphiumid salamanders. *PLoS ONE* 4 (5): e5615, 19 pp.

BONETT, R. M., M. A. STEFFEN, S. M. LAMBERT, J. J. WIENS, AND P. T. CHIPPINDALE. 2013a. Evolution of paedomorphosis in plethodontid salamanders: ecological correlates and re-evolution of metamorphosis. *Evolution* 2013: 1-17.

BONETT, R. M., A. L. TRUJANO-ALVAREZ, M. J. WILLIAMS, AND E. K. TIMPE. 2013b. Biogeography and body size shuffling of aquatic salamander communities on a shifting refuge. *Proceedings of the Royal Society B* 280: 1-8.

BONNATERRE, L. A. 1789. Erpétologie. *In*: Tableau encyclopédique et méthodique des trois Règnes de la nature, dédié et présenté a M. Necker, Ministre d'Ètat, & Directeur Général des Finances. Paris, Chez Panckouke: i-xxviii, 1-70, pl. 1-35 +A.

BORY DE SAINT-VINCENT, [J. B.]. 1842. Traité élémentaire d'erpétologie ou d'histoire naturelle des reptiles [including Iconographie des reptiles ou collection de figures]. Paris, Mairet et Fournier: i-viii, 1-292, [i-ii], 1-20, pl. 1-52.

BOSC, L. A. G. 1803. Le grenouille criarde, *Rana calamitans*. *In*: Nouveau dictionnaire d'histoire naturelle, appliqué aux arts, principalement à l'agriculture et à l'économie rurale et domestique, par une société de naturalistes et d'agriculteurs, avec des figures tirées des trois règnes de la nature. Tome 10. Paris, Crapelet: 436.

BOULENGER, G. A. 1878. Quelques mots sur les *Euproctes*. *Bulletin de la Société Zoologique de France* 3: 304-308.

BOULENGER, G. A. 1881. *Leptodactylus caliginosus* Girard et *L. albilabris* Günther. *Bulletin de la Société Zoologique de France* 6: 30-35.

BOULENGER, G. A. 1882a. Catalogue of the Batrachia Salientia s. Ecaudata in the collection of the British Museum. Second edition. London, Taylor & Francis: i-xvi, 1-503, pl. 1-30.

BOULENGER, G. A. 1882b. Catalogue of the Batrachia Gradientia s. Caudata and Batrachia Apoda in the collection of the British Museum. London, Taylor & Francis: i-xvi, 1-503, pl. 1-30.

BOULENGER, G. A. 1882c. Description of a new genus and species of frogs of the family Hylidae. *Annals and Magazine of Natural History* (5) 10: 326-328.

BOULENGER, G. A. 1883a. *Bufo beldingi*. *Zoologcal Record* 19: 26.

BOULENGER, G. A. 1883b. Notes on little known species of frogs. *Annals and Magazine of Natural History* 16-19.

BOULENGER, G. A. 1883c. Descriptions of new species of lizards and frogs collected by Herr A. Forrer in Mexico. *Annals and Magazine of Natural History* (5) 11: 342-344.

BOULENGER, G. A. 1887a. Description of a new tailed batrachian from Corea. *Annals and Magazine of Natural History* (5) 19: 67.

BOULENGER, G. A. 1887b. Descriptions of new Reptiles and Batrachians in the British Museum (Natural History). Part III. *Annals and Magazine of Natural History* (5) 20: 50-53.

BOULENGER, G. A. 1888a. On a rare American newt, *Molge meridionalis* Cope. *Annals and Magazine of Natural History* (6) 1: 24.

BOULENGER, G. A. 1888b. Note on the classification of the Ranidae. *Proceedings of the Zoological Society of London* 1888: 204-206.

BOULENGER, G. A. 1890a. Second report on additions to the batrachian collection in the Natural-History Museum. *Proceedings of the Zoological Society of London* 1890: 323-328 + 2 pl.

BOULENGER, G. A. 1890b. *Hypopachus cuneus*. *Zoologcal Record* 26: 21.

BOULENGER, G. A. 1891. Notes on American batrachians. *Annals and Magazine of Natural History* (6) 8: 453-457.

BOULENGER, G. A. 1893b. *Rana octoplicata*. *Zoologcal Record* 29: 37.

BOULENGER, G. A. 1894b. On the genus *Phryniscus* of Wiegmann. *Annals and Magazine of Natural History* (6) 14: 374-375.

BOULENGER, G. A. 1900. *Zoologcal Record* 36: 29.

BOULENGER, G. A. 1917. Descriptions of new frogs of the genus *Rana*. *Annals and Magazine of Natural History* (8) 20: 413-418.

BOULENGER, G. A. 1919. Synopsis of the American species of *Rana*. *Annals and Magazine of Natural History* (9) 3: 408-416.

BOULENGER, G. A. 1920. A monograph of the American frogs of he genus *Rana*. *Proceedings of the American Academy of Arts and Sciences* 55 (9): 413-480.

BOUNDY, J. 2000. *Batrachoseps attenuatus*. *Catalogue of American Amphibians and Reptiles*. Society for the Study of Amphibians and Reptiles (701): 1-6.

BOURRET, R. 1927. La faune de l'Indochine: vertébrés. Hanoi, Société de Geographie de Hanoi: [i-ix], 1-453, pl.I-IX, map.

BRADFORD, D. F., R. D. JENNINGS, AND J. R. JAEGER. 2005. *Rana onca*. Relict Leopard Frog. *In*: LANNOO, M. J. (ed.), Amphibian Declines: The Conservation Status of United States Species, Part 2: Species Accounts. Berkeley, University of California Press: 567-568.

BRADY, M. K., AND F. HARPER. 1935. A Florida subspecies of *Pseudacris nigrita* (Hylidae). *Proceedings of the Biological Society of Washington* 48: 107-110.

BRAGG, A. N. 1944. The spadefoot toads in Oklahoma with a summary of our knowledge of the group. *American Naturalist* 78: 517-533.

BRAGG, A. N. 1945. The spadefoot toads in Oklahoma with a summary of our knowledge of the group. II. *American Naturalist* 79: 52-72.

BRAGG, A. N. 1954. *Bufo terrestris charlesmithi*, a new subspecies from Oklahoma. *Wasmann Journal of Biology* 12 (2): 245-254.

BRAGG, A. N., AND O. SANDERS. 1951. A new subspecies of the *Bufo woodhousii* group of toads. *Wasmann Journal of Biology* 9 (3): 363-378.

BRAME, A. H., JR. 1957. List of the Living Caudata of the World. mimeographed, unpublished. [*fide* Frost, 2009-13; not seen]

BRAME, A. H., JR. 1959. Status of the salamander *Ambystoma tremblayi* Comeau. *Herpetologica* 15 (1): 20.

BRAME, A. H., JR. 1967. A list of the world's Recent and fossil salamanders. *Herpeton: The Journal of the Southwestern Herpetologists Society* 2 (1): 1-26.

BRAME, A. H., JR. 1970. A new species of *Batrachoseps* (slender salamander) from the desert of southern California. *Contributions in Science, Los Angeles County Museum of Natural History* (200): 1-11.

BRAME, A. H., JR. 1972. Checklist of Living and Fossil Salamanders of the World. 1-387. [unpublished manuscript *fide* Frost (2009-13); not seen]

BRAME, A. H., JR., AND K. F. MURRAY. 1968. Three new slender salamanders (*Batrachoseps*) with a discussion of relationships and speciation within the genus. *Science Bulletin of the Los Angeles County Museum of Natural History* 4: 1-35.

BRANCH, W. R., AND A. M. BAUER. 2005. The life and herpetological contributions of Andrew Smith, with an introduction, concordance of names and annotated bibliography. Villanova, Society for the Study of Amphibians and Reptiles: 1-19.

BRANDON, R. A. 1963. *Triturus* (*Gyrinophilus*) *lutescens* Rafinesque, 1832 (Amphibia, Caudata): proposed suppression under the plenary powers. Z.N.(S.) 1516. *Bulletin of Zoological Nomenclature* 20 (3): 210-211.

BRANDON, R. A. 1965a. A new race of the neotenic salamander *Gyrinophilus palleucus*. *Copeia* 1965 (3): 346-352.

BRANDON, R. A. 1965b. *Typhlotriton, T. nereus,* and *T. spelaeus. Catalogue of American Amphibians and Reptiles*. American Society of Ichthyologists and Herpetologists (20): 1-2.

BRANDON, R. A. 1966a. Systematics of the salamander genus *Gyrinophilus*. *Illinois Biological Monographs* (Urbana) 35: 1-86.

BRANDON, R. A. 1966b. A reevaluation of the status of the salamander, *Typhlotriton nereus* Bishop. *Copeia* 1966 (3): 555-561.

BRANDON, R. A. 1966c. *Phaeognathus* and *P. hubrichti*. *Catalogue of American Amphibians and Reptiles*. American Society of Ichthyologists and Herpetologists (26): 1-2.

BRANDON, R. A. 1967a. *Gyrinophilus*. *Catalogue of American Amphibians and Reptiles*. American Society of Ichthyologists and Herpetologists (31): 1-2.

BRANDON, R. A. 1967b. *Gyrinophilus palleucus*. *Catalogue of American Amphibians and Reptiles*. American Society of Ichthyologists and Herpetologists (32): 1-2.

BRANDON, R. A. 1967c. *Gyrinophilus porphyriticus. Catalogue of American Amphibians and Reptiles.* American Society of Ichthyologists and Herpetologists (33): 1-3.

BRANDON, R. A. 1967d. *Haideotriton* and *H. wallacei. Catalogue of American Amphibians and Reptiles.* American Society of Ichthyologists and Herpetologists (39): 1-2.

BRANDON, R. A. 1989. Natural history of the axolotl and its relationship to other ambystomatid salamanders. *In*: ARMSTRONG, J. B., AND G. M. MALACINSKI (eds.), Developmental Biology of the Axolotyl. New York, Oxford University Press: 12-21.

BRANDON, R. A., AND J. H. BLACK. 1970. The taxonomic status of *Typhlotriton braggi* (Caudata, Plethodontidae). *Copeia* 1970 (2): 388-391.

BRANDON, R. A., J. JACOBS, A. H. WYNN, AND D. M. SEVER. 1986. A naturally metamorphosed Tennessee Cave Salamander (*Gyrinophilus palleucus*). *Journal of the Tennessee Academy of Science* 61 (1): 1-2.

BRANDON-JONES, D., W. DUCKWORTH, P. D. JENKINS, A. B. RYLANDS, AND E. E. SARMIENTO. 2007. The genitive of species-group scientific names formed from personal names. *Zootaxa* (1541): 41-48.

BRANDT, B. B., AND C. F. WALKER. 1933. A new species of *Pseudacris* from the southeastern United States. *Occasional Papers of the Museum of Zoology, University of Michigan* (272): 1-7.

BRENNAN, T. C., AND A. T. HOLYCROSS. 2006. A field guide to amphibians and reptiles in Arizona. Phoenix, Arizona Game & Fish Department: i-vi, 1-150.

BRIMLEY, C. S. 1907. The salamanders of North Carolina. *Journal of the Elisha Mitchell Scientific Society* 23: 150-156.

BRIMLEY, C. S. 1912. Notes on the salamanders of the North Carolina mountains with descriptions of two new forms. *Proceedings of the Biological Society of Washington* 25 (28): 135-140.

BRIMLEY, C. S. 1915. List of reptiles and amphibians of North Carolina. *Journal of the Elisha Mitchell Scientific Society* 30 (4): 195-205.

BRIMLEY, C. S. 1917. The two forms of red *Spelerpes* occurring at Raleigh, N.C. *Proceedings of the Biological Society of Washington* 30 (21): 87-88.

BRIMLEY, C. S. 1924. The waterdogs (*Necturus*) of North Carolina. *Journal of the Elisha Mitchell Scientific Society* 40 (3-4): 166-168.

BRIMLEY, C. S. 1927. An apparently new salamander (*Plethodon clemsonae*) from S.C. *Copeia* (164): 73-75.

BRIMLEY, C. S. 1928a. Yellow-cheeked *Desmognathus* from Macon County, N.C. *Copeia* (166): 21-23.

BRIMLEY, C. S., AND W. B. MABEE. 1925. Reptiles, amphibians and fishes collected in eastern North Carolina in the autumn of 1923. *Copeia* (139): 14-16.

BRITCHER, H. W. 1903. Batrachia and reptilia of Onondaga County. *Proceedings of the Onondaga Academy of Science* (Syracuse) 1: 120-122.

BROCCHI, M. P. 1877a. Note sur quelques batrachiens hylaeformes recueillis au Mexique et au Guatemala. *Bulletin de la Société Philomathique de Paris* (7) 1: 122-132.

BROCCHI, M. P. 1877b. Sur quelques batraciens raniformes et bofoniformes de l'Amérique Centrale. *Bulletin de la Société Philomathique de Paris* (7) 1: 175-197.

BROCCHI, M. P. 1879. Sur divers batraciens anoures de l'Amérique Centrale. *Bulletin de la Société Philomathique de Paris* (7) 3: 19-24.

BROCCHI, M. P. 1881. Mission scientifique au Mexique et dans l'Amerique Centrale. *Ouvrage publié par ordre du Ministère de l'Instruction Publique. Recherches Zoologiques.* [*fide* Frost (2009-13); not seen]

BROCCHI, M. P. 1882. Étude des batraciens de l'Amérique Centrale. Paris, Imprimerie Nationale. [*fide* Frost (2009-13); not seen]

BRODIE, E. D., JR. 1970. Western salamanders of the genus *Plethodon*: systematics and geographic variation. *Herpetologica* 26 (4): 468-516.

BRODIE, E. D., JR. 1971. *Plethodon stormi. Catalogue of American Amphibians and Reptiles*. American Society of Ichthyologists and Herpetologists (91): 1-2.

BRODIE, E. D., JR., AND R. M. STORM. 1970. *Plethodon vandykei. Catalogue of American Amphibians and*

Reptiles. American Society of Ichthyologists and Herpetologists (103): 1-2.
BRODIE, E. D., JR., AND R. M. STORM. 1971. *Plethodon elongatus. Catalogue of American Amphibians and Reptiles*. American Society of Ichthyologists and Herpetologists (102): 1-2.
BRONGNIART, A. 1800. Essai d'une classification naturelle des reptiles, par le citoyen. Second Partie. Formation et disposition des genres. *Bulletin de la Science, Société Philomathique de Paris* 2 (36): 89-91 + pl. 6.
BRONN, H. G. 1849. Handbuch einer Geschichte der Natur. Volume 3. Stuttgart, E. Schwetzerbart: i-viii, I-LXXXIV, 1-1106.
BROOKES, J. 1828. A prodromus of a synopsis animalium, comprising a catalogue raisonné of the zootomical collection of Joshua Brookes. London, R. Taylor. [*fide* Frost (2009-13); not seen]
BROWN, A. E. 1908 ("1908-1909"). Generic types of nearctic reptilia and amphibia. *Proceedings of the Academy of Natural Sciences of Philadelphia* 60 (1908): 112-127.
BROWN, B. C. 1950. An annotated check list of the reptiles and amphibians of Texas. Waco, *Baylor University Studies*: i-xii, 1-259.
BROWN, B. C. 1967a. *Eurycea latitans. Catalogue of American Amphibians and Reptiles*. American Society of Ichthyologists and Herpetologists (34): 1-2.
BROWN, B. C. 1967b. *Eurycea nana. Catalogue of American Amphibians and Reptiles*. Society for the Study of Amphibians and Reptiles (35): 1-2.
BROWN, B. C. 1967c. *Eurycea neotenes. Catalogue of American Amphibians and Reptiles*. Society for the Study of Amphibians and Reptiles (36): 1-2.
BROWN, H. A. 1976. The status of California and Arizona populations of the Western Spadefoot Toads (genus *Scaphiopus*). *Contributions in Science, Los Angeles County Museum of Natural History* 286: 1-15.
BROWN, L. E. 1973. *Bufo houstonensis. Catalogue of American Amphibians and Reptiles*. American Society of Ichthyologists and Herpetologists (133): 1-2.

BROWN, L. E. 1992. *Rana blairi. Catalogue of American Amphibians and Reptiles.* Society for the Study of Amphibians and Reptiles (536): 1-6.

BROWN, L. E., H. M. SMITH, AND R. S. FUNK. 1977. Request for the conservation of *Rana sphenocephala* Cope 1886, and the suppression of *Rana utricularius* Harland, 1826 and *Rana virescens* Cope, 1889 (Amphibia: Salientia). Z.N.(S.) 2141. *Bulletin of Zoological Nomenclatur* 33 (3-4): 195-203.

BROWN, L. E., H. M. SMITH, AND R. S. FUNK. 1982. Comments on the proposed conservation of *Rana sphenocephala* Cope, 1886. Z.N.(S.) 2141 (5.). *Bulletin of Zoological Nomenclature* 39 (2): 84-90.

BROWN, W. C., AND S. C. BISHOP. 1947. A new species of Desmognathus from North Carolina. *Copeia* 1947: 163-166.

BRUCH, C. 1862. Beiträge zur Naturgeschichte und Classification der nackten Amphibien. *Würzburger Naturwissenschaftliche Zeitschrift* 3: 181-224.

BRYSON, R. W., JR., A. NIETO-MONTES DE OCA, J. R. JAEGER, AND B. R. RIDDLE. 2010. Elucidation of cryptic diversity in a widespread Nearctic treefrog reveals episodes of mitochondrial gene capture as frogs diversified across a dynamic landscape. *Evolution* 64 (8): 2315-2330.

BURGER, W. L., H. M. SMITH, AND F. E. POTTER, JR. 1950. Another neotenic *Eurycea* from the Edwards Plateau. *Proceedings of the Biological Society of Washington* 63: 51-57.

BURGER, W. L., P. W. SMITH, AND H. M. SMITH. 1949. Notable records of reptiles and amphibians in Oklahoma, Arkansas, and Texas. *Journal of the Tennessee Academy of Science* 24: 130-134.

BURNS, D. M. 1954. A new subspecies of the salamander *Plethodon vandykei. Herpetologica* 10: 83-87.

BURNS, D. M. 1962. The taxonomic status of the salamander *Plethodon vandykei larselli. Copeia* 1962 (1): 177-181.

BURNS, D. M. 1964. *Plethodon larselli. Catalogue of American Amphibians and Reptiles.* American Society of Ichthyologists and Herpetologists (13): 1.

BURT, C. E. 1932. Records of amphibians from the eastern and central United States. *American Midland Naturalist* 13 (702): 75-85.

BURT, C. E. 1936a. Contributions to the Herpetology of Texas I. Frogs of the genus *Pseudacris*. *American Midland Naturalist* 17 (4): 770-775.

BURT, C. E. 1938. Contributions to Texas herpetology. VI. Narrow-mouthed froglike toads (*Microhyla* and *Hypopachus*). *Papers of the Michigan Academy of Sciences, Arts, and Letters* 1937: 607-610.

CALDWELL, J. P. 1982. *Hyla gratiosa*. *Catalogue of American Amphibians and Reptiles*. Society for the Study of Amphibians and Reptiles (298): 1-2.

CAMERANO, L. 1879. Di alcune anfibii anuri esistenti nelle collezioni del R. Museo Zoologico di Torino. Atti dell'Accademia della Scienze di Torino. *Classe di Scienze Fisiche, Matimatiche, e Naturali* 14 (5): 866-898.

CAMP, C. D. 2004. *Desmognathus folkertsi*. *Catalogue of American Amphibians and Reptiles*. Society for the Study of Amphibians and Reptiles (782): 1-3.

CAMP, C. D., T. LAMB, AND J. R. MILANOVICH. 2012. *Urspelerpes* and *Urspelerpes brucei*. *Catalogue of American Amphibians and Reptiles*. Society for the Study of Amphibians and Reptiles (885): 1-3.

CAMP, C. D., W. E. PETERMAN, J. R. MILANOVICH, T. LAMB, J. C. MAERZ, AND D. B. WAKE. 2009. A new genus and species of lungless salamander (family Plethodontidae) from the Appalachian highlands of the south-eastern United States. *Journal of Zoology* 279: 86-94.

CAMP, C. D., Z. L. SEYMOUR, AND J. A. WOOTEN. 2013. Morphological variation in the cryptic species *Desmognathus quadramaculatus* (Black-belliied Salamander) and *Desmognathus folkertsi* (Dwarf Black-bellied Salamander). *Journal of Herpetology* 47 (3): 471-479.

CAMP, C. D., S. G. TILLEY, R. M. AUSTIN, JR., AND J. L. MARSHALL. 2002. A new species of Black-Bellied Salamander (genus *Desmognathus*) from the Appalachian Mountains of northern Georgia. *Herpetologica* 58 (4): 471-484.

CAMP, C. L. 1915. *Batrachoseps major* and *Bufo cognatus californicus*, new amphibia from Southern California. *University of California Publications in Zoology* 12 (12): 327-334.

CAMP, C. L. 1916a. *Spelerpes platycephalus*, a new alpine salamander from the Yosemite National Park, California. *University of California Publications in Zoology* 17 (3): 11-14.

CAMP, C. L. 1916c. Description of *Bufo canorus*, a new toad from Yosemite National Park. *University of California Publications in Zoology* 17 (6): 59-62.

CAMP, C. L. 1917. Notes on the systematic status of the toads and frogs of California. *University of California Publications in Zoology* 17 (9): 115-125.

CAMPBELL, B. 1931. Notes on *Batrachoseps*. *Copeia* 1931 (3): 131-134.

CARLIN, J. L. 1997. Genetic and morphological differentiation between *Eurycea longicauda longicauda* and *E. guttolineata* (Caudata: Plethodontidae). *Herpetologica* 53 (2): 206-217.

CARR, A. F., JR. 1939. *Haideotriton wallacei*, a new subterranean salamander from Georgia. *Occasional Papers of the Boston Society of Natural History* 8: 333-336.

CARR, A. F., JR. 1940. A contribution to the herpetology of Florida. *University of Florida Publication, Biological Science Series* 3 (1): i-iv, 1-118.

CARR, D. E. 1996. Morphological variation among species and populations of salamanders in the *Plethodon glutinosus* complex. *Herpetologica* 52 (1): 56-65.

CARVALHO, A. L. D. 1954. A preliminary synopsis of the genera of American microhylid frogs. *Occasional Papers of the Museum of Zoology University of Michigan* (555): 1-19.

CASE, S. M. 1978. Biochemical systematics of members of the genus *Rana* native to western North America. *Systematic Biology* 27 (3): 299-311.

CATESBY, M. 1743. The Natural history of Carolina, Florida, and the Bahama Islands: containing the figures of birds, beasts, fishes, serpents, insects, and plants. Volume 2. London, self published: [1-v], 1-101, i-xlx.

CHANNING, A. 1979 ("1978"). A new bufoid genus (Amphibia: Anura) from Rhodesia. *Herpetologica* 34 (4): 394-397.

CHAPARRO, J. C., J. B. PRAMUK, AND A. G. GLUESENKAMP. 2007a. A new species of arboreal *Rhinella* (Anura: Bufonidae) from cloud forest of southeastern Peru. *Herpetologica* 63 (2): 203-212.

CHE, J. J., J. PANG, H. ZHAO, G.-F. WU, E.-M. ZHAO, AND Y.-P. ZHANG. 2007. Phylogeny of Raninae (Anura: Ranidae) inferred from mitochondrial and nuclear sequences. *Molecular Phylogenetics and Evolution* 43 (1): 1-13.

CHERMOCK, R. L. 1952. A key to the amphibians and reptiles of Alabama. *Geological Survey of Alabama Museum Paper* (33): 1-88.

CHIPPINDALE, P. T. 2005a. *Eurycea chisholmensis*, Salado Salamander. *In*: LANNOO, M. J. (ed.), Amphibian Declines: The Conservation Status of United States Species. Part 2: Species Accounts. Berkeley, University of California Press: 739-740.

CHIPPINDALE, P. T. 2005b. *Eurycea latitans*, Cascade Caverns Salamander. *In*: LANNOO, M. J. (ed.), Amphibian Declines: The Conservation Status of United States Species. Part 2: Species Accounts. Berkeley, University of California Press: 746-747.

CHIPPINDALE, P. T. 2005c. *Eurycea naufragia*, Georgetown Salamander. *In*: LANNOO, M. J. (ed.), Amphibian Declines: The Conservation Status of United States Species. Part 2: Species Accounts. Berkeley, University of California Press: 756-757.

CHIPPINDALE, P. T. 2005d. *Eurycea neotenes*, Texas Salamander. *In*: LANNOO, M. J. (ed.), Amphibian Declines: The Conservation Status of United States Species. Part 2: Species Accounts. Berkeley, University of California Press: 757-758.

CHIPPINDALE, P. T. 2005e. *Eurycea robusta*, Blanco Blind Salamander. *In*: LANNOO, M. J. (ed.), Amphibian Declines: The Conservation Status of United States Species. Part 2: Species Accounts. Berkeley, University of California Press: 762.

CHIPPINDALE, P. T. 2005f. *Eurycea tonkawae*, Jollyville Plateau Salamander. *In*: LANNOO, M. J. (ed.), Amphibian Declines: The Conservation Status of United States

Species. Part 2: Species Accounts. Berkeley, University of California Press: 764-765.

CHIPPINDALE, P. T., R. M. BONETT, A. S. BALDWIN, AND J. J. WIENS. 2004. Phylogenetic evidence for a major reversal of life-history evolution in plethodontid salamanders. *Evolution* 58 (12): 2809-2822.

CHIPPINDALE, P. T., A. H. PRICE, AND D. M. HILLIS. 1993. A new species of perennibranchiate salamader (*Eurycea*: Plethodontidae) from Austin, Texas. *Herpetologica* 49: 248-259.

CHIPPINDALE, P. T., A. H. PRICE, J. J. WIENS, AND D. M. HILLIS. 2000. Phylogenetic relationships and systematic revision of central Texas hemidactyline plethodontid salamanders. *Herpetological Monographs* 14: 1-80.

CHRAPLIWY, P. S., AND E. V. MALNATE. 1961. The systematic status of the spadefoot toad *Spea laticeps* Cope. *Texas Journal of Science* 13 (2): 160-162.

CHRAPLIWY, P. S., K. WILLIAMS, AND H. M. SMITH. 1961. Noteworthy records of amphibians from Mexico. *Herpetologica* 17 (2): 85-90.

CLAY, W. M., R. B. CASE, AND R. CUNNINGHAM. 1955. On the taxonomic status of the slimy salamander, *Plethodon glutinosus* (Green), in southeastern Kentucky. *Transactions of the Kentucky Academy of Science* 16: 57-65.

CLINE, G. R. 2005a. *Hyla chrysoscelis*, Cope's Gray Treefrog. *In*: LANNOO, M. J. (ed.), Amphibian Declines: The Conservation Status of United States Species. Part 2: Species Accounts. Berkeley, University of California Press: 449-452.

CLINE, G. R. 2005b. *Hyla versicolor*, Eastern Gray Treefrog. *In*: LANNOO, M. J. (ed.), Amphibian Declines: The Conservation Status of United States Species. Part 2: Species Accounts. Berkeley, University of California Press: 458-461.

COCHRAN, D. M. 1961a. Type specimens of reptiles and amphibians in the United States National Museum. *Bulletin of the U.S. National Museum* 220: i-xv, 1-291.

COCHRAN, D. M. 1961b. Living amphibians of the world. Garden City, NY, Doubleday & Co.: 1-199.

COCHRAN, D. M., AND C. J. GOIN. 1970. The new field book of reptiles and amphibians. New York, G.P. Putnam's Sons: i-xxii, 1-359.

COCROFT, R. B. 1994. A cladistic analysis of Chorus Frog phylogeny (Hylidae: *Pseudacris*). *Herpetologica* 40 (4): 420-437.

COGGER, H. G., AND D. A. LINDNER. 1974. Frogs and reptiles. *In*: FRITH, H. J., AND J. H. CALABY (eds.), Faunal Survey of the Port Essington District, Cobourg Peninsula, Northern Territory of Australia. *CSIRO Division of Wildlife Research, Technical Paper* 28: 63-107.

COLLINS, J. P. 1981. Distribution, habitats, and life history variation in the tiger salamander, *Ambystoma tigrinum* in east-central and southeast Arizona. *Copeia* 1981 (4): 666-675.

COLLINS, J. T. 1989. New records of amphibians and reptiles in Kansas for 1989. *Kansas Herpetological Society Newsletter* (78): 16-21.

COLLINS, J. T. 1991. Viewpoint: A new taxonomic arrangement for some North American amphibians and reptiles. *Herpetological Review* 22 (2): 42-43.

COLLINS, J. T. 1997. Standard Common and Scientific Names for North American Amphibians and Reptiles. Fourth Edition. *Herpetological Circular* (25): i-iv, 1-40.

COLLINS, J. T., AND T. W. TAGGART. 2002. Standard common and current scientific names for North American amphibians, turtles, reptiles and crocodilians. 5th edition. Lawrence, Center for North American Herpetology: i-iv, 1-43.

COLLINS, J. T., AND T. W. TAGGART. 2009. Standard common and current scientific names for North American amphibians, turtles, reptiles, and crocodilians. Sixth edition. Lawrence, KS, Center of North Americna Herpetology: i-iv, 1-46.

COMEAU, N.-M. 1943. Une Ambystome nouvelle. *Annales de l'Association Canadienne-Francaise pour l'Avance du Science* 9: 124-125.

CONANT, R. 1958. A Field Guide to the Reptiles and Amphibians of the United States and Canada east of the 100th Meridian. Boston, Houghton Mifflin: i-xviii, 1-366, pl. 1-40.

CONANT, R. 1975. A field guide to reptiles and amphibians of eastern and central North America. Second editon. Boston, Houghton Mifflin Co.: 1-429.

CONANT, R., AND J. T. COLLINS. 1998. A field guide to the reptiles and amphibians, eastern and cetnral North America, third edition, expanded. Boston, Houghton Mifflin: i-xviii, 1-616.

CONDIT, J. M. 1958. True type locality of the salamander *Gyrinophilus prophyriticus inagnoscus*. *Copeia* 1958 (1): 46-47.

CONDIT, J. M. 1964. A list of the types of hylid frogs in the collection of the British Museum (Natural History). *Journal of the Ohio Herpetological Society* 4 (4): 85-98.

Conlon, J. M., L. Coquet, J. Leprince, T. Jouenne, H. Vaudry, J. Kolodzicjck, N. Nowotny, C. R. Bevier, and P. E. Moler. 2007. Peptidomic analysis of skin secretions from *Rana heckscheri* and *Rana okaloosae* provides insight into phylogenetic relationships among frogs of the *aquarana* species group. *Regulatory Peptides* 138: 87-93.

COOK, F. R. 1964. Additional records and a correction of the type locality for the Boreal Chorus Frog in northwestern Ontario. *Canadian Field Naturalist* 78: 186-192.

COOK, F. R. 1983. An analysis of toads of the *Bufo americanus* group in a contact zone in central northern North America. *National Museum of Natural Sciences (Ottawa). Publications in Natural Sciences* (3): i-vii, 1-89.

COOPER, J. G. 1860. Report upon the reptiles collected on the survey (Volume 12, Book 2). No. 4. *In*: Reports of Explorations and Surveys, to ascertain the most practicable and economical route for a railroad from the Mississippi River to the Pacific Ocean, made under the directon of the Secretary of War, in 1853-5, according to acts of Congress of March 3, 1853, May 31, 1854, and August 5, 1854. Vol. 12, Book 2, Part 3, Zoological Report. Washington, T.H. Ford, Printer: 292-306, pl. 12-16, 19-22, 24, 31.

COPE, E. D. 1859a. On the primary divisions of the Salamandridae with descriptions of two new species. *Proceedings of the Academy of Natural Sciences of Philadelphia* 11: 122-128.

Cope, E. D. 1861b. [Untitled. Concerning a Bahaman iguana, and Ohio and Pennsylvania salamanders]. *Proceedings of the Academy of Natural Sciences of Philadelphia* 13: 123-124.

Cope, E. D. 1861g. Descriptions of reptiles from tropical America and Asia. *Proceedings of the Academy of Natural Sciences of Philadelphia* 12 (1860): 368-374.

Cope, E. D. 1862a. On some new and little known American Anura. *Proceedings of the Academy of Natural Sciences of Philadelphia* 14 (2): 151-159 + 594.

Cope, E. D. 1862b. Notes upon some reptiles of the Old World. *Proceedings of the Academy of Natural Sciences of Philadelphia* 14: 337-344.

Cope, E. D. 1862c. Catalogues of the reptiles obtained during the explorations of the Parana, Paraguay, Vermejo and Uraguay [sic] rivers by Capt. Thos. J. Page, U.S.N., and of those procured by Lieut. N. Michler, U.S. Top. Eng., commander of the expedition conducting the survey of the Atrato River. *Proceedings of the Academy of Natural Sciences of Philadelphia* 14 (7-9): 346-369.

Cope, E. D. 1863. On *Trachycephalus*, *Scaphiopus* and other American Batrachia. *Proceedings of the Academy of Natural Sciences of Philadelphia* 15: 43-54.

Cope, E. D. 1864a. On the limits and relations of the Raniformes. *Proceedings of the Academy of Natural Sciences of Philadelphia* 16 (4): 181-183.

Cope, E. D. 1865a. Sketch of primary groups of Batrachia s. Salientia. *Natural History Review* (London) (New Series) 5: 97-120.

Cope, E. D. 1865b. Third contribution to the herpetology of tropical America. *Proceedings of the Academy of Natural Sciences of Philadelphia* 17: 185-198.

Cope, E. D. 1866a. On the structure and distribution of the genera of the Arciferous Anura. *Journal of the Academy of Natural Sciences of Philadelphia* (2) 6: 67-112, pl. 25.

Cope, E. D. 1866b. Fourth contribution to the herpetology of tropical America. *Proceedings of the Academy of Natural Sciences of Philadelphia* 1866: 123-132.

Cope, E. D. 1867a ("1866"). On the Reptilia and Batrachia of the Sonoran Province of the Nearctic Region. *Proceedings of the Academy of Natural Sciences of Philadelphia* 18: 300-314.

Cope, E. D. 1867b. On the families of raniform anura. *Journal of the Academy of Natural Sciences of Philadelphia* (2) 6: 189-206.

Cope, E. D. 1868a ("1867"). A review of the species of the Amblystomidae. *Proceedings of the Academy of Natural Sciences of Philadelphia* 19: 166-211.

Cope, E. D. 1868b. An examination of the Reptilia and Batrachia obtained by the Orton expedition to Equador [sic] and the upper Amazon, with notes on other species. *Proceedings of the Academy of Natural Sciences of Philadelphia* 20: 96-140.

Cope, E. D. 1869a ("1868"). Sixth contribution to the herpetology of tropical America. *Proceedings of the Academy of Natural Sciences of Philadelphia* 20: 305-313.

Cope, E. D. 1869b. A review of the species of the Plethodontidae and Desmognathidae. *Proceedings of the Academy of Natural Sciences of Philadelphia* 21: 93-118.

Cope, E. D. 1870b. Observations on the fauna of the southern Alleghanies. *American Naturalist* 4 (7): 392-402.

Cope, E. D. 1871a. Seventh contribution to the herpetology of tropical America. *Proceedings of the American Philosophical Society* 11 (1869-1870): 147-169, pl. 9-11.

Cope, E. D. 1871b (1965 reprint). Catalogue of Batrachia and Reptilia obtained by J. A. McNiel in Nicaragua. Catalogue of Reptilia and Batrachia obtained by C. J. Maynard in Florida. *Annual Report, Peabody Academy of Science* (Salem) 1871: 80-85. [Reprinted by Ohio Herpetological Society, Cincinnati]

Cope, E. D. 1871d. Ninth contribution to the herpetology of tropical America. *Proceedings of the Academy of Natural Sciences of Philadelphia* 1871: 200-224.

Cope, E. D. 1875a. The herpetology of Florida. *Proceedings of the Academy of Natural Sciences of Philadelphia* 1875: 10-11.

Cope, E. D. 1875b. Check-list of North American Batrachia and Reptilia; with a systematic list of the higher groups, and an essay on geographical distribution. Based on the specimens contained in the U.S. National Museum.

Bulletin of the United States National Museum 1: [i-iii], i-ii, 1-104.
COPE, E. D. 1875c ("1876"). "On the Batrachia and Reptilia of Costa Rica"; "On the Batrachia and Reptilia collected by Dr. John M. Branford during the Nicaraguan Canal Survey of 1874"; and "Report on the Reptiles brought by Professor James Orton from the middle and upper Amazon and western Peru". *Journal of the Academy of Natural Sciences of Philadelphia* N.S. 8: 93-183, pl. 1-6.
COPE, E. D. 1878a. A Texan cliff frog. *American Naturalist* 12 (3): 186.
COPE, E. D. 1878b. A new genus of Cystignathidae from Texas. *American Naturalist* 12 (4): 252-253.
COPE, E. D. 1878c ("1877"). Tenth contribution to the herpetology of tropical America. *Proceedings of the American Philosophical Society* 17 (1877-1878): 85-98.
COPE, E. D. 1879a. A contribution to the zoology of Montana. *American Naturalist* 13 (7): 432-441.
COPE, E. D. 1879b ("1880"). Eleventh contribution to the herpetology of tropical America. *Proceedings of the American Philosophical Society* 18 (1878-1880): 261-277.
COPE, E. D. 1880. On the zoological position of Texas. *Bulletin of the U.S. National Museum* (17): 1-51.
COPE, E. D. 1884 ("1883"). Notes on the geographical distribution of Batrachia and Reptilia in western North America. *Proceedings of the Academy of Natural Sciences of Philadelphia* 35: 10-35.
COPE, E. D. 1885. A contribution to the herpetology of Mexico. I. The collection of the Comisión Cientifica. *Proceedings of the American Philosophical Society* 22 (4): 379-404.
COPE, E. D. 1886b. Synonymic list of the North American species of *Bufo* and *Rana*, with some descriptions of some new species of Batrachia, from specimens in the National Museum. *Proceedings of the American Philosophical Society* 23: 514-526.
COPE, E. D. 1887a. Catalogue of the batrachians and reptiles of Central America and Mexico. *Bulletin of the U.S. National Museum* (32): 1-98.
COPE, E. D. 1887b. The hyoid structure in the amblystomid salamanders. *American Naturalist* 21 (1): 87-88.

COPE, E. D. 1888. Batrachia and Reptilia. *American Naturalist* 22 (253): 79-80.

COPE, E. D. 1889a ("1888"). Catalogue of Batrachia and Reptilia brought by William Taylor from San Diego, Tex. *Proceedings of the U.S. National Museum* (11): 395-398 + fig. 2, pl. xxxvi.

COPE, E. D. 1889b. The Batrachia of North America. *Bulletin of the U.S. National Museum* (34): 1-525, pl. 1-86.

COPE, E. D. 1890. On a new species of salamander from Indiana. *American Naturalist* 24 (286): 966-967.

COPE, E. D. 1891. A new species of frog from New Jersey. *American Naturalist* 25 (299): 1017-1019.

COPE, E. D. 1893a. On a new spade-foot from Texas. *American Naturalist* 27: 155-156.

COPE, E. D. 1894a ("1893"). On a collection of Batrachia and Reptilia from southwest Missouri. *Proceedings of the Academy of Natural Sciences of Philadelphia* 45: 383-385.

COPE, E. D. 1894c. Third additon to a knowledge of the Batrachia and Reptilia of Costa Rica. *Proceedings of the Academy of Natural Sciences of Philadelphia* 46: 194-206.

COPE, E. D. 1900. The crocodilians, lizards and snakes of North America. *Annual Report, U.S. National Museum* 1898: 153-1270.

COPE, E. D., AND A. S. PACKARD, JR. 1881. The fauna of Nickajack Cave. *American Naturalist* 15 (11): 877-882.

COPLAND, S. J. 1957. Australian tree frogs of the genus *Hyla*. *Proceedings of the Linnean Society of New South Wales* 82 (1): 9-108.

COUES, E. 1875. Synopsis of the reptiles and batrachians of Arizona, with critical and field notes, and an extensive synonymy. *In*: WHEELER, G. M. (ed.), Report upon geographical and geological explorations and surveys west of the one hundredth meridian, in charge of Lieut. Geo. M. Wheeler, under the direction of Brig. Gen. A.A. Humphreys, chief of engineers, U.S. Army. Vol. 5—Zoology. Washington, Government Printing Office: 585-633. pl. 16-25.

COVACEVICH, J. A. 1974. The status of *Hyla irrorata* De Vis, 1884 (Anura: Hylidae). *Memoirs of the Queensland Museum* 17: 49-53.

COX, C. L., J. W. STREICHER, C. M. I. I. I. SHEEHY, J. A. CAMPBELL, AND P. T. CHIPPINDALE. 2012. Patterns of genetic differentiation among populations of *Smilisca fodiens*. *Herpetologica* 68 (2): 226-235.

CRAWFORD, A. J., L. E. BROWN, AND C. W. PAINTER. 2005. *Rana blairi*. Plains Leopard Frog. *In*: LANNOO, M. J. (ed.), Amphibian Declines: The Conservation Status of United States Species, Part 2: Species Accounts. Berkeley, University of California Press: 532-534.

CRAWFORD, A. J., AND E. N. SMITH. 2005. Cenozoic biogeography and evolution in direct-developing frogs of Central America (Leptodactylidae: *Eleutherodactylus*) as inferred from a phylogenetic analysis of nuclear and mitochondrial genes. *Molecular Phylogenetics and Evolution* 35: 536-555.

CRAYON, J. J. 2005. *Xenopus laevis*, African Clawed Frog. *In*: LANNOO, M. J. (ed.), Amphibian Declines: The Conservation Status of United States. Species Accounts. Berkeley, University of California Press: 522-526.

CRESPI, E. J. 1996. Mountains as islands: genetic variability of the Pigmy Salamander (*Desmognathus wrighti*, family Plethodontidae) in the southern Appalachians. MS thesis, Wake Forest University. [not seen]

CRESPI, E. J., R. A. BROWNE, AND L. J. RISSLER. 2010. Taxonomic revision of *Desmognathus wrighti* (Caudata: Plethodontidae). *Herpetologica* 66 (3): 283-295.

Crother, B. I. (Ed.). 2012. Scientific and standard English names of amphibians and reptiles of North America north of Mexico, with comments regarding confidence in our understanding. Seventh edition. Society for the Study of Amphibians and Reptiles *Herpetological Circular* 39: [i-iii], 1-92.

CROTHER, B. I., J. A. BOUNDY, K. CAMPBELL, K. DE QUEIROZ, D. R. FROST, D. M. GREEN, R. HIGHTON, J. B. IVERSON, R. W. MCDIARMID, P. A. MEYLAN, T. W. REEDER, M. E. SEIDEL, J. W. SITES, JR., S. G. TILLEY, AND D. B. WAKE. 2003. Scientific and standard English names of amphibians and reptiles of North America north of Mexico: update. *Herpetological Review* 34: 196-203.

CROWHURST, R. S., K. M. FARIES, J. COLLANTES, J. T. BRIGGLER, J. B. KOPPELMAN, AND L. S. EGGERT. 2011. Genetic relationships of hellbenders in the Ozark highlinads of Missouri and conservation implications for the Ozark subspecies (*Cryptobranchus alleganiensis bishopi*). *Conservation Genetics* 25 (2): 637-646.

CUELLAR, O. 1996. Taxonomic status of the western spotted frog. *Biogeographica* (Paris) 72 (4): 145-150.

CUVIER, [G.]. 1817. Le règne animal distribué d'après son organisation, pour servir de base à l'histoire naturelle des animaux et d'introduction à l'anatomie comparée. Tome 1-4. Deterville, Paris, Tome 1: i-xxxvii, 1 page.

CUVIER, [G.] L. B. 1827. Sur le genre de reptiles batraciens; nommé *Amphiuma*, et sur une nouvelle espèce de ce genre (*Amphiuma tridactylum*). *Mémoires du Muséum National d'Histoire Naturelle*, Paris 14: 1-12, [i-ii], pl. 1-4.

CUVIER, [G.]. 1829. Le règne animal distribué d'après son organisation, pur sevir de base à l'histoire naturelle des animaux et d'introduction à l'anatomie comparée. 2nd edition, revue et augmentée par P.A. Latreille. Volume 2. Paris, Deterville & Crochard: i-xviii, 1-532.

CUVIER, [G.]. 1830. Le règne animal distribué d'après son organisation, pour servir de base à l'histoire naturelle des animaux et d'introduction à l'anatomie comparée, Ed. 2, Tome 3. Paris, Deterville & Crocher. [*fide* Frost (2009-13); not seen]

CUVIER, G. 1849. Le Règne animal distribué d'après son organisation, pour servir de base à l'histoire naturelle des animaux et d'introduction à l'anatomie comparée. Ed 3, Vol. 2. Paris, Deterville.

CUVIER, G., AND P. A. LATREILLE. 1831. The animal kingdom, arranged according to its organization, and the crustacea, arachnides and insecta. English edition. Volume II. New York, G & C & H Carvill: i-xvi, 1-475, pl. 1-10.

DAUDIN, F. M. 1800. Histoire naturelle des Quadrupèdes Ovipares. Livraison 1 & 2. Paris, Fuchs & Delalain: 29 pp., 12 pl. (unnumbered).

DAUDIN, F. M. 1802a. Histoire naturelle des rainettes, des grenouilles et des crapauds. Paris, Levrault: 1-108.

DAUDIN, F. M. 1802b. Histoire naturelle, générale et particuliére des reptiles. *In*: SONNINI, C. A., AND P. A. LATREILLE (eds.), Histoire naturelle des Reptiles, avec figures dissinées d'après nature. Tome 2. 1-432, 13 pl.

DAUDIN, F. M. 1803. Histoire naturelle, générale et particulière des reptiles. *In*: SONNINI, C. A., AND P. A. LATREILLE (eds.), Histoire naturelle des Reptiles, avec figures dissinées d'après nature. Tome 8. Paris, F. Dufart.

DAUGHERTY, C. H., F. W. ALLENDORF, W. W. DUNLAP, AND K. L. KNUDSEN. 1983. Systematic implications of geographic patterns of genetic variation in the genus *Dicamptodon*. *Copeia* 1983 (3): 679-691.

DAVID, A. 1871. Rapport adressé a MM. les professeurs-administrateur du Museum d'Histoire Naturelle. *Bulletin. Nouvelles Archives du Muséum d'Histoire Naturelle* (Paris) 7: 75-100.

DAVIS, D. D. 1935. A new generic and family position for *Bufo barbonica*. *Field Museum of Natural History Zoological Series* 20: 87-92.

DAVIS, D. R., AND G. B. PAULY. 2011. Morphological variation among populations of the Western Slimy Salamander on the Edwards Plateau of central Texas. *Copeia* 2011 (1): 103-112.

DAVIS, N. S., JR., AND F. L. RICE. 1883. List of Batrachia and Reptilia of Illinois. *Bulletin of the Chicago Academy of Science* 1 (3): 25-32.

DAVIS, W. B. 1955. A new sheep frog (genus *Hypopachus*) from Mexico. *Herpetologica* 11 (1): 71-72.

de BEAUVOIS, [A. M. F. J.]. 1799. On a new species of *Siren*. Translation of a memoir on a new species of *Siren*. *Transactions of the American Philosophical Society* 4: 277-281.

de BLAINVILLE, H. 1816a. Prodrome d'une nouvelle distribution systématique du règne animal. *Bulletin des Sciences par la Société Philomathique de Paris* 1816: 113-124.

de BLAINVILLE, H. 1816b. Prodrome d'une nouvelle distribution systématique du règne animal. *Journal de Physique, de Chimie, d'Histoire naturelle et des Arts* 83 (Octobre 1816): 244-267.

de BLAINVILLE, H. 1818. Vorläufige Anzeige einer neuen systematischen Eintheilung des Thierreichs. *Isis von Oken* 1818: 1365-1384.

de L'ISLE, A. 1877. Note sur un genre nouveau de batraciens bufoniformes du terrain a *Elephas meridionalis* de Dufort

(Gard) (Platosophus gervaisii). *Journal de Zoologie* (Paris) 6: 472-478.

de la Cepède, [B. G. E.], Conte. 1788b. Histoire naturelle des quadrupèdes ovipares et des serpens, des poisson et des cetaces. 16mo version. Volume 2. Paris, Hôtel de Thou: [i-ii], 1-462, [i-ii]. [This work and all subsequent editions ruled invalid and placed on official index of rejected and invalid works, as a non-binomial work. No names proposed here are available (ICZN, 2005)]

de la Espada, M. X. 1871. Faunae neotropicalis species quaedam nondum cognitae. *Jornal de Sciencias, Mathematicas, Physicas e Naturaes* (Academia Real das Sciencias de Lisboa) 3: 57-65.

De Vis, C. W. 1884. On a new species of *Hyla*. *Proceedings of the Royal Society of Queensland* 1: 128-130. [*fide* Frost (2009-13); not seen]

DeKay, J. E. 1842. Zoology of New York, or The New York Fauna; comprising detailed descriptions of all the animals hitherto observed within the state of New York, with brief notices of those occasionally found near its borders, and accompanied by approiate illustrations. Part 3. Reptiles and Amphibians. Albany, T. Weed: i-viii, 1-98 + 23 pl.

Desmarest, E. 1856. Reptiles et poissons. *In*: Chenu, D. (ed.), Encyclopédie d'histoire naturelle ou traité complet de cette science. Paris, Maresq & Co. & Gustave Havard: [i-x], 1-360, pl. 1-48.

Desor, E. 1870. Communique encore quelques details sur les metamorphoses du *Sirenodon*. *Bulletin de la Société des Sciences Naturelles de Neuchatel* 8: 266-269.

Devitt, T. J., S. J. E. Baird, and C. Moritz. 2011. Asymmetric reproductive isolation between terminal forms of the salamander ring species *Ensatina eschscholtzii* revealed by fine-scale genetic analysis of a hybrid zone. *BMC Evolutionary Biology* 11: 245-258.

Devitt, T. J., S. E. C. Devitt, B. D. Hollingsworth, J. A. McGuire, and C. Moritz. 2013. Montane refugia predict population genetic structure in the Large-blotched Ensatina salamander. *Molecular Ecology* 22: 1650-1665.

Dickerson, M. C. 1907. The Frog Book: North American toads and frogs with a study of the habits and life histories of those of the northeastern states. New York, Doubleday, Page & Co.: i-xviii, 1-253, pl. 1-96.

DIXON, J. R. 2000. Amphibians and Reptiles of Texas: With Keys, Taxonomic Synopses, Bibliography, and Distribution Maps. 2nd ed. College Station, Texas A&M University Press: i-viii, 1-421.

Dodd, C. K., Jr. 2004. *The amphibians of Great Smoky Mountains National Park.* Knoxville, University of Tennessee Press: 283 pp.

Dodd, C. K., Jr. 2013. *Frogs of the United States and Canada.* Baltimore, Johns Hopkins University Press, 2 volumes: i-xxxi, 1-460; [i-ix], 461-982.

DOWLING, H. G., AND W. E. DUELLMAN. 1974-1978. Systematic Herpetology: a synopsis of families and higher categories. New York, HISS Publications: i-vii, + 118 "issues," each with one or more pages.

DUBOIS, A. 1981. Liste des genres et sous-genres nominaux de Ranoidea (Amphibiens, Anores) du monde, avec identification de leurs espèces-types: conséquences nomenclaturales. *Monitore Zoologico Italiano N.S. Supplemento* 15: 225-284.

DUBOIS, A. 1982. Les notions de genre, sous-genre et groupe despèces en zoologie à la lumière de la systématique évolutive. *Monitore Zoologico Italiano N.S. Supplemento* 16: 9-65.

DUBOIS, A. 1983. Classification et nomenclature supragénérique des Amphibiens Anoures. *Bulletin mensuel de la Société linnéenne de Lyon* 52: 270-276.

DUBOIS, A. 1984a. Miscellanea nomenclatorica batrachologica (III). *Alytes* 3: 85-89.

DUBOIS, A. 1984b. Miscellanea nomenclatorica batrachologica (IV). *Alytes* 3: 103-110.

Dubois, A. 1984c. Note sur les Grenouilles brunes (groupe de *Rana temporaria* Linné, 1758). III. Un critère méconnu pour distinguer *Rana dalmatina* et *Rana temporaria. Alytes* 3: 117-124.

DUBOIS, A. 1987a ("1986"). Miscellanea taxinomica batrachologica (I). *Alytes* 5 (1-2): 7-95.

DUBOIS, A. 1987b. Living amphibians of the world: a first step towards a comprehensive checklist. *Alytes* 5 (3): 99-149.

DUBOIS, A. 1987c. Again on the nomenclature of frogs. *Alytes* 6 (1-2): 27-55.

DUBOIS, A. 1992. Notes sur la classification des Ranidae (Amphibiens Anoures). *Bulletin Mensuel de la Societe Linnéenne de Lyon* 61 (10): 305-352.

Dubois, A. 1995. Comments on the proposed conservation of Hemidactyliini Hallowell, 1856 (Amphibia, Caudata). *Bulletin of Zoological Nomenclature* 52 (4): 337-338.

DUBOIS, A. 1999. Miscellanea nomenclatorica batrachologica. *Alytes* 17: 81-100.

DUBOIS, A. 2000. Synonymies and related lists in zoology: general proposals, with examples in herpetology. *Dumérilia* 4 (2): 33-98.

DUBOIS, A. 2004a. The higher nomenclature of recent amphibians. *Alytes* 22 (1): 1-14.

DUBOIS, A. 2005a. Proposed rules for the incorporation of nomina of higher-ranked zoological taxa in the International Code of Zoological Nomenclature. 1. Some general questions, concepts and terms of biological nomenclature. *Zoosystema* 27 (2): 365-426.

DUBOIS, A. 2005b. Amphibia Mundi. 1.1. An ergotaxonomy of recent amphibians. *Alytes* 23 (1-2): 1-24.

DUBOIS, A. 2005c. Proposals for the incorporation of nomina of higher-ranked taxa into the Code. *Bulletin of Zoological Nomenclature* 62: 200-209.

DUBOIS, A. 2006a. Proposed Rules for the incorporation of nomina of higher-ranked zoological taxa in the International Code of Zoological Nomenclature. 2. The proposed Rules and their rationale. *Zoosystema* 28 (1): 165-258.

DUBOIS, A. 2006b. Incorporation of nomina of higher-ranked taxa into the International Code of Zoological Nomenclature: some basic questions. *Zootaxa* (1337): 1-37.

DUBOIS, A. 2006c. New proposals for naming lower-ranked taxa within the frame of the International Code of Zoological Nomenclature. *Comptes Rendus des l'Academie des Sciences, Paris Biologies* 329: 823-840.

DUBOIS, A. 2007a. Naming taxa from cladograms: some confusions, misleading statements, and necessary clarifications. *Cladistics* 23 (1): 390-402.

DUBOIS, A. 2007b. Genitives of species and subspecies nomina derived from personal names should not be emended. *Zootaxa* (1550): 49-68.

DUBOIS, A. 2008a. Authors of zoological publications and nomina are signatures, not persons. *Zootaxa* (1771): 63-68.

DUBOIS, A. 2008b. Phylogenetic hypotheses, taxa and nomina in zoology. *IN*: Minelli, A., L. Bonato, & G. Fusco (eds.), Updating the Linnaean heritage: names as tools for thinking about animals and plants. *Zootaxa* (1950): 51-86.

DUBOIS, A. 2009a. Incorporation of nomina of higher-ranked taxa into the International Code of Zoological Nomenclature: the nomenclatural status of class-series zoological nomina published in a non-Latinized form. *Zootaxa* (2106): 1-12.

DUBOIS, A. 2009b. Miscellanea nomenclatorica batrachologica. 20. Class-series nomina are nouns in the nominative plural: Terrarana Hedges, Duellman & Heinicke, 2008 must be emended. *Alytes* 26 (1-4): 167-175.

Dubois, A. 2010. Retroactive changes should be introduced in the *Code* only with great care: problems related to the spellings of nomina. *Zootaxa* (2426): 1-42.

Dubois, A. 2011. The *International Code of Zoological Nomenclature* must be drastically improved before it is too late. *Bionomina* 2: 1-104.

Dubois, A. 2012. The distinction between introduction of a new nomen and subsequent use of a previously introduced nomen in zoological nomenclature. *Bionomina* 5: 57-80.

DUBOIS, A., AND R. BOUR. 2010b. The nomenclatural status of the nomina of amphibians and reptiles created by Garsault (1764), with a parsimonious solution to an old nomenclatural problem regarding the genus *Bufo* (Amphibia, Anura), comments on the taxonomy of this genus, and comments on some nomina created by Laurenti (1768). *Zootaxa* 2447: 1-52.

Dubois, A., P.-A. Crochet, E. C. Dickinson, A. Nemésio, E. Aescht, A. M. Bauer, V. Blagoderov, R. Bour, M. R. de Carvalho, L. Desutter-Grandcolas, T. Frétey, P. Jäger. V. Koyamba, E. O. Lavilla, I. Löbl, A. Louchart, V. Malécot, H. Schatz, and A. Ohler. 2013. Nomenclatural and taxonomic problems related to the electronic publication of new nomina and nomenclatural acts in zoology, with brief comments on optical discs and on the situation in botany. *Zootaxa: Monograph* (3735): 001-094.

Dubois, A., R. I. Crombie, and F. Glaw. 2005. *Amphibia Mundi.* 1.2. Recent amphibians: generic and infrageneric taxonomic additions (1981-2002). *Alytes* 23 (1-2): 25-69.

Dubois, A., and W. R. Heyer. 1992. *Leptodactylus labialis,* the valid name for the American white-lipped frog (Amphibia: Leptodactylidae). *Copeia* 1992: 584-585.

Dubois, A., and A. Ohler. 1995. Frogs of the subgenus *Pelophylax* (Amphibia, Anura, genus Rana): a catalogue of available and valid scientific names, with comments on name-bearing types, complete synonymies, proposed common names, and maps showing all type localities. *Zoologica Poloniae* 39 (3-4): 139-204.

Dubois, A. and A. Ohler. 2001. Systematics of the genus *Philautus* Gistel, 1848 (Amphibia, Anura, Ranidae, Rhacophorinae): some historical and metataxonomic comments. *Journal of.South Asian Natural History* 5 (2): 173-186.

Dubois, A., and J. Raffaëlli. 2009. A new ergotaxonomy of the family Salamandridae Goldfuss, 1820 (Amphibia, Urodela). *Alytes* 26 (1-4): 1-85.

Dubois, A., and J. Raffaëlli. 2012. A new ergotaxonomy of the order Urodela Duméril, 1805 (Amphibia, Batrachia). *Alytes* 28 (3-4): 77-161.

Duellman, W. E. 1961. The amphibians and reptiles of Michoacán. *University of Kansas Publications, Museum of Natural History* 15 (1): 1-148, pl. 1-6.

Duellman, W. E. 1968a. *Smilisca. Catalogue of American Amphibians and Reptiles.* American Society of Ichthyologists and Herpetologists (58): 1-2.

Duellman, W. E. 1968b. *Smilisca baudinii. Catalogue of American Amphibians and Reptiles.* American Society of Ichthyologists and Herpetologists (59): 1-2.

Duellman, W. E. 1970. Hylid frogs of Middle America. 2 volumes. Monographs of the Museum of Natural History, University of Kansas, 1-753, pl. 1-72.

Duellman, W. E. 1975. On the classification of frogs. *Occasional Papers of the Museum of Natural History, University of Kansas* 42: 1-14.

DUELLMAN, W. E. 1977. Liste der rezenten Amphibien und Reptilien: Hylidae, Centrolenidae, Pseudidae. *In*: *Das Tierreich*, Lieferung 95. Berlin & New York, W. de Guyter: i-xix, 1-225.

DUELLMAN, W. E. 1993. Amphibian species of the world: additions and corrections. Lawrence, *University of Kansas Museum of Natural HIstory. Special Publication* 21: i-iii, 1-372.

DUELLMAN, W. E. 2001a. Hylid frogs of Middle America, 2nd ed. Volume 1. Ithaca, Society for the Study of Amphibians and Reptiles: i-xvi, 1-694.

DUELLMAN, W. E. 2001b. Hylid frogs of Middle America, 2nd ed. Volume 2. Ithaca, Society for the Study of Amphibians and Reptiles: i-x, 695-115

DUELLMAN, W. E., AND R. I. CROMBIE. 1970. *Hyla septentrionalis*. *Catalogue of American Amphibians and Reptiles*. American Society of Ichthyologists and Herpetologists (92): 1-4. 9, pl. 1-92.

DUELLMAN, W. E., AND J. D. LYNCH. 1969. Descriptions of *Atelopus* tadpoles and their relevance to atelopodid classification. *Herpetologica* 25 (3): 231-240.

DUELLMAN, W. E., AND L. TRUEB. 1966. Neotropical hylid frogs, genus *Smilisca*. *University of Kansas Publications, Museum of Natural History* 17 (7): 281-375, pl. 1-12.

DUMÉRIL, A. 1863. Catalogue méthodique de la collection des batraciens du Muséum d'Histoire Naturelle de Paris. *Mémoires de la Société Impériale des Sciences Naturelles de Cherbourg* 9: 295-321.

DUMÉRIL, A. M. C. 1805. Zoologie analytique, ou méthode naturelle de classification des animaux, rendue plus facile à l'aide de tableaux synoptiques. Paris, Allais: i-xxxii, 1-344.

DUMÉRIL, A. M. C., AND G. BIBRON. 1841. Erpétologie générale ou histoire naturelle complète des Reptiles. Tome 8. Paris, Roret: i-vii, 1-792.

DUMÉRIL, A. M. C., G. BIBRON, AND A. DUMÉRIL. 1854c. Erpétologie générale: ou, Histoire naturelle complète des reptiles. Tome 9. Paris, Roret: i-xx, 1-440, 12 pl.

DUNCAN, R., AND R. HIGHTON. 1979. Genetic relationships of the eastern large *Plethodon* of the Ouachita Mountains. *Copeia* 1979 (1): 95-110.

DUNDEE, H. A. 1950. Notes on the type locality of *Eurycea multiplicata* (Cope). *Herpetologica* 6 (2): 27-28.

DUNDEE, H. A. 1965a. *Eurycea multiplicata. Catalogue of American Amphibians and Reptiles*: American Society of Ichthyologists and Herpetologists (21): 1-2.

DUNDEE, H. A. 1965b. *Eurycea tynerensis. Catalogue of American Amphibians and Reptiles:* American Society of Ichthyologists and Herpetologists (22): 1-2.

DUNDEE, H. A. 1972 ("1971"). *Cryptobrahchus and C. alleganiensis. Catalogue of American Amphibians and Reptiles.* American Society of Ichthyologists and Herpetologists (101): 1-4.

DUNDEE, H. A. 1996. Some reallocations of type localities of reptiles and amphibians described from the Major Stephen H. Long expedition to the Rocky Mountains, with comments on some of the statements made in the account written by Edwin James. *Tulane Studies in Zoology and Botany* 30 (2): 75-89.

DUNDEE, H. A. 1998. *Necturus punctatus. Catalogue of American Amphibians and Reptiles:* Society for the Study of Amphibians and Reptiles (663): 1-5.

DUNLAP, D. G., AND J. E. PLATZ. 1981. Geogaphic variation of proteins and call in *Rana pipiens* from the northcentral United States. *Copeia* 1981 (4): 876-879.

DUNN, E. R. 1916. Two new salamanders of the genus *Desmognathus. Proceedings of the Biological Society of Washington* 29: 73-76.

DUNN, E. R. 1917a. The salamanders of the genera *Desmognathus* and *Leurognathus. Proceedings of the U.S. National Museum* 53: 393-433.

DUNN, E. R. 1917b. Reptile and amphibian collections from the North Carolina mountains, with especial reference to salamanders. *Bulletin of the American Museum of Natural History* 37: 593-634.

DUNN, E. R. 1918a. A preliminary list of the reptiles and amphibians of Virginia. *Copeia* (53): 16-27.

DUNN, E. R. 1918b. The collection of the Amphibia Caudata of the Museum of Comparative Zoology. *Bulletin of the Museum of Comparative Zoology* 62 (9): 443-471.

DUNN, E. R. 1920b. Notes on two Pacific coast Ambystomatidae. *Proceedings of the New England Zoological Club* (Cambridge) 7 (1919-1921): 55-59.

DUNN, E. R. 1920c. Some reptiles and amphibians from Virginia, North Carolina, Tennessee and Alabama. *Proceedings of the Biological Society of Washington* 33: 129-137.

DUNN, E. R. 1922. The sound-transmitting apparatus of salamanders and the phylogeny of the caudata. *American Naturalist* 56 (646): 418-427.

DUNN, E. R. 1923. Mutanda herpetologica. *Proceedings of the New England Zoological Club* (Cambridge) 8: 39-40.

DUNN, E. R. 1926b. The salamanders of the family Plethodontidae. Northampton, MA, Smith College: i-xii, 1-441.

DUNN, E. R. 1927a. A new mountain race of *Desmognathus*. *Copeia* (164): 84-86.

DUNN, E. R. 1928. A new genus of salamanders from Mexico. *Proceedings of the New England Zoological Club* (Cambridge) 10: 85-86.

DUNN, E. R. 1929. A new salamander from southern California. *Proceedings of the U.S. National Museum* 74 (25): 1-3.

DUNN, E. R. 1931. New frogs from Panama and Costa Rica. *Occasional Papers of the Boston Society of Natural History* 5: 385-401.

DUNN, E. R. 1938. Notes on frogs of the genus *Acris*. *Proceedings of the Academy of Natural Sciences of Philadelphia* 90: 153-154.

DUNN, E. R. 1939. *Bathysiredon*, a new genus of salamanders, from Mexico. *Notulae Naturae* (36): 1.

DUNN, E. R. 1940. The races of *Ambystoma tigrinum*. *Copeia* 1940 (3): 154-162.

DUNN, E. R. 1941. Notes on *Dendrobates auratus*. *Copeia* 1941: 88-93.

DUNN, E. R. 1944. Notes on the salamanders of the *Ambystoma gracile* group. *Copeia* 1944: 129-130.

DUNN, E. R., AND M. T. DUNN. 1940. Generic names proposed in herpetology by E.D. Cope. *Copeia* 1940 (2): 69-76.

DUNN, E. R., AND J. T. EMLEN, JR. 1932. Reptiles and amphibians from Honduras. *Proceedings of the Academy of Natural Sciences of Philadelphia* 84: 21-32.

DUNN, E. R., AND A. A. HEINZE. 1933. A new salamander from the Ouachita Mountains. *Copeia* 1933 (3): 121-122.

EASTEAL, S. 1986. *Bufo marinus. Catalogue of American Amphibians and Reptiles.* Society for the Study of Amphibians and Reptiles (395): 1-4.

EDWARDS, J. L. 1976. Spinal nerves and their bearing on salamander phylogeny. *Journal of Morphology* 148 (3): 305-328.

EICHWALD, C. E. V. 1831. Zoologia specialis quam expositis animalibus tum vivis, tum fossilibus potissimum rossiae in universum, et poliniae in specie, in usum lectionum publicarum in Universitate Caesarea Vilnensi habendarum. Pars posterior. Volume 3. Vilnius, J. Zawadski: 1-404, pl. 1.

EIGENMANN, C. H. 1901. Description of a new cave salamander, *Spelerpes stejnegeri,* from the caves of southwestern Missouri. *Transactions of the American Microscopical Society* 22: 189-192, pl 27-28.

ELLIOTT, L., C. GERHARDT, AND C. DAVIDSON. 2009. The frogs and toads of North America: a comprehensive guide to their identification, behavior, and calls. Boston & New York, Houghton Mifflin Harcourt: 1-343.

ELLIS, M. M., AND J. HENDERSON. 1913. The Amphibia and Reptilia of Colorado. *University of Colorado Studies* 10 (2): 39-129, pl. 1-8.

ENGBRECHT, N. J., S. J. LANNOO, J. O. WHITAKER, AND M. J. LANNOO. 2011. Comparative morphometrics in ranid frogs (subenus *Nenirana*): are apomorphic elongation and a blunt snout responses to small-bore burrow dwelling in Crawfish Frogs (*Lithobates areolatus*)? *Copeia* 2011 (2): 285-295.

ERMAN, A. 1835. Reise um die Erde. Naturhistorischer Atlas. Berlin, G. Riemer: 64 pp. [*fide* Frost (2009-13); not seen)

ESCHSCHOLTZ, F. 1833. Zoologischer Atlas, enthaltend Abbildungen und Beschreibungen neuer Thierarten wärend des Flottcapitains von Kotzebue zweiter Reise um die Welt, auf der Russisch-Kaiserlichen Kriegsschlupp

Predpriaetie in den Jahren 1823-1826. Vol. 5. Berlin, G. Riemer: i-viii, 1-28, pl. 1-25.

ESPINOZA, F. A., J. E. DEACON, AND A. SIMMIN. 1970. An economic and biostatistical analysis of the bait fish industry in the Lower Colorado River. *Special Publication, University of Nevada Las Vegas* 1-87.

ESTES, R. 1965. Fossil salamanders and salamander origins. *American Zoologist* 5: 319-334.

FAIVOVICH, J., C. F. B. HADDAD, P. C. A. GARCIA, D. R. FROST, J. A. CAMPBELL, AND W. C. WHEELER. 2005. Systematic review of the frog family Hylidae, with special reference to Hylinae: phylogenetic analysis and taxonomic revision. *Bulletin of the American Museum of Natural History* (294): 1-240.

FATIO, V. 1872. Faune des Vertebrates de la Suisse. Vol. 3. Histoire naturelle des reptiles et des batraciens. Geneve, H. Georg: [i-vi], 1-602.

FEDER, J. H., G. Z. WURST, AND D. B. WAKE. 1978. Genetic variation in western salamanders of the genus *Plethodon*, and the status of *Plethodon gordoni*. *Herpetologica* 34 (1): 64-69.

FEI, L., C. YE, AND Y. HUANG. 1990. Zhongguo liang qi dong wu jian suo [Key to Chinese amphibians]. Chongqing, China, Chongqing Shi. Publishing house for Scientific and Technological Literature: [i-ii], i-vi, 1-364. [*fide* Frost (2009-13), not seen]

FEI, L., C. YE, AND JIANG. 2000. A new genus of the subfamily Amolopinae *Pseudoamolops*, and its relationship to related genera. *Acta Zoologica Sinica* (Dong uw xue bao, Beijing) 46: 19-26. [*fide* Frost (2009-13); not seen]

FEJÉRVÁRY, G. G. 1917. Fosszilis békák a püspökfürdői praeglaciális rétegekből. Különős tekintettel az anurák sacrumának phyletikaai feilődésére. *Földtani Közlöny* (Budapest) 47: 25-199.

FEJÉRVÁRY, G. G. 1920. Liste des batraaciens et reptiles recueillis dans la Vallée du Haut-Rhône. *Bulletin de la Société Vaudoise des Sciences Naturelles* (Lausanne) 53 (198): 187-193.

FEJÉRVÁRY, G. G. 1923. Ascaphidae, a new family of the tailless batrachians. *Annales Historico-Naturales Musei Nationalis Hungarici* 20: 178-181.

FELLERS, G. M. 2005a. *Rana boylii*. Foothill Yellow-legged Frog. *In*: LANNOO, M. J. (ed.), Amphibian Declines: The Conservation Status of United States Species, Part 2: Species Accounts. Berkeley, University of California Press: 534-536.

FELLERS, G. M. 2005b. *Rana draytonii*. California Red-legged Frog. *In*: LANNOO, M. J. (ed.), Amphibian Declines: The Conservation Status of United States Species, Part 2: Species Accounts. Berkeley, University of California Press: 552-554.

FERGUSON, D. E. 1961. The geographic variation of *Ambystoma macrodactylum* Baird, with the description of two new subspecies. *American Midland Naturalist* 65: 311-338.

FERGUSON, D. E. 1963. *Ambystoma macrodactylum*. *Catalogue of American Amphibians and Reptiles*. American Society of Ichthyologists and Herpetologists (4): 1-2.

FERGUSON, J. H., AND C. H. LOWE, JR. 1969. Evolutionary relationships of the *Bufo punctatus* group. *American Midland Naturalist* 81 (2): 435-446.

FIRSCHEIN, I. L. 1950. A new record of *Spea bombifrons* from northern Mexico and remarks on the status of the *hammondii* group of spadefoot anurans. *Herpetologica* 6 (1): 75-77.

FIRSCHEIN, I. L. 1954. Definition of some little understood members of the leptodactylid genus *Syrrhophus*, with a description of a new species. *Copeia* 1954 (1): 48-58.

Fischer, G. 1813. Zoognosia tabulis synopticis illustrata: in Usum Praelectionum Academiae Imperialis Medico-Chirurgicae Mosquensis Edita. Editio tertia: classium, ordinum, generum illustratione perpetua aucta. Volumen primum. Nicolai Sergeidis Vsevolozsky, Moscow, i-xiv, 1 page.

FISHER, C., A. JOYNT, AND R. J. BROOKS. 2007. Reptiles and amphibians of Canada. Edmonton, Lone Pine Publishing: 1-208.

FITCH, H. S. 1938a. An older name for *Triturus similans* Twitty. *Copeia* 1938 (3): 148-149.

FITZGERALD, K. T., H. M. SMITH, AND L. J. GUILLETTE, JR. 1981. Nomenclature of the diploid species of the diploid-tetraploid *Hyla versicolor* complex. *Journal of Herpetology* 15 (3): 356-360.

FITZINGER, L. 1826. Neue Classification der Reptilien nach ihren natürlichen Verwandtschaften nebst einer Verwandtschafts-Tafel und einem Verzeichnisse der Reptilien-Sammlung des k.k. zoologisch Museum's zu Wien. Wien, J.G. Heubner: i-viii, 1-66 + 1 foldout table.

FITZINGER, L. J. F. J. 1827. Neue Classification der Reptilien, nach ihren naturlichen Verwandtschaften, nebst einer Verwandtschaftstafel und einem Vergleichniss der Reptiliensammlung des k. k. zoolog. Museums zu Wien. *Isis von Oken* 20: 262-267.

FITZINGER, L. 1828. Erwiederung an Herrn Schlegel, Conservator am königlichen naturhistorischen Museum zu Leyden, und Herrn Dr. Wagler, Professor an der königlichen Universität zu München, in Betreff ihrer Angriffe gegen meine neue Classification der Reptilien. *Isis von Oken* 21: 3-24.

FITZINGER, L. 1843. Systema reptilium. Fasciculus primus. Amblyglossae. Vindobonae, Braumüller & Seidel: i-ii, 1-106, i-ix.

FITZINGER, L. J. 1861. Die Ausberete der österreichische Naturforsher an Säugethieren und Reptilien während der Weltumsegelung Sr. Majestät Fregatte Novara. *Sitzungsberichte der Akademie der Mathematisch-Naturwissenschaftliche Classe der Kaiserlichen Akademie der Wissenschaften* (Wien) 42 (25/26): 383-416.

FITZINGER, L. J. 1864. Bilder-Atlas zur wissenschaftlich-populären Naturgeschichte der Amphibien in ihren Sämmtlichen hauptformen. Wien, Kaiserl. Konigl. Hof-& Staatsdruckerei: 109 pl. [*fide* Frost (2009-13); not seen]

FLEMING, J. 1822. The philosophy of zoology; or, a general view of the structure, functions, and classification of animals. Vol. 2. Edinburgh, A. Constable & Co.: [i-vi], 1-618.

FLORES-VILLELA, O. 1993. Herpetofauna of Mexico: distribution and endemism. *In*: RAMAMOORTHY, R., R. BYE, A. LOT, AND J. FA (eds.), Biological Diversity of Mexico: Origins and Distribution. New York & Oxford, Oxford University Press: 253-280.

FLORES-VILLELA, O., AND R. A. BRANDON. 1992. *Siren lacertina* (Amphibia: Caudata) in Northeastern Mexico and Southern Texas. *Annals of the Carnegie Museum* 61 (4): 289-291.

Folt, B., T. Pierson, J. Goessling, S. M. Goetz, D. Laurencio, D. Thompson, and S. P. Graham. 2013. Amphibians and reptiles of Jasper County, Mississippi, with comments on the potentially extinct Bay Springs Salamander (*Plethodon ainsworthi*). *Herpetological Review* 44 (2): 283-286.

Ford, L. S., and D. C. Cannatella. 1993. The major clades of frogs. *Herpetological Monographs* 7: 94-117.

Fouquette, M. J., Jr. 1968a. Remarks on the type specimen of *Bufo alvarius* Girard. *Great Basin Naturalist* 28 (2): 70-72.

Fouquette, M. J., Jr. 1969. *Rhinophrynidae, Rhinophrynus, R. dorsalis. Catalogue of American Amphibians and Reptiles*: American Society of Ichthyologists and Herpetologists (78): 1-2.

Fouquette, M. J., Jr. 1970. *Bufo alvarius. Catalogue of American Amphibians and Reptiles*: American Society of Ichthyologists and Herpetologists (93): 1-4.

Fouquette, M. J., Jr. 2005. *Rhinophrynus dorsalis*, Burrowing Toad. *In*: Lannoo, M. J. (ed.), Amphibian Declines: The Conservation Status of United States Species. Part 2: Species Accounts. Berkeley, University of California: 599-600.

Fouquette, M. J., Jr., C. W. Painter, and P. Nanjappa. 2005. *Bufo alvarius*, Colorado River Toad. *In*: Lannoo, M. J. (ed.), Amphibian Declines: The Conservation Status of United States Species. Part 2: Species Accounts. Berkeley, University of California: 384-386.

Fowler, H. W. 1906b ("1905-1906"). The sphagnum frog of New Jersey, *Rana virgatipes. Proceedings of the Academy of Natural Sciences of Philadelphia* 57 (1905): 662-664, pl.40.

Fowler, H. W. 1907. The amphibians and reptiles of New Jersey. *Annual Report of the New Jersey State Museum* (Trenton) 1906 (Part 2): 23-250 + 351-385 + 402-408, pl. 1-69.

Fowler, H. W. 1925a. Records of amphibians and reptiles for Delaware, Maryland and Virginia. I. Delaware. *Copeia* (145): 57-61.

Fowler, H. W., and E. R. Dunn. 1917. Notes on salamanders. *Proceedings of the Academy of Natural Sciences of Philadelphia* 69 (1917): 7-28, pl. 3-4.

FRANK, N., AND E. RAMUS. 1995. A complete guide to scientific and common names of reptiles and amphibians of the world. Pottsdale, PA, NG Publishing: 1-377.

FRANZ, R., AND C. J. CHANTELL. 1978. *Limnaoedus. L. ocularis. Catalogue of American Amphibians and Reptiles.* Society for the Study of Amphibians and Reptiles (209): 1-2.

FREEMAN, T., AND F. CUSTIS (Eds.). 1807. An Account of the Red River in Louisiana, drawn up from the returns of Messrs. Freeman and Custis, to the War Office of the United States, who explored same in the year 1806 U.S. Department of War Office, Washington: 65 pp + 2 tab.

FREYTAG, G. E. 1959. Zur Anatomie und systematischen Stellung von *Ambystoma schmidti* Taylor 1938 und verwandten Arten. *Vierteljahrschrift der Naturforschenden Gesselschaft in Zürich* 104: 79-89.

FROST, D. R. (ed.). 1985. Amphibian species of the world: a taxonomic and geographic reference. Lawrence, KS, Allen Press & Association of Systematics Collections: i-v, 1-732.

FROST, D. R. 2009-2013. Amphibian Species of the World: an Online Reference. Version 5.3 and later. Database accessible at http://research.amnh.org/herpetology/amphibia/.

FROST, D. R., T. GRANT, J. FAIVOVICH, R. H. BAIN, A. HAAS, C. F. B. HADDAD, R. O. DE SÁ, A. CHANNING, M. WILKINSON, S. C. DONNELLAN, C. J. RAXWORTHY, J. A. CAMPBELL, B. L. BLOTTO, P. E. MOLER, R. C. DREWES, R. A. NUSSBAUM, J. D. LYNCH, D. B. GREEN, AND W. C. WHEELER. 2006a. The amphibian tree of life. *Bulletin of the American Museum of Natural History* (297): 1-370.

FROST, D. R., T. GRANT, AND J. R. MENDELSON, III. 2006b. *Ollotis* Cope, 1875 is the oldest name for the genus currently referred to as *Cranopsis* Cope, 1875 (Anura: Hyloides: Bufonidae). *Copeia* 2006: 558.

FROST, D. R., AND D. M. HILLIS. 1990. Species in concept and practice: herpetological applications. *Herpetologica* 46 (1): 87-104.

FROST, D. R., R. W. MCDIARMID, AND J. R. MENDELSON, III. 2008. Anura: frogs. *In*: Scientific and standard English names of amphibians and reptiles of North America north of Mexico, with comments regarding confidence in our understanding. Sixth Edition. *Edited by* CROTHER,

B. I. Society for the Study of Amphibians and Reptiles, *Herpetological Circular* 37, 2-12.

Frost, D. R., R. W. McDiarmid, and J. R. Mendelson, III. 2009a. Response to the *Point of View* of Gregory B. Pauly, David M. Hillis, and David C. Cannetella, by the Anuran Subcommittee of the SSAR/HL/ASIH Scientific and Standard English Names List. *Herpetologica* 65: 136-153.

Frost, D. R., R. W. McDiarmid, J. R. Mendelson, III, and D. M. Green. 2012. Anura: frogs. *In*: Crother, B. I. (ed.), Scientific and standard English names of amphibians and reptiles of North America north of Mexico, with comments regarding confidence in our understanding. Seventh Edition. Society for the Study of Amphibians and Reptiles, *Herpetological Circular* 39: 11-22.

Frost, D. R., J. R. Mendelson, III., and J. B. Pramuk. 2009b. Further notes on the nomenclature of Middle American toads (Bufonidae). *Copeia* 2009: 418.

Funk, W. C., C. A. Pearl, H. M. Drabeim, M. J. Adams, T. D. Mullins, and S. M. Haig. 2008. Range-wide phylogeographic analysis of the spotted frog complex (*Rana luteiventris* and *Rana pretiosa*) in northwestern North America. *Molecular Phylogenetics and Evolution* 49: 198-210.

Gadow, H. 1901. Amphibia and Reptiles. Volume 8. London, Macmillan & Co.: i-xiv, 1-668.

Gage, S. H. 1891. Life history of the vermilion-spotted newt (*Diemyctylus viridescens* Raf.). *American Naturalist* 25 (300): 1084-1110 + pl. 23.

Gaige, H. T. 1917. Description of a new salamander from Washington. *Occasional Papers of the Museum of Zoology, University of Michigan* (40): 1-3.

Gaige, H. T. 1932. The status of *Bufo copei*. *Copeia* 1932 (3): 134.

Gallardo, J. M. 1965. A propósito de los Leptodactylidae (Amphibia, Anura). *Papéis Avulsos de Zoologia* (São Paulo) 17 (8): 77-87.

Gamble, T., P. B. Berendzen, H. B. Shaffer, D. E. Starkey, and A. M. Simons. 2008. Species limits and phylogeography of North American cricket frogs (*Acris*: Hylidae). *Molecular Phylogenetics and Evolution* 48: 112-125.

GAO, K., AND N. H. SHUBIN. 2001. Late Jurassic salamanders from northern China. *Nature* 410: 574-577.

GARCÍA-PARIS, M., D. R. BUCHHOLZ, AND G. PARRA-OLEA. 2003. Phylogenetic relationships of Pelobatoidea re-examined using mtDNA. *Molecular Phylogenetics and Evolution* 28 (1): 12-23.

GARMAN, H. 1890. Notes on Illinois reptiles and amphibians, including several species not before recorded from the northern states. *Bulletin of the Illinois State Laboratory of Natural History* 3: 185-190.

GARMAN, H. 1894. A preliminary list of the vertebrate animals of Kentucky. *Bulletin of the Essex Institute* 26: 1-63.

GARMAN, S. W. 1876. On a variation in the colors of animals. *Proceedings of the American Academy for the Advancement of Science* 25 (1877): 187-204.

GARMAN, S. 1883. The reptiles and batrachians of North America. Frankfort, KY, Kentucky Geological Survey: [i-vii], i-xxxiv, 1-185, pl. I-IX.

GARMAN, S. 1884a. The North American reptiles and batrachians. A list of the species occurring north of the Isthmus of Tehuantepec, with references. *Bulletin of the Essex Institute* 16: 1-46.

GARMAN, S. 1888b ("1887"). Reptiles and batrachians from Texas and Mexico. *Bulletin of the Essex Institute* 19: 119-138.

GARMAN, S. 1896. *Diemyctylus viridescens* var. *vittatus*, a new variety of the red-spotted triton. *Journal of the Cincinnati Society of Natural History* 19: 49-51.

GARNIER, [J. H.]. 1888. Synopsis of a paper read before the biological section by Dr. Garnier on a new species of *Menobranchus*. *Proceedings of the Canadian Institute* (Toronto) (3) 5: 218-219.

GARRETT, J. M., AND D. G. BARKER. 1987. A Field Guide to Reptles and Amphibians of Texas. Austin, Texas Monthly Press: i-xi, 1-225, pl. 1-48.

GARSAULT, [F. A.] D. 1764-1765. Les figures des plantes et animaux d'usage en medecine, décrits dans la Matiere Medicale de Geoffroy Medecin. Paris, Desprez: i-v, pl. 1-118.

GASSÓ MIRACLE, M. E., L. W. VAN DEN HOEK OSTENDE, AND J. W. ARMTZEN. 2007. Type specimens of amphibians in the

National Museum of Natural History, Leiden. *Zootaxa* (1482): 25-68.

GATES, W. R. 1988. *Pseudacris nigrita. Catalogue of American Amphibians and Reptiles.* Society for the Study of Amphibians and Reptiles (416): 1-3.

GAUDIN, A. J. 1979. *Hyla cadaverina. Catalogue of American Amphibians and Reptiles.* Society for the Study of Amphibians and Reptiles (225): 1-2.

GEHLBACH, F. R. 1966. Types and type-localities of some taxa in the synonymy of *Ambystoma tigrinum* (Green). *Copeia* 1966(4): 881-882.

GEHLBACH, F. R. 1967. *Ambystoma tigrinum. Catalogue of American Amphibians and Reptiles.* American Society of Ichthyologists and Herpetologists (52): 1-4.

GERGUS, E. W. A., T. W. REEDER, AND B. K. SULLIVAN. 2004. Geographic variation in *Hyla wrightorum*: Advertisement calls, allozymes, mtDNA, and morphology. *Copeia* 2004 (4): 758-769.

GIBBES, L. R. 1845. Description of a new species of salamander. *Boston Journal of Natural History* 5: 89-90.

GIBBES, L. R. 1850. On a new species of *Menobranchus*, from South Carolina. *Proceedings of the American Association for the Advancement of Science* 1850: 159.

GILL, T. 1907. *Diemictylus* or *Notophthalmus* as names of a salamander. *Science* 26 (660): 256.

GILLIAMS, J. 1818. Description of two new species of Linnaean *Lacerta. Journal of the Academy of Natural Sciences of Philadelphia* 1: 460-462.

GIRARD, C. 1856a ("1854"). A list of the North American bufonids, with diagnoses of new species. *Proceedings of the Academy of Natural Sciences of Philadelphia* 7: 86-88. [Sometimes cited as Baird & Girard, but only Girard indicated as author]

GIRARD, C. 1856b ("1854"). Abstract of a report to Lieut. James M. Gilliss, U.S.N., upon the reptiles collected during the U.S.N. astronomical expedition to Chili. *Proceedings of the Academy of Natural Sciences of Philadelphia* 7: 226-227.

GIRARD, C. 1857a ("1856"). On a new genus and species of Urodela, from the collections of the U.S. Expl. Exped. under Comm. Charles Wilkes, U.S.N. *Proceedings of the Academy of Natural Sciences of Philadelphia* 8 (4): 140-141.

GIRARD, C. 1858. United States exploring expedition, during the years 1838, 1839, 1840, 1841, 1842 under the command of Charles Wilkes, U.S.N. Vol. 20: Herpetology. Philadelphia, Lippincott & Co.: i-xvi, 1-496, pl. 1-32. [prepared under the superintendence of S. F. Baird. Note: on p. v, Introduction, Baird states that Girard was responsible for the entire work]

GIRARD, C. 1859. *Bufo alvarius* Grd. *In*: BAIRD, S. F. (ed.), Reptiles of the Boundary (viz.): 26.

GIRARD, C. 1860 ("1859"). Herpetological notices. *Proceedings of the Academy of Natural Sciences of Philadelphia* 11: 169-170.

GISTEL, J. 1848. Naturgeschiechte des Thierreichs für höhere Schulen bearbeitet durch Johannes Gistel.mit einem Atlas von 32 Tafein (darstelland 617 illuminierte Figuren) und mehrem dem Texte eingednruckten Xylographien. Stuttgart, Hoffman: i-xvi, 1-216.

GISTEL, J. AND G. TILESIUS. 1868. Die Lurch Europas. Ein Beitrag zur Lehre von der geographischen Verbreitung derselben. *In*: Blicke in das Leben der Natur und des Menschen. Ein Taschenbuch zur Verbreitung gemeinnutziger Kenntiniss insebesondere des Natur-Lander-un Volkerkunde, Kunste und Gewerbe. Leipzig, Verlag Gb. Wartig: 144-167.

GLASS, B. P. 1946. A new *Hyla* from south Texas. *Herpetologica* 3 (2): 101-103.

GLAW, F., AND M. FRANZEN. 2006. Type catalogue of amphibians in the Zoologische Staatssammlung Muenchen. *Spixiana* 29 (2): 153-192.

GLOGER, C. 1827. Assembly of the eastern North American herpetofauna: new evidence from lizards and frogs. *Notizen aus dem Gebiete der Natur-und Heilkunde, gesammelt und mitgebeilt* 16 (18): 277-280.

GLORIOSO, B. M. 2010. *Pseudacris ornata. Catalogue of American Amphibians and Reptiles.* Society for the Study of Amphibians and Reptiles (866): 1-8.

GMELIN, J. F. 1789 ("1788"). Pars 3. Amphibia, Pisces. *In*: LINNAEUS, C. (ed.), Systema Naturae per regna tria naturae secundum classes, ordines, genera, species, cum characteribus, differentis, synonymis, locis. Editio 13. Tom. 1. Leipzig, G.E. Beer: [i-ii], 1033-1516: Classis III Amphibia 1033-1125; Classis IV. Pisces 1126-1516.

GOEBEL, A. M., T. A. RANKER, P. S. CORN, AND R. G. OLMSTEAD. 2009. Mitochondrial DNA evolution in the *Anaxyrus boreas* [Amphibia: Bufonidae] species group. *Molecular Phylogenetics and Evolution* 50: 209-225.

GOIN, C. J. 1938. The status of *Amphiuma tridactylum* Cuvier. *Herpetologica* 1: 127-130.

GOIN, C. J. 1942. Description of a new race of *Siren intermedia* Le Conte. *Annals of the Carnegie Museum* 29: 211-217.

GOIN, C. J. 1950. A study of the salamander *Ambystoma cingulatum*, with a description of a new subspecies. *Annals of the Carnegie Museum* 31 (14): 299-321.

GOIN, C. J. 1957. Description of a new salamander of the genus *Siren* from the Rio Grande. *Herpetologica* 13 (1): 37-42.

GOIN, C. J., AND J. W. CRENSHAW, JR. 1949. Description of a new race of the salamander *Pseudobranchus striatus* (Le Conte). *Annals of the Carnegie Museum* 31 (10): 277-280.

GOIN, C. J., AND O. B. GOIN. 1962. Introduction to Herpetology. San Francisco, Freeman:

GOIN, C. J., AND M. G. NETTING. 1940. A new gopher frog from the Gulf coast, with comments upon the *Rana areolata* group. *Annals of the Carnegie Museum* 27: 137-168, pl. 12.

GOLDBERG, C. S., K. F. FIELD, AND M. J. SREDL. 2004a. Mitochondrial DNA sequences do not support species status of the Ramsey Canyon Leopard Frog (*Rana subaquavocalis*). *Journal of Herpetology* 38 (3): 313-319.

GOLDBERG, C. S., B. K. SULLIVAN, J. H. MALONE, AND C. R. SCHWALBE. 2004b. Divergence among Barking Frogs (*Eleutherodactylus augusti*) in the southwestern United States. *Herpetologica* 60 (3): 312-320.

GOLDFUSS, G. A. 1820. Handbuch der Zoologie. Dritter Theil, zweite Abtheilung. Nürnberg, J.L. Schrang: i-xxiv, 1-512, pl. 3-4.

GOOD, D. A. 1989. Hybridization and cryptic species in *Dicamptodon* (Caudata: Dicamptodontidae). *Evolution* 43: 728-744.

GOOD, D. A., AND D. B. WAKE. 1992. Geographic variation and speciation in the torrent salamanders of the genus

Rhyacotriton (Caudata: Rhyacotritonidae). *University of California Publications in Zoology* 126: 1-91.

GORDON, R. E. 1967. *Aneides aeneus. Catalogue of American Amphibians and Reptiles.* American Society of Ichthyologists and Herpetologists (30): 1-2.

GORHAM, S. W. 1970. The amphibians and reptiles of New Brunswick. *Publications of the New Brunswick Museum. Monographic Series* (6): i-ix, 1-30.

GORHAM, S. W. 1974. Checklist of world amphibians up to January 1, 1970. Liste des amphibiens du monde d'apres l'etat du 1er Janvier 1970. Saint John, New Brunswick Museum: 1-173.

GORMAN, J. 1954. A new species of salamander from central California. *Herpetologica* 10: 153-158.

GORMAN, J. 1960. Treetoad studies. 1. *Hyla californiae*, new species. *Herpetologica* 16: 214-222.

GORMAN, J. 1964a. *Hydromantes. Catalogue of American Amphibians and Reptiles.* American Society of Ichthyologists and Herpetologists (10): 1-2.

GORMAN, J. 1964b. *Hydromantes brunus, H. platycephalus, and H. shastae. Catalogue of American Amphibians and Reptiles.* American Society of Ichthyologists and Herpetologists (11): 1-2.

GORMAN, J., AND C. L. CAMP. 1953. A new cave species of salamander of the genus *Hydromantes* from California, with notes on habits and habitat. *Copeia* 1953 (1): 39-43.

GORZULA, S., AND J. C. SEÑARIS. 1999 ("1998"). 1. A data base. *IN*: Contribution to the herpetofauna of the Venezuelan Guayana. *Scientia Guaianae* (Caracas) (8): i-xvii, 1-268, pl. x-xx.

GOSNER, K. L., AND I. H. BLACK. 1967. *Hyla andersonii. Catalogue of American Amphibians and Reptiles.* American Society of Ichthyologists and Herpetologists (54): 1-2.

GOSNER, K. L., AND I. H. BLACK. 1968. *Rana virgatipes. Catalogue of American Amphibians and Reptiles.* American Society of Ichthyologists and Herpetologists (67): 1-2.

GRANDISON, A. G. C. 1980. Aspects of breeding morphology in *Mertensophryne microanotis* (Anura: Bufonidae): secondary sexual characters, eggs and tadpole. *Bulletin of the British Museum (Natural History). Zoology* 39: 299-304.

GRAVENHORST, J. L. C. 1807. Vergliechende Uebersicht des Linneischen und einer neuern zoologischen Systeme von J.L.C Gravenhorst ... nebst dem Eingeschalteten Verzeichniss der Zoologischen Sammlung des Verfassers und den Beschreibungen neuer Thierarten, die in Derselben Vorhanden Sind. Göttingen, H. Dieterich: i-xx, 1-476.

GRAVENHORST, J. L. C. 1829. Deliciae Musei Zoologici Vratislaviensis. Recensita et descripta. Fasciculus primus. Chelonios et Batrachia. Voss, Leipzig, i-xiv, 1-20, Tab. I-V.

GRAVENHORST, J. L. C. 1843. Vergleichende Zoologie. Breslau, Grass, Barth & Co.: i-xx, 1-686.

GRAVENHORST, J. L. C. 1845. Das Thiereich nach den Verwandtschaften und Uebergangen in den Klassen und Ordnungen desselben dargestellt. Breslau, Grass, Barth & Co.: i-x, 1-254, pl. 1-12.

GRAY, J. E. 1825. A synopsis of the genera of Reptiles and Amphibia, with a description of some new species. *Annals of Philosophy* (London) (2) 10: 193-217.

GRAY, J. E. 1829. Synopsis generum Reptilium et Amphibiorum. *Isis von Oken* 22 (2): 187-206.

GRAY, J. E. 1831a. A synopsis of the species of Class Reptilia. *In*: GRIFFITH, E., AND E. PIDGEON (ed.s.), The animal kingdom arranged in conformity with its organisation by the Baron Cuvier with additional descriptions of all the species hitherto named, and of many before noticed. Volume 9. Reptilia. Supplement. London, Whittaker, Treacher & Co.: 1-110.

GRAY, J. E. 1831b. Synopsis reptilium: or short descriptions of the species of reptiles. Part 1.—Cataphracta: Tortoises, crocodiles, and enaliosaurians. London, Treuttel, Wurtz & Co.: [i-v], i-vii, 1-85.

GRAY, J. E. 1839. Reptiles. *In*: RICHARDSON, J., N. A. VIGORS, G. T. LAY, E. T. BENNETT, R. OWEN, J. E. GRAY, W. BUCKLAND, AND G. B. SOWERBY (eds.), The zoology of Captain Beechey's Voyage to the Pacific and Behring's Straits performedin His Majesty's ship Blossom. London, H.G. Bohn: 93-97.

GRAY, J. E. 1842a. The Northern Zoological Gallery. *In*: Synopsis of the contents of the British Museum, 44th edition. London, British Museum: 97-157.

GRAY, J. E. 1850. Catalogue of the specimens of Amphibians in the Collection of the British Museum. Part II, Batrachia Gradientia, etc. London, Spottiswoodes & Shaw: [i-v], 1-72, pl. I-IV.

GRAY, J. E. 1853a. On a new species of salamander from California. *Proceedings of the Zoological Society of London* 1853 (21): 11, pl. 7.

GRAY, J. E. 1859. Descriptions of new species of salamanders from China and Siam. *Proceedings of the Zoological Society of London* 1859 (27): 229-230.

GREEN, D. M. 1986a. Systematics and evolution of western North American frogs allied to *Rana aurora* and *Rana boylii*: karyological evidence. *Systematic Zoology* 35 (3): 273-282.

GREEN, D. M. 1986b. Systematics and evolution of western North American frogs allied to *Rana aurora* and *Rana boylii*: electrophoretic evidence. *Systematic Zoology* 35 (3): 283-296.

GREEN, D. M., H. KAISER, T. F. SHARBEL, J. KEARSLEY, AND K. R. MCALLISTER. 1997. Cryptic species of spotted frogs, *Rana pretiosa* complex, in western North America. *Copeia* 1997 (1): 1-8.

GREEN, D. M., T. F. SHARBEL, J. KEARSLEY, AND H. KAISER. 1996. Postglacial range fluctuation, genetic subdivision and speciation in the western North American spotted frog complex, *Rana pretiosa*. *Evolution* 50: 374-390.

GREEN, J. 1818. Descriptions of several species of North American Amphibia, accompanied with observations. *Journal of the Academy of Natural Sciences of Philadelphia* 1 (Part 2): 348-359.

GREEN, J. 1825. Description of a new species of salamander. *Journal of the Academy of Natural Sciences of Philadelphia* 5: 116-118.

GREEN, J. 1827. An account of some new species of salamanders. *Contributions of the Maclurian Lyceum* 1 (1): 3-8.

GREEN, J. 1831. Descriptions of two new species of salamander. *Journal of the Academy of Natural Sciences of Philadelphia* 6 (Part 2): 253-255.

GREEN, N. B. 1938a. A new salamander, *Plethodon nettingi*, from West Virginia. *Annals of the Carnegie Museum* 27 (19): 295-299.

GRIFFITH, E., AND E. PIDGEON. 1831. The class Reptilia arranged by the Baron Cuvier, with specific descriptions. *In*: GRIFFITH, E. (ed.), The animal kingdom arranged in conformity with its organization, by the Baron Cuvier, member of the Institute of France, &c. &c. &c., with additional descriptions of all the species hitherto named, and of many not before noticed, Volume the ninth. London, Whittaker, Treacher & Co.: [i-vi] + 1-481 + 1-110, pl. 1-55.

GRINNELL, J., AND C. L. CAMP. 1917. A distributional list of the reptiles and amphibians of California. *University of California Publications in Zoology* 17 (10): 127-208.

GRINNELL, J., AND T. I. STORER. 1924. Animal life in the Yosemite. Berkeley, University of California Press, Museum of Vertebrate Zoology: [1-viii], v-xviii, 1-752, pl. 1-60 + 2 maps.

GROBMAN, A. B. 1943. Notes on salamanders, with a description of a new species of *Cryptobranchus*. *Occasional Papers of the Museum of Zoology, University of Michigan* 1-12, 1 pl.

GROBMAN, A. B. 1944. The distribution of the salamanders of the genus *Plethodon* in eastern United States and Canada. *Annals of the New York Academy of Science* 45 (7): 261-316.

GROBMAN, A. B. 1945. The identity of *Desmognathus phoca* (Matthes) and of *Desmognathus monticola* Dunn. *Proceedings of the Biological Society of Washington* 58: 39-43.

GROBMAN, A. B. 1949. Some recent collections of *Plethodon* from Virginia with the description of a new form. *Proceedings of the Biological Society of Washington* 62: 135-140.

GROBMAN, A. B. 1959. The anterior cranial elements of the salamanders *Pseudotriton* and *Gyrinophilus*. *Copeia* 1959 (1): 60-63.

GUÉRIN-MÉNEVILLE, F. E. 1838 (1829-1844). Iconographie du règne animal de G. Cuvier: ou, Repésentation d'après nature de l'une des espèces les plus et souvent non encore figurées de chaque genre d'animaux. 3 vol. Paris & London, J.B Baillière. [*fide* Frost (2009-13); not seen]

GUETTARD, M. 1770. Mémoires sur différentes parties des sciences et arts. Tome second. Paris, Pault. [*fide* Frost (2009-13); not seen]

GUIBÉ, J. 1950. Catalogue des types d'amphibiens du Muséum National d'Histoire Naturelle. Paris, Imprimerie nationale: 1-71.

GUNTER, G., AND W. E. BRODE. 1964. *Necturus* in the state of Mississippi, with notes on adjacent areas. *Herpetologica* 20: 114-126.

GÜNTHER, A. 1858a. On the systematic arrangement of the tailless batrachians and the structure of *Rhinophrynus dorsalis*. *Proceedings of the Zoological Society of London* 1858: 61-74.

GÜNTHER, A. 1859a ("1858"). Catalogue of the Batrachia Salientia in the collection of the British Museum. London, Taylor and Francis: [i-iii], i-xvi, 1-160, pl. I-XII.

GÜNTHER, A. 1868. *Zoologcal Record*: Reptilia 4: 145.

GÜNTHER, A. 1870. Second account of species of tailless batrachians added to the collection of the British Museum. *Proceedings of the Zoological Society of London* 1870: 401-402, pl. 30.

GÜNTHER, A. 1900. Reptilia and Batrachia. Part 155. *In*: SALVIN, O., AND F. D. GODMAN (eds.), Biologia Centrali Americana. Volume 7. London, Porter, Dulau & Co.: 213-220.

GÜNTHER, A. 1901a. Reptiles and batrachia. Part 162. *In*: SALVIN, O., AND F. D. GEDMANN (eds.), Biologia Centrali-Americana. Vol. 7. London, Porter, Dulau & Co.: 237-252.

GÜNTHER, A. C. L. G. 1901b. Reptilia and batrachia. Part 164. *In*: SALVIN, O., AND F. D. GODMAN (ed.s.), Biologia Centrali-Americana. Vol. 7. London, Porter, Dulau & Co.: 253-260.

GÜNTHER, A. C. L. G. 1901c. Reptilia and Batrachia. Part 169. *In*: SALVIN, O., AND F. D. GODMAN (eds.), Biologia

Centrali-Americana. Vol. 7. London, Porter Dulau & Co.: 293-300.

GUTTMAN, S. I., A. A. KARLIN, AND G. M. LABANICK. 1978. A biochemical and morphological analysis of the relationship between *Plethodon longicrus* and *Plethodon yonahlossee* (Amphibia, Urodela, Plethodontidae). *Journal of Herpetology* 12 (4): 445-454.

GUTTMAN, S. I., L. A. WEIGT, P. E. MOLER, R. E. ASHTON, JR., B. W. MANTELL, AND J. PEAVY. 1990. An electrophoretic analysis of *Necturus* from the southeastern United States. *Journal of Herpetology* 24 (2): 163-175.

HAECKEL, E. 1866. Generelle Morphologie der Organismen. 2 vol. Berlin, G Reimer: Vol. 1: I-XXXII, 1-574; v2: I-CLX, 1-462.

HAIRSTON, N. G. 1950. Intergradation in Appalachian salamanders of the genus *Plethodon*. *Copeia* 1950: 262-273.

HAIRSTON, N. G. 1993. On the validity of the name *teyahalee* as applied to a member of the *Plethodon glutinosus* complex (Caudata: Plethodontidae): a new name. *Brimleyana* 18: 65-69.

HAIRSTON, N. G., AND C. H. POPE. 1948. Geographic variation and speciation in Appalachian salamanders (*Plethodon jordani* group). *Evolution* 2 (3): 266-278.

HALDEMAN, [S. S.]. 1848. [untitled extract from letter: on identity of two *Salamandra*]. *Proceedings of the Academy of Natural Sciences of Philadelphia* 3: 315.

HALE, S. F., P. C. ROSEN, J. L. JARCHOW, AND G. A. BRADLEY. 2005. Effects of the chytrid fungus on the Tarahumara Frog (*Rana tarahumarae*) in Arizona and Sonora, Mexico. *USDA Forest Service, Rocky Mountain Research Station Proceedings* 36: 407-411.

HALL, J. A. 1998. *Scaphiopus intermontanus*. *Catalogue of American Amphibians and Reptiles*. Society for the Study of Amphibians and Reptiles (650): 1-17.

HALLOWELL, E. 1850. Description of a new species of salamander from Upper California. *Proceedings of the Academy of Natural Sciences of Philadelphia* 4 (1848-1849): 126.

HALLOWELL, E. 1852a. Descriptions of new species of reptiles inhabiting North America. *Proceedings of the Academy of Natural Sciences of Philadelphia* 6: 177-182.
HALLOWELL, E. 1854d. Descriptions of new species of reptiles inhabiting North America. *Proceedings of the Academy of Natural Sciences of Philadelphia* 6 (1852-1853): 177-182.
HALLOWELL, E. 1854e. Descriptions of new species of reptiles from Oregon. *Proceedings of the Academy of Natural Sciences of Philadelphia* 6 (1852-1853): 182-183.
HALLOWELL, E. 1854f. On a new genus and three new species of reptiles inhabiting North America. *Proceedings of the Academy of Natural Sciences of Philadelphia* 6 (1852-1853): 206-209.
HALLOWELL, E. 1855a ("1854"). Descriptions of new reptiles from California. *Proceedings of the Academy of Natural Sciences of Philadelphia* 7 (1854-1855): 91-97.
HALLOWELL, E. 1857a. Descriptions of several species of Urodela with remarks on the geographical distribution of the Caducibranchiate division of these animals and their classification. *Proceedings of the Academy of Natural Sciences of Philadelphia* 8 (1856): 6-12.
HALLOWELL, E. 1857b. [untitled report: On some anurans and a new species of *Rana*]. *Proceedings of the Academy of Natural Sciences of Philadelphia* 8 (1856): 141-143.
HALLOWELL, E. 1857c. Description of two new species of urodeles, from Georgia. *Proceedings of the Academy of Natural Sciences of Philadelphia* 8 (1856): 130-131.
HALLOWELL, E. 1857d. Notes on a collection of reptiles from Kansas and Nebraska, presented to the Academy of Natural Sciences, by Dr. Hammond, U.S.A. *Proceedings of the Academy of Natural Sciences of Philadelphia* 8 (1856): 238-253.
HALLOWELL, E. 1857e. [untitled: notes on salamanders of the Philadelphia area]. *Proceedings of the Academy of Natural Sciences of Philadelphia* 8: 101.
HALLOWELL, E. 1857f. Note on the collection of reptiles from the neighborhood of San Antonio, Texas, recently presented to the Academy of Natural Sciences by Dr. A. Heermann. *Proceedings of the Academy of Natural Sciences of Philadelphia* 8 (1856): 306-310.

HALLOWELL, E. 1858a. Description of several new North American reptiles. *Proceedings of the Academy of Natural Sciences of Philadelphia* 9 (1857): 215-216.

HALLOWELL, E. 1858b. On the caducibranchiate urodele batrachians. *Journal of the Academy of Natural Sciences of Philadelphia* (2) 3: 337-366.

HALLOWELL, E. 1858c. Notice on a collection of reptiles from the Gaboon country, West Africa, recently presented to the Academy of Natural Sciences of Philadelphia, by Dr. Henry A. Ford. *Proceedings of the Academy of Natural Sciences of Philadelphia* 9 (1857): 48-72.

HALLOWELL, E. 1859b. Report upon reptiles of the route (No. 1). *In*: Reports of explorations in California for a railroad route to connect with the route near the 35th and 32d parallel of north latitude by Lieutenant R.S. Williamson, Corps of Topographical Engineers. In: Reports of explorations and surveys to axcertain the most practicable and economical route for a railroad from the Mississippi River to the Pacific Ocean, made under the direction of the Secretary of War, in 1853-5, according to acts of Congress of March 3, 1853, May 31, 1854, and August 5, 1854. Volume 10, Part 4, Zoological Report. Washington, 1-27, pl. 1-10.

HANSEN, R. W., R. H. GOODMAN, JR., AND D. B. WAKE. 2005. *Batrachoseps gabrieli*, San Gabriel Mountains Slender Salamander. *In*: LANNOO, M. J. (ed.), Amphibian Declines: The Conservation Status of United States Species. Part 2: Species Accounts. Berkeley, University of California Press: 672-673.

HANSEN, R. W., AND D. B. WAKE. 2005a. *Batrachoseps aridus*, Desert Slender Salamander. *In*: LANNOO, M. J. (ed.), Amphibian Declines: The Conservation Status of United States Species, Part 2: Species Accounts. Berkeley, University of California Press: 666-667.

HANSEN, R. W., AND D. B. WAKE. 2005b. *Batrachoseps diabolicus*, Gregarious Slender Salamander. *In*: LANNOO, M. J. (ed.), Amphibian Declines: The Conservation Status of United States Species, Part 2: Species Accounts. Berkeley, University of California Press: 671-672.

HANSEN, R. W., AND D. B. WAKE. 2005c. *Batrachoseps gregarius*, Hell Hollow Slender Salamander. *In*: LANNOO,

M. J. (ed.), Amphibian Declines: The Conservation Status of United States Species, Part 2: Species Accounts. Berkeley, University of California Press: 675-676.

HANSEN, R. W., AND D. B. WAKE. 2005d. *Batrachoseps incognitus*, San Simeon Slender Salamander. *In*: LANNOO, M. J. (ed.), Amphibian Declines: The Conservation Status of United States Species, Part 2: Species Accounts. Berkeley, University of California Press: 677-678.

HANSEN, R. W., AND D. B. WAKE. 2005e. *Batrachoseps kawia*, Sequoia Slender Salamander. *In*: LANNOO, M. J. (ed.), Amphibian Declines: The Conservation Status of United States Species, Part 2: Species Accounts. Berkeley, University of California Press: 678-679.

HANSEN, R. W., AND D. B. WAKE. 2005f. *Batrachoseps luciae*, Santa Lucia Mountains Slender Salamander. *In*: LANNOO, M. J. (ed.), Amphibian Declines: The Conservation Status of United States Species, Part 2: Species Accounts. Berkeley, University of California Press: 679-680.

HANSEN, R. W., AND D. B. WAKE. 2005g. *Batrachoseps minor*, Lesser Slender Salamander. *In*: LANNOO, M. J. (ed.), Amphibian Declines: The Conservation Status of United States Species, Part 2: Species Accounts. Berkeley, University of California Press: 682-683.

HANSEN, R. W., AND D. B. WAKE. 2005h. *Batrachoseps regius*, Kings River Slender Salamander. *In*: LANNOO, M. J. (ed.), Amphibian Declines: The Conservation Status of United States Species, Part 2: Species Accounts. Berkeley, University of California Press: 686-688.

HANSEN, R. W., AND D. B. WAKE. 2005i. *Batrachoseps relictus*, Relictual Slender Salamander. *In*: LANNOO, M. J. (ed.), Amphibian Declines: The Conservation Status of United States Species, Part 2: Species Accounts. Berkeley, University of California Press: 688-690.

HANSEN, R. W., AND D. B. WAKE. 2005j. *Batrachoseps robustus*, Kern Plateau Salamander. *In*: LANNOO, M. J. (ed.), Amphibian Declines: The Conservation Status of United States Species, Part 2: Species Accounts. Berkeley, University of California Press: 690-691.

HANSEN, R. W., AND D. B. WAKE. 2005k. *Batrachoseps simatus*, Kern Canyon Slender Salamander. *In*: LANNOO, M. J. (ed.), Amphibian Declines: The Conservation

Status of United States Species, Part 2: Species Accounts. Berkeley, University of California Press: 691-693.

HANSEN, R. W., AND D. B. WAKE. 2005l. *Batrachoceps attenuatus*. California Slender Salamander. *In*: LANNOO, M. J. (ed.), Amphibian Declines: The Conservation Status of United States Species, Part 2: Species Accounts. Berkeley, University of California Press: 667-669.

HANSEN, R. W., AND D. B. WAKE. 2005m. *Batrachoceps gavilanensis*. Gavilan Mountains Slender Salamander. *In*: LANNOO, M. J. (ed.), Amphibian Declines: The Conservation Status of United States Species, Part 2: Species Accounts. Berkeley, University of California Press: 673-675.

HANSEN, R. W., AND D. B. WAKE. 2005n. *Batrachoceps major*. Garden Slender Salamander. *In*: LANNOO, M. J. (ed.), Amphibian Declines: The Conservation Status of United States Species, Part 2: Species Accounts. Berkeley, University of California Press: 680-682.

HANSEN, R. W., AND D. B. WAKE. 2005o. *Batrachoceps nigriventris*. Black-bellied Slender Salamander. *In*: LANNOO, M. J. (ed.), Amphibian Declines: The Conservation Status of United States Species, Part 2: Species Accounts. Berkeley, University of California Press: 683-685.

HANSEN, R. W., AND D. B. WAKE. 2005p. *Batrachoceps pacificus*. Channel Islands Slender Salamander. *In*: LANNOO, M. J. (ed.), Amphibian Declines: The Conservation Status of United States Species, Part 2: Species Accounts. Berkeley, University of California Press: 685-686.

HARDY, J. D., JR. 1964. A new frog, *Rana palustris mansuetii,* subsp. nov. from the Atlantic Coastal Plain. *Chesapeake Science* 5: 91-100.

HARDY, J. D., JR., AND J. D. ANDERSON. 1970. *Ambystoma maybeei. Catalogue of American Amphibians and Reptiles*. American Society of Ichthyologists and Herpetologists (81): 1-2.

HARDY, J. D., JR., AND R. J. BURROWS. 1986. Systematic status of the Spring Peeper, *Hyla crucifer* (Amphibia: Hylidae). *Bulletin of the Maryland Herpetological Society* 22 (2): 68-89.

HARLAN, R. 1825b. Description of a variety of the *Coluber fulvius*, Linn., a new species of *Scincus*, and two new species of *Salamandra*. *Journal of the Academy of Natural Sciences of Philadelphia* 5 (1): 154-158.

HARLAN, R. 1825c. Observations on the genus *Salamandra*, with the anatomy of the *Salamandra gigantea* (Barton) or *S. alleghaniensis* (Michaux) and two new genera proposed. *Annals of the Lyceum of Natural History of New York* 1 (2): 222-234, pl. 16-18.

HARLAN, R. 1825d. Note to a paper entitled "Observations on the genus *Salamandra*" at p. 222 of this volume. *Annals of the Lyceum of Natural History of New York* 1 (2): 270-271.

HARLAN, R. 1825e. Description of a new species of *Salamandra*. *Journal of the Academy of Natural Sciences of Philadelphia* 5 (1): 136.

HARLAN, R. 1826a. Descriptions of several new species of batrachian reptiles, with observations on the larvae of frogs. *American Journal of Science and Arts* 10 (7): 53-65.

HARLAN, R. 1826b. Notice of a new species of salamander, (inhabiting Pennsylvania). *American Journal of Sciences and Arts* 10: 286-287.

HARLAN, R. 1827. American herpetology, or genera of North American Reptilia, with a synopsis of the species. *Journal of the Academy of Natural Sciences of Philadelphia* 5 (2): 317-404.

HARLAN, R. 1829. Description of a new species of *Salamandra*. *Journal of the Academy of Natural Sciences of Philadelphia* 6 (1): 101.

HARLAN, R. 1835. Genera of North American Reptilia, and a synopsis of the species. *In*: (ed.), *Medical and Physical Research or Original Memoires in Medicine, Surgery, Physiology, Geology, Zoology, and Comparative Anatomy*. Philadelphia, L.R. Bailey: 84-160.

HARPER, F. 1935. Records of amphibians in the southeastern states. *American Midland Naturalist* 16 (3): 275-310.

HARPER, F. 1939a. A southern subspecies of the spring peeper (*Hyla crucifera*). *Notulae Naturae* (Philadelphia) (27): 1-4.

HARPER, F. 1939b. Distribution, taxonomy, nomenclature, and habits of the little tree-frog (*Hyla ocularis*). *American Midland Naturalist* 22: 134-149.

HARPER, F. 1940. Some works of Bartram, Daudin, LaTreille, and Sonnini, and their bearing upon North American herpetological nomenclature. *American Midland Naturalist* 23 (3): 692-723.

HARPER, F. 1947. A new cricket frog (*Acris*) from the middle western states. *Proceedings of the Biological Society of Washington* 60: 39-40.

HARPER, F. 1955a. A new chorus frog (*Pseudacris*) from the eastern United States. *Natural History Miscellany, Chicago Academy of Science* (150): 1-6.

HARPER, F. 1955b. The type locality of *Hyla triseriata* Wied. *Proceedings of the Biological Society of Washington* 68: 155-156.

HARRISON, J. R. 1992. *Desmognathus aeneus. Catalogue of American Amphibians and Reptiles*. Society for the Study of Amphibians and Reptiles (534): 1-4.

HARRISON, J. R. 2000. *Desmognathus wrighti. Catalogue of American Amphibians and Reptiles*. Society for the Study of Amphibians and Reptiles (704): 1-7.

HARRISON, J. R. 2005. *Eurycea chamberlaini,* Chamberlain's Dwarf Salamander. *In*: LANNOO, M. J. (ed.), Amphibian Declines: The Conservation Status of United States Species, Part 2: Species Accounts. Berkeley, University of California Press: 738-739.

HARRISON, J. R., AND S. I. GUTTMAN. 2003. A new species of *Eurycea* (Caudata: Plethodontidae) from North and South Carolina. *Southeastern Naturalist* 2 (2): 159-178.

HARTMANN, R. 1879. Über die Umwaldlung des *Siredon lichenoides* Baird in *Amblystoma* (*Ambystoma*) *mavortium* Baird. *Sitzungsberichte der Gesellschaft Naturforschender Freunde zu Berlin* 1879: 76-78.

HAY, O. P. 1885. Description of a new species of *Amblystoma* (*Amblystoma copeianum*) from Indiana. *Proceedings of the U.S. National Museum* 8: 209-213, pl. 14.

HAY, O. P. 1891. Note on *Gyrinophilus maculicaudus* Cope. *American Naturalist* 25 (300): 1133-1135.

HAY, O. P. 1892. The batrachians and reptiles of the state of Indiana. *Indiana Department of Geology and Natural Resources, Annual Report* (for 1891) 17: 409-602.

HECHT, M. K. 1957. A case of parallel evolution in salamanders. *Proceedings of the Zoological Society of Calcutta* 1957: 283-292.

HECHT, M. K. 1958. A synopsis of the mud puppies of eastern North America. *Proceedings of the Staten Island Institute of Arts and Sciences* 21: 5-38.

HECHT, M. K., AND B. L. MATALAS. 1946. A review of middle North American toads of the genus *Microhyla*. *American Museum Novitates* (1315): 1-21.

HECK, J. G., AND S. F. BAIRD. 1851. Iconographic Encyclopaedia of Science, Literature and Art systematically arranged by J.G. Heck (translated from the German, with additions). Volume 2: Botany, Zoology, Anthropology, and Surgery. New York, R. Garrigue: i-xxiv, Botany 1-202, Zoology1-502, Anthropology & Surgery 1-219, i-xii, i-xvi, i-v.

HEDEEN, S. E. 1977. *Rana septentrionalis*. *Catalogue of American Amphibians and Reptiles*. Society for the Study of Amphibians and Reptiles (202): 1-2.

HEDGES, S. B. 1986. An electrophoretic analysis of Holarctic hylid frog evolution. *Systematic Zoology* 35 (1): 1-21.

HEDGES, S. B. 1989. Evolution and biogeography of West Indian frogs of the genus *Eleutherodactylus*: slow-evolving loci and the major groups. *In*: WOODS, C. (ed.), Biogeography of the West Indies: Past, Present, and Future. Gainesville, Sandhill Crane Press: 305-370.

HEDGES, S. B. 1996. The origin of West Indian amphibians and reptiles. In: POWELL, R., AND R.W. HENDERSON (eds.), Contributions to West Indian Herpetology: A tribute to Albert Schwartz. *Contributions to herpetology*. Ithaca, Society for the study of Amphibians and Reptiles: 95-128.

HEDGES, S. B., W. E. DUELLMAN, AND M. P. HEINICKE. 2008. New World direct-developing frogs (Anura: Terrarana): molecular phylogeny, classification, biogeography, and conservation. *Zootaxa* (1737): 1-182.

HEKKALA, E. R., R. A. SAUMURE, J. R. JAEGER, H.-W. HERRMANN, M. J. SREDL, D. F. BRADFORD, D. DRABECK, AND M. J. BLUM. 2011. Resurrecting an extinct species: archival DNA, taxonomy, and conservation of the Vegas Valley leopard frog. *Conservation Genetics* 12 (5): 1379-1385.

HEMPRICH, W. 1820. Grundniss der Naturgeschichte für höhere Lehranstalten. Berlin, August Rucker: i-viii,1-432. [fide (Frost 2009-13); not seen]

HENDERSON, R. W., AND R. POWELL. 2003. Some historical perspectives. *In*: HENDERSON, R. W., AND R. POWELL (ed.s.),

Islands and the sea. Essays on herpetological exploration in the West Indies. *Contributions to herpetology* 20. Salt Lake City, Society for the Study of Amphibians and Reptiles: 9-20.

HENDRICKSON, J. R. 1954. Ecology and systematics of salamanders of the genus *Batrachoseps*. *University of California Publications in Zoology* 17: 127-208.

HERRE, W. 1934. Die systematische Stellung von *Taricha torosa* Eschscholtz. *Blätter für Aquarien-und Terrarienkunde* (Stuttgart) 45: 250-252.

HERRE, W. 1936. Ueber Rasse und Artbildung. Studien an Salamandriden. *Abhhandlungen und Berichte aus dem Museum für Natur-und Heimatkunde zu Magdeburg* 6 (3): 193-221.

HEYER, M. M., W. R. HEYER, AND R. O. DE SÁ. 2006. *Leptodactylus fragilis*. *Catalogue of American Amphibians and Reptiles*. Society for the Study of Amphibians and Reptiles (830): 1-26.

HEYER, W. R. 1971. *Leptodactylus labialis*. *Catalogue of American Amphibians and Reptiles*. American Society of Ichthyologists and Herpetologists (104): 1-3.

HEYER, W. R. 1974. *Vanzolinius*, a new genus proposed for *Leptodactylus discodactylus* (Amphibia, Leptodactylidae). *Proceedings of the Biological Society of Washington* 87 (11): 81-90.

HEYER, W. R. 1978. Systematics of the *fuscus* group of the frog genus *Leptodactylus* (Amphibia, Leptodactylidae). *Natural History Museum of Los Angeles County, Science Bulletin* (29): [i], 1-85.

HEYER, W. R. 2002. *Leptodactylus fragilis*, the valid name for the Middle American and northern South American White-lipped Frog (Amphibia: Leptodactylidae). *Proceedings of the Biological Society of Washington* 115 (2): 321-322.

HIGHTON, R. 1961. A new genus of lungless salamander from the Coastal Plain of Alabama. *Copeia* 1961 (1): 65-68.

HIGHTON, R. 1962. Revision of North American salamanders of the genus *Plethodon*. *Bulletin of the Florida State Museum* 6 (3): 235-367.

HIGHTON, R. 1970. Evolutionary interactions between species of North American salamanders of the genus *Plethodon*.

Part 1. Genetic and ecological relationships of *Plethodon jordani* and *P. glutinosus* in the southern Appalachian Mountains. *Evolutionary Biology* 4: 211-241.

HIGHTON, R. 1972 ("1971"). Distributional interactions among eastern North American salamanders of the genus *Plethodon*. *In*: HOLT, P. C., et al. (eds.), The distributional history of the biota of the southern Appalachians. Part III: Vertebrates. Blackburg, *Research Division Monograph 4, Virginia Polytechnic Institute*: 139-188.

HIGHTON, R. 1973. *Plethodon jordani. Catalogue of American Amphibians and Reptiles*. Society for the Study of Amphibians and Reptiles (130): 1-2.

HIGHTON, R. 1977. The endemic salamander, *Plethodon shenandoah*, of Shenandoah National Park. *First Annual Shenandoah Research Symposium, Research Reports* (11): 15-17.

HIGHTON, R. 1979. A new cryptic species of salamander of the genus *Plethodon* from the southeastern United States (Amphibia: Plethodontidae). *Brimleyana* (1): 31-36.

HIGHTON, R. 1984 ("1983"). A new species of woodland salamander of the *Plethodon glutinosus* group from the southern Appalachian Mountains. *Brimleyana* (9): 1-20.

HIGHTON, R. 1986a. *Plethodon aureolus. Catalogue of American Amphibians and Reptiles*. Society for the Study of Amphibians and Reptiles (381): 1.

HIGHTON, R. 1986b. *Plethodon kentucki. Catalogue of American Amphibians and Reptiles*. Society for the Study of Amphibians and Reptiles (382): 1-2.

HIGHTON, R. 1986c. *Plethodon nettingi. Catalogue of American Amphibians and Reptiles*. Society for the Study of Amphibians and Reptiles (383): 1-2.

HIGHTON, R. 1986d. *Plethodon websteri. Catalogue of American Amphibians and Reptiles*. Society for the Study of Amphibians and Reptiles (384): 1-2.

HIGHTON, R. 1986e. *Plethodon fourchensis. Catalogue of American Amphibians and Reptiles*. Society for the Study of Amphibians and Reptiles (391): 1.

HIGHTON, R. 1986f. *Plethodon hoffmani. Catalogue of American Amphibians and Reptiles*. Society for the Study of Amphibians and Reptiles (392): 1-2.

HIGHTON, R. 1986g. *Plethodon hubrichti. Catalogue of American Amphibians and Reptiles.* Society for the Study of Amphibians and Reptiles (393): 1-2.
HIGHTON, R. 1986h. *Plethodon serratus. Catalogue of American Amphibians and Reptiles.* Society for the Study of Amphibians and Reptiles (394): 1-2.
HIGHTON, R. 1987a. *Plethodon teyahalee. Catalogue of American Amphibians and Reptiles.* Society for the Study of Amphibians and Reptiles (401): 1-2.
HIGHTON, R. 1987b. *Plethodon wehrlei. Catalogue of American Amphibians and Reptiles.* Society for the Study of Amphibians and Reptiles (402): 1-3.
HIGHTON, R. 1988a. *Plethodon shenandoah. Catalogue of American Amphibians and Reptiles.* Society for the Study of Amphibians and Reptiles (413): 1-2.
HIGHTON, R. 1988b. *Plethodon punctatus. Catalogue of American Amphibians and Reptiles.* Society for the Study of Amphibians and Reptiles (414): 1-2.
HIGHTON, R. 1989. Part I. Geographic protein variation. In: HIGHTON, R., G. C. MAHA, AND L. R. MAXSON (ed.s.), Biochemical evolution in the slimy salamanders of the *Plethodon glutinosus* complex in the eastern United States. *Illinois Biological Monographs* 57: 1-78.
HIGHTON, R. 1997. Geographic protein variation and speciation in the *Plethodon dorsalis* complex. *Herpetologica* 53: 345-356.
HIGHTON, R. 1999. Geographic protein variation and speciation in the salamanders of the *Plethodon cinereus* group with the description of two new species. *Herpetologica* 55 (1): 43-90. [have pdf]
HIGHTON, R. 2004. A new species of Woodland Salamander of the *Plethodon cinereus* group from the Blue Ridge Mountains of Virginia. *Jeffersoniana* 14: 1-22.
HIGHTON, R., AND A. H. BRAME, JR. 1965. *Plethodon stormi* species nov. Amphibia: Urodela: Plethodontidae. *Pilot Register of Zoology* (Card No. 20):
HIGHTON, R., AND A. B. GROBMAN. 1956. Two new salamanders of the genus *Plethodon* from the southeastern United States. *Herpetologica* 12 (3): 186-188.

HIGHTON, R., AND A. LARSON. 1979. The genetic relationships of the salamanders of the genus *Plethodon*. *Systematic Zoology* 28 (4): 579-599.

HIGHTON, R., AND J. R. MACGREGOR. 1983. *Plethodon kentucki* Mittleman: a valid species of Cumberland Plateau woodland salamander. *Herpetologica* 39 (3): 189-200.

HIGHTON, R., AND R. B. PEABODY. 2000. Geographic variation and speciation in the *Plethodon jordani* and *Plethodon glutinosus* complexes in the southern Appalachian Mountains with the description of four new species. *In*: BRUCE, R. C., R. G. JAEGER, AND L. D. HOUCK (ed.s.), The Biology of Plethodontid Salamanders. New York, Kluwer Academic/Plenum: 31-93.

HIGHTON, R., S. G. TILLEY, AND D. B. WAKE. 2001 ("2000"). Salamanders. In: CROTHER, B. I., J. BOUNDY, B. CAMPBELL, K. DE QUEIROZ, D. R. FROST, R. HIGHTON, J. B. IVERSON, P. A. MEYLAN, T. W. REEDER, SEIDEL, J. SITES, T. W. TAGGART, S. G. TILLEY, AND D. B. WAKE (eds.), Scientific and standard English names of amphibians and reptiles of North America north of Mexico, with comments regarding the confidence in our understanding. *Herpetological Circular* 29: 21-37.

HIGHTON, R., AND T. P. WEBSTER. 1976. Geographic protein variation and divergence in populations of the salamander *Plethodon cinereus*. *Evolution* 30 (1): 33-45.

HIGHTON, R., AND R. D. WORTHINGTON. 1967. A new salamander of the genus *Plethodon* from Virginia. *Copeia* 1967 (4): 617-626.

HILL, I. R. 1954. The taxonomic status of the mid-Gulf Coast *Amphiuma*. *Tulane Studies in Zoology* 1 (12): 189-215.

HILLIS, D. M. 2007. Constraints in naming parts of the Tree of Life. *Molecular Phylogenetics and Evolution* 42: 331-338.

HILLIS, D. M., D. A. CHAMBERLAIN, T. P. WILCOX, AND P. T. CHIPPINDALE. 2001. A new species of subterranean blind salamander (Plethodontidae: Hemidactyliini: *Eurycea*: *Typhlomolge*) from Austin, Texas, and a systematic revision of central Texas paedomorphic salamanders. *Herpetologica* 57 (3): 266-280.

HILLIS, D. M., AND S. K. DAVIS. 1986. Evolution of ribosomal DNA: fifty million years of recroded history in the frog genus *Rana*. *Evolution* 40 (6): 1275-1288.

HILLIS, D. M., J. S. FROST, AND D. A. WRIGHT. 1983. Phylogeny and biogeography of the *Rana pipiens* complex: a biochemical evaluation. *Systematic Zoology* 32 (2): 132-143.

HILLIS, D. M., AND T. P. WILCOX. 2005. Phylogeny of the New World true frogs (*Rana*). *Molecular Phylogenetics and Evolution* 34: 299-314.

HINCKLEY, M. H. 1882 ("1881"). On some differences in the mouth-structure of tadpoles of the anourous Batrachians found in Milton, Mass. *Proceedings of the Boston Society of Natural History* 21: 307-314.

HODGKINS, J. G. 1856. Remarks on a Canadian specimen of the Proteus of the Lakes. *Canadian Journal of Industry, Science, and Art* 1: 19-23.

HOFFMAN, R. L. 1945. Notes on the herpetological fauna of Alleghany County, Virginia. *Herpetologica* 2: 199-205.

HOFFMAN, R. L. 1951. A new subspecies of salamander from Virginia. *Journal of the Elisha Mitchell Scientific Society* 67 (1): 249-254.

HOFFMAN, R. L. 1980. *Pseudacris brachyphona. Catalogue of American Amphibians and Reptiles.* Society for the Study of Amphibians and Reptiles (244): 1-2.

HOFFMAN, R. L. 1983. *Pseudacris brimleyi.* Catalogue of American Amphibians and Reptiles. Society for the Study of Amphibians and Reptiles 1-2.

HOFFMAN, R. L. 1988. *Hyla femoralis. Catalogue of American Amphibians and Reptiles.* Society for the Study of Amphibians and Reptiles (436): 1-3.

HOFFMANN, C. K. 1878. Klassen und Ordnungen der Amphibien wissenschaftlich dargestellt in Wort un Bild. Band 6, Abt. 2. In: BRONN, H. G. (ed.), Die Klassen und Ordnungen des Thier-Reichs wissenschaftlich dargestellt in Wort und Bild. Leipzig & Heidelberg, C.F. Winter: [i-ii], 1-726, pl. 1-52.

HOGG, J. 1838. [On the classification of the Amphibia]. *Annals of Natural History or, Magazine of Zoology, Botany, and Geology* (London) (1) 1: 152.

HOGG, J. 1839. On the classification of the Amphibia. *The Magazine of Natural History* (London) (New Series) 3: 265-274, 367-378.

HOLBROOK, J. E. 1835. American herpetology; or, a description of the reptiles inhabiting the United States [advertising brochure for forthcoming volumes]. Charleston, J.E. Holbrook, printed by E.J. Van Brunt: [unnumbered pages]. [fide Adler, 1976: xlii; not seen]

HOLBROOK, J. E. 1836. North American herpetology; or, A description of the reptiles inhabiting the United States. Vol. 1 [first edition, first version]. Philadelphia, J. Dobson: [i-v], i-vii, [viii], 9-120, pl 1-23.

HOLBROOK, J. E. 1838a. North American herpetology; or, a description of the reptiles inhabiting the United States. Vol. 2 (first edition, first version). Philadelphia, Dobson: [i-iv], 5-125, pl. 1-28, [i-ii].

HOLBROOK, J. E. 1838b. North American herpetology; or, A description of the reptiles inhabiting the United States. Vol. 3 (first edition). Philadelphia, Dobson: i-viii, 9-122, pl. 1-30.

HOLBROOK, J. E. 1839a ("1836"). North American herpetology; or, a description of the reptiles inhabiting the United States. Vol. 1 (first edition, second version). Philadelphia, J. Dobson: i-viii, 9-132, pl. 1-27.

HOLBROOK, J. E. 1839b ("1838"). North American herpetology; or, A description of the reptiles inhabiting the United States. Vol. 2 [first edition, second version]. Philadelphia, J Dobson: [i-v], i-iv, 5-130, pl. 1-30.

HOLBROOK, J. E. 1840. North American herpetology; or, A description of the reptiles inhabiting the United States. Vol. 4 (first edition). Philadelphia, Dobson: [i-iii], i-x, 9-126, pl. 1-28.

HOLBROOK, J. E. 1842. North American herpetology; or, A description of the reptiles inhabiting the United States. Vol. 5 (2nd Ed.). Philadelphia, Dobson: [i-v], i-vi, 5-119, pl. 1-38.

HOLBROOK, J. E. 1842. North American herpetology; or, A description of the reptiles inhabiting the United States. Five volumes. Second edition. (SSAR reprint, 1976). Philadelphia, Dobson: 1: [i-vi], vii-xv, 17-152, pl. 1-24; 2: i-vi, iii-iv, 9-142, pl. 1-20; 3: i-ii, 3-128, pl. 1-30; 4: i-vi, 7-138, pl. 1-35; 5: i-vi, 5-118, pl. 1-38.

HOLLOWAY, A. K., D. C. CANNATELLA, H. C. GERHARDT, AND D. M. HILLIS. 2006. Polyploids with different origins and ancestors form a single sexual polyploid species. *American Naturalist* 167 (4): E88-E101.

HOOGMOED, M. S. 1978. An annotated review of the salamander types described in the Fauna Japonica. *Zoologische Mededelingen, Leiden* 53: 91-105.

HOOGMOED, M. S., AND U. GRUBER. 1983. Spix and Wagler type specimens of reptiles and amphibians in the Natural History Musea in Munich (Germany) and Leiden (The Netherlands). *Spixiana* 9 (Supplement): 319-415.

HOSER, R. 2009. A reclassification of the rattlesnakes; species formerly exclusively referred to the genera *Crotalus* and *Sistrurus*. *Australasian Journal of Herpetology* 6: 1-21.

HOUTTUYN, M. 1782. Het onderscheid der salamanderen van de haagdissen in 't algemeen, en van de gekkoos in 't byzonder aangetoond. *Verhandelingen Uitgegeven door het Zeeuwsch Genootschap derWetenschappen te Vlissingen* 9: 305-336, +1 pl.

HOWE, R. H., JR. 1899. North American wood frogs. *Proceedings of the Boston Society of Natural History* 28 (14): 369-374.

HSÜ, H.-F. 1930. A new giant frog of Amoy. *Contributions from the Biological Laboratory of the Science Society of China. Zoological Series* 6 (3): 19-23.

HUA, X., C. FU, J. LI, A. NIETO-MONTES DE OCA, AND J. J. WIENS. 2009. A revised phylogeny of Holarctic treefrogs (genus *Hyla*) based on nuclear and mitochondrial DNA sequences. *Herpetologica* 65 (3): 246-259.

HULSE, A. C. 1978. *Bufo retiformis*. *Catalogue of American Amphibians and Reptiles*. Society for the Study of Amphibians and Reptiles (207): 1-2.

HUNTINGTON, C., T. N. STUHLMAN, AND D. J. CULLEN. 1993. *Plethodon sequoyah*. *Catalogue of American Amphibians and Reptiles*. Society for the Study of Amphibians and Reptiles (557): 1-2.

HUTCHINSON, V. H. 1966. *Eurycea lucifuga*. *Catalogue of American Amphibians and Reptiles*. American Society of Ichthyologists and Herpetologists (24): 1-2.

INTERNATIONAL COMMISSION ON ZOOLOGICAL NOMENCLATURE. 1926. Opinion 92. Sixteen generic names of Pisces, Amphibia, and Reptilia placed in the Official List of

Generic Names. *Smithsonian Miscellaneous Collection* (2873): 3-4.

INTERNATIONAL COMMISSION ON ZOOLOGICAL NOMENCLATURE. 1956a. Opinion 377. Suppression under the plenary powers of the specific name "*tereticauda*" Eschscholtz, 1833, as published in the combination "*Triton tereticauda*", for the purpose of rendering the specific name "*lugubris*" Hallowell, 1849, as published in the combination "*Salamandra lugubris*", the oldest available name for the species concerned (Class Amphibia, Order Caudata). *Opinions and declarations rendered by the International Commission on Zoological Nomenclature* 11(27): 401-410.

INTERNATIONAL COMMISSION ON ZOOLOGICAL NOMENCLATURE. 1956b. Direction 57. Addition to the Official List of Specific Names in Zoology (a) of the specific names of forty-seven species belonging to the Classes Cyclostomata, Pisces, Amphibia and Reptilia, each of which is the type species of a genus, the name of which was placed on the Official List of Generic Names in Zoology in the period up to the end of 1936 and (b) of the specific name of one species of the Class Amphibia which is currently treated as a senior subjective synonym of the name of such a species. *Opinions and Declarations rendered by the International Commission on Zoological Nomenclature* 1. Section D. Part D.18: 367-388.

INTERNATIONAL COMMISSION ON ZOOLOGICAL NOMENCLATURE. 1958. Opinions and declarations rendered by the international commission on zoological nomenclature. Volume 1. Section B. Facsimile edition of Opinions 1-133. London, International Trust for Zoological Nomenclature: i-xix, 1-508.

INTERNATIONAL COMMISSION ON ZOOLOGICAL NOMENCLATURE. 1962. Opinion 635. *Notophthalmus* Rafinesque, 1820 (Amphibia): addition to the official list as the name to be used for the eastern North-American newt. *Bulletin of Zoological Nomenclature* 19: 152-155.

INTERNATIONAL COMMISSION ON ZOOLOGICAL NOMENCLATURE. 1963b. Opinion 665. *Salamandra erythronota* Rafinesque, 1818 (Amphibia): Suppression under the plenary powers. *Bulletin of Zoological Nomenclature* 20: 199-200.

INTERNATIONAL COMMISSION ON ZOOLOGICAL NOMENCLATURE. 1963c. Opinion 649. *Ambystoma* Tschudi, 1838 (Amphibia): validation under the plenary powers. *Bulletin of Zoological Nomenclature* 20: 102-104.

INTERNATIONAL COMMISSION ON ZOOLOGICAL NOMENCLATURE. 1963d. Opinion 662. *Salamandra tigrina* Green, 1825 (Amphibia): validation under the plenary powers. *Bulletin of Zoological Nomenclature* 20: 193-194.

INTERNATIONAL COMMISSION ON ZOOLOGICAL NOMENCLATURE. 1965. Opinion 738. *Triturus* (*Gyrinophilus*) *lutescens* Rafinesque, 1832 (Amphibia): suppressed under the plenary powers. *Bulletin of Zoological Nomenclature* 22 (3): 167-168.

INTERNATIONAL COMMISSION ON ZOOLOGICAL NOMENCLATURE. 1970. Opinion 921. Plethodontidae in Pisces and Amphibia: removal of homonymy under the plenary powers. *Bulletin of Zoological Nomenclature* 27(2): 79-80.

INTERNATIONAL COMMISSION ON ZOOLOGICAL NOMENCLATURE. 1974. Opinion 1024. *Epirhexis* Cope, 1866 (Amphiba Salientia): suppressed under the plenary powers. *Bulletin of Zoological Nomenclature* 31 (3): 130-132.

INTERNATIONAL COMMISSION ON ZOOLOGICAL NOMENCLATURE. 1977. Opinion 1071. Emendation under the plenary powers of Liopelmatina to Leiopelmatidae (Amphibia Salientia). *Bulletin of Zoological Nomenclature* 33 (3-4): 167-169.

INTERNATIONAL COMMISSION ON ZOOLOGICAL NOMENCLATURE. 1978. Opinion 1104. Relative precedence of *Cornufer* Tschudi, 1888,and *Platymantis* Günther, 1858 (Amphibia Salientia). *Bulletin of Zoological Nomenclature* 34 (4): 222-233.

INTERNATIONAL COMMISSION ON ZOOLOGICAL NOMENCLATURE. 1992. Opinion 1685. *Rana sphenocephala* Cope, 1886 (Amphibia, Anura): given precedence over *Rana utricularia* Harlan, 1826. *Bulletin of Zoological Nomenclature* 49 (2): 171-173.

INTERNATIONAL COMMISSION ON ZOOLOGICAL NOMENCLATURE. 1993a. Opinion 1716. *Hyla chrysoscelis* Cope, 1880 (Amphibia, Anura): specific name conserved by the designation of a neotype. *Bulletin of Zoological Nomenclature* 50 (1): 94-95.

INTERNATIONAL COMMISSION ON ZOOLOGICAL NOMENCLATURE. 1997a. Opinion 1866. *Hydromantes* Gistel, 1848 (Amphibia, Caudata): Spelerpes platycephalus Camp, 1916 designated as the type species. *Bulletin of Zoological Nomenclature* 54 (1): 72-74.

INTERNATIONAL COMMISSION ON ZOOLOGICAL NOMENCLATURE. 1997b. Opinion 1873. Hemidactyliini Hallowell, 1856 (Amphibia, Caudata): conserved. *Bulletin of Zoological Nomenclature* 54 (2): 140-141.

INTERNATIONAL COMMISSION ON ZOOLOGICAL NOMENCLATURE. 1999. International Code of Zoological Nomenclature. 4th edition. London, International Trust for Zoological Nomenclature: i-xxix, 1-306.

INTERNATIONAL COMMISSION ON ZOOLOGICAL NOMENCLATURE. 2005. Opinion 2104. Lacepède, B.G.E., de la V., 1788, Histoire naturelle des quadrupèes ovipares: rejected as non-binomial work. *Bulletin of Zoological Nomenclature* 62 (1): 55.

INTERNATIONAL COMMISSION ON ZOOLOGICAL NOMENCLATURE. 2009. Opinion 2223 (Case 3345) Dendrobatidae Cope, 1865 (1850) (Amphibia, Anura): family-group name conserved. *Bulletin of Zoological Nomenclature* 66 (1): 103-105.

IRELAND, P. H. 1979. *Eurycea longicauda. Catalogue of American Amphibians and Reptiles*. Society for the Study of Amphibians and Reptiles (221): 1-4.

IRSCHICK, D. J., AND H. B. SHAFFER. 1997. The polytypic species revisited: morphological differentiation among Tiger Salamaders (*Ambystoma tigrinum*) (Amphibia: Caudata). *Herpetologica* 53 (1): 30-49.

IVES, J. E. 1892 ("1891"). Reptiles and batrachians from northern Yucatan and Mexico. *Proceedings of the Academy of Natural Sciences of Philadelphia* 43 (1891): 458-463.

JACKMAN, T. R. 1998. Molecular and historical evidence for the introduction of clouded salamanders (genus *Aneides*) to Vancouver Island, British Columbia, Canada, from California. *Canadian Journal of Zoology* 76: 1570-1580.

JACKMAN, T. R., G. APPLEBAUM, AND D. B. WAKE. 1997. Phylogenetic relationships of bolitoglossine salamanders: a demonstration of the effects of combining morphological and molecular data sets. *Molecular Biology and Evolution* 14 (8): 883-891.

JACKSON, N. D. 2005. Phylogenetic history, morphological parallelism, and speciation in a complex of Appalachian salamanders (genus *Desmognathus*). M.S. thesis, Dept. of Biology, Brigham Young University, Provo, UT. [*fide* Camp *et al.*, 2012; not seen]

JACOBS, J. F. 1987. A preliminary investigation of geographic genetic variation and systematics of the 2-lined salamander, *Eurycea bislineata* (Green). *Herpetologica* 43 (4): 423-446.

JAEGER, J. R., B. R. RIDDLE, R. D. JENNINGS, AND D. F. BRADFORD. 2001. *Rana onca*: evidence for phylogenetically distinct leopard frogs from the border region of Nevada, Utah, and Arizona. *Copeia* 2001 (2): 339-354.

JAMES, E. 1822. Account of an expedition from Pittsburgh to the Rocky Mountains, performed in the years 1819 and '20. By order of the Hon. J. C. Calhoun, Sec'y of War; under the command of Major Stephen H. Long,. From the notes of Major Long, Mr. T. Say, and other gentlemen of the exploring party. Compiled by Edwin James, botanist and geologist for the expedition. Volume 1. Philadelphia, Carey & Lea: 1-503.

JAMES, E. 1823. Account of an expedition from Pittsburgh to the Rocky Mountains, performed in the years 1819, 1820. By order of the Hon. J. C. Calhoun, Secretary of War; under the command of Maj. S. H. Long, of the U.S. Top. Engineers. Compiled from the notes of Major Long, Mr. T. Say, and other gentlemen of the party by Edwin James, botanist and geologist to the expedition. Volume 1. London, Longman, Hurst, Rees, Orme, and Brown: i-viii, 1-344, map.

JAMES, P. 1966. The Mexican burrowing toad, *Rhinophrynus dorsalis*, an addition to the vertebrate fauna of the United States. *Texas Journal of Science* 18: 272-276.

JAMESON, D. L., J. P. MACKEY, AND R. C. RICHMOND. 1966. The systematics of the Pacific tree frog, *Hyla regilla*.

Proceedings of the California Academy of Sciences 33 (19): 551-619.

JAN, G. 1857. Cenni sul Museo Civico di Milano ed indice sistematico dei rettili ed anfibi esposti nel medesimo. Milano, L. di Giacomo Pirola: [i-vii], 1-61, [i-xii].

JAROCKIEGO, F. P. 1822. Zoologiia czyli Zwiérzetopismo Ogólne podlug náynowszego systematu. Tom [Volume] 3. Gady i Plazy [Reptilia and Sirenia (=Amphibia)]. Warsaw, Daukarni Latriewicza przy Ulicy Senatorskiéy: [i-viii], 1-184, [i-ix], pl. 1-4.

JAUME, M. L. 1966. Catalogo de los anfibios de Cuba. *Trabajo Divulg. de la Sociedad Cubana de Historia Natural "Felipe Poey"* 35: 1-21.

JENNINGS, M. R. 1988a ("1987"). A biography of Dr. Charles Elisha Boyle, with notes on his 19th century natural history collection from California. *Wasmann Journal of Biology* 45 (1-2): 59-68.

JENNINGS, M. R. 1988b. *Rana onca. Catalogue of American Amphibians and Reptiles.* Society for the Study of Amphibians and Reptiles (417): 1-2.

JENNINGS, M. R., AND M. P. HAYES. 1994. Amphibian and repitle species of special concern in California. Sacramento, California Department of Fish and Game: i-iii, 1-255.

JENSEN, J. B., AND C. D. CAMP. 2004. *Plethodon petraeus. Catalogue of American Amphibians and Reptiles.* Society for the Study of Amphibians and Reptiles (783): 1-2.

JENSEN, J. B., AND S. C. RICHTER. 2005. *Rana capito.* Gopher Frog. *In*: LANNOO, M. J. (ed.), Amphibian Declines: The Conservation Status of United States Species, Part 2: Species Accounts. Berkeley, University of California Press: 536-538.

JIANG, L. FEI, C. YE, ZENG, AHEN, Z. XIE, AND B.-Y. CHEN. 1997. Studies on the taxonomics of species of *Pseudorana* and discussions of the phylogenetical relationships with its relative genera. *Cultum Herpetologica Sinica* (Liang qi pa xing dong wu xue yan jiu) 6-7: 67-74. [*fide* Frost (2009-13); not seen]

JIMÉNEZ DE LA ESPADA, M. 1875a. *Urotropis platensis. Anales de la Sociedad Española de Historia Natural* (Madrid) 4: 69-73, + 1 pl.

Jiménez De La Espada, M. 1875b. Vertebrados del viaje al Pacifico verificado de 1862 a 1865 por una comision de naturalistas enviada por el Gobierno Español. Batracios. Madrid, M. Ginesta: 1-208, pl. 1-6.

Jockusch, E. L. 2001. *Batrachoseps campi. Catalogue of American Amphibians and Reptiles*. Society for the Study of Amphibians and Reptiles (722): 1-2.

Jockusch, E. L., I. Martínez-Solano, R. W. Hansen, and D. Wake. 2012. Morphological and molecular diversification of slender salamanders (Caudata: Plethodontidae: *Batrachoseps*) in the southern Sierra Nevada of California with descriptons of two new species. *Zootaxa* (3190): 1-30.

Jockusch, E. L., and D. B. Wake. 2002. Falling apart and merging: diversification of slender salamanders (Plethodontidae: *Batrachoseps*) in the American west. *Biological Journal of the Linnaean Society* 76 (3): 361-391.

Jockusch, E. L., D. B. Wake, and K. P. Yanev. 1998. New species of *Batrachoseps* (Caudata: Plethodontidae) from the Sierra Nevada, California. *Contributions in Science, Natural History Museum, Los Angeles County* (472): 1-17.

Jockusch, E. L., K. P. Yanev, and D. B. Wake. 2001. Molecular phylogenetic analysis of slender salamanders, genus *Batrachoseps* (Amphibia: Plethodontidae), from central coastal California with descriptions of four new species. *Herpetological Monographs* 15: 54-99.

Johnson, C. R. 1966. Species recognition in the *Hyla versicolor* complex. *Texas Journal of Science* 18: 361-364.

Jones, M. T., S. R. Voss, M. B. Ptacek, D. W. Weisrock, and D. W. Tonkyn. 2006. River drainages and phylogeography: an evolutionary significant lineage of shovel-nosed salamander (*Desmognathus marmoratus*) in the southern Appalachians. *Molecular Phylogenetics and Evolution* 38 (1): 280-287.

Jones, T. R., A. G. Kluge, and A. J. Wolf. 1993. When theories and methodologies clash: a phylogenetic reanalysis of the North American ambystomatid salamanders (Caudata: Ambystomatidae). *Systematic Biology* 42 (1): 92-102.

Jordan, D. S. 1878. Manual of Vertebrates of the Northern United States, including the district east of the Mississippi River and north of North Carolina and Tennessee, exclusive of marine species. 2nd Editon. Chicago, Mansen, McClurg: 1-407.

KAISER, H., AND J. D. HARDY, JR. 1994. *Eleutherodactylus martinicensis*. *Catalogue of American Amphibians and Reptiles*. Society for the Study of Amphibians and Reptiles (582): 1-4.

KALM, P. 1761. En resa til Norra America. Tom. 3. Stockholm, L. Salvii: [i-ii],1-538, [i-xiv].

KARLSTROM, E. L. 1973. *Bufo canorus*. *Catalogue of American Amphibians and Reptiles*. Society for the Study of Amphibians and Reptiles (132): 1-2.

KAUFFELD, C. F. 1936. New York the type locality of *Rana pipiens* Schreber. *Herpetologica* 1 (1): 11.

KAUFFELD, C. F. 1937. The status of the leopard frogs, *Rana brachycephala* and *Rana pipiens*. *Herpetologica* 1 (3): 84-87.

KEFERSTEIN, W. 1867. Ueber einige neue oder seltene Batrachier aus Australien und dem tropischen Amerika. *Nachrichten von der Köngliche Gesellschaft der Wissenschaften* (Göttingen) 18: 341-361.

KEFERSTEIN, W. 1868a. Über die Batrachier Australiens. *Archiv für Naturgeschichte* 34 (1): 253-290, pl. 5-8.

KELLOGG, R. 1932. Mexican tailless amphibians in the United States National Museum. *Bulletin of the U.S. National Museum* 160: [i-iv], 1-224, pl. 1.

KING, W. 1936. A new salamander (*Desmognathus*) from the southern Appalachians. *Herpetologica* 1 (1): 57-60.

KING, W. 1939. A survey of the herpetology of Great Smoky Mountains National Park. *American Midland Naturalist* 21 (3): 531-582.

KIRK, J. J. 1991. *Batrachoseps wrighti*. *Catalogue of American Amphibians and Reptiles*. Society for the Study of Amphibians and Reptiles (506): 1-3.

KLUGE, A. G. 1966. A new pelobatine frog from the Lower Miocene of South Dakota with a discussion of the evolution of the *Scaphiopus-Spea* complex. *Contributions in Science, Natural History Museum, Los Angeles County* 113: 1-26.

KLUGE, A. G. 1983a. Type-specimens of amphibians in the University of Michigan Museum of Zoology. *Miscellaneous*

Publications, Museum of Zoology University of Michigan (166): i-iv, 1-68.

KNAUER, F. K. 1878. Naturgeschichte der Lurche. (Amphibiologie). Eine umfassendere Darlegung unserer Kenntnisse von dem anatomischen Bau, der Entwicklung und systematischen Eintheilung der Amphibien sowie eine eingenende Schilderung des Lebens dieser Thiere. Wien, A. Pichler's Witwe & Sohn: I-XX, 1-340 +4 maps, 2 tables.

KNEELAND, S. 1857. [On a supposed new species of Siredon]. *Proceedings of the Boston Society of Natural History* 6 (1856-1859): 152-1154.

KOBEL, H. R., B. BARANDUN, AND C. H. THIEBAUD. 1998. Mitochondrial rDNA phylogeny in *Xenopus*. *Herpetological Journal* 8 (1): 13-17.

KORKY, J. K. 1999. *Bufo punctatus. Catalogue of American Amphibians and Reptiles.* Society for the Study of Amphibians and Reptiles (689): 1-5.

KOZAK, K. H., A. LARSON, R. M. BONETT, AND L. J. HARMON. 2005. Phylogenetic analysis of ecomorphological divergence, community structure, and diversification rates in Dusky Salamanders (Plethodontidae: *Desmognathus*). *Evolution* 59 (9): 2000-2016.

KOZAK, K. H., D. W. WEISROCK, AND A. LARSON. 2006. Rapid lineage accumulation in a non-adaptive radiation: phylogenetic analysis of diversification rates in eastern North American woodland salamanders (Plethodontidae: *Plethodon*). *Proceedings of the Royal Society B* 273: 539-546.

KRAUS, F. 1985. A new unisexual salamander from Ohio. *Occasional Papers of the Museum of Zoology University of Michigan* (709): 1-24.

KRAUS, F. 1996. *Ambystoma barbouri. Catalogue of American Amphibians and Reptiles.* Society for the Study of Amphibians and Reptiles (621): 1-4.

Kraus, F. 2009. *Alien reptiles and amphibians: A scientific compendium and analysis*. Springer, Dordrecht: i-xii, 1-563.

Kraus, F., and E. W. Campbell III. 2002. Human-mediated escalation of a formerly eradicable problem: the invasion of Caribbean frogs in the Hawaiian Islands. *Biological Invasions* 4: 327-332.

KRAUS, F., E. W. CAMPBELL, A. ALLISON, AND T. PRATT. 1999. *Eleutherodactylus* frog introductions to Hawaii. *Herpetological Review* 30: 21-25.

KRAUS, F., AND R. NUSSBAUM. 1989. The status of the Mexican salamander, *Ambystoma schmidti* Taylor. *Journal of Herpetology* 23 (1): 78-79.

KRAUS, F., AND J. W. PETRANKA. 1989. A new sibling species of *Ambystoma* from the Ohio River drainage. *Copeia* 1989 (1): 94-110.

KRUPA, J. J. 1990. *Bufo cognatus. Catalogue of American Amphibians and Reptiles.* Society for the Study of Amphibians and Reptiles (457): 1-8.

Krysko, K. L., J. P. Burgess, M. R. Rochford, C. R. Gillette, D. Cueva, K. M. Enge, L. A. Somma, J. L. Stabile, D. C. Smith, J. A. Wasilewski, G. N. Kieckhefer III, M. C. Granatosky and S. V. Nielsen. 2011. Verified non-indigenous amphibians and reptiles in Florida from 1863 through 2010: outlining the invasion process and identifying invasion pathways and states. *Zootaxa* (3028): 1-64.

KUCHTA, S. R., D. S. PARKS, R. L. MUELLER, AND D. B. WAKE. 2009b. Closing the ring: historical biogeography of the salamander ring species *Ensatina eschscholtzii. Journal of Biogeography* 36: 982-995.

KUCHTA, S. R., D. S. PARKS, AND D. B. WAKE. 2009a. Pronounced phylogeographic structure on a small spatial scale: geomorphological evolution and lineage history in the salamander ring species *Ensatina eschscholtzii* in central coastal California. *Molecular Phylogenetics and Evolution* 50: 240-255.

KUHN, O. 1965. Die Amphibien; System und Stammesgeschichte. Krailling bei München, Verlag Oeben: 1-102.

KUZMIN, S. L. 1996. The taxonomic position of amphibian species from "Zoographia Rosso-Asiatica" by P.S. Pallas. *Advances in Amphibian Research in the Former Soviet Union* 1: 47-65.

LANGEBARTEL, D. A., AND H. M. SMITH. 1954. Summary of the Norris collection of reptiles and amphibians from Sonora, Mexico. *Herpetologica* 10: 125-136.

LANNOO, M. J. 2005. *Desmognathus abditus,* Cumberland Dusky Salamander. (Editor's note). *In*: LANNOO, M. J. (ed.).

2005. Amphibian Declines: The Conservation Status of United States Species, Part 2: Species Accounts. Berkeley, University of California Press: 696.

LANNOO, M. J., AND C. A. PHILLIPS. 2005. *Ambystoma tigrinum*, Tiger Salamander. *In*: LANNOO, M. J. (ed.), Amphibian Declines: The Conservation Status of United States Species, Part 2: Species Accounts. Berkeley, University of California Press: 636-639.

LANZA, B., AND S. VANNI. 1981. On the biogeography of Plethodontid salamanders (Amphibia Caudata) with a description of a new genus. *Monitore Zoologico Italiano N.S. Supplemento* 15: 117-121.

LATREILLE, P. A. 1800. Histoire naturelle des salamandres de France, précédée d'un tableau méthodique des autres reptiles indigènes. Paris, Villier: i-xlvii, 1-63, pl. 1-6.

LAURENT, R. F. 1942. Note sur les procoeliens firmisternes (Batrachia Anura). *Bulletin du Musée Royal d'Histoire Naturelle de Belgique* 18 (43): 1-20.

LAURENT, R. F. 1967. Taxonomia de los anuros. *Acta Zoologica Lilloana* 22: 207-210.

LAURENT, R. F. 1975. La distribution des amphibiens et les translations continentales. *Mémoires du Museum National d'Histoire Naturelle, Paris. A-Zoologie* 88: 176-191.

LAURENTI, J. N. 1768. Specimen medicum, exhibens synopsin Reptilium emendatam cum experimentis circa venena et antidota Reptilium austriacorum. Vienna, Joan, Thomae Nob. de Trattnern: i-ii, 1-215, pl. 1-5.

LAZELL, J. D., JR. 1998. New salamander of the genus *Plethodon* from Mississippi. *Copeia* 1998 (4): 967-970.

Lazell, J. D., Jr. 2005. *Plethodon ainsworthi,* Bay Springs Salamander. *In:* LANNOO, M. J. (ed.), Amphibian Declines: The Conservation Status of United States Species, Part 2: Species Accounts. Berkeley, University of California Press: 787-788.

LAZELL, J. D., JR., AND R. A. BRANDON. 1962. A new stygian salamander from the southern Cumberland Plateau. *Copeia* 1962 (2): 300-306.

LE CONTE, J. 1824. Description of a new species of *Siren*, with some observations on animals of a similar nature. *Annals of the Lyceum of Natural History* (New York) 1 (1): 52-58, pl. 4.

Le Conte, J. 1825. Remarks on the American species of the genera *Hyla* and *Rana*. *Annals of the Lyceum of Natural History of New York* 1 (2): 278-282.

Le Conte, J. 1827. *Siren intermedia, In*: Harlan, R. 1827. American herpetology, or genera of North American Reptilia, with a synopsis of the species. *Journal of the Academy of Natural Sciences of Philadelphia* 5 (2): 322.

Le Conte, J. 1828. Description of a new species of *Siren*. *Annals of the Lyceum of Natural History* (New York) 2: 133-134, pl. 1.

Le Conte, J. 1855b ("1856"). Descriptive catalog of the Ranina of the United States. *Proceedings of the Academy of Natural Sciences of Philadelphia* 7 (1854-1855): 423-431, pl. 5.

Le Conte, J. 1856 ("1857"). Description of a new species of *Hyla* from Georgia. *Proceedings of the Academy of Natural Sciences of Philadelphia* 8 (1856): 146.

Leach, W. E. 1814. Crustaceology. *In*: *Brewster's Edinburgh Encyclopaedia* 7 (2): 343-437.

Lemos-Espinal, J. A., H. M. Smith, and D. Chiszar. 2004. Introducción a los amphibios y reptiles del estado de Chihuahua. Mexico, D.F., Comisión Nacional para el Coocimiento y Uso de la Biodiversidad.

Lemos-Espinal, J. A., and H. M. Smith. 2007a. Ampibians and reptiles of the state of Coahuila, Mexico. Anfibios y reptiles del estado de Coahuila, México. Mexico, DF, Universidad Nacional Autónoma de México (and CONABIO): i-xii, 1-550.

Lemos-Espinal, J. A., and H. M. Smith. 2007b. Amphibians and reptiles of the state of Chihuahua, Mexico. Mexico, DF, Universidad Nacional Autonóma de México (and CONABIO): 613pp.

Leuckart, F. S. 1821. Einiges über die fischartigen Amphibien. *Isis von Oken* 9: 257-265, pl 5.

Leunis, J. 1844. Synopsis der drei Naturreiche. Ein Handbuch für höhere Lehranstalten und für Alle, welche sich wissenschaftlich mit Naturgeschichte beschäftigen wollen. Erster Theil. Zoologie. Hannover, Hahn'sche Hofbuchhandlung: i-xxxi, 1-476.

Leunis, J. 1860. Synopsis der drei Naturreiche. Ein Handbuch für höhere Lehranstalten und für Alle, welche sich wissenschaftlich mit Naturgeschichte beschäftigen wollen. Theil 2. Zoologie. Hannover, Hahn'sche Hofbuchhandlung.

LEVER, C. 2003. Naturalized reptiles and amphibians of the world. Oxford, Oxford University Press: i-xx, 1-318.

LINDHOLM, W. A. 1924. A forgotten description of a North American frog. *Copeia* (129): 46-47.

LINER, E. A. 1994. Scientific and common names for the amphibians and reptiles of Mexico in English and Spanish. Nombres Cientificos y comunes en ingles y español de los anfibios y los reptiles de México. Society for the Study of Amphibians and Reptiles, *Herpetological Circular* 23: 1-113.

LINNAEUS, C. 1758. Systema Naturae per regna tria naturae, secundum classes, ordines, genera, species, cum characteribus, differentiis, synonymis, locis. Editio decima, reformata. Tomus I. Stockholm, Holmiae, Laurentii Salvii: [i-iv], 1-824.

LINNAEUS, C. 1766. Systemae Naturae per regna tria naturae, secundum classes, ordines, genera, species, cum characteribus, differentiis, synonymis, locis. 12th ed. Stockholm, Salvius: 532 pp.

LINNAEUS, C. 1767. Systema Naturae. Tome I. Pars II. Editio duodecima reformata. Salvius, Stockholm, 533-1327, [i-xxxv, appendices], [xxxv-xxxvii, addendum].

LINSDALE, J. M. 1940. Amphibians and reptiles in Nevada. *Proceedings of the American Academy of Arts and Sciences* 73 (8): 197-257.

LIU, W., A. LATHROP, J. FU, D. YANG, AND R. W. MURPHY. 2000. Phylogeny of East Asian bufonids inferred from mitochondrial DNA sequences (Anura: Amphibia). *Molecular Phylogenetics and Evolution* 14 (3): 423-435.

LÖDING, C. P. 1922. A preliminary catalogue of Alabama amphibians and reptiles. *Geological Survey of Alabama, Museum Papers, Alabama Museum of Natural History* (5): 1-59.

LOFTUS-HILLS, J., AND M. J. LITTLEJOHN. 1992. Reinforcement and reproductive character displacement in *Gastrophryne carolinensis* and *G. olivacea* (Anura: Microhylidae): a reexamination. *Evolution* 46: 896-906.

LOHMAN, K., AND R. B. BURY. 2005. *Dicamptodon aterrimus*, Idaho Giant Salamander. *In*: LANNOO, M. J. (ed.),

Amphibian Declines: The Conservation Status of United States Species, Part 2: Species Accounts. Berkeley, University of California Press: 651-652.

LONG, C. A. 1982. Rare gigantic toads, *Bufo americanus*, from Lake Michigan isles. *University of Wisconsin Museum of Natural History, Reports on the Fauna and Flora of Wisconsin* 18: 16-19.

LONGLEY, G. 1978. Status of *Typhlomolge* (= *Eurycea*) *rathbuni*, the Texas blind salamander. U.S. Fish and Wildlife Service, Albuquerque, New Mexico, Endangered Species Report 2: 1-11.

LÖNNBERG, E. 1895. Notes on reptiles and batrachians collected in Florida in 1892 and 1893. *Proceedings of the U.S. National Museum* 17 (1894): 317-339.

LÖNNBERG, E. 1896. Linnean type specimens of birds, reptiles, batrachians, and fishes in the Zoological Museum of the R. University in Upsala. *Bihang till Konglika Svenska Vetenskaps-Academien Handligar* 22 (4): 1-45.

LOVERIDGE, A. 1932. New races of a skink (*Siaphos*) and frog (*Xenopus*) from the Uganda protectorate. *Proceedings of the Biological Society of Washington* 45: 113-115.

LOWCOCK, L. A., L. E. LICHT, AND J. P. BOGART. 1987. Nomenclature in hybrid complexes of *Ambystoma* (Urodela: Ambystomatidae): no case for the erection of hybrid "species". *Systematic Zoology* 36: 328-336.

LOWE, C. H., JR. 1950a. The systematic status of the salamander *Plethodon hardii,* with discussion of biogeographical problems in *Aneides. Copeia* 1950 (2): 92-99.

Lowe, C. H., Jr. 1950b. Speciation and ecology in salamanders of the genus *Aneides. PhD dissertation, UCLA, Los Angeles.* [not seen]

LOWE, C. H., JR. 1954. A new salamander (Genus *Ambystoma*) from Arizona. *Proceedings of the Biological Society of Washington* 67: 243-245.

LOWE, C. H., JR. 1955a. The salamanders of Arizona. *Transactions of the Kansas Academy of Science* 58: 237-251.

LOWE, C. H., JR. 1964. An annotated checklist of the amphibians and reptiles of Arizona. *In*: LOWE, C. H., JR (ed.), The vertebrates of Arizona. Tucson, University of Arizona Press: 153-174.

LUTZ, A. 1930. Segunda memória sobre espécies brasileiras do gênero *Leptodactylus*, incluindo outras aliadas. Second paper on Brazilian and some closely related species of the genus *Leptodactylus*. *Memorias do Instituto Oswaldo Cruz* 23: 1-20 (Portuguese), 21-34 (English), pl. 1-5.

LUTZ, B. 1954. Anfibios Anuros do Distrito Federal. *Memorias do Instituto Oswaldo Cruz* 52: 155-238.

LUTZ, B. 1969. Adaptações, especializações e linhagens nos anuros neotropicais. *Acta Zoologica Lilloana* 24: 267-292.

LYNCH, J. D. 1965a. A review of the eleutherodactylid frog genus *Microbatrachylus* (Leptodacylidae). *Natural History Miscellanea* (182): 1-12.

LYNCH, J. D. 1965b. The races of the microhylid frog, *Gastrophryne usta*, in México. *Transactions of the Kansas Academy of Science* 68: 369-400.

LYNCH, J. D. 1967. *Epirhexis* Cope, 1866 (Amphibia: Salientia): Request for suppression under the plenary powers. Z.N.(S.) 1824. *Bulletin of Zoological Nomenclature* 24: 313-315.

LYNCH, J. D. 1968. Genera of leptodactylid frogs in Mexico. *University of Kansas Publications, Museum of Natural History* 17 (11): 503-515.

LYNCH, J. D. 1970. A taxonomic revision of the leptodactylid frog genus *Syrrhophus* Cope. *University of Kansas Publications, Museum of Natural History* 20 (1): 1-45.

LYNCH, J. D. 1971. Evolutionary relationships, osteology, and zoogeography of leptodactylid frogs. *Miscellaneous Publications, Museum of Natural History, University of Kansas* 53: 1-238.

LYNCH, J. D. 1973. The transition from archaic to advanced frogs. *In*: VIAL, J. L. (ed.), Evolutionary Biology of the Anurans: Contemporary research on major problems. Columbia, University of Missouri Press: 132-182.

LYNCH, J. D. 1974. *Aneides flavipunctatus*. *Catalogue of American Amphibians and Reptiles*. Society for the Study of Amphibians and Reptiles (158): 1-2.

LYNCH, J. D. 1996. Replacement names for three homonyms in the genus *Eleutherodactylus* (Anura: Leptodactylidae). *Journal of Herpetology* 30 (2): 278-280.

LYNCH, J. D., AND W. E. DUELLMAN. 1997. Frogs of the genus *Eleutherodactylus* in western Ecuador: systematics, ecology and biostratigraphy. *University of Kansas Publications, Museum of Natural History* 23: 1-236.

LYNCH, J. D., AND C. M. FUGLER. 1965. A survey of the frogs of Honduras. *Journal of the Ohio Herpetological Society* 5: 5-18.

LYNCH, J. D., AND D. B. WAKE. 1974. *Aneides lugubris*. *Catalogue of American Amphibians and Reptiles*. Society for the Study of Amphibians and Reptiles (159): 1-2.

MACARTNEY, J. 1802. Lectures on comparative anatomy. Translated from the French of G. Cuvier by William Ross, under the inspection of James Macartney. London, Longman & Rees: i-xl, 1-542, tab. i-ix.

MACEY, J. R. 2005. Plethodontid salamander mitochondrial genomics: a parsimony evaluation of character conflict and implications for historical biogeography. *Cladistics* 21: 194-202.

MACEY, J. R., J. L. STRASBURG, J. A. BRISSON, V. T. VREDENBURG, M. JENNINGS, AND A. LARSON. 2001. Molecular phylogenetics of western North American frogs of the *Rana boylii* species group. *Molecular Phylogenetics and Evolution* 19: 131-143.

MAHONEY, M. J. 2001. Molecular systematics of *Plethodon* and *Aneides* (Caudata: Plethodontidae: Plethodontini): phylogenetic analysis of an old and rapid radiation. *Molecular Phylogenetics and Evolution* 18 (2): 174-188.

MAKOWSKY, R. J., J. CHESSER, AND L. J. RISSLER. 2009. A striking lack of genetic diversity across the wide-ranging amphibian *Gastrophryne carolinensis* (Anura: Microhylidae). *Genetica* 135 (2): 169-183.

MALNATE, E. V. 1971. A catalog of primary types in the herpetological collections of the Academy of Natural Sciences, Philadelphia (ANSP). *Proceedings of the Academy of Natural Sciences of Philadelphia* 123 (9): 345-351.

MANEYRO, R., AND A. KWET. 2008. Amphibians in the border region between Uruguay and Brazil: updated species list with comments on taxonomy and natural history (Part I: Bufonidae). *Stuttgarter Beiträguzer Naturkunde A* (n.s. 1): 95-121.

Marlow, R. W., J. M. Brode, and D. B. Wake. 1979. A new salamander, genus *Batrachoseps*, from the Inyo Mountains of California, with a discussion of relationships in the genus. *Contributions in Science, Los Angeles County Museum of Natural History* (308): 1-17.

Martin, P. S. 1958. A biogeography of reptiles and amphibians in the Gomez Farias region, Tamaulipas, Mexico *Miscellaneous Publications of the Museum of Zoology, University of Michigan* (101): 1-117.

Martínez-Solano, I., E. L. Jockusch, and D. B. Wake. 2007. Extreme population subdivision throughout a continuous range: phylogeography of *Batrachoseps attenuatus* (Caudata: Plethodontidae) in western North America. *Molecular Ecology* 16 (20): 4335-4355.

Martínez-Solano, I., A. Peralta-García, E. L. Jockusch, D. B. Wake, E. Vázquez-Dominguez, and G. Parra-Olea. 2012. Molecular systematics of *Batrachoseps* (Caudata, Plethodontidae) in southern California and Baja California: mitochondrial-nuclear DNA discordance and the evolutionary history of *B. major*. *Molecular Phylogenetics and Evolution* 63: 131-149.

Martof, B. S. 1956. Three new subspecies of *Leurognathus marmorata* from the southern Appalachian mountains. *Occasional Papers of the Museum of Zoology University of Michigan* (575): 1-14, pl. I-IV.

Martof, B. S. 1962. Some aspects of the life history and ecology of the salamander *Leurognathus*. *American Midland Naturalist* 67: 1-35.

Martof, B. S. 1963. *Leurognathus and L. marmoratus*. *Catalogue of American Amphibians and Reptiles*. American Society of Ichthyologists and Herpetologists (3): 1-2.

Martof, B. S. 1968. *Ambystoma cingulatum*. *Catalogue of American Amphibians and Reptiles*. American Society of Ichthyologists and Herpetologists (57): 1-2.

Martof, B. S. 1970. *Rana sylvatica*. *Catalogue of American Amphibians and Reptiles*. American Society of Ichthyologists and Herpetologists (86): 1-4.

Martof, B. S. 1972. *Pseudobranchus, P. striatus*. *Catalogue of American Amphibians and Reptiles*. Society for the Study of Amphibians and Reptiles (118): 1-4.

MARTOF, B. S. 1973a. *Siren intermedia. Catalogue of American Amphibians and Reptiles.* Society for the Study of Amphibians and Reptiles (127): 1-3.
MARTOF, B. S. 1973b. *Siren lacertina. Catalogue of American Amphibians and Reptiles.* Society for the Study of Amphibians and Reptiles (128): 1-2.
MARTOF, B. S. 1974a. Sirenidae. *Catalogue of American Amphibians and Reptiles.* Society for the Study of Amphibians and Reptiles (151): 1-2.
MARTOF, B. S. 1974b. *Siren. Catalogue of American Amphibians and Reptiles.* Society for the Study of Amphibians and Reptiles (152): 1-2.
MARTOF, B. S. 1975a. *Pseudotriton. Catalogue of American Amphibians and Reptiles.* Society for the Study of Amphibians and Reptiles (165): 1-2.
MARTOF, B. S. 1975b. *Pseudotriton montanus. Catalogue of American Amphibians and Reptiles.* Society for the Study of Amphibians and Reptiles (166): 1-2.
MARTOF, B. S. 1975c. *Pseudotriton ruber. Catalogue of American Amphibians and Reptiles.* Society for the Study of Amphibians and Reptiles (167): 1-4.
MARTOF, B. S. 1975d. *Hyla squirella. Catalogue of American Amphibians and Reptiles.* Society for the Study of Amphibians and Reptiles (168): 1-2.
MARTOF, B. S., AND R. L. HUMPHRIES. 1959. Geographic variation in the woodfrog, *Rana sylvatica. American Midland Naturalist* 61 (2): 350-389.
MARTOF, B. S., AND F. L. ROSE. 1962. The taxonomic status of the plethodontid salamander, *Desmognathus planiceps. Copeia* 1962: 215-216.
MARTOF, B. S., AND F. L. ROSE. 1963. Geographic variation in southern populations of *Desmognathus ochrophaeus. American Midland Naturalist* 69: 376-425.
MARX, H. 1976. Supplementary catalogue of type specimens of reptiles and amphibians in the Field Museum of Natural History. *Fieldiana: Zoology* 69 (2): 1-94.
MASSALONGO, A. 1853. Sopra un nuovo genere di rettili della provincia Padovana. *Nuovi Annali della Scienze Naturali* (Bologna) (3) 7: 5-16, pl. 1.
MASTA, S. E., B. K. SULLIVAN, T. LAMB, AND E. J. ROUTMAN. 2002. Molecular systematics, hybridization, and

phylogeography of the *Bufo americanus* complex in eastern North America. *Molecular Phylogenetics and Evolution* 24: 302-314.

MATSON, T. O. 2005. *Necturus maculosus,* Mudpuppy. *In* Amphibian Declines: The conservation status of United States species, Part 2: Species accounts. *Edited by* LANNOO, M. J. University of California Press, Berkeley, pp 870-871.

MATTHES, B. 1855. Die Hemibatrachier im Allgemeinen und die Hemibatrachier von Nordamerika im Speciellem. *Allegemeine deutsche naturhistorische* (n.s.) 1: 249-280.

MAXSON, L. R., P. E. MOLER, AND B. W. MANSELL. 1988. Albumin evolution in salamanders of the genus *Necturus. Journal of Herpetology* 22 (2): 231-235.

MAYER, A. F. J. C. 1835. Analecten für vergleichende Anatomie. Thiel 2. Bonn, E. Weber: [fide Frost, 2009-13; not seen]

Mazerolle, M. J., Y. Dubois, C. Fontenot, P. Galois, D. Lesbarrères, M. Ouellet, and D. M. Green. 2012. Noms Français standardisés des amphibiens. *In:* Moriarty, J. J. (Ed.) Noms Français standardisés des amphibiens et des reptiles d'Amérique du Nord au nord du Mexique. Standard French names of amphibians and reptiles of North America north of Mexico. *SSAR Herpetological Circular* 40: 7-24.

MCCOY, C. J. 1992. Rediscovery of the mud salamander (*Pseudotriton montanus*, Amphibia, Plethodontidae) in Pennsylvania, with restriction of the type-locality. *Journal of the Pennsylvania Academy of Science* 66: 92-93.

MCCRADY, E. 1954. A new species of *Gyrinophilus* (Plethodontidae) from Tennessee caves. *Copeia* 1954 (2): 200-206.

MCCRANIE, J. R., L. D. WILSON, AND K. L. WILLIAMS. 1989. A new genus and species of toad (Anura: Bufonidae) with and extraordinary stream-adapted tadpole from northern Honduras. *Occasional Papers of the Museum of Natural History, University of Kansas* (129): 1-18.

McKeown, S. 1996. *A field guide to reptiles and amphibians in the Hawai'ian Islands.* Los Osos, California, Diamond Head Publishing: i-iv, 1-173.

MCMURTRIE, H. 1834. Cuvier's Animal Kingdon: arranged according to its organizatin, translagted frm the French, and abridged for the use of students. London, Orr & Smith: i-xx, 1-508. [Translation of Cuvier, 1831, slightly revised, with occasional nomenclatural or taxonomic modification]

MEAD, L. S., D. R. CLAYTON, R. S. NAUMAN, D. H. OLSON, AND M. E. PFRENDER. 2005. Newly discovered populations of salamanders from Siskiyou County California represent a species distinct from *Plethodon stormi*. *Herpetologica* 61 (2): 158-177.

MEANS, D. B. 1974. The status of *Desmognathus brimleyorum* Stejneger and an analysis of the genus *Desmognathus* (Amphibia: Urodela) in Florida. *Bulletin of the Florida State Museum, Biological Science* 18 (1): 1-100.

MEANS, D. B. 1993. *Desmognathus apalachicolae*. *Catalogue of American Amphibians and Reptiles*. Society for the Study of Amphibians and Reptiles (556): 1-2.

MEANS, D. B. 1996. *Amphiuma pholeter*. *Catalogue of American Amphibians and Reptiles*. Society for the Study of Amphibians and Reptiles (622): 1-2.

MEANS, D. B. 1999a. *Desmognathus auriculatus*. *Catalogue of American Amphibians and Reptiles*. Society for the Study of Amphibians and Reptiles (681): 1-6.

MEANS, D. B. 1999b. *Desmognathus brimleyorum*. *Catalogue of American Amphibians and Reptiles*. Society for the Study of Amphibians and Reptiles (682): 1-4.

MEANS, D. B. 2005. *Hyla andersonii*, Pine Barrens Treefrog. *In*: LANNOO, M. J. (ed.), Amphibian Declines: The Conservation Status of United States Species, Part 2: Species Accounts. Berkeley, University of California Press: 445-447.

MEANS, D. B., AND R. M. BONETT. 2005. *Desmognathus conanti*, Spotted Dusky Salamander. *In*: LANNOO, M. J. (ed.), Amphibian Declines: The Conservation Status of United States Species, Part 2: Species Accounts. Berkeley, University of California Press: 705-706.

MEANS, D. B., AND A. A. KARLIN. 1989. A new species of *Desmognathus* from the eastern Gulf Coastal Plain. *Herpetologica* 45 (1): 37-46.

MECHAM, J. S. 1967a. *Notophthalmus perstriatus*. *Catalogue of American Amphibians and Reptiles*. American Society of Ichthyologists and Herpetologists (38): 1-2.

Mecham, J. S. 1967b. *Notophthalmus viridescens. Catalogue of American Amphibians and Reptiles.* American Society of Ichthyologists and Herpetologists (53): 1-4.

MECHAM, J. S. 1968a. *Notophthalmus meridionalis. Catalogue of American Amphibians and Reptiles.* American Society of Ichthyologists and Herpetologists (74): 1-2.

MECHAM, J. S. 1968b. On the relationships between *Notophthalmus meridionalis* and *Notophthalmus kallerti. Journal of Herpetology* 2 (3/4): 121-127.

MECHAM, J. S., M. J. LITTLEJOHN, R. S. OLDHAM, L. E. BROWN, AND J. R. BROWN. 1973. A new species of leopard frog (*Rana pipiens* complex) from the plains of the central United States. *Occasional Papers of the Museum of Texas Tech University* (18): 1-11.

MERREM, B. 1820. Versuch eines Systems der Amphibien. Tentamen systematis amphibiorum. Marburg, J.C. Krieger: (i-xv, 1-191) x 2, 1 pl.

MERTENS, R. 1936. Eine übersehene "Herpetologia europaea". *Senckenbergiana. Biologica* (Frankfurt am Main) 18: 75-78.

MERTENS, R. 1938. Amphibien und Reptilien aus Santo Domingo, gesammest von Prof. Dr. H. Böker. *Senckenbergiana. Biologica* (Frankfurt am Main) 20 (5): 332-342.

MESHAKA, W. E., JR. 2005. *Eleutherodactylus planirostris*, Greenhouse Frog. *In*: LANNOO, M. J. (ed.), Amphibian Declines: The Conservation Status of United States Species, Part 2: Species Accounts. Berkeley, University of California Press: 499-500.

MESHAKA, W. E., B. P. BUTTERFIELD, AND J. B. HAUGE. 2004. The exotic amphibians and reptiles of Florida. Malabar, FL, Krieger Publishing: i-x, 1-155.

MESHAKA, W. E., AND S. E. TRAUTH. 2006. *Plethodon angusticlavius. Catalogue of American Amphibians and Reptiles.* Society for the Study of Amphibians and Reptiles (801): 1-3.

METCALF, M. M. 1923. The opalinid ciliate infusorians. *Bulletin of the U.S. National Museum* 120: i-vii, 1-484.

METTER, D. E. 1964. A morphological and ecological comparison of two populations of the tailed frog *Ascaphus truei* Stejneger. *Copeia* 1964 (1): 181-195.

METTER, D. E. 1968. *Ascaphus and A. truei. Catalogue of American Amphibians and Reptiles.* American Society of Ichthyologists and Herpetologists (69): 1-2.

MICHAHELLES, [M.]. 1833. Waglers Synonymie der Sebaischen Amphibien. *Isis von Oken* 1833 (9): 884-905.

MICHELS, J. P., AND A. M. BAUER. 2004. Some corrections to the scientific names of amphibians and reptiles. *Bonner zoologische Beiträge* 52 (1/2): 83-94.

MILLER, B. T., AND M. L. NIEMILLER. 2012. *Gyrinophius palleucus. Catalogue of American Amphibians and Reptiles.* Society for the Study of Amphibians and Reptiles (884): 1-7.

MILLER, G. S. 1901. A new treefrog from the District of Columbia. *Proceedings of the Biological Society of Washington* 13 (1899-1900): 75-78.

Miller, M. P., S. M. Haig and R. S. Wagner. 2005. Conflicting patterns of genetic structure produced by nuclear and mitochondrial markers in the Oregon slender salamander (*Batrachoseps wrighti*): implications for conservation efforts ad species management. *Conservation Genetics* 6: 275-287.

MILSTEAD, W. W., J. S. MECHAM, AND H. MCCLINTOCK. 1950. The amphibians and reptiles of the Stockton Plateau in northern Terrell County, Texas. *Texas Journal of Science* 2 (4): 543-562.

MINTON, S. A., JR. 1954. Salamanders of the *Ambystoma jeffersonianum* complex in Indiana. *Herpetologica* 10 (3): 173-179.

MIRANDA-RIBEIRO, A. 1926. Notas para servirem ao estudo do Gymnobatrachios (Anura) brasileiros. *Archivos do Museu Nacional do Rio de Janeiro* 27: 1-227, pl. 1-22.

MITCHELL, J. C. 2005. *Ambystoma mabeei*, Mabee's Salamander. *In*: Amphibian Declines: The Conservation Status of United States Species, Part 2: Species Accounts. *Edited by* LANNOO, M. J. University of California Press, Berkeley: 616-617.

MITCHELL, R. W., AND J. R. REDDELL. 1965. *Eurycea tridentifera*, a new species of troglobitic salamander from Texas and a reclassification of *Typhlomolge rathbuni*. *Texas Journal of Science* 17 (1): 12-27.

MITCHELL, R. W., AND R. E. SMITH. 1972. Some aspects of the osteology and evolution of the neotenic spring and cave salamanders (*Eurycea*, Plethodontidae) of central Texas. *Texas Journal of Science* 23 (3): 343-362.

MITCHILL, S. L. 1822. Description of a batracian animal from Georgia, different from the reptiles of that order hitherto known. *American Medical Recorder* 5 (3): 499-503.

MITTLEMAN, M. B. 1942a. Notes on salamanders of the genus *Gyrinophilus*. *Proceedings of the New England Zoological Club* 20: 25-42, pl. 1-4.

MITTLEMAN, M. B. 1942b. A new longtailed *Eurycea* from Indiana, and notes on the *longicauda* complex. *Proceedings of the New England Zoological Club* 21: 101-105.

MITTLEMAN, M. B. 1945a. The status of *Hyla phaeocrypta* with notes on its variation. *Copeia* 1945 (1): 31-37.

MITTLEMAN, M. B. 1946. Nomenclatural notes on two southeastern frogs. *Herpetologica* 3 (2): 57-60.

MITTLEMAN, M. B. 1947a. American Caudata. I. Geographic variation in *Manculus quadridigitatus*. *Herpetologica* 3 (3): 209-224.

MITTLEMAN, M. B. 1947b. Miscellaneous notes on Indiana amphibians and reptiles. *American Midland Naturalist* 38 (2): 466-484.

MITTLEMAN, M. B. 1948a. American Caudata II. Geographic variation in *Ambystoma macrodactylum*. *Herpetologica* 4 (2): 81-95.

MITTLEMAN, M. B. 1948b. American Caudata, V: notes on certain Appalachian salamanders of the genus *Plethodon*. *Journal of the Washington Academy of Science* 38: 416-419.

MITTLEMAN, M. B. 1949. American caudata. VI. The races of *Eurycea bislineata*. *Proceedings of the Biological Society of Washington* 62: 89-96.

MITTLEMAN, M. B. 1951. American Caudata. VII. Two new salamanders of the genus *Plethodon*. *Herpetologica* 7: 105-112.

MITTLEMAN, M. B. 1967a. *Manculus and M. quadridigitatus*. *Catalogue of American Amphibians and Reptiles*. American Society of Ichthyologists and Herpetologists (44): 1-2.

MITTLEMAN, M. B. 1967b. *Eurycea bislineata. Catalogue of American Amphibians and Reptiles.* American Society of Ichthyologists and Herpetologists (45): 1-4.

MITTLEMAN, M. B., AND H. T. GIER. 1942. Notes on leopard frogs. *Proceedings of the New England Zoological Club* 20: 7-15.

MITTLEMAN, M. B., AND H. G. M. JOPSON. 1941. A new salamander of the genus *Gyrinophilus* from the southern Appalachians. *Smithsonian Miscellaneous Collections* 101 (2): 1-5, pl. 1.

MITTLEMAN, M. B., AND J. C. LIST. 1953. The generic differentiation of the swamp treefrogs. *Copeia* 1953 (1): 80-83

MITTLEMAN, M. B., AND G. S. MYERS. 1949. Geographic variation in the ribbed frog *Ascaphus truei. Proceedings of the Biological Society of Washington* 62: 57-66.

MIVART, S. G. 1868 ("1867"). On *Plethodon persimilis* of Gray. *Proceedings of the Zoological Society of London* 1867: 695-699.

MIVART, S. G. 1869. On the classification of the anurous batrachians. *Proceedings of the Zoological Society of London* 1869: 280-295.

MOCQUARD, M. F. 1899. Reptiles et batraciens recueillis au Mexique par M. León Diguet en 1896 et 1897. *Bulletin des Sciences, par la Société Philomatique de Paris* (9) 1: 154-169, pl. 1.

MOLER, P. E. 1985. A new species of frog (Ranidae: *Rana*) from northwestern Florida. *Copeia* 1985 (2): 379-383.

MOLER, P. E. 1993. *Rana okaloosae. Catalogue of American Amphibians and Reptiles.* Society for the Study of Amphibians and Reptiles (561): 1-3.

MOLER, P. E., AND J. KEZER. 1993. Karyology and systematics of the salamander genus *Pseudobranchus. Copeia* 1993 (1): 39-47.

MONTANUCCI, R. R. 2006. A review of the amphibians of the Jim Timmerman Natural Resources Area, Oconee and Pickens Counties, South Carolina. *Southeastern Naturalist* 5 (Monograph 1): 1-58.

MOORE, G. A., AND R. C. HUGHES. 1939. A new plethodontid from eastern Oklahoma. *American Midland Naturalist* 22 (3): 696-699.

MOORE, G. A., AND R. C. HUGHES. 1941. A new plethodont salamander from Oklahoma. *Copeia* 1941: 139-142.

MOORE, J. P. 1899. *Leurognathus marmorata*, a new genus and species of salamander of the family Desmognathidae. *Proceedings of the Academy of Natural Sciences of Philadelphia* 51 (1899): 316-323, pl. 14.

MORIARTY, E. C., AND D. C. CANNATELLA. 2004. Phylogenetic relationships of the North American chorus frogs (*Pseudacris*: Hylidae). *Molecular Phylogenetics and Evolution* 30: 409-420.

MORIARTY LEMMON, E., A. R. LEMMON, AND D. C. CANNATELLA. 2007a. Geological and climatic forces driving speciation in the continentally distributed trilling chorus frogs (*Pseudacris*). *Evolution* 61 (9): 2086-2103.

MORIARTY LEMMON, E., A. R. LEMMON, J. T. COLLINS, AND D. C. CANNATELLA. 2008. A new North American chorus frog species (Amphibia: Hylidae: *Pseudacris*) from the south-central United States. *Zootaxa* (1675): 1-30.

MORIARTY LEMMON, E., A. R. LEMMON, J. T. COLLINS, J. A. LEE-YAW, AND D. C. CANNATELLA. 2007b. Phylogeny-based delimitation of species boundaries and contact zones in the trilling chorus frogs (*Pseudacris*). *Molecular Phylogenetics and Evolution* 44: 1068-1082.

MORITZ, C., C. J. SCHNEIDER, AND D. B. WAKE. 1992. Evolutionary relationships within the *Ensatina eschscholtzii* complex confirm the ring species interpretation. *Systematic Biology* 41 (3): 273-291.

MOUNT, R. H. 1975. The reptiles and amphibians of Alabama. University of Alabama Press: i-vii, 1-347.

MUCHMORE, W. B. 1955. Brassy flecking in the salamander *Plethodon c. cinerreus*, and the validity of *Plethodon huldae*. *Copeia* 1955 (3): 170-172.

MUELLER, R. L., J. R. MACEY, M. JACKEL, D. B. WAKE, AND J. L. BOORE. 2004. Morphological homoplasy, life history evolution, and historical biogeography of plethodontid salamanders inferred from complete mitochondrial genomes. *Proceedings of the National Academy of Sciences (USA)* 101 (38): 13820-13825.

MULCAHY, D. G., AND J. R. MENDELSON, III. 2000. Phylogeography and speciation of the morphologically variable, widespread species *Bufo valliceps*, based on molecular evidence from mtDNA. *Molecular Phylogenetics and Evolution* 17 (2): 173-189.

MÜLLER, L., AND W. HELLMICH. 1936. Wissenschaftliche Ergebnisse der Deutschen Gran Chaco-Expedition, vol. 1. Amphibien und Reptilien. Band 4, Teil 1: Amphibia, Chelonia, Loricata. Stuttgart, Strecker & Schroder: i-xvi, 1-120.

MUTHS, E., AND P. NANJAPPA. 2005. *Bufo boreas*, Western Toad. *In*: LANNOO, M. J. (ed.), Amphibian Declines: The Conservation Status of United States Species, Part 2: Species Accounts. Berkeley, University of California Press: 392-396.

MYERS, G. S. 1926 ("1925"). A synopsis for the identification of the amphibians and reptiles of Indiana. *Proceedings of the Indiana Academy of Sciences* 35: 277-294.

MYERS, G. S. 1927 ("1926"). Notes on Indiana amphibians and reptiles. *Proceedings of the Indiana Academy of Sciences* 36: 337-340.

MYERS, G. S. 1930a. Notes on some amphibians in western North America. *Proceedings of the Biological Society of Washington* 43: 55-64.

MYERS, G. S. 1930b. The status of the Southern California Toad, *Bufo californicus* (Camp). *Proceedings of the Biological Society of Washington* 43: 73-77.

MYERS, G. S. 1942a. Notes on Pacific coast *Triturus*. *Copeia* 1942 (2): 77-82.

MYERS, G. S. 1942b. The black toad of Deep Springs valley, Inyo County, California. *Occasional Papers of the Museum of Zoology University of Michigan* (480): 1-13, pl. 1-3.

MYERS, G. S. 1950. The systematic status of *Hyla septentrionalis*, the large tree frog of the Florida Keys, the Bahamas and Cuba. *Copeia* 1950 (3): 203-214.

MYERS, G. S. 1962. The American leptodactylid frog genera *Eleutherodactylus*, *Hylodes* (=*Elosia*), and *Caudiverbera* (=*Calyptocephalus*). *Copeia* 1962 (1): 195-202.

MYERS, G. S., AND A. L. D. CARVALHO. 1945. Notes on some new or little-known Brazilian amphibians, with an examination of the history of the Plata salamander, *Ensatina platensis*. *Boletim do Museu Naconal. Nova Série. Zoologia* 35: 1-24.

MYERS, G. S., AND T. P. MASLIN. 1948. The California plethodont salamander, *Aneides flavipunctatus* (Strauch), with description of a new subspecies and notes on other western *Aneides*. *Proceedings of the Biological Society of Washington* 61: 127-135.

NAKAMURA, K., AND S. I. UENO. 1963. Japanese reptiles and amphibians in color. Tokyo, Ie-no Hikari Kiyokai: 1-206. [in Japanese]

NASH, C. W. 1905. Check list of the vertebrates of Ontario and Catalogue of specimens in the Biological Section of the Provincial Museum. Toronto, L.K. Cameron.

NEILL, W. T. 1948a. Salamanders of the genus *Pseudotriton* from Georgia and South Carolina. *Copeia* 1948 (2): 134-136.

NEILL, W. T. 1948b. A new subspecies of tree-frog from Georgia and South Carolina. *Herpetologica* 4 (3): 175-179.

NEILL, W. T. 1948c. The status of the salamander *Desmognathus quadramaculatus amphileucus*. *Copeia* 1948 (3): 218.

NEILL, W. T. 1949a. The status of *Hyla flavigula*. *Copeia* 1949 (1): 78.

NEILL, W. T. 1949b. The status of Baird's Chorus-Frog. *Copeia* 1949 (3): 227-228.

NEILL, W. T. 1950a. A new species of salamander, genus *Desmognathus*, from Georgia. *Publication of the Research Division of Ross Allen's Reptile Institute* 1 (1): 1-6.

NEILL, W. T. 1950b. Taxonomy, nomenclature, and distribution of southeastern cricket frogs, genus *Acris*. *American Midland Naturalist* 43 (1): 152-156.

NEILL, W. T. 1951b. A new subspecies of Dusky Salamander, genus *Desmognathus*. *Publication of the Research Division of Ross Allen's Reptile Institute* 1: 25-38.

NEILL, W. T. 1951c. A new subspecies of salamander, genus *Pseudobranchus*, from the Gulf hammock region of Florida. *Publication of the Research Division of Ross Allen's Reptile Institute* 1 (4): 39-46.

NEILL, W. T. 1952. Remarks on salamander voices. *Copeia* 1952 (3): 195-196.

NEILL, W. T. 1954. Ranges and taxonomic allocations of amphibians and reptiles in the southeastern United States. *Publication of the Research Division of Ross Allen's Reptile Institute* 1 (7): 75-96.

NEILL, W. T. 1957a. The status of *Rana capito stertens* Schwartz and Harrison. *Herpetologica* 13: 47-52.

NEILL, W. T. 1957b. Objections to wholesale revision of type localities. *Copeia* 1957 (2): 140-141.

NEILL, W. T. 1963a. Notes on the Alabama waterdog, *Necturus alabamensis* Viosca. *Herpetologica* 19 (2): 166-174.

NEILL, W. T. 1963b. *Hemidactylium. Catalogue of American Amphibians and Reptiles.* American Society of Ichthyologists and Herpetologists (1): 1.

NEILL, W. T. 1963c. *Hemidactylium scutatum. Catalogue of American Amphibians and Reptiles.* American Society of Ichthyologists and Herpetologists (2): 1-2.

NEILL, W. T. 1964. A new species of salamander, genus *Amphiuma,* from Florida. *Herpetologica* 20: 62-66.

NELSON, C. E. 1972a. *Gastrophryne carolinensis. Catalogue of American Amphibians and Reptiles.* Society for the Study of Amphibians and Reptiles (120): 1-4.

NELSON, C. E. 1972b. *Gastrophryne olivacea. Catalogue of American Amphibians and Reptiles.* Society for the Study of Amphibians and Reptiles (122): 1-4.

NELSON, C. E. 1972c. Systematic studies of the North American microhylid genus *Gastrophryne. Journal of Herpetology* 6 (2): 111-137.

NELSON, C. E. 1973a. Systematics of the Middle American upland populations of *Hypopachus* (Anura: Microhylidae). *Herpetologica* 29 (1): 6-17.

NELSON, C. E. 1973b. Mating calls of the Microhylinae: descriptions and phylogenetic and ecological considerations. *Herpetologica* 29 (2): 163-176.

NELSON, C. E. 1973c. *Gastrophryne. Catalogue of American Amphibians and Reptiles.* Society for the Study of Amphibians and Reptiles (134): 1-2.

NELSON, C. E. 1974. Further studies on the systematics of *Hypopachus* (Anura: Microhylidae). *Herpetologica* 30 (3): 250-274.

Nemésio, A. and A. Dubois. 2012. The endings of specific nomina dedicated to persons should not be emended: nomenclatural issues in Phalangopsidae (Hexapoda: Grylloidea). *Zootaxa* (3270): 67-68.

NETTING, M. G., AND C. J. GOIN. 1942. Description of two new salamanders from peninsular Florida. *Annals of the Carnegie Museum* 29: 175-196.

NETTING, M. G., AND C. J. GOIN. 1945. The cricket-frog of peninsular Florida. *Quarterly Journal of the Florida Academy of Sciences* 8 (4): 304-310.

NETTING, M. G., AND C. J. GOIN. 1946. *Acris* in Mexico and Trans-Pecos Texas. *Copeia* 1946 (4): 107.

NETTING, M. G., AND M. B. MITTLEMAN. 1938. Description of *Plethodon richmondi,* a new salamander from West Virginia and Ohio. *Annals of the Carnegie Museum* 27 (18): 287-293.

NEWCOMER, R. 1961. Investigations into the status of *Gyrinophilus lutescens* (Rafinesque). *Bulletin of the Association of Southeastern Biologists* 8 (2): 21. [*fide* Brandon, 1966a; not seen]

Newman, C. E., J. A. Feinberg, L. J. Rissler, J. Burger, and H. B. Shaffer. 2012. A new species of Leopard Frog (Anura: Ranidae) from the urban northeastern U.S. *Molecular Phylogenetics and Evolution* 63: 445-455.

NEWMAN, W. B. 1954. A new plethodontid salamander from southwestern Virginia. *Herpetologica* 10 (1): 9-14.

NEWMAN, W. B. 1955. *Desmognathus planiceps,* a new salamander from Virginia. *Journal of the Washington Academy of Science* 45: 83-86.

Ng, P. K. L. 1994. The citation of species names and the role of the author's name. *Raffles Bulletin of Zoology* 42 (3): 509-513.

NICHOLLS, J. C., JR. 1949. A new salamander of the genus *Desmognathus* from east Tennessee. *Journal of the Tennessee Academy of Science* 24 (2): 127-129.

NICKERSON, M. A., AND C. E. MAYS. 1973. The Hellbenders: North American "giant salamanders". *Milwaukee Public Museum Publications in Biology and Geology* 1: i-viii, 1-106.

NIEDEN, F. 1923. Anura I. Subordo Aglossa und Phaerglossa Sectio 1 Arcifera. *Das Tierreich* 46. Prubischen Akademie der Wissenschaften zu Berlin: i-xxxii, 1-584.

NIEDEN, F. 1926. Anura II. Engystomatidae. Prubischen Akademie der Wissenschaften zu Berlin: i-xvi, 1-110.

NIELSON, M., K. LOHMAN, AND J. SULLIVAN. 2001. Phylogeography of the Tailed Frog (*Ascaphus truei*): implications for the biogeography of the Pacific Northwest. *Evolution* 55 (1): 147-160.

NIEMILLER, M. L., AND B. T. MILLER. 2010. *Gyrinophilus gulolineatus*. *Catalogue of American Amphibians and Reptiles*. Society for the Study of Amphibians and Reptiles (862): 1-4.

NIKOLSKY, A. M. 1918. Zemnovodnye I presmykayushcheesya v serii “Fauna Rossii sopredel-nyukh stran [Amphibia and Reptiles. *IN*: ”Fauna of Russia and adjacent countries.”]. Petrograd, Izdatel’stvo Academii Nauk. [cited by Frost, 1985; not seen]

NOBLE, G. K. 1921. The anterior cranial elements of *Oedipus* and certain other salamanders. *Bulletin of the American Museum of Natural History* 44 (1): 1-6, pl. 1-2.

NOBLE, G. K. 1923. The generic and genetic relations of *Pseudacris*, the swamp tree frogs. *American Museum Novitates* (70): 1-6.

NOBLE, G. K. 1924. A new spadefoot toad from the Oligocene of Mongolia with a summary of the evolution of the Pelobatidae. *American Museum Novitates* (132): 1-14.

NOBLE, G. K. 1931. The biology of the Amphibia. New York, McGraw-Hill: i-xiv, 1-577.

NUSSBAUM, R. A. 1970. *Dicamptodon copei,* n. sp., from the Pacific Northwest, U.S.A. (Amphibia: Caudata: Ambystomatidae). *Copeia* 1970 (3): 506-514.

NUSSBAUM, R. 1976. Geographic variation and systematics of salamanders of the genus *Dicamptodon* (Ambystomatidae). *Miscellaneous Publications, Museum of Zoology University of Michigan* (149): 1-94.

NUSSBAUM, R. 1983. *Dicamptodon copei*. *Catalogue of American Amphibians and Reptiles*. Society for the Study of Amphibians and Reptiles (334): 1-2.

NUSSBAUM, R., AND E. D. BRODIE, JR. 1981a. *Taricha*. *Catalogue of American Amphibians and Reptiles*. Society for the Study of Amphibians and Reptiles (271): 1-2.

NUSSBAUM, R., AND E. D. BRODIE, JR. 1981b. *Taricha granulosa*. *Catalogue of American Amphibians and Reptiles*. Society for the Study of Amphibians and Reptiles (272): 1-4.

NUSSBAUM, R., AND E. D. BRODIE, JR. 1981c. *Taricha torosa. Catalogue of American Amphibians and Reptiles*. Society for the Study of Amphibians and Reptiles (273): 1-4.

ODUM, B. A., AND P. S. CORN. 2005a. *Ambystoma tigrinum*, Tiger Salamander. *In*: LANNOO, M. J. (ed.), Amphibian Declines: The Conservation Status of United States Species, Part 2: Species Accounts. Berkeley, University of California Press: 390-392.

ODUM, B. A., AND P. S. CORN. 2005b. *Bufo baxteri*, Wyoming Toad. *In*: (eds.), Amphibian Declines: The Conservation Status of United States Species, Part 2: Species Accounts: 390-392.

OKEN, L. von. 1815-1817. Lehrbuch der Naturgeschichte. Th. 3. Zoologie. Abthielung 1, 2, atlas. Leipzig, C.H. Reclam: i-xv, 1-1270.

OLIVER, J. A., AND J. R. BAILEY. 1939. Amphibians and reptiles of New Hampshire. *In*: New Hampshire Fish and Game Department: 195-221.

Oliver, J. A., and C. E. Shaw. 1953. The amphibians and reptiles of the Hawaiian Islands. *Zoologica* 38: 65-95.

OPPEL, M. 1811. Die Ordnungen, Familien und Gattungen der Reptilien als Prodrom einer Naturgeschichte derselben. München, Joseph Lindauer: i-xii, 1-87.

ORGAN, J. A. 1961. The eggs and young of the spring salamander, *Pseudotriton porphyriticus. Herpetologica* 17: 53-56.

O'SHAUGNESSY, A. W. E. 1879. Zoologcal Record 14: 13.

ÖSTERDAM, A. 1766. *Siren lacertina. Amoenitates Academicae* 7: 311-325, + 1 pl.

OWEN, [R.]. 1844. Characters of a new species of axolotl. *ILAR Journal* 14: 23.

PACE, A. E. 1974. Systematic and biological studies of the leopard frogs (*Rana pipiens* complex) of the United States. *Miscellaneous Publications of the Museum of Zoology, University of Michigan* (148): [i-vii], 1-140.

PACKARD, G. C. 1971. Inconsistency in application of the biological species concept to disjunct populations of anurans in southeastern Wyoming and north-central Colorado. Journal of Herpetology 5 (3/4): 191-193.

PALLAS, P. 1831. Zoographia rosso-asiatica, sistens omnium animalium in extenso Imperio rossico, et adjacentibus maribus observatorum recensionem, domicilia, mores et descriptiones, anatomen atque icones plurimorum. Volumen tertium: Imperii rossici animalia monocardia seu frigidi sanguinis (1811; 1831 edition). Petropoli [St. Petersburg], Caes. Academiae Scientiarum Impress: i-viii, 1-428, I-CXXV.

PARKER, H. W. 1933. A list of the frogs and toads of Trinidad. *Tropical Agriculture* (Trinidad) 10: 8-12. [*fide* Frost (2009-13); not seen]

PARKER, H. W. 1934. A monograph of the frogs of the family Microhylidae. London, British Museum of Natural History: i-vii, 1-266.

PARKER, H. W. 1936. Reptiles and amphibians collected by the Lake Rudolph Rift Valley Expedition, 1934. *Annals and Magazine of Natural History* (10) 18: 594-609. [*fide* Frost, 2009; not seen]

PARKER, H. W. 1940. The Australasian frogs of the family Leptodactylidae. *Novitates Zoologicae* (London) 42 (1): 1-106.

PARRIS, M. J., AND M. REDMER. 2005. *Rana areolata*. Crawfish Frog. *In*: LANNOO, M. J. (ed.), Amphibian Declines: The Conservation Status of United States Species, Part 2: Species Accounts. Berkeley, University of California Press: 526-528.

PAULY, G. B., D. M. HILLIS, AND D. C. CANNATELLA. 2004. The history of Nearctic colonization: molecular phylogenetics and biogeography of the Nearctic toads (*Bufo*). *Evolution* 58: 2517-2535.

PAULY, G. B., O. PISKUREK, AND B. SHAFFER. 2007. Phylogeographic concordance in the southeastern United States: the flatwoods salamander, *Ambystoma cingulatum*, as a test case. *Molecular Ecology* 16: 415-429.

PAULY, G. B., D. M. HILLIS, AND D. C. CANNATELLA. 2009. Taxonomic freedom and the role of official lists of species names. *Herpetologica* 65 (2): 115-128.

PEABODY, R. B. 1978. Electrophoretic analysis of geographic variation and hybridization of two Appalachian salamanders, *Plethodon jordani* and *Plethodon glutinosus*. PhD dissertation, University of Maryland, i-v, 1-111.

PEARL, C. A. 2005. *Rana aurora*. Northern Red-Legged Frog. *In*: Lannoo, M. J.(ed.), Amphibian Declines: The Conservation Status of United States Species, Part 2: Species Accounts. 528-530.

PEARL, C. A., AND M. J. ADAMS. 2005. *Rana cascadae*. Cascade Frog. *In*: LANNOO, M. J. (ed.), Amphibian Declines: The Conservation Status of United States Species, Part 2: Species Accounts. Berkeley, University of California Press: 538-540.

PEARL, C. A., AND M. P. HAYES. 2005. *Rana pretiosa*. Oregon Spotted Frog. *In*: LANNOO, M. J. (ed.), Amphibian Ddeclines: The Conservation Status of United States Species, Part 2: Species Accounts. Berkeley, University of California Press: 577-580.

Pemberton, C. E. 1933. Introduction to Hawaii of the tropical American toad *Bufo marinus*. *Hawaiian Planters' Record* 37: 15-16.

Pemberton, C. E. 1934. Local investigations on the introduced tropical American toad *Bufo marinus*. *Hawaiian Planters' Record* 38: 186-191.

PENNANT, T. 1792. Arctic Zoology Second Edition. London, Robert Faulder: [i-xv], i-cccxxxiv, [i-viii].

PEREIRA, R. J., W. B. MONAHAN, AND D. B. WAKE. 2011. Predictors for reproductive isolation in a ring species complex following genetic and ecological divergence. *BMC Evolutionary Biology* 11: 194-207.

PEREIRA, R. J., AND D. B. WAKE. 2009. Genetic leakage after adaptive and nonadaptive divergence in the *Ensatina eschscholtzii* ring species. *Evolution* 63: 2288-2301.

PERRET, J.-L. 1966. Les amphibiens du Cameroun. Zoologische Jahrbücher. *Abteilug für Systematik, Ökologie und Geographie der Tiere. Jena* 93: 289-494.

PETERS, J. A. 1952. Catalogue of type specimens in the herpetological collections of the University of Michigan Museum of Zoology. *Occasional Papers of the Museum of Zoology University of Michigan* (539): 1-55.

PETERS, W. C. H. 1859. Eine neue Gattung und eine neue Art von Fröschen aus Caracas. *Monatsberichte der königlich Akademie der Wissenschaften zu Berlin* 1859: 402-403.

PETERS, W. C. H. 1862a. Eine neue Gattung von Laubfröschen, *Plectromantis*, aus Ecuador. *Monatsberichte der*

Preussischen Akademie der Wissenschaften zu Berlin 1862: 232-233.
PETERS, W. C. H. 1863a. Bemerkungen über verchiedene Batrachier, namentlich über die Original-exemplare der von Schneider und Wiegmann beschriebenen Arten des zoologischen Museums zu Berlin. *Monatsberichte der königlich Akademie der Wissenschaften zu Berlin* 1863: 76-82.
PETERS, W. C. H. 1867. Herpetologische Notisen. *Monatsberichte der Preussischen Akademie der Wissenschaften zu Berlin* 1867: 13-37.
PETERS, W. C. H. 1870a. Über neue Amphibien (*Hemidactylus, Urosaura, Tropidolepisma, Geophis, Uriechis, Scaphiophis, Hoplocephalus, Rana, Entomoglossus, Cystignathus, Hylodes, Arthroleptis, Phyllobates, Cophomantis*) des Königlich zoologischen Museums. *Monatsberichte der Preussischen Akademie der Wissenschaften zu Berlin* 1870: 641-652, pl. 1.
PETERS, W. C. H. 1872. Über die von Spix in Brassilien gesammelten Batrachier des Königl. Naturalienkabinets zu Muunchen. *Monatsberichte der königlich Akademie der Wissenschaften zu Berlin* 1872: 196-227.
PETERS, W. C. H. 1873b. Über eine neue Schildkrötenart, *Cinosternon effeldtii* und einige andere neue oder weniger bekannte Amphibien. *Monatsberichte der königlich Akademie der Wissenschaften zu Berlin* 1873: 603-618.
PETERS, W. C. H. 1873c. Über die von Dr. J. J. v. Tschudi beschriebenen Batrachier aus Perú. *Monatsberichte der königlich Akademie der Wissenschaften zu Berlin* 1873: 622-624.
PETERS, W. C. H. 1874 ("1873"). Über neue Saurier (*Spæriodactylus, Anolis, Phrynosoma, Tropidolepisma, Lygosoma, Ophioscincus*) aus Centralamerica, Mexico und Australien. *Monatsberichte der königlich Akademie der Wissenschaften zu Berlin* 1873: 738-747.
PETERS, W. C. H. 1880. Über neue oder weniger bekannte Amphibien des Berliner Zoologischen Museums (*Leposoma dispar, Monopeltis (Phractogonus) jugularis, Typhlops depressus, Leptocalamus trilineatus, Xenodon punctatus, Elapomorphus erythronotus, Hylomantis*

fallax). *Monatsberichte der Preussischen Akademie der Wissenschaften zu Berlin* 1880: 217-224.

PETERS, W. C. H. 1882a. Über neue Batrachier der gattungen *Hyperolius* und *Limnodytes* (*Hylorana*) aus Africa. *Sitzungsberichte der Gesellschaft Naturforschender Freunde zu Berlin* 1882 (1): 8-10.

PETERS, W. C. H. 1882b. Verlegung dreier neuen Batrachier. *Sitzungsberichte der Gesellschaft Naturforschender Freunde zu Berlin* 10: 145-148.

PETERS, W. C. H. 1882c. Zoologie III. Amphibien. Naturwissenschaftlichte Reise nach Mossambique auf Befehl seiner Majestät des Königs Friedrich Wilhelm IV: In den Jahren 1842 bis 1848 ausgeführt von Wilhelm C. H. Peters. Berlin, G. Reimer: i-xv, 1-191, pl. 1-33.

PETERSON, H. W. 1952. A new salamander from the everglades of southern Florida. *Herpetologica* 8 (3): 103-106.

PETERSON, H. W., R. GARRETT, AND J. P. LANTZ. 1952. The mating period of the giant tree frog *Hyla dominicensis*. *Herpetologica* 8 (3): 63.

PETRANKA, J. W. 1998. Salamanders of the United States and Canada. Washington, Smithsonian Institution Press: i-xiv, 1-587, Pl. 1-172.

PFEILER, E., AND T. A. MARKOW. 2008. Phylogenetic relationships of leopard frogs (*Rana pipiens* complex) from an isolated coastal mountain range in southern Sonora, Mexico. *Molecular Phylogenetics and Evolution* 49 (1): 343-348.

PHILIPPI, R. A. 1902. Suplemento a los Batraquios Chilenos descritos en la Historia Fisica y Politica de Chile de Don Claudio Gay. Santiago: 1-161.

PIATT, J. 1934. The systematic status of *Eleutherodactylus latrans* (Cope). *American Midland Naturalist* 15 (1): 89-91.

PIERCE, B. A., AND P. H. WHITEHURST. 1990. *Pseudacris clarkii*. *Catalogue of American Amphibians and Reptiles*. Society for the Study of Amphibians and Reptiles (458): 1-3.

PIMENTEL, R. A. 1958. On the validity of *Taricha granulosa* Skilton. *Herpetologica* 14 (3): 165-168.

PLATZ, J. E. 1984. Status report for *Rana onca* Cope. Prepared for Office of Endangered Species, U.S. Fish and Wildlife Service, Albuquerque, New Mexico: i-iv, 1-27.

PLATZ, J. E. 1988. *Rana yavapaiensis. Catalogue of American Amphibians and Reptiles.* Society for the Study of Amphibians and Reptiles (418): 1-2.

PLATZ, J. E. 1989. Speciation within the Chorus Frog *Pseudacris triseriata:* Morphometric and mating call analyses of the boreal and western subspecies. *Copeia* 1989 (3): 704-712.

PLATZ, J. E. 1991. *Rana berlandieri. Catalogue of American Amphibians and Reptiles.* Society for the Study of Amphibians and Reptiles (508): 1-4.

PLATZ, J. E. 1993. *Rana subaquavocalis*, a remarkable new species of Leopard Frog (*Rana pipiens* complex) from southeastern Arizona that calls under water. *Journal of Herpetology* 27 (2): 154-162.

PLATZ, J. E., AND J. S. FROST. 1984. *Rana yavapaiensis,* a new species of Leopard Frog (*Rana pipiens* complex). *Copeia* 1984 (4): 940-948.

PLATZ, J. E., AND J. S. MECHAM. 1979. *Rana chiricahuensis,* a new species of Leopard Frog (*Rana pipiens* complex) from Arizona. *Copeia* 1979 (3): 383-390.

PLATZ, J. E., AND J. S. MECHAM. 1984. *Rana chiricahuensis. Catalogue of American Amphibians and Reptiles.* Society for the Study of Amphibians and Reptiles (347): 1-2.

POPE, C. H. 1924. Notes on North Carolina salamanders with especial reference to the egg-laying habits of *Leurognathus* and *Desmognathus. American Museum Novitates* (153): 1-15.

POPE, C. H. 1928. Some plethodontid salamanders from North Carolina and Kentucky, with the description of a new race of *Leurognathus. American Museum Novitates* (306): 1-19.

POPE, C. H. 1949. The salamander *Desmognathus quadramaculatus amphileucus* reduced to synonymy. *Natural History Miscellanea* (44): 1-4.

POPE, C. H. 1964. *Plethodon caddoensis. Catalogue of American Amphibians and Reptiles.* American Society of Ichthyologists and Herpetologists (14): 1.

POPE, C. H. 1965. *Plethodon yonahlossee. Catalogue of American Amphibians and Reptiles.* American Society of Ichthyologists and Herpetologists (18): 1-2.

POPE, C. H., AND H. W. FOWLER. 1949. A new species of salamander (*Plethodon*) from southwestern Virginia. *Natural History Miscellanea* 1-4.
POPE, C. H., AND N. G. HAIRSTON. 1947. Distribution of *Leurognathus*, a southern Appalachian genus of salamanders. *Fieldiana: Zoology* 31 (20): 155-162.
POPE, C. H., AND N. G. HAIRSTON. 1948. Two new subspecies of the salamander *Plethodon shermani*. *Copeia* 1948 (1): 106-107.
POPE, C. H., AND S. H. POPE. 1951. A study of the salamander *Plethodon ouachitae* and the description of an allied form. *Bulletin of the Chicago Academy of Science* 9: 129-152.
PORTER, K. R. 1968. Evolutionary status of a relict population of *Bufo hemiophrys* Cope. *Evolution* 22 (3): 583-594.
PORTER, K. R. 1969. Description of *Rana maslini*, a new species of wood frog. *Herpetologica* 25 (3): 212-215.
PORTER, K. R. 1970. *Bufo valliceps*. *Catalogue of American Amphibians and Reptiles*. American Society of Ichthyologists and Herpetologists (94): 1-4.
PORTIS, A. 1885. Resti di batraci fossili italiani. *Atti della reale Accademia delle Scienze di Torino* 20: 935-962 + 1 pl.
POTTER, F. E., JR. 1963. Gross morphological variation in the genus *Typhlomolge* with a description of a new species. Master of Arts thesis, Department of Zoology, University of Texas, Austin.
POTTER, F. E., JR., AND S. S. SWEET. 1981. Generic boundaries in Texas cave salamanders, and a redescription of *Typhlomolge robusta* (Amphibia: Plethodontidae). *Copeia* 1981 (1): 64-75.
POUGH, H. F. 2007. Amphibian biology and husbandry. *ILAR Journal* 48 (3): 203-213.
POYNTON, J. C. 1964. The Amphibia of southern Africa: a faunal study. *Annals of the Natal Museum* 17: 1-334.
PRAMUK, J. B. 2000. Prenasal bones and snout morphology in West Indian bufonids and the *Bufo granulosus* species group. *Journal of Herpetology* 34 (2): 334-340.
PRAMUK, J. B., AND E. LEHR. 2005. Taxonomic status of *Atelophryniscus chrysophorus* McCranie, Wilson, and Williams, 1989 (Anura: Bufonidae) inferred from phylogeny. *Journal of Herpetology* 39 (4): 610-618.

PRAMUK, J. B., T. ROBERTSON, J. W. SITES, JR, AND B. P. NOONAN. 2008. Around the world in 10 million years: biogeography of the nearly cosmopolitan true toads (Anura: Bufonidae). *Global Ecology and Biogeography* 17 (1): 72-83.

PRATT, H. S. 1935. A Manual of Land and Fresh Water Vertebrate Animals of the United States (Exclusive of Birds). Philadelphia, P. Blakiston's Son & Co: i-xv, 1-422.

PREGILL, G. 1981. Cranial morphology and the evolution of West Indian toads (Salientia: Bufonidae): resurrection of the genus *Peltophryne* Fitzinger. *Copeia* 1981 (2): 273-285.

PRICE, A. H., AND B. K. SULLIVAN. 1988. *Bufo microscaphus. Catalogue of American Amphibians and Reptiles.* Society for the Study of Amphibians and Reptiles (415): 1-3.

PROVANCHER, M. 1875. Description d'une salamandre nouvelle. *Le Nauraliste Canadien* 7 (8): 251-252.

PULIS, J. G., AND D. B. MEANS. 2005. *Ambystoma cingulatum*, Flatwoods Salamander. *In*: LANNOO, M. J. (ed.), Amphibian Declines: The Conservation Status of United States Species, Part 2: Species Accounts. Berkeley, University of California Press: 608-609.

PYRON, R. A., AND J. J. WIENS. 2011. A large-scale phylogeny of Amphibia including over 2800 species, and a revidsed classification of extant frogs, salamanders, and caecilians. *Molecular Phylogenetics and Evolution* 61: 543-583.

PYTEL, B. A. 1986. Biochemical systematics of the eastern North American frogs of the genus *Rana*. *Herpetologica* 42 (3): 273-282.

RABB, G. B. 1966. *Stereochilus* and *S. marginatus*. *Catalogue of American Amphibians and Reptiles.* American Society of Ichthyologists and Herpetologists (25): 1-2.

RADDI, G. 1823. Continuazione della descrizione dei rettili Brasiliani. *Memorie Matematica et di Fisica della Società Italiana delle Scienze residente in Modena* 19 (1): 58-73.

RAFFAËLLI, J. 2007. Les Urodèles du Monde. Plumelec, Penclen édition: [i-vi], 1-377.

RAFINESQUE, C. S. 1814b. Fine del Prodromo d'Erpetologia Siciliana. *Specchio delle Scienze o Giornale enciclopedico di Sicilia* 2: 102-104.

RAFINESQUE, C. S. 1815. Analyse de nature, ou Tableau de l'universe et des corps organisés. Palermo, Jean Barravecchia: 1-224, 1 pl.

RAFINESQUE, C. S. 1818a. Description of a new American salamander—the red-backed salamander from the Highlands. *Scientific Journal of New York* 1 (2): 25-26.

RAFINESQUE, C. S. 1818b. Farther accounts of discoveries in natural history, in the western states. *Amercan Monthly Magazine and Critical Reviews* 4: 39-42.

RAFINESQUE, C. S. 1819. Prodrome de 70 nouveaux genres d'animaux découverts dans l'intérieur des Etats-Unis d'Amérique, durant l'année 1818. *Journal de Physique, de Chimie, d'Histoire Naturelle et des Arts* 88: 417-429.

RAFINESQUE, C. S. 1820. Annals of nature or annual synopsis of new genera and species of animals, plants, etc., discovered in North America. Lexington, Thomas Smith: 1-16.

RAFINESQUE, C. S. 1822a. On two new salamanders of Kentucky. *Kentucky Gazette (Lexington)*(new series) 1 (9): 3.

RAFINESQUE, C. S. 1832a. Description of the Spelerpes, or salamander of the caves of Kentucky. *Atlantic Journal and Friend of Knowledge (Philadelphia)* 1 (2): 22.

RAFINESQUE, C. S. 1832c. On three new water salamanders of Kentucky. *Atlantic Journal and Friend of Knowledge (Philadelphia)* 1 (3): 121.

RAFINESQUE, C. S. 1832d. On the salamander of the hills of east Kentucky. S. lurida. *Atlantic Journal and Friend of Knowledge (Philadelphia)* 1 (2): 63-64.

RAO, D., AND D. YANG. 1994. The study of early development and evolution of *Torrentophryne aspinia. Zoological Research* 15: 142-157.

REASER, J. K., AND D. S. PILLIOD. 2005. *Rana luteiventris.* Columbia Spotted Frog. *In*: LANNOO, M. J. (ed.), Amphibian Declines: The Conservation Status of United States Species, Part 2: Species Accounts. Berkeley, University of California Press: 559-563.

RECUERO, E., I. MARTINEZ-SOLANO, G. PARRA-OLEA, AND M. GARCIA-PARIS. 2006a. Phylogeography of *Pseudacris regilla* (Anura: Hylidae) in western North America, with a proposal for a new taxonomic rearrangement. *Molecular Phylogenetics and Evolution* 39 (2): 293-304.

RECUERO, E., I. MARTINEZ-SOLANO, G. PARRA-OLEA, AND M. GARCIA-PARIS. 2006b. Corrigendum to "Phylogeography of *Pseudacris regilla* (Anura: Hylidae) in western North America, with a proposal for a new taxonomic rearrangement". *Molecular Phylogenetics and Evolution* 41 (2): 511.

REDMER, M. 2005. *Hyla avivoca*, Bird-voiced Treefrog. *In*: LANNOO, M. J. (ed.), Amphibian Declines: The Conservation Status of United States Species, Part 2: Species Accounts. Berkeley, University of California Press: 448-449.

REDMER, M., AND R. A. BRANDON. 2003. *Hyla cinerea. Catalogue of American Amphibians and Reptiles.* Society for the Study of Amphibians and Reptiles (766): 1-14.

REED, C. F. 1956. H*yla cinerea* in Maryland, Delaware, and Virginia, with notes on the taxonomic status of *Hyla cinerea evittata. Journal of the Washington Academy of Science* 46: 328-332.

REESE, R. W. 1950. The identity of the salamander *Gyrinophilus danielsi*, with a description of a new subspecies. *Natural History Miscellanea* (63): 1-7.

REGAL, P. 1966. Feeding specialisations and the classification of terrestrial salamanders. *Evolution* 20 (3): 392-407.

REGESTER, K. J. 2000a. *Plethodon electromorphus. Catalogue of American Amphibians and Reptiles.* Society for the Study of Amphibians and Reptiles (706): 1-3.

REGESTER, K. J. 2000b. *Plethodon richmondi. Catalogue of American Amphibians and Reptiles.* Society for the Study of Amphibians and Reptiles (707): 1-3.

REID, H. A. 1895. History of Pasadena. Pasadena, Pasadena History Co.: 1-675.

Reilly, S. B., M. F. Mulks, J. M. Reilly, W. B. Jennings, and D. B. Wake. 2013. Genetic diversity of Black Salamanders (*Aneides flavipunctatus*) across watersheds in the Klamath Mountains. *Diversity* 5: 657-679.

REINHARDT, J., AND C. F. LÜTKEN. 1863 ("1862"). Bidrag til der vestindiske Öriges og navnilgen til de dansk-vestindiske Örs herpetologie. *Videnskabelige Meddelelser fra den Naturhistoriske Forening i Kjøbenhavn* (2) 4 (10-18): 153-291.

RHOADS, S. N. 1895 ("1896"). Contributions to the zoology of Tennessee. No. 1, amphibians and reptiles. *Proceedings of the Academy of Natural Sciences of Philadelphia* 47 (1895): 376-407.

RICE, F. L., AND N. S. DAVIS. 1878. Addendum 2 (b), for p. 188. *R. circulosa. In*: JORDAN, D. S. (Ed.), Manual of vertebrates of the northern United States, including the district east of the Mississippi River, and north of North Carolina and Tennessee, exclusive of marine species. Second edition. Jansen, McClurg & Company, Chicago: 355.

RICHTER, S. C., AND J. B. JENSEN. 2005. *Rana sevosa*. Dusky Gopher Frog. *In*: LANNOO, M. J. (ed.), Amphibian Declines: The Conservation Status of United States Species, Part 2: Species Accounts. Berkeley, University of California Press: 584-586.

RIEMER, W. J. 1958. Variation and systematic relationships within the salamander genus *Taricha*. *University of California Publications in Zoology* 56 (3): 301-390.

Rissler, L. J., and J. J. Apodaca. 2007. Adding more ecology into species delimitation: ecological niche models and phylogeography help define cryptic species in the Black Salamanders (*Aneides flavipunctatus*). *Systematic Biology* 56 (6): 924-942.

RISSLER, L. J., AND D. R. TAYLOR. 2003. The phylogenetics of desmognathine salamander populations across the southern Appalachians. *Molecular Phylogenetics and Evolution* 27: 197-211.

RITGEN, F. A. V. 1828. Versuch einer Natürichen Eintheilung der Amphibien. *Nova Acta physico-medica Academiae Caesareae Leopoldino-Carolinae Naturae Curiosorum* (Halle) 14 (1): 245-284.

RIVERO, J. A. 1961. Salientia of Venezuela. *Bulletin of the Museum of Comparative Zoology* 126 (1): 1-207.

Robertson, A. V., C. Ramsden, J. Niedzwiecki, J. Fu, and J. P. Bogart. 2006. An unexpected recent ancestor of unisexual *Ambystoma*. *Molecular Ecology* 15: 3339-3351.

ROCEK, Z. 1981 ("1980"). Cranial anatomy of frogs of the family Pelobatidae Stannius, 1856, with outlines of their phylogeny and systematics. Acta Universitatis Carolinae-Biologica 1980: 1-164.

ROELANTS, K., D. J. GOWER, M. WILKINSON, S. P. LOADER, S. D. BIJU, K. GUILLAUME, L. MORIAU, AND F. BOSSUYT. 2007. Global patterns of diversification in the history of modern amphibians. *Proceedings of the National Academy of Science (US)* 104: 887-892.

ROMER, A. S. 1945. Vertebrate Paleontology. 2nd edition. Chicago, University of Chicago Press: i-x, 1-687.

RORABAUGH, J. C. 2005. *Rana pipiens*. Northern Leopard Frog. *In*: LANNOO, M. J. (ed.), Amphibian Declines: The Conservation Status of United States Species, Part 2: Species Accounts. Berkeley, University of California Press: 570-577.

ROSE, F. L. 1971. *Eurycea aquatica. Catalogue of American Amphibians and Reptiles.* American Society of Ichthyologists and Herpetologists (116): 1-2.

ROSE, F. L., AND F. M. BUSH. 1963. A new species of *Eurycea* (Amphibia: Caudata) from the southeastern United States. *Tulane Studies in Zoology* 10 (2): 121-128.

ROSE, F. L., T. R. SIMPSON, M. R. J. FORSTNER, D. J. MCHENRY, AND J. WILLIAMS. 2006. Taxonomic status of *Acris crepitans palludicola*: in search of the pink frog. *Journal of Herpetology* 40: 428-434.

ROSSMAN, D. A. 1958. A new race of *Desmognathus fuscus* from the south-central United States. *Herpetologica* 14 (3): 158-160.

ROSSMAN, D. A. 1959. Ecosystematic relationships of the salamanders *Desmognathus fuscus auriculatus* Holbrook and *Desmognathus fuscus carri* Neill. *Herpetologica* 15 (2): 149-155.

ROVITO, S. M. 2010. Lineage divergence and speciation in the Web-toed Salamanders (Plethodontidae: *Hydromantes*) of the Sierra Nevada, California. Molecular Ecology 19: 4554-4571.

RUBENSTEIN, N. M. 1971. Ontogenetic allometry in the salamander genus *Desmognathus*. *American Midland Naturalist* 85 (2): 329-348.

RUSSELL, R. W., AND J. D. ANDERSON. 1956. A disjunct population of the long-nosed salamander from the coast of California. *Herpetologica* 12 (2): 137-140.

RUTHVEN, A. G. 1912a. The amphibians and reptiles collected by the University of Michigan-Walker Expedition in southern Vera Cruz, Mexico. *Zoologische Jahrbücher Abteilung für Systematik, Ökologie und Geographie der Tiere (Jena)* 32: 295-332 + pl. 6-11.

RUTHVEN, A. G. 1912b. Description of a new salamander from Iowa. *Proceedings of the U.S. National Museum* 41: 517-519.

RYDER, J. A. 1879 ("1880"). Morphological notes on the limbs of Amphiumidae, as indicating a possible synonymy of the supposed genera. *Proceedings of the Academy of Natural Sciences of Philadelphia* 31 (1879): 14-15.

SAGER, A. 1839. On American Amphibia. *American Journal of Sciences and Arts* 36: 320-324.

SAGER, A. 1858. Description of a new genus of perennibranchiate amphibians. *Peninsular Journal of Medicine* 5: 428-429.

SALTHE, S. N. 1973a. Amphiumidae, *Amphiuma*. *Catalogue of American Amphibians and Reptiles*. Society for the Study of Amphibians and Reptiles (147): 1-4.

SALTHE, S. N. 1973b. *Amphiuma means*. *Catalogue of American Amphibians and Reptiles*. Society for the Study of Amphibians and Reptiles (148): 1-2.

SALTHE, S. N. 1973c. *Amphiuma tridactylum*. *Catalogue of American Amphibians and Reptiles*. Society for the Study of Amphibians and Reptiles (149): 1-4.

SANCHIZ, B. 1998. Handbuch der Paläoherpetologie / Encyclopedia of Paleoherptology. Part 4, Salientia. Mūnchen, F. Pfeil.

SANDERS, O. 1953. A new species of toad, with a discussion of morphology of the bufonid skull. *Herpetologica* 9 (1): 25-47.

SANDERS, O. 1973. A new leopard frog (*Rana berlandieri brownorum*) from southern Mexico. *Journal of Herpetology* 7 (2): 87-92.

SANDERS, O. 1984. *Rana heckschedri*. *Catalogue of American Amphibians and Reptiles*. Society for the Study of Amphibians and Reptiles (348): 1-2.

SANDERS, O. 1986. The heritage of *Bufo woodhousei* Girard in Texas. *Occasional Papers of the Strecker Museum (Baylor Univ.)* (1): 1-28.

SANDERS, O. 1987. Evolutionary hybridization and speciation in North American indigenous bufonids. Dallas, (privately published): i-viii, 1-110.

SANDERS, O., AND H. M. SMITH. 1951. Geographic variation in toads of the *debilis* group of *Bufo*. *Field and Laboratory* 19: 141-160, pl. 1-3.

SATTLER, P. W. 1980. Genetic relationships among selected species of North American *Scaphiopus*. *Copeia* 1980 (4): 605-610.

SAVAGE, J. M. 1959. A preliminary biosystematic analysis of toads of the *Bufo boreas* group in Nevada and California. *Year Book of the American Philosophical Society* 1959: 251-254.

SAVAGE, J. M. 1967. The systematic status of the toad *Bufo politus* Cope, 1862. *Copeia* 1967 (1): 225-226.

SAVAGE, J. M. 1973. The geographic distribution of frogs: patterns and predictions. *In*: VIAL, J. L. (ed.), Evolutionary Biology of the Anurans: Contemporary research on major problems. Columbia, University of Missouri Press: 351-445.

SAVAGE, J. M. 1974. Type localities for species of amphibians and reptiles described from Costa Rica. *Revista de Biologia Tropical* 22 (1): 71-122.

SAVAGE, J. M. 1986. Nomenclatural notes on the Anura (Amphibia). *Proceedings of the Biological Society of Washington* 99 (1): 42-45.

SAY, T. 1818. Notes on Professor Green's paper on the amphibia, published in the September number of this journal. *Journal of the Academy of Natural Sciences of Philadelphia* 1: 405-407.

SCHAAF, R. T., JR, AND P. W. SMITH. 1970. Geographic variation in the pickerel frog. *Herpetologica* 26 (2): 240-254.

SCHAAF, R. T., JR, AND P. W. SMITH. 1971. *Rana palustris*. *Catalogue of American Amphibians and Reptiles*. Society for the Study of Amphibians and Reptiles (117): 1-3.

SCHARLINSKI, H. 1939. Nachtrag zum Katalog der Wolterstorff-Sammlun im Museum fürNaturkunde und Vorgeschichte zu Magdeburg. *Abhandlungen und Berichte aus dem Museum für Natur-und Heimatkunde zu Magdeburg* 7: 31-57.

SCHINZ, H. R. 1822. Das Thierreich eingetheilt nach dem Bau der Thiere als Grundlage ihrer Naturgeschichte und der vergleichenden Anatomie. Mit vielen Zusätzenversehen. Volume 2. Stuttgart & Tubingen, J.G. Cotta: i-xvi, 1-835. [fide Frost (2009-13); not seen]

SCHINZ, H. R. 1833. Naturgeschichte und Abbildungen der Reptilien: nach den neuesten Systemen zum gemeinnützigen Gebrauche entworfen und mit Berücksichtigung für der Unterricht der Jugend. Schaffhausen, Brodtmann's lithographischer Kunstanstalt: [i-iv], i-iv, 1-240, pl. 1-102.

SCHLEGEL, H. 1826. Notice sur l'erpétologie de l'île de Java; par M. Boie (ouvrage manuscript). *Bulletin des Sciences Naturelles et de Géologie* (2) 9: 233-240.

SCHLEGEL, H. 1858. Handleiding tot de beoefening der dierkunde. II Deel [Vol. 2]. Breda, Koninklijke Militaire Akademie: i-xx + 1-628 + [i-ii], pl. 1-27.

SCHMIDT, K. P. 1924. A list of amphibians and reptiles collected near Charleston, S.C. *Copeia* (132): 67-69.

SCHMIDT, K. P. 1925. New Chinese amphibians and reptiles. *American Museum Novitates* (175): 1-3.

SCHMIDT, K. P. 1932. Reptiles and amphibians of the Mandel Venezuelan Expedition. *Field Museum of Natural History Zoological Series* 18 (7): 159-163.

SCHMIDT, K. P. 1938. A geographic variation gradient in frogs. *Field Museum of Natural History Zoological Series* 20 (29): 377-382.

SCHMIDT, K. P. 1939. New Central American frogs of the genus Hypopachus. *Field Museum of Natural History Zoological Series* 24: 1-5.

SCHMIDT, K. P. 1941. The amphibians and reptiles of British Honduras. *Field Museum of Natural History Zoological Series* 22: 475-510.

SCHMIDT, K. P. 1953. A checklist of North American Amphibians and Reptiles. 6th edition. Chicago, University of Chicago Press: i-viii, 1-280.

SCHMIDT, K. P., AND W. L. NECKER. 1935. Amphibians and reptiles of the Chicago region. *Bulletin of the Chicago Academy of Science* 5 (4): 55-77.

SCHMIDT, K. P., AND T. F. SMITH. 1944. Amphibians and reptiles of the Big Bend region of Texas. *Field Museum of Natural History Zoological Series* 29: 79-96.

SCHNEIDER, J. G. 1799. Historiae Amphibiorum naturalis et literariae. Fasciculus primus, continens Ranas, Calamitas, Bufones, Salamandras et Hydros, in genera et species descriptos notisque suis distinctos. Jena, Frommann: i-xv,1-264, pl. 1-2.

SCHREBER, [J. C. D.]. 1782. Beytrag zur Naturgeschicte der Froesche. *Der Naturforscher* (Halle) 18: 182-193.

SCHULTZE, E. 1891. Amphibia Europaea. Jahresbericht und Abhandlungen des Naturwissenschaftlichen Vereins in Magdeburg 1890: 163-178.

SCHWALBE, C. R., AND C. S. GOLDBERG. 2005. *Eleutherodactylus augusti*, Barking Frog. In: LANNOO, M. J. (ed.), Amphibian Declines: The Conservation Status of United States Species, Part 2: Species Accounts. Berkeley, University of California Press: 491-492.

SCHWARTZ, A. 1952. A new race of *Pseudobranchus striatus* from southern Florida. *Natural History Miscellanea* (115): 1-9.

SCHWARTZ, A. 1957. Chorus frogs (*Pseudacris nigrita* LeConte) in South Carolina. *American Museum Novitates* (1838): 1-12.

SCHWARTZ, A. 1967. Frogs of the genus *Eleutherodactylus* in the Lesser Antilles. *Studies of the Fauna of Curaçao and other Caribbean Islands* 24 (91): 1-62.

SCHWARTZ, A. 1974. *Eleutherodactylus planirostris. Catalogue of American Amphibians and Reptiles*. Society for the Study of Amphibians and Reptiles (154): 1-4.

SCHWARTZ, A., AND W. E. DUELLMAN. 1952. The taxonomic status of the newts, *Diemictylus viridescens*, of Peninsular Florida. *Bulletin of the Chicago Academy of Science* 9 (12): 219-227.

SCHWARTZ, A., AND J. R. HARRISON. 1956. A new subspecies of gopher frog (*Rana capito* LeConte). *Proceedings of the Biological Society of Washington* 69: 135-144.

SCHWARTZ, A., AND R. W. HENDERSON. 1991. Amphibians and reptiles of the West Indies. Gainesville, University of Florida Press: 720 pp.

SCOPOLI, I. A. 1777. Introductio ad historiam naturalem, sistens genera lapidium, planatarum, et animalium hactenus detecta, caracteribus essentialibus donata, in tribus divisa, subinde ad leges naturae. Prague, Gerle: [i-vii] + 1-506 + i-xxxiv.

SCOTT, E. 2005. A phylogeny of ranid frogs (Anura: Ranoidea: Ranidae), based on a simultaneous analysis of morphological and molecular data. *Cladistics* 21: 507-574.

SEBA, A. 1734. Locupletissimi Rerum naturalium Thesauri accurata Descriptio, et Iconibus artificiosissimus Expressio, per universam Physices Historium. Opus cui in hoc Rerum Generi, nullum par exstitit. Vol. 1. Janssonio-Waesburgios, Amsterdam, 1-178 + pl. 1-111.

SESSIONS, S. K., AND A. LARSON. 1987. Developmental correlates of genome size in plethodontid salamanders and their implications for genome evolution. *Evolution* 41 (6): 1239-1251.

SESSIONS, S. K., AND J. E. WILEY. 1985. Chromosome evolution in salamanders of the genus *Necturus*. *Brimleyana* 1985 (10): 1-12.

SEVER, D. M. 1972. Geographic variation and taxonomy of *Eurycea bislineata* (Caudata: Plethodontidae) in the upper Ohio River valley. *Herpetologica* 28 (4): 314-324.

SEVER, D. M. 1983. *Eurycea junaluska*. *Catalogue of American Amphibians and Reptiles*. Society for the Study of Amphibians and Reptiles (321): 1-2.

SEVER, D. M. 1999a. *Eurycea bislineata*. *Catalogue of American Amphibians and Reptiles*. Society for the Study of Amphibians and Reptiles (683): 1-5.

SEVER, D. M. 1999b. *Eurycea cirrigera*. *Catalogue of American Amphibians and Reptiles*. Society for the Study of Amphibians and Reptiles (684): 1-6.

SEVER, D. M. 1999c. *Eurycea wilderae*. *Catalogue of American Amphibians and Reptiles*. Society for the Study of Amphibians and Reptiles (685): 1-4.

SEVER, D. M. 2005. *Eurycea wilderae*, Blue Ridge Two-Lined Salamander. *In*: LANNOO, M. J. (ed.), Amphibian Declines: The Conservation Status of United States Species. Part 2: Species Accounts. Berkeley, University of California Press: 770-772.

SEVER, D. M., H. A. DUNDEE, AND C. D. SULLIVAN. 1976. A new *Eurycea* (Amphibia: Plethodontidae) from southwestern North Carolina. *Herpetologica* 32 (1): 26-29.

SHAFFER, H. B. 2005. *Ambystoma gracile*, Northwestern Salamander. *In*: LANNOO, M. J. (ed.), Amphibian Declines: The Conservation Status of United States Species, Part 2: Species Accounts. Berkeley, University of California Press: 609-611.

SHAFFER, H. B., J. M. CLARK, AND E. KRAUS. 1991. When molecules and morphology clash: a phylogenetic analysis of North American ambystomatid salamanders (Caudata: Ambystomatidae). *Systematic Zoology* 40 (3): 284-303.

SHAFFER, H. B., G. M. FELLERS, R. VOSS, J. C. OLIVER, AND G. B. PAULY. 2004a. Species boundaries, phylogeography and conservation genetics of the red-legged frog (*Rana aurora / draytonii*) complex. *Molecular Ecology* 13 (10): 2667-2677.

SHAFFER, H. B., AND M. L. MCKNIGHT. 1996. The polytypic species revisited: differentiation and molecular phylogenetics of the Tiger Salamander *Ambystoma tigrinum* (Amphibia: Caudata) complex. *Evolution* 50 (1): 417-433.

SHANNON, F. A. 1949. A western subspecies of *Bufo woodhousii* hitherto erroneously associated with *Bufo compactilis*. *Bulletin of the Chicago Academy of Science* 8 (15): 301-312.

SHANNON, F. A., AND F. L. HUMPHREY. 1958. A discussion of the polytypic species *Hypopachus oxyrrhinus*, with a description of a new subspecies. *Herpetologica* 14: 85-95.

SHANNON, F. A., AND C. H. LOWE, JR. 1955. A new subspecies of *Bufo woodhousei* from the inland Southwest. *Herpetologica* 11 (3): 185-190.

SHANNON, F. A., AND J. E. WERLER. 1955. Notes on the amphibians of the Los Tuxtlas range of Veracruz, Mexico. *Transactions of the Kansas Academy of Science* 58 (3): 360-381.

SHAW, G. 1802a. General Zoology or Systematic Natural History—Vol. 3 (Part 1) Amphibia. London, G. Kearsley: i-viii, 1-312, + pl. 1-86.

Shen, X. X., D. Liang, Y. J. Feng, M. Y. Chen, and P. Zhang. 2013. A versatile and highly efficient toolit including 102 nuclear markers for vertebrate phylogenomics, tested by resolving the higher level relationships of the Caudata. *Molecular Biology and Evolution* 30 (10): 2235-2248.

SHEPARD, D. B., AND L. E. BROWN. 2005. *Bufo houstonensis*. Houston Toad. *In*: LANNOO, M. J. (ed.), Amphibian Declines: The Conservation Status of United States Species, Part 2: Species Accounts. Berkeley, University of California Press: 415-417.

SHERBORN, C. D. 1891. Note on the authors of the specific names in John White's 'Journal of a Voyage to New South Wales' 1790. *Annals and Magazine of Natural History* (6) 7: 535.

SHOOP, C. R. 1964. *Ambystoma talpoideum. Catalogue of American Amphibians and Reptiles*. American Society of Ichthyologists and Herpetologists (8): 1-2.

SHREVE, B. 1945. Application of the name *Eleutherodactylus ricordii*. *Copeia* 1945: 117.

SHREVE, B. 1957. Reptiles and amphibians from the Selva Lacandona. *IN*: Biological investigations in the Selva Lacandona, Chiapas, Mexico (R.A. Paynter, Jr., Ed). *Bulletin of the Museum of Comparative Zoology* 116 (4): 242-248.

SILVA, H. R. D. 1997. Two character states new for hylines and the taxonomy of the genus *Pseudacris*. *Journal of Herpetology* 31 (4): 609-613.

SILVERSTONE, P. A. 1975. A revision of the poison-arrow frogs of the genus *Dendrobates* Wagler. *Science Bulletin, Natural History Museum of Los Angeles County* 21: 1-55.

SIMPSON, G. G. 1940. Types in modern taxonomy. *American Journal of Science* 238: 413-431.

SINCLAIR, R. M. 1955. Larval development of *Gyrinophilus warneri*. *Journal of the Tennessee Academy of Science* 30: 133. [fide ASW]

SITES, J. W., JR, M. MORANDO, R. HIGHTON, F. HUBER, AND R. E. JUNG. 2004. Phylogenetic relationships of the endangered Shenandoah Salamander (*Plethodon shenandoah*) and other salamanders of the Plethodon cinereus group (Caudata: Plethodontidae). *Journal of Herpetology* 38 (1): 96-105.

SKILTON, A. J. 1849. Description of two reptiles from Oregon. *American Journal of Science and Arts* (2) 7: 202.

SLATER, J. R. 1939. Description and life history of a new *Rana* from Washington. *Herpetologica* 1: 145-149.

SLATER, J. R. 1940. Salamander records from British Columbia. *Occasional Papers, Department of Biology, College of Puget Sound* (9): 43-44.

SLATER, J. R., AND J. W. SLIPP. 1940. A new species of *Plethodon* from northern Idaho. *Occasional Papers, Department of Biology, College of Puget Sound* (38): 38-43.

SLEVIN, J. R. 1928. The amphibians of western North America: an account of the species known to inhabit California, Alaska, British Columbia, Washington, Oregon, Idaho, Utah, Nevada, Arizona, Sonora, and Lower California. *Occasional Papers of the California Academy of Sciences* 16: 1-152, pl. 1-23.

SLEVIN, J. R., AND A. E. LEVITON. 1956. Holotype specimens of reptiles and amphibians in the collection of the California Academy of Sciences. *Proceedings of the California Academy of Sciences* (4) 28 (14): 529-560.

SMITH, A. 1831. Contributions to the natural history of South Africa, &c. *South African Quarterly Journal* (1) 2 (5): 9-24.

SMITH, C. C. 1968. A new *Typhlotriton* from Arkansas (Amphibia Caudata). *Wasmann Journal of Biology* 26 (1): 155-159.

SMITH, H. M. 1933b. On the proper name for the brevicipitid frog *Gastrophryne texensis* (Girard). *Copeia* 1933 (4): 217.

SMITH, H. M. 1934a. The amphibians of Kansas. *American Midland Naturalist* 15: 377-528.

SMITH, H. M. 1947a. Subspecies of the Sonoran toad (*Bufo compactilis* Wiegman). *Herpetologica* 4 (1): 7-13.

SMITH, H. M. 1947b. Notes on Mexican amphibians and reptiles. *Journal of the Washington Academy of Science* 37 (11): 408-412.

SMITH, H. M. 1950. Handbook of amphibians and reptiles of Kansas. *University of Kansas Museum of Natural History, Miscellaneous Publications* (2): 1-336

SMITH, H. M. 1953a. The generic name of the newts of eastern North America. *Herpetologica* 9 (2): 95-99.

SMITH, H. M. 1978. Amphibians of North America. New York, Golden Press: 1-160.

SMITH, H. M., AND B. C. BROWN. 1947. The Texan subspecies of the treefrog, *Hyla versicolor*. *Proceedings of the Biological Society of Washington* 60: 47-49.

SMITH, H. M., AND D. CHIZAR. 2006. Dilemma of name-recognition: why and when to use new combinations of scientific names. *Herpetological Conservation and Biology* 1 (1): 6-8.

SMITH, H. M., G. E. FAWCETT, J. D. FAWCETT, AND R. B. SMITH. 1970. J. G. Wood and the Mexican axolotl. *Journal of the Society for the Bibliography of Natural History* 5: 362-365.

SMITH, H. M., K. T. FITZGERALD, AND L. J. GUILLETTE, JR. 1992. Note on the proposed designation of a neotype for *Hyla chrysoscelis* Cope, 1880, and the designation of a neotype for *H. versicolor* Le Conte, 1825 (Amphibia, Anura). *Bulletin of Zoological Nomenclature* 49: 151-153.

SMITH, H. M., AND W. L. NECKER. 1943. Alfredo Dugès' types of Mexican reptiles and amphibians. *Annales de la Escuela Nacional de Ciencias Biologicas* 3 (1/2): 179-219.

SMITH, H. M., AND F. E. POTTER, JR. 1946. A third neotenic salamander of the genus *Eurycea* from Texas. *Herpetologica* 3 (4): 105-109.

SMITH, H. M., AND O. SANDERS. 1952. Distributional data on Texas amphibians and reptiles. *Texas Journal of Science* 4 (2): 204-219.

SMITH, H. M., T. SCHNEIDER, AND R. B. SMITH. 1977. An overlooked synonym of the giant toad *Bufo marinus* (Linnaeus) (Amphibia, Anura, Bufonidae). *Journal of Herpetology* 11 (4): 423-425.

SMITH, H. M., AND E. H. TAYLOR. 1948. An annotated checklist and key to the amphibia of Mexico. *Bulletin of the U.S. National Museum* (194): i-iv, 1-118.

SMITH, H. M., AND E. H. TAYLOR. 1950a. Type localities of Mexican reptiles and amphibians. *University of Kansas Science Bulletin* 33 (2): 313-380.

SMITH, H. M., AND J. A. TIHEN. 1961. *Tigrina* (*Salamandra*) Green, 1825: Proposed validation under the plenary powers (Amphibia, Caudata). Z. N. (S.) 1460. *Bulletin of Zoological Nomenclature* 18 (3): 214-216.

SMITH, H. M., R. T. ZAPPALORTI, A. R. BREISCH, AND D. L. MCKINLEY. 1995. The type locality of the frog *Acris crepitans*. *Herpetological Review* 26 (1): 14.

SMITH, J. E. 1821. A selection of the correspondence of Linnaeus and other naturalists. Vol. 1. London, Longman, Hurst, Rees, Orme and Brown: i-xxxi, 1-605.

SMITH, M. A. 1931. Fauna of British India, second edition. Reptilia and Amphibia 1: Loricata, Testudines. London, Taylor & Francis: i-xxviii, 1-185, 2 Pl.

SMITH, P. W. 1951. A new frog and a new turtle from the western Illinois sand prairies. *Bulletin of the Chicago Academy of Science* 9 (10): 189-199.

SMITH, P. W. 1953. A reconsideration of the status of *Hyla phaeocrypta*. *Herpetologica* 9 (4): 169-173.

SMITH, P. W. 1956. The status, correct name, and geographic range of the boreal chorus frog. *Proceedings of the Biological Society of Washington* 69: 169-176.

SMITH, P. W. 1961. The amphibians and reptiles of Illinois. *Illinois Natural History Survey Bulletin* 28: 1-298.

SMITH, P. W. 1963. *Plethodon cinereus*. *Catalogue of American Amphibians and Reptiles*. American Society of Ichthyologists and Herpetologists (5): 1-3.

SMITH, P. W. 1966a. *Pseudacris streckeri*. *Catalogue of American Amphibians and Reptiles*. American Society of Ichthyologists and Herpetologists (27): 1-2.

SMITH, P. W. 1966b. *Hyla avivoca*. *Catalogue of American Amphibians and Reptiles*. American Society of Ichthyologists and Herpetologists (28): 1-2.

SMITH, P. W., AND D. M. SMITH. 1952. The relationship of the chorus frogs, *Pseudacris nigrita feriarum* and *Pseudacris n. triseriata*. *American Midland Naturalist* 48 (1): 165-180.

SMITH, S. A., P. R. STEPHENS, AND J. J. WIENS. 2005. Replicate patterns of species richness, historical biogeography, and phylogeny in Holarctic treefrogs. *Evolution* 59 (11): 2433-2450.

SMITH, W. H. 1877. The tailed amphibians including the caecilians. A thesis. Detroit, Herald Publishing: 1-158.

SNYDER, R. C. 1963. *Ambystoma gracile*. *Catalogue of American Amphibians and Reptiles*. American Society of Ichthyologists and Herpetologists (6): 1-2.

SOLLAS, I. B. J. 1906. Reptilia and Batrachia. *Zoological Record* 42: 1-40.

SONNINI, C. S., AND P. A. LATREILLE. 1801. Histoire naturelle des reptiles; avec figures dessinées d'après nature. Tome 2. Paris, Chez Deterville: i-vi, 1-332, pl. 1-19.

SONNINI, C. S., AND P. A. LATREILLE. 1802. Histoire naturelle des reptiles; avec figures dessinées d'après nature. Tome 4. Paris, Deterville: i-iii, 1-410, pl. 1-14.

SONNINI, C. S., AND P. A. LATREILLE. 1830a. Histoire naturelle des reptiles, avec figures dessinée d'après nature (Nouvelle Éditon). Tome premiere. Paris, Chez Raynal: v. 1: i-xxii, 1-280; v. 2: i-iv, 1-332; v. 3: i-iv, 1-335; v. 4: i-iv, 1-408.

SONNINI, C. S., AND P. A. LATREILLE. 1830d. Histoire naturelle des reptiles, avec figures dessinée d'après nature (Nouvelle Éditon). Tome quatrième. Paris, Chez Raynal: i-iv, 1-408.

SPINAR, Z. V. 1972. Tertiary frogs from central Europe. The Hague, W. Junk N.V: 1-286, pl. 1-184.

SPIX, J. B. de. 1824. Animalia nova siva species novae Testudinum et Ranarum quas in itinere per Brasiliam annis MDCCCXVII-MDCCCXX jussu et auspiciis Maximiliana Josephi I Bavariae Regis suscepto collegit et descripsit Dr. J. B. de Spix. Munchen, Hübschmann: i-iii, 1-53, pl. 1-22.

STANNIUS, H. 1856. Zootomie der Amphibien. *In*: VON SIEBOLD, C. T., AND H. STANNIUS (ed.s.), Handbuch der Zootomie der Wirbelthiere. Volume 2. Berlin, Von Viet & Co.: i-vi + 1-270. [fide Frost (2009-13), not seen]

STAUB, N. L., AND D. B. WAKE. 2005. *Aneides vagrans*, Wandering Salamander. *In*: LANNOO, M. J. (ed.), Amphibian Declines: The Conservation Status of United States Species, Part 2: Species Accounts. Berkeley, University of California Press: 664-666.

STEBBINS, R. C. 1949. Speciation in salamanders of the plethodontid genus *Ensatina*. *University of California Publications in Zoology* 48: 377-526.

STEBBINS, R. C. 1951. The Amphibians of Western North America. Berkeley, University of California Press: i-ix, 1-539.

STEBBINS, R. C. 1985. A Field Guide to Western Reptiles and Amphibians, 2nd edition, revised. Boston, Houghton Mifflin: i-xvi, 1-336, 48 pl.

STEBBINS, R. C. 2003. A field guide to western reptiles and amphibians, third edition. Boston, Houghton Mifflin: i-xiii, 1-533, 56 pl.

STEBBINS, R. C., AND C. H. LOWE, JR. 1949. The systematic status of *Plethopsis* with a discussion of speciation in the genus *Batrachoseps*. *Copeia* 1949 (2): 116-129.

STEBBINS, R. C., AND C. H. LOWE, JR. 1951. Subspecific differentiation in the Olympic salamander *Rhyacotriton olympicus*. *University of California Publications in Zoology* 50 (4): 465-484.

STEBBINS, R. C., AND W. J. RIEMER. 1950. A new species of plethodontid salamander from the Jemez Mountains of New Mexico. *Copeia* 1950 (2): 73-80.

STEELE, C. A., B. C. CARSTENS, A. STORFER, AND J. SULLIVAN. 2005. Testing hypotheses of speciation timing in *Dicamptodon copei* and *Dicamptodon aterrimus* (Caudata: Dicamptodontidae). *Molecular Phylogenetics and Evolution* 36 (1): 90-100.

STEINDACHNER, F. 1864. Batrachologische Mittheilungen. *Verhandlungen des Zoologisch-Botanischen Vereins in Wien* 14: 239-288, pl. 9-17.

STEINDACHNER, F. 1867b. Reise der Oesterreichischen Fregatte Novara um die Erde om dem Kajrem 1857. 1858, 1859, unter den Befehlen des Commodore B. von Wüllersdorf-Urbair. Zoologischer Theil. Vol. 2 Amphibien. Wien, K.K. Hof-& Staatsdruckerei: 1-98.

STEJNEGER, L. 1890a. Annotated list of reptiles and batrachians collected by Dr. C. Hart Merriam and Vernon Bailey on the San Francisco Mountain Plateau and Desert of the Little Colorado, Arizona, with descriptions of new species. *North American Fauna* 3: 103-118.

STEJNEGER, L. 1893a. Preliminary description of a new genus and species of blind cave salamander from North America. *Proceedings of the U.S. National Museum* 15 (1892): 115-117, pl. 9.

STEJNEGER, L. 1893b. Annotated list of the reptiles and batrachians collected by the Death Valley expedition in 1891, with descriptions of new species. *North American Fauna* 7: 159-228.

STEJNEGER, L. 1895. Description of a new salamander from Arkansas with notes on *Ambystoma annulatum*.

Proceedings of the U.S. National Museum 17 (1894): 597-599.

STEJNEGER, L. 1896. Description of a new genus and species of blind tailed batrachian from the subterranean waters of Texas. *Proceedings of the U.S. National Museum* 18 (1895): 619-621.

STEJNEGER, L. 1899a. Description of a new genus and species of discoglossid toad from North America. *Proceedings of the U.S. National Museum* 21: 899-901, pl. 89.

STEJNEGER, L. 1902a. A new species of bullfrog from Florida and the Gulf Coast. *Proceedings of the U.S. National Museum* 24: 211-215.

STEJNEGER, L. 1902b. A salamander new to the District of Columbia. *Proceedings of the Biological Society of Washington* 15: 239-240.

STEJNEGER, L. 1904. The herpetology of Porto Rico. *Annual Report of the U.S. National Museum for 1902*: 549-724.

STEJNEGER, L. 1906. A new salamander from North Carolina. *Proceedings of the U.S. National Museum* 30: 559-562.

STEJNEGER, L. 1907. Herpetology of Japan and adjacent territory. *Bulletin of the U.S. National Museum* 58: i-xx, 1-577, pl. 1-35.

STEJNEGER, L. 1910. The amphibian generic name *Engystoma* untenable. *Proceedings of the Biological Society of Washington* 23: 165-167.

STEJNEGER, L. 1915a. A new species of tailless batrachian from North America. *Proceedings of the Biological Society of Washington* 28: 131-132.

STEJNEGER, L. 1916. New generic name for a tree-toad from New Guinea. and notes on amphisbaenian nomenclature. *Proceedings of the Biological Society of Washington* 29: 85.

STEJNEGER, L., AND T. BARBOUR. 1917. A check list of North American amphibians and reptiles. Cambridge, Harvard University Press: i-iv, 5-126.

STEJNEGER, L., AND T. BARBOUR. 1923. A check list of North American amphibians and reptiles. 2nd edition. Cambridge, Harvard University Press: i-x, 1-171.

STEJNEGER, L., AND T. BARBOUR. 1933. A check list of North American amphibians and reptiles. 3rd edition. Cambridge, Harvard University Press: i-xiv, 1-185.

STEJNEGER, L., AND T. BARBOUR. 1939. A check list of North American amphibians and reptiles. 4th edition. Cambridge, Harvard University Press: i-xvi, 1-207.

STEJNEGER, L., AND T. BARBOUR. 1943. A check list of North American amphibians and reptiles. 5th edition. *Bulletin of the Museum of Comparative Zoology* 93: i-xix, 1-260.

STEPHENSON, N. G. 1951. Observations on the development of the amphicoelous frogs, *Liopelma* and *Ascaphus*. *Journal of the Linnean Society of London (Zoology)* 42: 18-28.

STEVENSON, D., AND D. CROWE. 1992. *Pseudacris crucifer bartramiana* (Southern Spring Peeper). *Herpetological Review* 23 (3): 86.

STEWART, M. M. 1968. *Rana clamitans*. *Catalogue of American Amphibians and Reptiles*. Society for the Study of Amphibialns and Reptles (337): 1-4.

STEWART, M. M. 1984. Redescription of the type of *Rana clamitans*. *Copeia* 1984 (2): 210-213.

STEWART, M. M., AND M. J. LANNOO. 2005. *Eleutherodactylus coqui*, Coqui. *In*: LANNOO, M. J. (ed.), Declining Amphibians: The Conservation Status of United States Species, Part 2: Species Accounts. 492-494.

STORER, D. H. 1839. Reports of the fishes, reptiles and birds of Massachusetts. Boston, Dutton & Wentworth: i-xvi, 1-426, pl. 1-4.

STORER, D. H. 1840. A report on the reptiles of Massachusetts. *Boston Journal of Natural History* 3 (1): 1-64.

STORER, T. I. 1925. A synopsis of the Amphibia of California. *University of California Publications in Zoology* 27: 1-342, pl. 1-18.

STORER, T. I. 1929. Notes on the genus *Ensatina* in California, with description of a new species from the Sierra Nevada. *University of California Publications in Zoology* 30: 443-452.

STORM, R. M., AND E. D. BRODIE, JR. 1970a. *Plethodon dunni*. *Catalogue of American Amphibians and Reptiles*. American Society of Ichthyologists and Herpetologists (82): 1-2.

STORM, R. M., AND E. D. BRODIE, JR. 1970b. *Plethodon vehiculum*. *Catalogue of American Amphibians and Reptiles*. American Society of Ichthyologists and Herpetologists (83): 1-3.

STRAUCH, A. 1870. Revision der Salamandriden-Gattungen nebst Beschreibung einiger neuen oder weniger bekannten Arten dieser Familie. *Mémoires de l'Académie Imperiale des Sciences de St. Pétersbourg* (7) 16 (4): [i-ii], 1-110, pl. 1-2.

STRECKER, J. K., JR. 1908. The reptiles and batrachians of Victoria and Refugio Counties, Texas. *Proceedings of the Biological Society of Washington* 21: 47-52.

STRECKER, J. K., JR. 1909. Notes on the narrow-mouthed toads (*Engystoma*) and the description of a new species from southeastern Texas. *Proceedings of the Biological Society of Washington* 22: 115-120.

STRECKER, J. K., JR. 1910. Description of a new solitary spadefoot (*Scaphiopus hurterii*) from Texas, with other herpetological notes. *Proceedings of the Biological Society of Washington* 23: 115-122.

STRECKER, J. K. 1915. Reptiles and amphibians of Texas. *Baylor Bulletin* 18 (4): 1-82.

STRECKER, J. K., AND W. J. WILLIAMS. 1928. Field notes on the herpetology of Bowie County, Texas. *Contributions of the Baylor University Museum* (12): 3-19.

STUART, L. C. 1940. A new *Hypopachus* from Guatemala. *Proceedings of the Biological Society of Washington* 53: 19-21.

STUART, L. C. 1963. A checklist of the herpetofauna of Guatemala. *Miscellaneous Publications of the Museum of Zoology, University of Michigan* (122): 1-150.

STUART, S. N., M. HOFFMAN, J. S. CHANSON, N. A. COX, R. J. BERRIDGE, P. RAMANI, AND B. E. YOUNG (Eds.). 2008. Threatened Amphibians of the World. (IUCN, Gland, Switzerland, and Conservation International, Arlington, Virginia) Lynx Edicions, Barcelona: i-xv, 1-758.

SULLIVAN, B. K. 1986. Advertisement call variation in the Arizona Tree Frog, *Hyla wrightorum* Taylor, 1938. *Great Basin Naturalist* 46 (2): 378-381.

SULLIVAN, B. K., R. W. BOWKER, K. B. MALMOS, AND E. W. A. GERGUS. 1996a. Arizona distribution of three Sonoran Desert anurans: *Bufo retiformis, Gastrophryne olivacea*, and *Pternohyla fodiens*. Great Basin Naturalist 56 (1): 38-47.

SULLIVAN, B. K., K. B. MALMOS, AND M. F. GIVEN. 1996b. Systematics of the *Bufo woodhousii* complex (Anura:

Bufonidae): advertisement call variation. *Copeia* 1996 (2): 274-280.

Svihla, A. 1936. *Rana rugosa* (Schlegel): notes on the life history of this interesting frog. *Mid-Pacific Magazine* 49 124-125.

Swainson, W. 1839. On the Natural History and Classification of Fishes, Amphibians and Reptiles, or Monocardian Animals, Volume 2. London, Longman & Co.: i-vi, 1-452.

Sweet, S. S. 1978. On the status of *Eurycea pterophila* (Amphibia: Plethodontidae). *Herpetologica* 34 (2): 101-108.

Sweet, S. S. 1984. Secondary contact and hybridization in the Texas cave salamanders *Eurycea neotenes* and *E. tridentifera*. *Copeia* 1984 (4): 428-441.

Sweet, S. S., and B. K. Sullivan. 2005. *Bufo californicus*. Arroyo Toad. *In*: Lannoo, M. J. (ed.), Amphibian Declines: The Conservation Status of United States Species, Part 2: Species Accounts. Berkeley, University of California Press: 396-400.

Tanner, V. M. 1929. A distributional list of the amphibians and reptiles of Utah, no. 3. *Copeia* (171): 46-52.

Tanner, V. M. 1933. *Bufo lamentor* Girard. *Copeia* 1933: 42.

Tanner, V. M. 1939. A study of the genus *Scaphiopus*. *Great Basin Naturalist* 1 (1): 3-26.

Tanner, W. W. 1989a. Amphibians of western Chihuahua. *Great Basin Naturalist* 49 (1): 38-70.

Tanner, W. W. 1989b. Status of *Spea stagnalis* Cope (1875), *Spea intermontanus* Cope (1889), and a systematic review of *Spea hammondii* Baird (1839) (Amphibia: Anura). *Great Basin Naturalist* 49 (4): 503-510.

Taylor, D. R., and H. M. Smith. 1945. Summary of the collections of amphibians made in Mexico under the Walter Rathbone Bacon traveling scholarship. *Proceedings of the U.S. National Museum* 95 (3185): 521-647.

Taylor, E. H. 1932a. *Leptodactylus albilabris* (Günther): a species of toad new to the fauna of the United States. *University of Kansas Science Bulletin* 20 (11): 243-245.

Taylor, E. H. 1938b ("1936"). Notes on the herpetological fauna of the Mexican state of Sinaloa. *University of Kansas Science Bulletin* 24 (20): 505-537.

TAYLOR, E. H. 1939a ("1938"). Concerning Mexican salamanders. *University of Kansas Science Bulletin* 25 (14): 259-313.

TAYLOR, E. H. 1939b ("1938"). New species of Mexican tailless Amphibia. *University of Kansas Science Bulletin* 25 (17): 385-405.

TAYLOR, E. H. 1939c ("1938"). Frogs of the *Hyla eximia* group in Mexico, with descriptions of two new species. *University of Kansas Science Bulletin* 25 (19): 421-445.

TAYLOR, E. H. 1940a. Two new anuran amphibians from Mexico. *Proceedings of the U.S. National Museum* 89 (3093): 43-46, pl. 1-3.

TAYLOR, E. H. 1940b ("1939"). Herpetological miscellany No. 1. *University of Kansas Science Bulletin* 26 (15): 489-571.

TAYLOR, E. H. 1941a. A new plethodont salamander from New Mexico. *Proceedings of the Biological Society of Washington* 54: 77-79.

TAYLOR, E. H. 1941b. Extinct toads and salamanders from Middle Pliocene beds of Wallace and Sherman counties, Kansas. *Bulletin of the State Geological Survey of Kansas. Reports of Studies* 38 (6): 177-196.

TAYLOR, E. H. 1942a. New Caudata and Salientia from Mexico. *University of Kansas Science Bulletin* 28 (14): 295-323.

TAYLOR, E. H. 1942b. Extinct toads and frogs from the upper Pliocene deposits of Meade County, Kansas. *University of Kansas Science Bulletin* 28 (10): 199-235, pl. 14-20.

TAYLOR, E. H. 1943a. Herpetological novelties from Mexico. *University of Kansas Science Bulletin* 29 (Part 2) (8): 343-361.

TAYLOR, E. H. 1952a. Third contribution to the herpetology of the Mexican state of San Luis Potosi. *University of Kansas Science Bulletin* 34 (pt.2) (13): 793-815.

TAYLOR, E. H. 1952b. A review of the frogs and toads of Costa Rica. *University of Kansas Science Bulletin* 35 (5): 577-942.

TAYLOR, E. H. 1954. Additions to the known herpetological fauna of Costa Rica with comments on other species. No. I. *University of Kansas Science Bulletin* 36 (9): 597-639.

TAYLOR, D. R., AND H. M. SMITH. 1945. Summary of the collections of amphibians made in Mexico under the

Walter Rathbone Bacon traveling scholarship. *Proceedings of the U.S. National Museum* 95: 521-647.

TEMMINCK, C. J., AND H. SCHLEGEL. 1838. Les batraciens. *In*: DE SIEBOLD, P. F., AND W. DE HAHN (eds.), Fauna Japonica sive descriptio animalium, quae in itinere per Japoniam jussu et auspiciis superiorum, qui summum in India Batava Imperium tenent, suscepto, annis 1823-1830 colleget, notis observationibus et adumbrationibus illustratis. 6 volumes, 1833-1850. Volume 3 (Chelonia, Ophidia, Sauria Batrachia. Part III Saurii et Batrachii): 67-124. Leiden, J.G. Laiau.

TENNESSEN, J. A., AND M. S. BLOUIN. 2010. A revised leopard frog phylogeny allows a more detailed examination of adaptive evolution at ranatuerin-2 antimicrobial peptide loci. *Immunogenetics* 62: 333-343.

TEST, F. C. 1893. Annotated list of reptiles and batrachians collected. *Bulletin of the U.S. Fish Commission* 11: 57-59. [*fide* Frost (2009-13); not seen]

TEST, F. C. 1899. A contribution to the knowledge of the variations of the tree frog *Hyla regilla*. *Proceedings of the U.S. National Museum* 21: 477-492, pl. 39.

THIREAU, M. 1986. Catalogue des types des Urodeles du Muséum Natiional d'Histoire Naturelle, Paris. Paris, Muséum National d'Histoire Naturelle: 1-96.

THOMAS, R. 1966 ("1965"). New species of Antillean *Eleutherodactylus*. *Quarterly Journal of the Florida Academy of Sciences* 28 (4): 375-391.

THOMPSON, H. B. 1914. Description of a new subspecies of *Rana pretiosa* from Nevada. *Proceedings of the Biological Society of Washington* 26 (1913): 53-55, pl. 3.

THUROW, G. R. 1956a. Comparisons of two species of salamanders, *Plethodon cinereus* and *Plethodon dorsalis*. *Herpetologica* 12 (2): 177-182.

THUROW, G. R. 1956b. A new subspecies of *Plethodon welleri*, with notes on other members of the genus. *American Midland Naturalist* 55 (2): 343-346.

THUROW, G. R. 1957. A new *Plethodon* from Virginia. *Herpetologica* 13 (1): 59-66.

THUROW, G. R. 1964. *Plethodon welleri*. *Catalogue of American Amphibians and Reptiles*. American Society of Ichthyologists and Herpetologists (12): 1-2.

THUROW, G. R. 1966. *Plethodon dorsalis. Catalogue of American Amphibians and Reptiles.* American Society of Ichthyologists and Herpetologists (29): 1-3.

THUROW, G. R. 1968. On the small black *Plethodon* problem. *Western Illinois University, Series in Biological Science* (6): 1-48.

TIHEN, J. A. 1958. Comments on the osteology and phylogeny of ambystomatid salamanders. *Bulletin of the Florida State Museum* 3 (1): 1-50.

TIHEN, J. A. 1960a. On *Neoscaphiopus* and other Pliocene pelobatid frogs. *Copeia* 1960 (2): 89-94.

TIHEN, J. A. 1960b. Two new genera of African bufonids, with remarks on the phylogeny of related genera. *Copeia* 1960 (3): 225-233.

TIHEN, J. A. 1969. *Ambystoma. Catalogue of American Amphibians and Reptiles.* American Society of Ichthyologists and Herpetologists: 1-4.

TIHEN, J. A. 1974. Two new North American Miocene salamandrids. *Journal of Herpetology* 8: 211-218.

TILLEY, S. G. 1973. *Desmognathus ochrophaeus. Catalogue of American Amphibians and Reptiles.* Society for the Study of Amphibians and Reptiles (121): 1-4.

TILLEY, S. G. 1981. A new species of *Desmognathus* (Amphibia: Caudata: Plethodontidae) from the southern Appalachian Mountains. *Occasional Papers of the Museum of Zoology University of Michigan* (695): 1-23.

TILLEY, S. G. 1985. *Desmognathus imitator. Catalogue of American Amphibians and Reptiles.* Society for the Study of Amphibians and Reptiles (359): 1-2.

TILLEY, S. G. 2000. *Desmognathus santeetlah. Catalogue of American Amphibians and Reptiles.* Society for the Study of Amphibians and Reptiles (703): 1-3.

TILLEY, S. G. 2010. *Desmognathus abditus. Catalogue of American Amphibians and Reptiles.* Society for the study of Amphibians and Reptiles (861): 1-??

TILLEY, S. G., R. L. ERIKSEN, AND L. A. KATZ. 2008. Systematics of dusky salamanders, *Desmognathus* (Caudata: Plethodontidae), in the mountain and Piedmont regions of Virginia and North Carolina, USA. *Zoological Journal of the Linnean Society* 152: 115-130.

TILLEY, S. G., R. HIGHTON, AND D. B. WAKE. 2012. Caudata: Salamanders. *In*: CROTHER, B. I. (ed.), Scientific and standard English names of amphibians and reptiles of North America north of Mexico, with comments regarding confidence in our understanding. Seventh edition. Society for the Study of Amphibians and Reptiles *Herpetological Circular* 39: 23-31.

TILLEY, S. G., AND M. J. MAHONEY. 1996. Patterns of genetic differentiation in salamanders of the *Desmognathus ochrophaeus* complex (Amphibia: Plethodontidae). *Herpetological Monographs* 10: 1-42.

TILLEY, S. G., R. B. MERRITT, B. WU, AND R. HIGHTON. 1978. Genetic differentiation in salamanders of the *Desmognathus ochrophaeus* complex (Plethodontidae). *Evolution* 32 (1): 93-115.

TIMPE, E. K., S. P. GRAHAM, AND R. M. BONETT. 2009. Phylogeography of the Brownback Salamander reveals patterns of local endemism in southern Appalachian springs. *Molecular Phylogenetics and Evolution* 52: 368-376.

TITUS, T. A. 1990. Genetic variation in two subspecies of *Ambystoma gracile* (Caudata: Ambystomatidae). *Journal of Herpetology* 24: 107-111.

TITUS, T. A., AND A. LARSON. 1996. Molecular phylogenetics of desmognathine salamanders (Caudata: Plethodontidae): a reevaluation of evolution in ecology, life history, and morphology. *Systematic Biology* 45 (4): 451-472.

TRAPIDO, H., AND R. T. CLAUSEN. 1938. Amphibians and reptiles of eastern Quebec. *Copeia* 1938 (3): 117-125.

TRAUTH, S. E. 2005a. *Ambystoma annulatum*, Ringed Salamander. *In*: LANNOO, M. J. (ed.), Amphibian Declines: The Conservation Status of United States Species, Part 2: Species Accounts. Berkeley, University of California Press: 602-603.

TRAUTH, S. E. 2005b. *Ambystoma talpoideum*, Mole Salamander. *In*: LANNOO, M. J. (ed.), Amphibian Declines: The Conservation Status of United States Species, Part 2: Species Accounts. Berkeley, University of California Press: 632-634.

TROOST, G. 1840. Geological report to the twenty-first twenty-seventh General Assembly of the state of

Tennessee. S. Nye & Co., Nashville. [*fide* Frost, 2009-2013, not seen]

TRUEB, L. 1969. *Pternohyla, P. dentata, P. fodiens. Catalogue of American Amphibians and Reptiles.* American Society of Ichthyologists and Herpetologists (77): 1-4.

TRUEB, L. 1971. Phylogenetic relationships of certain neotropical toads with the description of a new genus (Anura: Bufonidae). *Natural History Museum of Los Angeles County, Contributions to Science* 218: 1-40.

TRUEB, L., AND M. J. TYLER. 1974. Systematics and evoluton of the greater Antillean hylid frogs. *Occasional Papers of the Museum of Natural History, University of Kansas* 24: 1-60.

TSCHUDI, J. J. von. 1838. Classification der Batrachier mit Berücksichtigung der fossilen Thiere dieser Abtheilung der Reptilien. Neuchâtel, Petitpierre: i-ii, 1-98, pl. 1-6.

TSCHUDI, J. J. D. 1845. Reptilium conspectus quae in Republica Peruana reperiuntur et pleraquae observata vel collecta sunt in itinere a Dr. J.J. de Tschudi. *Archiv für Naturgeschichte* 11 (1): 150-170.

TURBOTT, E. G. 1942. The distribution of the genus *Leiopelma* in New Zealand with a description of a new species. *Transactions and Proceedings of the Royal Society of New Zealand* 71: 247-253.

TURNER, F. B., AND P. C. DUMAS. 1972. *Rana pretiosa. Catalogue of American Amphibians and Reptiles.* Society for the Study of Amphibians and Reptiles (119): 1-4.

TWITTY, V. C. 1935. Two new species of *Triturus* from California. *Copeia* 1935 (3): 211-218.

TWITTY, V. C. 1942. The species of California *Triturus*. *Copeia* 1942 (2): 65-76.

TWITTY, V. C. 1964b. *Taricha rivularis. Catalogue of American Amphibians and Reptiles.* American Society of Ichthyologists and Herpetologists (9): 1-2.

TYLER, M. J. 1971. The phylogenetic significance of vocal sac structure in hylid frogs. *University of Kansas Publications, Museum of Natural History* 19 (4): 321-360.

TYLER, M. J. 1974. The systematic position and geographic distribution of the Australian frog *Chiroleptes alboguttatus* Günther. *Proceedings of the Royal Society of Queensland* 85: 27-32.

TYMOWSKA, J., AND M. FISCHBERG. 1973. Chromosome complements of the genus *Xenopus*. *Chromosoma* 44 (3): 335-342.

UETZ, P., J. GOLL, AND J. HALLERMANN. 1995-. The TIGR Reptile Database (accessed during 2009-2013). [Database accessible at http://www.reptile-database.org/ (Also available on CD)]

UZZELL, T. M., JR. 1964. Relations of the diploid and triploid species of the *Ambystoma jeffersonianum* complex (Amphibia, Caudata). *Copeia* 1964 (2): 257-300.

UZZELL, T. 1967a. *Ambystoma jeffersonianum*. *Catalogue of American Amphibians and Reptiles*. American Society of Ichthyologists and Herpetologists (47): 1-2.

UZZELL, T. 1967b. *Ambystoma laterale*. *Catalogue of American Amphibians and Reptiles*. American Society of Ichthyologists and Herpetologists (48): 1-2.

UZZELL, T. 1967c. *Ambystoma platineum*. *Catalogue of American Amphibians and Reptiles*. American Society of Ichthyologists and Herpetologists (49): 1-2.

UZZELL, T. 1967d. *Ambystoma tremblayi*. *Catalogue of American Amphibians and Reptiles*. American Society of Ichthyologists and Herpetologists (50): 1-2.

UZZELL, T. 1982. Comments on the proposed conservation of *Rana sphenocephala* Cope, 1886. Z.N.(S.) 2141 (3.). *Bulletin of Zoological Nomenclature* 39 (2): 83.

VALENTINE, B. D. 1961. Variation and distribution of *Desmognathus ocoee* Nicholls (Amphibia: Plethodontidae). *Copeia* 1961: 315-322.

VALENTINE, B. D. 1963. The salamander genus *Desmognathus* in Mississippi. *Copeia* 1963 (1): 130-139.

VALENTINE, B. D. 1964. *Desmognathus ocoee*. *Catalogue of American Amphibians and Reptiles*. American Society of Ichthyologists and Herpetologists (7): 1-2.

VALENTINE, B. D. 1974. *Desmognathus quadramaculatus*. *Catalogue of American Amphibians and Reptiles*. Society for the Study of Amphibians and Reptiles (153): 1-4.

Van Bocxlaer, I., S. P.Loader, K. Roelants, S. D. Bijou, M. Menegon, and F. Bossuyt. 2010. Gradual adaptation toward a range-expansion phenotype initiated the global radiation of toads. *Science* (327): 679-682.

VAN DENBURGH, J. 1905. The reptiles and amphibians of the islands of the Pacific coast of North America, from the Farallons to Cape San Lucas and the Revilla Gigedos. *Proceedings of the California Academy of Sciences, Ser. 3, Zoology* 4 (1): 1-41, pl 1-8.

VAN DENBURGH, J. 1906. Description of a new species of the genus *Plethodon* (*Plethodon vandykei*) from Mount Rainier, Washington. *Proceedings of the California Academy of Sciences* (3) 4 (4): 61-63.

VAN DENBURGH, J. 1916. Four species of salamanders new to the state of California, with a description of *Plethodon elongatus*, a new species, and notes on other salamanders. *Proceedings of the California Academy of Sciences* (4) 6 (7): 215-221.

VAN DER HOEVEN, J. 1833. Handboek der Dierkunde, of grondbeginsels der natuurlijke geschiedenis van het dierenrijk. Tweeden deels. Amsterdam, Sulpke: [i-v], i-ii, i-x, i-v, 1-698.

VAN DER HOEVEN, J. 1838 ("1837"). Iets over den grooten zoogenoemden den Salamander von Japan. *Tidjschrift Voor Natuurlijke Geschiedenis en Physiologie* (Leiden) 4: 375-386, + 2 pl.

VAN DER MEIJDEN, A., M. VENCES, S. HOEGG, R. BOISTEL, A. CHANNING, AND A. MEYER. 2007. Nuclear gene phylogeny of narrow-mouthed toads (Family: Microhylidae) and a discussion of competing hypotheses concerning their biogeographical origins. *Molecular Phylogenetics and Evolution* 44 (3): 1017-1030.

VAN DIJK, D. E. 1966. Systematic and field keys to the families, genera and described species of southern African anuran tadpoles. *Annals of the Natal Museum* 18 (2): 231-286.

VERRILL, A. E. 1865a. Catalogue of the reptiles and batrachians found in the vicinity of Norway, Oxford Co., Maine. *Proceedings of the Boston Society of Natural History* 9 (1862-1863): 195-199.

VERRILL, A. E. 1865b. Notice of the eggs and young of a salamander, Desmognathus fusca Baird, from Maine. *Proceedings of the Boston Society of Natural History* 9 (1962-1963): 253-255.

VIEITES, D. R., M.-S. MIN, AND D. B. WAKE. 2007. Rapid diversification and dispersal during periods of global warming by plethodontid salamanders. *Proceedings of the National Academy of Science (US)* 104 (50): 19903-19907.

VIEITES, D. R., N. ROMÁN, M. H. WAKE, AND D. B. WAKE. 2011. A multigenic perspective on phylogenetic relationships in the largest family of salamanders, the Plethodontidae. *Molecular Phylogenetics and Evolution* 59: 623-635.

VIEITES, D. R., P. ZHANG, AND D. B. WAKE. 2009. Salamanders (Caudata). *In*: HEDGES, S. B., AND S. KUMAR (eds.), The Timetree of Life. 365-368.

VINCENT, W. S. 1947. A check list of amphibians and reptiles of Crater Lake National Park. *Nature Notes from Crater Lake* 13: 20-22.

Viosca, P., Jr. 1923. Notes on the status of *Hyla phaeocrypta* Cope. *Copeia* (122): 96-99.

VIOSCA, P., JR. 1928. A new species of *Hyla* from Louisiana. *Proceedings of the Biological Society of Washington* 41: 89-91, pl. 11-12.

VIOSCA, P., JR. 1937. A tentative revision of the genus *Necturus*, with descriptions of three new species from the southern Gulf drainage area. *Copeia* 1937 (2): 120-138.

VIOSCA, P., JR. 1938. A new water dog from central Louisiana. *Proceedings of the Biological Society of Washington* 51: 143-145, pl. 1-2.

VIOSCA, P., JR. 1949. Amphibians and Reptiles of Louisiana. *Popular Science Bulletin, Louisiana Academy of Science* (1): 1-12.

VITT, L. J., AND J. P. CALDWELL. 2009. Herpetology, 3rd edition. Burlington, San Diego, London, Academic Press: i-xiv, 1-697.

VOSS, S. R., D. G. SMITH, C. K. BEACHY, AND D. G. HECKEL. 1995. Allozyme variation in neighboring isolated populations of the plethodontid salamander *Leurognathus marmoratus*. *Journal of Herpetology* 29: 493-497.

VREDENBURG, V. T., R. BINGHAM, R. KNAPP, J. A. T. MORGAN, C. MORITZ, AND D. WAKE. 2007. Concordant molecular and phenotypic data delineate new taxonomy and conservation priorities for the endangered mountain yellow-legged frog. *Journal of Zoology* (London) 271 (4): 361-374.

VREDENBURG, V. T., G. M. FELLERS, AND C. DAVIDSON. 2005. *Rana muscosa*. Mountain Yellow-legged Frog. In: LANNOO, M. J. (ed.), Amphibian Declines: The Conservation

Status of United States Species, Part 2: Species Accounts. Berkeley, University of California Press: 563-566.

WAGLER, [J. G.]. 1827. H. Boie an Wagler. Buitenzorg aus Java ben 25. Aug. 1825. *Isis von Oken* 20: 724-726.

WAGLER, [J. G.]. 1828. Systema Amphibiorum. *Isis von Oken* 21 (8): 859-861.

WAGLER, J. 1830. Natürliches System der Amphibien, mit vorangehender Classificaiton der Säugethiere und Vögel. Ein Beitrag zur vergleichenden Zoologie. München, Stuttgart & Tübingen, J.G. Cotta: i-vi, 1-354.

WAGLER, J. 1933a. Descriptiones et icones amphibiorum. Part 2. Munich, J.G. Cotta: 36 pl.

WAHLERT, G. V. 1952. On the systematic position of the salamandrid genus *Taricha* and its species. *Copeia* 1952 (2): 29-30.

WAKE, D. 1961. The distribution of the Sinaloa narrow-mouthed toad, *Gastrophryne mazatlanensis* (Taylor). *Bulletin of the Southern California Academy of Science* 60 (2): 88-92.

Wake, D. 1963. Comparative osteology of the plethodontid salamander genus *Aneides*. *Journal of Morphology* 113: 77-118.

WAKE, D. B. 1965a. *Aneides ferreus. Catalogue of American Amphibians and Reptiles.* American Society of Ichthyologists and Herpetologists (16): 1-2.

WAKE, D. B. 1965b. *Aneides hardii. Catalogue of American Amphibians and Reptiles.* American Society of Ichthyologists and Herpetologists (17): 1-2.

WAKE, D. B. 1966. Comparative osteology and evolution of the lungless salamanders, family Plethodontidae. *Memoirs of the Southern California Academy of Sciences* 4: 1-111.

WAKE, D. B. 1974. *Aneides. Catalogue of American Amphibians and Reptiles. Society for the Study of Amphibians and Reptiles* (157): 1-2.

WAKE, D. B. 1993. Phylogenetic and taxonomic issues relating to salamanders of the family Plethodontidae. *Herpetologica* 49 (2): 229-237.

WAKE, D. B. 1996. A new species of *Batrachoseps* (Amphibia: Plethodontidae) from the San Gabriel Mountains, Southern California. *Contributions in Science, Natural History Museum, Los Angeles County* (463): 1-12.

WAKE, D. B 2012. Taxonomy of salamanders of the family Plethodontidae (Amphibia, Caudata). *Zootaxa* (3484): 75-82.

WAKE, D. B., AND T. R. JACKMAN. 1999 ("1998"). Description of a new species of plethodontid salamander from California [appendix to Jackman: Molecular and historical evidence for the introduction of clouded salamnaders (genus *Aneides*) to Vancouver Island, British Columbia, Canada, from California: 1570-1580]. *Canadian Journal of Zoology* 76: 1579-1580.

WAKE, D. B., AND E. L. JOCKUSCH. 2000. Detecting species borders using diverse data sets: plethodontid salamanders in California. *In*: BRUCE, R. C., R. G. JAEGER, AND L. D. HOUCK (ed.s.), The biology of plethodontid salamanders. New York, Kluwer Academic/Plenum Publishers: 95-119.

WAKE, D., E. L. JOCKUSCH, AND T. J. PAPENFUSS. 1998. Does *Batrachoseps* occur in Alaska? *Herpetological Review* 29 (1): 12-14.

WAKE, D. B., A. SALVADOR, AND M. A. ALONZO-ZARAZAGA. 2005. Taxonomy of the plethodontid salamander genus *Hydromantes* (Caudata: Plethodontidae). *Amphibia-Reptilia* 26: 543-548.

WAKE, D. B., K. P. YANEV, AND M. M. FRELOW. 1989. Sympatry and hybridization in a 'ring species': the plethodontid salamander *Ensatina eschscholtzii. In:* OTTE, D., AND J. A. ENDLER (eds.), Speciation and its consequences. Sunderland, MA, Sinauer: 134-157.

WAKE, D. B., K. P. YANEV, AND R. W. HANSEN. 2002. New species of slender salamander, genus *Batrachoseps*, from the southern Sierra Nevada of California. *Copeia* 2002: 1029-1036.

WALBAUM, J. J. 1784. Beschreibung eies Meerfrosches. *Schriften der Gesellschaft Naturforschender Freunde zu Berlln* 5: 239. [*fide* Frost (2009-13); not seen]

WALKER, C. F. 1931. Description of a new salamander from North Carolina. *Proceedings of the Junior Society of Natural Science, Cincinnati* 2: 48-51.

WALKER, C. F. 1932. *Pseudacris brachyphona* (Cope) a valid species. *Ohio Journal of Science* 32: 379-384.

WALKER, C. F. 1938. The structure and systematic relationships of the genus *Rhinophrynus*. *Occasional Papers of the Museum of Zoology University of Michigan* (372): 1-11.
WALKER, C. F., AND W. H. WELLER. 1932. The identity and status of *Pseudotriton duryi*. *Copeia* 1932 (2): 81-83.
WALLACE, J. E. 2005. *Eleutherodactyus (= Syrrhophus) cystignathoides*, Rio Grande Chirping Frog. In: LANNOO, M. J. (ed.), Amphibian Declines: The Conservation Status of United States Species, Part 2: Species Accounts. Berkeley, University of California Press: 494-495.
WASSERMAN, A. O. 1968. *Scaphiopus holbrookii*. *Catalogue of American Amphibians and Reptiles*. American Society of Ichthyologists and Herpetologists (70): 1-4.
WASSERMAN, A. O. 1970. *Scaphiopus couchii*. *Catalogue of American Amphibians and Reptiles*. American Society of Ichthyologists and Herpetologists (85): 1-4.
WASSERSUG, R. 1982. Comments on the proposed conservation of *Rana sphenocephala* Cope, 1886. *Bulletin of Zoological Nomenclature* 39 (2): 81, 83.
WASSERSUG, R. J., AND W. P. PYBURN. 1987. The biology of the Pe-ret toad, *Otophryne robusta* (Microhylidae), with special consideration of its fossorial larva and systematic relationships. *Zoological Journal of the Linnean Society* 91: 137-169.
WATSON, M. B., AND T. E. PAULEY. 2005. *Ambystoma barbouri*, Streamside Salamander. In: LANNOO, M. J. (ed.), Amphibian Declines: The Conservation Status of United States Species, Part 2: Species Accounts. Berkeley, University of California Press: 603-605.
WEED, A. C. 1922. New frogs from Minnesota. *Proceedings of the Biological Society of Washington* 35: 107-110.
WELLBORN, V. 1936. Beschreibung eines neuen Molches der Gattung *Cryptobranchus*. *Zoologischer Anzeiger* 114 (3-4): 63-64.
WELLER, W. F., AND D. M. GREEN. 1997. Checklist and current status of Canadian amphibians. *In*: Amphibians in decline: Canadian studies of a global problem. *Edited by* GREEN, D. M. *Herpetological Conservation* No. 1. Society for the Study of Amphibians and Reptiles, St. Louis: 309-328.
WELLER, W. H. 1930. Notes on amphibians collected in Carter County, Kentucky. *Proceedings of the Junior Society of*

Natural Science, Cincinnati 1 (5-6): 4 pp (unnumbered, 6-9).

WELLER, W. H. 1931a. Corrections to herpetological notices. *Proceedings of the Junior Society of Natural Science, Cincinnati* 2 (1): 7-9.

WELLS, M. R. 1964. A study of the paper electrophoretic serum protein patterns of the subspecies of *Rana pipiens*. *Journal of the Tennessee Academy of Science* 39: 50-63.

WELLS, R. W., AND C. R. WELLINGTON. 1985. A classification of the Amphibia and Reptilia of Australia. *Australian Journal of Herpetology Supplement Series* 1: 1-61.

WELTER-SCHULTES, F. W., R. KLUG, AND A. LUTZE. 2008. Les figures des plantes et animaux d'usage en médecine, a rare work published by F.A.P. de Garsault in 1764. *Archives of Natural History* 35 (1): 118-127.

WERNER, F. 1893. Herpetologische Nova. *Zoologischer Anzeiger* 16 (414): 81-84.

WERNER, F. 1896b. Beiträge zur Kenntniss der Reptilien und Batrachier von Centralamerika und Chile, sowie einiger seltenerer Schlangenarten. *Verhandlungen des Zoolgisch-Botanischen Veriens in Wien* 46: 344-365. [*fide* Frost (2009-13); not seen]

WERNER, F. 1897. Über einige noch unbeschriebene Reptilien und Batrachier. *Zoologischer Anzeiger* 20 (537): 261-267.

WERNER, F. 1899a. Beschreibung einiger neuer Schlangen und Batrachier. *Zoologischer Anzeiger* 22: 114-117.

WERNER, F. 1899b. Ueber Reptilien und Batrachier aus Columbien und Trinidad. *Verhandlungen der K. K. zoologisch-botanischen Gesellschaft in Wien* 49: 470-484.

WERNER, F. 1909. Amphibien und Reptilien I. Stuttgart, Strecker & Schröder: i-viii, 1-104.

WERNER, F. 1917. Über einige neue Reptilien und einen neuen Frosch des Zoologischen Museums in Hamburg. *Mitteilungen aus dem Zoologie Museum in Hamburg* 34: 31-36.

WHITE, J. 1790. Journal of a voyage to New South Wales with sixty-five plates of non-descript animals, birds, lizards, serpents, curious cones of trees and other natural productions. London, J. Debrett. [*fide* Frost, 2009-13; not seen]

WIED-NEUWIED, M. A. P. 1838. Reise in das Innere Nord-Amerika in den Jaren 1832 bis 1834. Vol. 1. Coblenz, J. Hoelscher: i-xvi, 1-653 + [1], pl. 1-20. [*fide* Frost, 2009-13; not seen]

WIED [-NEUWIED], [A. P.] M. Z. 1865. Verzeichniss der Reptilien, welche auf einer Reise im nördlichen America beobachtet wurden. *Nova Acta Academiae Caesareae Leopoldino-Carolinae Germanicae Naturae Curiosorum* (Halle) 32: i-viii, 1-141, [i-ii], pl. 1-7.

WIEDERSHEIM, R. 1877. Das Kopfskelet der Urodelen (Fortsetzung). Morphologisches Jahrbuch 3 (4): 459-548 + pl. 24-27.

WIEGMANN, A. F. A. 1832. Klasse Amphibien. Amphibia. *In*: WIEGMANN, A. F. A., AND J. F. RUTHE (eds.), Handbuch der Zoologie. Berlin, C. G. Lüderitz: 160-205.

WIEGMANN, A. F. A. 1833. Herpetologischen Beyträge. I. Ueber die mexicanischen Kröten nebst bemerkungen üuber ihren verwandte Arten anderer Weltgegenden. *Isis von Oken* 26: 651-662.

WIEGMANN, A. F. A. 1834b. Amphibien, *IN*: Meyen, F.J.F. (ed.), Reise um die Erde ausgeführt auf dem Königlisch Preussischen Sechandlungs-Schiffe Prinzess Louise, comandieri von Capitan W. Wendt, in den Jahren 1830, 1831 und 1832, von Dr. F.J.F. Meyen. Vol. 3. Zoologisher Bericht. Berlin, Sanderschen Buchhandlung: 433-522.

WIENS, J. J., R. M. BONETT, AND P. T. CHIPPINDALE. 2005a. Ontogeny discombobulates phylogeny: paedomorphosis and higher-level salamander relationships. *Systematic Biology* 54: 91-110.

WIENS, J. J., T. N. ENGSTROM, AND P. T. CHIPPINDALE. 2006. Rapid diversification, incomplete isolation, and the "speciation clock" in North American salamanders (genus Plethodon): testing the hybrid swarm hypothesis of rapid radiation. *Evolution* 60 (12): 2585-2603.

WIENS, J. J., AND T. A. TITUS. 1991. A phylogenetic analysis of *Spea* (Anura: Pelobatidae). *Herpetologica* 47 (1): 21-28.

WILCZYNSKI, W., A. S. RAND, AND M. J. RYAN. 2002 ("2001"). Evolution of calls and auditory tuning in the *Physalaemus pustulosus* species group. *Brain, Behavior and Evolution* 58: 137-151.

WILD, E. R. 1995. New genus and species of Amazonian microhylid frog with a phylogenetic analysis of New World genera. *Copeia* 1995 (4): 837-849.

WILDER, H. H. 1894 ("1893"). Lungenlose Salamandriden. *Anatomischer Anzeiger* 9: 216-220.

WILLIAMS, S. R. 1973. *Plethodon neomexicanus. Catalogue of American Amphibians and Reptiles.* Society for the Study of Amphibians and Reptiles (131): 1-2.

WILSON, A. G., AND P. OHANJANIAN. 2002. *Plethodon idahoensis. Catalogue of American Amphibians and Reptiles.* Society for the Study of Amphibians and Reptiles (741): 1-4.

WITSCHI, E. 1954 ("1953"). The Cherokee frog, *Rana sylvatica cherokiana* nom. nov., of the Appalachian Mountain region. *Proceedings of the Iowa Academy of Science* 60: 764-769.

WOLTERSTORFF, W. 1914. Ueber *Diemyctylus viridescens* Raf. subsp. *louisianenesis* n. subsp. *Abhandlungen und Berichte aus dem Museum für Natur-und Hiematkunde zu Magdeburg* 2 (4): 383-392, pl. 8.

WOLTERSTORFF, W. 1925. Katalog der Amphibien-Sammlung im Museum fur Natur-und Heimatkunde zu Magdeburg. *Abhandlungen und Berichte aus dem Museum für Natur-und Hiematkunde zu Magdeburg* 4 (2): 231-310.

WOLTERSTORFF, W. 1930. Beitrage zur Herpetologie Mexikos. II. Zur Systematik un Biologie der Urodelen Mexikos. *Abhandlungen und Berichte aus dem Museum für Natur-und Hiematkunde zu Magdeburg* 6 (2): 129-149.

WOLTERSTORFF, W. 1935. Über eine eigentumliche Form des kalifonischen Wassermolches, *Taricha torosa* (Rathke). *Blätter für Aquarien-und Terrarienkunde* (Stuttgart) 46 (8): 176-184.

WOOD, J. G. 1863. Illustrated Natural History: Animals and Birds. Volume 3, Reptiles, Fishes, Molluscs, etc. New York, Routledge & Sons: 1-146.

WOOD, W. F. 1940. A new race of salamander, *Ensantina eschscholtzii picta,* from northern California and southern Oregon. *University of California Publications in Zoology* 42 (10): 425-428.

Wood, W. W. 1825. Description of a new species of salamander. *Journal of the Academy of Natural Sciences of Philadelphia* 4: 306-307.

Woodman, N. 2010. History and dating of the publication of the Philadelphia (1822) and London (1823) editions of Edwin James's account of an expedition from Pittsburgh to the Rocky Mountains. *Archives of Natural History* 37 (1): 28-38.

Worthington, R. D., and P. H. Worthington. 1976. John Edwards Holbrook, father of American herpetology. In: Adler, K. (ed.), Holbrook's North American Herpetology. Ithaca, xiii-xxvii.

Wright, A. H. 1924. A new bullfrog (*Rana heckscheri*) from Georgia and Florida. *Proceedings of the Biological Society of Washington* 37: 141-152, pl. 11-12.

Wright, A. H., and A. A. Wright. 1932. Life-histories of frogs of Okefinokee Swamp, Georgia, North American Salientia (Anura) No. 2. New York, Macmillan Co.: i-xv, 1-497.

Wright, A. H., and A. A. Wright. 1933. Handbook of frogs and toads of the United States and Canada (1st edition). Ithaca, Comstock: i-xi, 1-231.

Wright, A. H., and A. A. Wright. 1938. Amphibians of Texas. *Transactions of the Texas Academy of Science* 21: 5-35, pl. 1-3, [i-ii].

Wright, A. H., and A. A. Wright. 1942. Handbook of frogs and toads of the United Staes and Canada. Second edition. Ithaca, NY, Comstock.

Wright, A. H., and A. A. Wright. 1949. Handbook of frogs and toads of the United States and Canada. Third Editon. Ithaca, NY, Comstock.

Wynn, A. H., R. Highton, and J. F. Jacobs. 1988. A new species of rock-crevice dwelling *Plethodon* from Pigeon Mountain, Georgia. *Herpetologica* 44: 133-143.

Yanev, K. P. 1980. Biogeography and distribution of three parapatric salamander species in coastal and borderland California. In: Power, D. M. (ed.), The California Islands: Proceedings of a Multidisciplinary Symposium. Santa Barbara, CA, Santa Barbara Museum of Natural History: 531-550.

Yang, D., W. Liu, and D. Rao. 1996. A new toad genus of Bufonidae—*Torrentophryne* from Transhimalaya mountain of Yunnan of China with its biology. *Zoological Research*. Kunming Institute of Zoology, Kunming, Yunnan 17 (4): 353-359.

Yang, D.-T. (ed.). 1991. Amphibian-fauna of Yunnan [in Chinese]. Bejiing, China Forestry Publishing House: i-iv, 1-260. [*fide* Frost (2009-13); not seen]

YARROW, H. C. 1875. Report upon the collections of Batrachians and Reptiles made in portions of Nevada, Utah, California, Colorado, New Mexico, and Arizona during the years 1871, 1872, 1873, and 1874 (Chapter 4). *In*: WHEELER, G. M. (ed.), Report upon geographical and geological explorations and surveys west of the one hundredth meridian, in charge of Lieut. Geo. M. Wheeler, under the direction of Brig. Gen. A.A. Humphreys, chief of engineers, U.S. Army. Vol. 5—Zoology. Washington, Government Printing Office: 509-584.

YARROW, H. C. 1883a ("1882"). Check List of North American Reptilia and Batrachia, with Catalogue of Specimens in U. S. National Museum. *Bulletin of the U.S. National Museum* (24): i-vi 1-249.

YARROW, H. C. 1883b. Descriptions of new species of reptles and amphibians in the United States National Museum. *Proceedings of the U.S. National Museum* 5 (1882): 438-443.

YARROW, H. C., AND H. W. HENSHAW. 1878. Appendix L. Report upon the reptiles and batrachians, collected during the years of 1875, 1876, and 1877, in California, Arizona, and Nevada. In: WHEELER, G. M. (ed.), Annual Report of the Chief of Engineers, War Department, for 1878. Washington DC, Government Printing Office: 206-226.

YOUNG, J. E., AND B. I. CROTHER. 2001. Allozyme evidence for the separation of *Rana areolata* and *Rana capito* and for the resurrection of *Rana sevosa*. *Copeia* 2001 (2): 382-388.

ZEYL, C. 1993. Allozyme variation and divergence among populations of *Rana sylvatica*. *Journal of Herpetology* 27 (2): 233-236.

ZHANG, P., T. J. PAPENFUSS, M. H. WAKE, L. QU, AND D. B. WAKE. 2008. Phylogeny and biogeography of the family Salamandridae (Amphibia: Caudata) inferred from complete mitochondrial genomes. *Molecular Phylogenetics and Evolution* 49: 586-597.

ZHANG, P., AND D. B. WAKE. 2009. Higher-level salamander relationships and divergence dates inferred from complete mitochondrial genomes. *Molecular Phylogenetics and Evolution* 53: 492-508.

ZHANG, Y.-X., AND Y.-T. WEN. 2000. Amphibians in Guangxi. Guangxi Biodiversity Studies. Guangxi shi fan da xue chu ban she. [in Chinese]. Guilin, China, Guangxi Normal University Press: 183 pp.

ZHOU, E.-M., AND K. ADLER. 1993. Herpetology of China. *Contributions to Herpetology* 10. Oxford, Ohio, Society for the Study of Amphibians and Reptiles.

ZITTEL, K. A. 1888. Handbuch der Palaeontologie. I Abtheilung. Palaeozoologie. III Band. Vertebrata (Pisces, Amphibia, Reptilia, Aves). 2nd edition. München & Leipzig, Oldenbourg: 257-436.

ZUG, G. R. 1982. Comments on the proposed conservation of *Rana sphenocephala* Cope, 1886. Z.N. (S.) 2141 (1.). *Bulletin of Zoological Nomenclature* 39 (2): 80-82.

ZWEIFEL, R. G. 1955. Ecology, distribution, and systematics of frogs of the *Rana boylei* group. *University of California Publications in Zoology* 54 (4): 207-292.

ZWEIFEL, R. G. 1956a. Two pelobatid frogs from the Tertiary of North America and their relationships to fossil and Recent forms. *American Museum Novitates* (1762): 1-45.

ZWEIFEL, R. G. 1956b. A survey of the frogs of the *augusti* group, genus *Eleutherodactylus*. *American Museum Novitates* (1813): 1-35.

ZWEIFEL, R. G. 1958. Results of the Puritan-American Museum of Natural History Expedition to western Mexico. 2. Notes on reptiles and amphibians from the Pacific coastal islands of Baja California. *American Museum Novitates* (1895): 1-17.

ZWEIFEL, R. G. 1966. *Cornufer unicolor* Tschudi, 1838 (Amphibia, Salientia): request for suppression under the penary powers, Z.H. (S) 1749. *Bulletin of Zoological Nomenclature* 23: 167-168.

ZWEIFEL, R. G. 1967. *Eleutherodactylus augusti*. *Catalogue of American Amphibians and Reptiles*. American Society of Ichthyologists and Herpetologists (41): 1-4.

ZWEIFEL, R. G. 1968a. *Rana muscosa. Catalogue of American Amphibians and Reptiles.* American Society of Ichthyologists and Herpetologists (65): 1-2.
ZWEIFEL, R. G. 1968b. *Rana tarahumarae. Catalogue of American Amphibians and Reptiles.* American Society of Ichthyologists and Herpetologists (66): 1-2.
ZWEIFEL, R. G. 1968c. *Rana boylii. Catalogue of American Amphibians and Reptiles.* American Society of Ichthyologists and Herpetologists (71): 1-2.

APPENDICES

1. Abbreviations Used for Museum Numbers of Type Specimens Listed

(this is duplicated in both volumes)

AMNH: American Museum of Natural History (New York)
AMS: Australian Museum (Sydney)
ANSP: Academy of Natural Sciences, Philadelphia
ARB: Academie Royale de Belgique (Brussels)
ASSP: Academy of Science of St. Petersburg, Russia
BM: British Museum (London)
BYU: Brigham Young University, Monte Bean Museum (Provo, Utah)
CAS: California Academy of Science (San Francisco)
CAS-SU: Stanford University collections, transferred to California Academy of Science
CHAS: Chicago Academy of Sciences
CM: Carnegie Museum
CHM: Charleston Museum (Charleston, South Carolina)
CNHM: Chicago (Field) Natural History Museum
CPS: College/University of Puget Sound, Museum of Natural History, Tacoma
CR: Charleston Museum, Charleston, South Carolina.
CSNM: Cincinnati Society of Natural History, now Museum of Natural History and Science
CU: Cornell University, Division of Biological Sciences
EHT-HMS: Private collection of Edward H. Taylor and Hobart M. Smith; now in UIMNH and FMNH
ERA-WTN: Private collection of E. Ross Allen and Wilfred T. Neill (now in FSM)
FAS: Private collection of Frederick A. Shannon (mostly transferred to UIMNH after his death)
FMNH: Field Museum of Natural History (called Chicago Natural History Museum for several years)
FPC: Floyd Potter Collection (disposition following his death unknown)
FSM: Florida State Museum (Biological Sciences, University of Florida), Gainesville

GRT: G.R. Thurow, personal collection
GSF: Geological Survey of Florida
GSMNP: Great Smoky Mountains National Park
IM: Indian Museum (Calcutta)
INHS: Illinois Natural History Survey (Champaign)
LACM: Los Angeles County Museum
LMK: Lawrence M. Klauber, personal collection
MCZ: Museum of Comparative Zoology, Harvard University
MCSN: Museu Civico Storia Naturale, Venezia, Italia
MDUG: Universidad de Guanajuato, Musei "Alfredo Dugès", Guanajuato, Mexico
MH: Museum of Hildesheim (Hildesheim, Germany)
MHNG: Museum d'Histoire Naturelle, Geneva
MKUB: Museum der Königlichen, Universität zu Berlin
MM: Madgeburg Museum, Berlin
MMS: Macleay Museum, University of Sydney
MNCN: Museo Nacional de Ciencias Naturales, Madrid
MNHN: Muséum National d'Histoire Naturelle, Paris
MNHNC: Museo Nacional de Historia Natural, Santiago, Chile
MNHNM: Museo Nacional de Historia Natural, Cd. Mexico
MSB: Museum of Southwestern Biology, University of New Mexico, Santa Fe
MVZ: Museum of Vertebrate Zoology, University of California, Berkeley (has also been called University of California Museum of Zoology [UCMZ]).
MZUP: Museo Zoologico, Instituto di Zoologia, Universitá di Padova, Italy
MZUT: Museo di Zoologia, Università di Torino, Italy
MZT: Zoological Museum, Tarta, Estonia
NMW: Naturhistorisches Museum, Zoologische Abtheilung, Wien, Austria
NRM: Naturhistoriska Riksmuseet, Stockholm, Sweden
NWU: Collections of Northwestern University, Evanston, Illinois
OKMNH: University of Oklahoma Museum of Natural History, now Stovall Museum of Science and History
OKSU: Oklahoma State University collections
OS: Ottys Sanders personal collection; transferred to UIMNH
OSUS: Oklahoma State University, Stillwater
PAS: Peabody Academy of Science, Salem, Massachusetts

PBS: Philippine Bureau of Science
QM: Queensland Museum
RCSM: Royal College of Surgeons, London
RMNH: Rijksmuseum van Natuurlijke Historie, Leiden
RWA: Private collection of Ralph W. Axtell.
ROM: Royal Ontario Museum, Toronto
SDNHM: San Diego Natural History Museum
SDSNH: San Diego Society of Natural History
SM: Strecker Museum, Baylor University, Waco, Texas
SMF: Senckenberg Museum, Frankfurt
SU: Stanford University, Stanford, California
TAMU: Texas A&M University collection
TCU: Texas Christian University, Fort Worth
TCWC: Texas Cooperative Wildlife Collection at Texas A&M University
TMM: Texas Memorial Museum, University of Texas, Austin
TNHC: Texas Natural History Collection, University of Texas, Austin
TU: Tulane University, New Orleans
UAZ: University of Arizona, Tucson
UBIPRO: Unidad de Biología, Tecnología y Prototipos, UNAM.
UCLA: University of California at Los Angeles; now at CAS
UCM: University of Colorado Museum, Herpetology Collection, Boulder
UCPM: University of California Paleontology Museum
UIMNH: University of Illinois Museum of Natural History; now maintained by INHS
UKMNH: University of Kansas Museum of Natural History
UMMP: University of Michigan Museum of Paleontology
UMMZ: University of Michigan Museum of Zoology
USNM: United States National Museum, Smithsonian Institution, Washington DC
UTEP: University of Texas at El Paso
UWZM: University of Wroclaw (formerly Univ. of Breslau), Zoological Museum, Wrocaw, Poland
UUMZ: University of Utah Muzeum of Zoology
WBN: W.B. Newman, private collection
WMNH: Wisconsin Museum of Natural History
ZFMK: Zoologisches Forschungsinstitut und Museum Alexander Koenig (Bonn, Germany)

ZISP: Zoological Institute, Academy of Sciences, St. Petersburg (Russia)
ZIUG: Zoological Institute, University of Geneva
ZMA: Zoological Museum of Amsterdam (now part of the Dutch Center for Biodiversity)
ZMB: Museum für Naturkunde, Universität Humboldt, Berlin
ZMG: Zoologisches Museum, Göttingen, Germany
ZMH: Zoologisches Institut und Museum, Universitäät Hamburg
ZMUC: Universitets København, Zoologisk Museum, Copenhagen
ZMUU: Zoologiska Museet, Uppsala Universitet (Uppsala, Sweden)
ZSI: Zoological Survey of India (Port Blair)
ZSM: Zoologisches Sammlung des Bayerischen Staates, München

2. Postal Codes for U.S. States and Canadian Provinces

U.S. States:

AK Alaska
AL Alabama
AR Arkansas
AZ Arizona
CA California
CO Colorado
CT Connecticut
DC Distr. of Columbia
DE Delaware
FL Florida
GA Georgia
HI Hawaii
IA Iowa
ID Idaho
IL Illinois
IN Indiana
KS Kansas
KY Kentucky
LA Louisiana
MA Massachusetts
MD Maryland
ME Maine
MI Michigan
MN Minnesota
MO Missouri
MS Mississippi
MT Montana
NC North Carolina
ND North Dakota
NE Nebraska
NH New Hampshire
NJ New Jersey
NM New Mexico
NV Nevada
NY New York
OH Ohio
OK Oklahoma
OR Oregon
PA Pennsylvania

RI Rhode Island
SC South Carolina
SD South Dakota
TN Tennessee
TX Texas
UT Utah
VA Virginia
VT Vermont
WA Washington
WI Wisconsin
WV West Virginia
WY Wyoming

Canadian Provinces
AB Alberta
BC British Columbia
MB Manitoba
NB New Brunswick
NL Newfoundland & Labrador
NS Nova Scotia
NT Northwest Territories
NU Nunavut
ON Ontario
PE Prince Edward Island
QC Quebec
SK Saskatchewan
YT Yukon

Index to Generic and Species Names

(in **bold** if the name is treated as valid here)

(includes nomina of genus, subgenus, species and subspecies)

Index to Common Names

(including French Canadian names for Canadian species)

Salamanders (Sal. = Salamander)

Frogs and Toads

Lightning Source UK Ltd.
Milton Keynes UK
UKOW01n2035080218
317519UK00003BB/50/P

9 781493 170340